FUNDAMENTALS OF MOLECULAR SPECTROSCOPY

FUNDAMENTALS OF MOLECULAR SPECTROSCOPY

WALTER S. STRUVE
Department of Chemistry
Iowa State University
Ames, Iowa

WILEY
A WILEY-INTERSCIENCE PUBLICATION
JOHN WILEY & SONS
New York / Chichester / Brisbane / Toronto / Singapore

Library of Congress Cataloging in Publication Data:

Struve, Walter S.
Fundamentals of molecular spectroscopy.

"A Wiley–Interscience publication."
Bibliography: p.
1. Molecular spectroscopy. I. Title.
QC454.M6S87 1989 539′.6 88-5459
ISBN 0-471-85424-7

Printed in the United States of America

10 9 8 7 6 5 4 3 2 1

To Helen and my family

PREFACE

This book grew out of lecture notes for a graduate-level molecular spectroscopy course that I developed at Iowa State University between 1974 and 1987. It is intended to fill a pressing need for a concise introduction to the spectroscopy of atoms and molecules. I have tried to stress logical continuity throughout, with a view to developing readers' confidence in their physical intuition and problem-solving techniques. A suitable quantum mechanical background is furnished by the first seven and a half chapters of P. W. Atkins' *Molecular Quantum Mechanics*, 2d ed. (Oxford University Press, London, 1983): The Schrödinger equation for simple systems, angular momentum, the hydrogen atom, stationary state perturbation theory, and the variational theorem are all presumed in this book. Group theory is used extensively from Chapter 3 on; it is not developed here, because many excellent texts are available on this subject. A one-semester undergraduate course in electromagnetism is helpful but not strictly necessary: The concepts of vector and scalar potentials are introduced in Chapter 1. Other requisite material, such as time-dependent perturbation theory and second quantization, is developed in the text.

Eight or nine of the eleven chapters in this book can be comfortably accommodated within a one-semester course. The underlying time-dependent perturbation theory for molecule–radiation interactions is emphasized early, revealing the hierarchies of multipole and multiphoton transitions that can occur. Several of the chapters are introduced using illustrative spectra from the literature. This technique, extensively used by Herzberg in his classic series of monographs, avoids excessive abstraction before spectroscopic applications are reached. Diatomic rotations and vibrations are introduced explicitly in the context of the Born-Oppenheimer principle. Electronic band spectra are examined with careful attention to electronic structure, angular momentum coupling, and rotational fine structure. The treatment of polyatomic rotations hinges on a physically transparent demonstration of the commutation rules for molecule-fixed and space-fixed angular momenta. From these, all of the energy levels and selection rules that govern microwave spectroscopy are accessible without recourse to detailed rotational eigenstates. The chapter on polyatomic electronic spectra focuses on triatomic molecules and aromatic hydrocarbons—the former for their environmental and astrophysical interest, and the latter for their illustrations of vibronic coupling and radiationless relaxation phenomena. Population inversion criteria, specific laser systems, and the principles of ultrahigh-resolution lasers and ultrashort pulse generation are outlined in a chapter on lasers, which have emerged as a ubiquitous tool in spectroscopy laboratories. Some of the higher order terms in the time-dependent perturbation expansion are fleshed out for several multiphoton spectroscopies (Raman, two-photon absorption, second-harmonic generation, and CARS) in the final two chapters. The reader is guided through the powerful diagrammatic perturbation techniques in a discussion designed to enable facile determination of transition probabilities for arbitrary multiphoton processes of the reader's choice.

It is my great pleasure to acknowledge the people who made this book possible. I am particularly indebted to my teachers, Dudley Herschbach and Roy Gordon, who communicated to me the inherent beauty and cohesiveness of molecular quantum mechanics. The original suggestion for writing this book came from Cheuk-Yiu Ng. David Hoffman made seminal contributions to the chapter on polyatomic rotations. I am grateful to numerous anonymous referees for valuable suggestions for improving the manuscript, though, of course, the responsibility for errors is still mine. The line drawings were supplied by Linda Emmerson, and Sandra Bellefeuille, Klaus Ruedenberg, and Gregory Atchity generated the computer graphics. Finally, I must express deep appreciation to my wife, Helen, whose moral support was essential to completing this work.

WALTER S. STRUVE

Ames, Iowa
June 1988

CONTENTS

FUNDAMENTALS OF MOLECULAR SPECTROSCOPY

1

RADIATION–MATTER INTERACTIONS

In its broadest sense, spectroscopy is concerned with interactions between light and matter. Since light consists of electromagnetic waves, this chapter begins with classical and quantum mechanical treatments of molecules subjected to static (time-independent) electric fields. Our discussion identifies the molecular properties that control interactions with electric fields: the electric multipole moments and the electric polarizability. Time-dependent electromagnetic waves are then described classically using vector and scalar potentials for the associated electric and magnetic fields **E** and **B**, and the classical Hamiltonian is obtained for a molecule in the presence of these potentials. Quantum mechanical time-dependent perturbation theory is finally used to extract probabilities of transitions between molecular states. This powerful formalism not only covers the full array of multipole interactions that can cause spectroscopic transitions, but also reveals the hierarchies of multiphoton transitions that can occur. This chapter thus establishes a framework for multiphoton spectroscopies (e.g., Raman spectroscopy and coherent anti-Stokes Raman spectroscopy, which are discussed in Chapters 10 and 11) as well as for the one-photon spectroscopies that are described in most of this book.

1.1 CLASSICAL ELECTROSTATICS OF MOLECULES IN ELECTRIC FIELDS

Consider a molecule composed of N electric charges e_n (electrons and nuclei) located at positions $\mathbf{r}_n$ referenced to an arbitrary origin in space. The total molecular charge is

$$q = \sum_{n=1}^{N} e_n \tag{1.1}$$

and the electric dipole moment is

$$\boldsymbol{\mu} = \sum_{n=1}^{N} e_n \mathbf{r}_n \tag{1.2}$$

The latter expression reduces to a familiar expression for the dipole moment in a neutral "molecule" consisting of two point charges, $e_1 = +Q$ and $e_2 = -Q$ (Fig. 1.1). In this case, we have $\boldsymbol{\mu} = +Q\mathbf{r}_1 + (-Q)\mathbf{r}_2 = Q(\mathbf{r}_1 - \mathbf{r}_2) = Q\mathbf{R}$, which is the conventional expression for the dipole moment of a pair of opposite charges $\pm Q$ separated by the vector $\mathbf{R}$. By convention, $\mathbf{R}$ points toward the positive charge. For molecules characterized by electric charge distributions $\rho(\mathbf{r})$ instead of point charges, the expressions for the molecular charge and dipole moment are superseded by

$$q = \int_V \rho(\mathbf{r}) d\mathbf{r} \tag{1.3}$$

and

$$\boldsymbol{\mu} = \int_V \mathbf{r}\rho(\mathbf{r}) d\mathbf{r} \tag{1.4}$$

where the integration volume encloses the entire charge distribution.

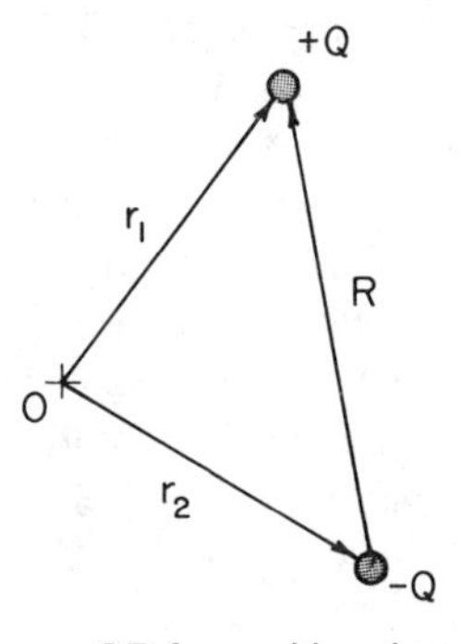

Figure 1.1 Dipole moment $\boldsymbol{\mu} = Q\mathbf{R}$ formed by charges $\pm Q$ separated by the vector $\mathbf{R}$.

*Numbers in brackets are citations of references at the end of the chapter.

For a point charge e located at position $\mathbf{r}$ under an external electrostatic potential $\phi(\mathbf{r})$, the energy of interaction with the potential is [1]*

$$W = e\phi(\mathbf{r}) \tag{1.5}$$

$\phi(\mathbf{r})$ can be expanded in a Taylor series about $\mathbf{r} = 0$ (whose location is arbitrary) as

$$\begin{aligned}
\phi(\mathbf{r}) &= \phi(0) + x\frac{\partial\phi}{\partial x}(0) + y\frac{\partial\phi}{\partial y}(0) + z\frac{\partial\phi}{\partial z}(0) \\
&\quad + \frac{1}{2}\left[x^2\frac{\partial\phi^2}{\partial x^2}(0) + y^2\frac{\partial^2\phi}{\partial y^2}(0) + z^2\frac{\partial^2\phi}{\partial z^2}(0)\right. \\
&\quad \left. + 2xy\frac{\partial^2\phi}{\partial x\partial y}(0) + 2xz\frac{\partial^2\phi}{\partial x\partial z}(0) + 2yz\frac{\partial^2\phi}{\partial y\partial z}(0)\right] \\
&\quad + \cdots \\
&\equiv \phi(0) + \mathbf{r}\cdot\nabla\phi(0) + \frac{1}{2}\sum_{ij} x_i x_j\frac{\partial^2\phi(0)}{\partial x_i\partial x_j} + \cdots
\end{aligned} \tag{1.6}$$

where the components of $\mathbf{r}$ are expressed as either (x, y, z) or (x_1, x_2, x_3). Since the electric field $\mathbf{E}(0)$ at the origin is related to $\phi(0)$ by

$$\mathbf{E}(0) = -\nabla\phi(0) \tag{1.7}$$

this implies that

$$\begin{aligned}
\phi(\mathbf{r}) &= \phi(0) - \mathbf{r}\cdot\mathbf{E}(0) + \frac{1}{2}\sum_{ij} x_i x_j\frac{\partial}{\partial x_i}\frac{\partial\phi}{\partial x_j}(0) + \cdots \\
&= \phi(0) - \mathbf{r}\cdot\mathbf{E}(0) - \frac{1}{2}\sum_{ij} x_i x_j\frac{\partial E_j}{\partial x_i}(0) + \cdots
\end{aligned} \tag{1.8}$$

The interaction energy is then

$$W = e\phi(\mathbf{r}) = e\phi(0) - e\mathbf{r}\cdot\mathbf{E}(0) - \tfrac{1}{2}e\sum_{ij} x_i x_j\frac{\partial E_j}{\partial x_i}(0) + \cdots \tag{1.9}$$

For a molecule consisting of N point charges e_n at locations $\mathbf{r}_n$, this becomes

$$\begin{aligned}
W &= \left(\sum_n^N e_n\right)\phi(0) - \left(\sum_n^N e_n\mathbf{r}_n\right)\cdot\mathbf{E}(0) \\
&\quad - \frac{1}{2}\sum_n^N e_n\sum_{ij}(r_n)_i(r_n)_j\frac{\partial E_i(0)}{\partial(r_n)_j} + \cdots \\
&= q\phi(0) - \boldsymbol{\mu}\cdot\mathbf{E}(0) - \frac{1}{6}\sum_n\sum_{ij} Q_{ij}^{(n)}\frac{\partial E_i(0)}{\partial(r_n)_j} + \cdots
\end{aligned} \tag{1.10}$$

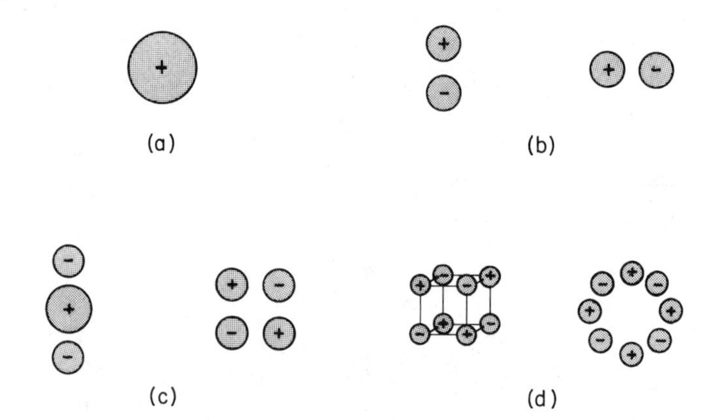

Figure 1.2 Examples of charge distributions with (a) nonzero charge, (b) nonzero dipole moment, (c) nonzero quadrupole moment, and (d) nonzero octupole moment.

The first two terms in W arise from the interaction of the molecular charge with the scalar potential ϕ and the interaction of the molecular dipole moment with the electric field $\mathbf{E}$, respectively. The next terms in W are due to interactions between the various electric field gradients $\partial E_i/\partial(r_n)_j$ and the corresponding components [2]

$$Q_{ij}^{(n)} = \tfrac{1}{2}e_n[3(r_n)_i(r_n)_j - \mathbf{r}_n^2\delta_{ij}] \tag{1.11}$$

of the electric quadrupole moment tensor. (Note that the term $-\mathbf{r}_n^2\delta_{ij}$ in $Q_{ij}^{(n)}$ drops out when Eq. 1.11 for the latter is used in Eq. 1.10, because $\nabla \cdot \mathbf{E} = 0$ for an external field in free space.) Hence the expansion (1.10) illustrates how the various electric multipoles interact with an external field. Examples of charge distributions exhibiting nonzero dipole, quadrupole, and octupole moments are shown in Fig. 1.2.

1.2 QUANTUM THEORY OF MOLECULES IN STATIC ELECTRIC FIELDS

We will be concerned almost exclusively with interactions between light and isolated molecules. The total Hamiltonian of the system is then given by

$$\hat{H} = \hat{H}_0 + \hat{H}_{\text{rad}} + W \tag{1.12}$$

$\hat{H}_0$ is the Hamiltonian for the isolated molecule, $\hat{H}_{\text{rad}}$ is the Hamiltonian of the radiation field, and W is the interaction term describing the coupling of the molecular states to the radiation field. We will denote the isolated unperturbed molecular states with $|\Psi_n\rangle$; they obey the time-independent and time-dependent

Schrödinger equations

$$\hat{H}_0|\Psi_n\rangle = E_n^{(0)}|\Psi_n\rangle \tag{1.13}$$

$$i\hbar \frac{\partial}{\partial t}|\Psi_n\rangle = \hat{H}_0|\Psi_n\rangle \tag{1.14}$$

The unperturbed molecular states $|\Psi_n\rangle$ depend on both position (**r**) and time (t); since they are eigenstates of $\hat{H}_0$, they can be factorized into spatial and time-dependent portions,

$$|\Psi_n(\mathbf{r}, t)\rangle = e^{-iE_n t/\hbar}|\psi_n(\mathbf{r})\rangle \tag{1.15}$$

If, instead of (1.12), the total Hamiltonian for the molecule in the presence of light were $\hat{H} = \hat{H}_0 + \hat{H}_{\mathrm{rad}}$, the eigenstates of the system would become simple products of molecular and radiation field states $|\Psi_n(\mathbf{r}, t)\rangle|\chi_{\mathrm{rad}}\rangle$, where the radiation field states $|\chi_{\mathrm{rad}}\rangle$ would depend on photon occupation numbers, energies and polarizations. Since no coupling of light with molecular states is implied in such a Hamiltonian, no transitions can occur between molecular states due to absorption or emission of light in this description. The simplest Hamiltonian that can account for spectroscopic transitions is therefore the one in Eq. 1.12.

Formally, the inclusion of the interaction term W requires that the Schrödinger equations (1.13) and (1.14) be solved again after replacing $\hat{H}_0 + \hat{H}_{\mathrm{rad}}$ with the total Hamiltonian $\hat{H}$ for the molecule in the presence of light. A simplification arises here because the interaction term W in Eq. 1.12 normally introduces only a small perturbation to the isolated molecular Hamiltonian $\hat{H}_0$. For example, the electric field of bright sunlight is on the order of 5 V/cm. By comparison, an electron spaced by a Bohr radius a_0 from the nucleus in a H atom experiences an electric field of $e^2/4\pi\varepsilon_0 a_0^2$, which is on the order of 5×10^9 V/cm. Hence, W can generally be treated as a perturbation to $\hat{H}_0 + \hat{H}_{\mathrm{rad}}$. This approach proves to be useful even for describing molecules subject to intense laser beams; in such cases, higher order perturbations assume unusual importance in comparison to the situation of molecules exposed to classical light sources.

For molecules subjected to *static* (time-independent) electromagnetic fields, the perturbed energies and eigenstates may be evaluated from *stationary-state* perturbation theory [3]. The full Hamiltonian may be written in terms of a perturbation parameter λ (which may be set to unity at the end of the calculation, after serving its usual purpose of keeping track of orders in the perturbation expansions for the energies and eigenstates) as

$$\hat{H} = \hat{H}_0 + \lambda W \tag{1.16}$$

($\hat{H}_{\mathrm{rad}}$ has been dropped here because we are interested primarily in how the applied field affects the molecular energy levels.) The time-independent eigenstates $|\psi_n\rangle$, which we use here instead of $|\Psi_n\rangle$ because we are dealing with the static problem, and the eigenvalues E_n are expanded as

$$|\psi_n\rangle = |\psi_n^{(0)}\rangle + \lambda|\psi_n^{(1)}\rangle + \lambda^2|\psi_n^{(2)}\rangle + \cdots \tag{1.17}$$

$$E_n = E_n^{(0)} + \lambda E_n^{(1)} + \lambda^2 E_n^{(2)} + \cdots \tag{1.18}$$

Substituting Eqs. 1.16–1.18 into $\hat{H}|\psi_n\rangle = E_n|\psi_n\rangle$ and using Eqs. 1.13 and 1.14 yields successive approximations to the perturbed energy

$$E_n = E_n^{(0)} + \lambda E_n^{(1)} + \lambda^2 E_n^{(2)} + \cdots \tag{1.19}$$

where

$$E_n^{(0)} = \langle\psi_n^{(0)}|\hat{H}_0|\psi_n^{(0)}\rangle \tag{1.20}$$

is the unperturbed energy in state $|\psi_n^{(0)}\rangle$ and

$$E_n^{(1)} = \langle\psi_n^{(0)}|W|\psi_n^{(0)}\rangle \tag{1.21}$$

$$E_n^{(2)} = \sum_{l \neq n} \frac{\langle\psi_n^{(0)}|W|\psi_l^{(0)}\rangle\langle\psi_l^{(0)}|W|\psi_n^{(0)}\rangle}{E_n^{(0)} - E_l^{(0)}} \tag{1.22}$$

are the leading terms in the energy corrections to $E_n^{(0)}$.

As an example of using stationary-state perturbation theory to compute the perturbed energies of a molecule in a static electromagnetic field, consider an uncharged molecule in a *uniform* (position-independent) static electric field **E**. In this case, the only nonvanishing term in the expansion (1.10) is $W = -\boldsymbol{\mu}\cdot\mathbf{E}(0)$. Substitution of this expression for the perturbation W in Eqs. 1.21 and 1.22 yields

$$E_n^{(1)} = -\mathbf{E}\cdot\langle\psi_n^{(0)}|\boldsymbol{\mu}|\psi_n^{(0)}\rangle \tag{1.23}$$

$$E_n^{(2)} = \sum_{l \neq n} \frac{\mathbf{E}\cdot\langle\psi_n^{(0)}|\boldsymbol{\mu}|\psi_l^{(0)}\rangle\langle\psi_l^{(0)}|\boldsymbol{\mu}|\psi_n^{(0)}\rangle\cdot\mathbf{E}}{E_n^{(0)} - E_l^{(0)}} \tag{1.24}$$

The first- and second-order corrections to the energy are linear and quadratic in the electric field **E**, respectively. It is interesting to compare these results with the classical energy of an uncharged molecule with permanent dipole moment $\boldsymbol{\mu}_0$ in a uniform, static **E** field: according to Eq. 1.10, this would be

$$E = E_0 - \boldsymbol{\mu}_0\cdot\mathbf{E} \tag{1.25}$$

Here E_0, the classical energy of the molecule in the absence of the field, can be identified with unperturbed energy $E_n^{(0)}$ in Eq. 1.20. Comparison of the terms linear in **E** in Eqs. 1.23 and 1.25 shows that the permanent dipole moment $\boldsymbol{\mu}_0$ is

Table 1.1 Permanent dipole moments of some molecules[a]

Molecule	μ_0
HCl	1.03
HBr	0.788
H_2O	1.81
CO	1.2
CH_3Cl	1.9
NO_2	0.399

[a]In units of debyes: 1 D = 3.33564×10^{-30} C·m.

the expectation value of the instantaneous dipole moment operator $\boldsymbol{\mu} = \Sigma\, e_n \mathbf{r}_n$,

$$\boldsymbol{\mu}_0 = \langle \psi_n^{(0)} | \boldsymbol{\mu} | \psi_n^{(0)} \rangle \tag{1.26}$$

(Values of permanent dipole moments are given for several small molecules in Table 1.1.) However, the classical energy 1.25 has no counterparts to the second- and higher order terms in the perturbation expansion (1.19) for E_n.

This situation arises because the multipole expansion for W per se (Eq. 1.10) has no provision for molecular polarization by the electric field. When an atom (or nonpolar molecule of sufficiently high symmetry) is subjected to an electric field **E**, the latter separates the centers of gravity of the species' positive and negative charges, creating an *induced dipole moment* $\boldsymbol{\mu}_{\text{ind}}$ which is parallel to **E** (Fig. 1.3). For sufficiently small **E**, $\boldsymbol{\mu}_{\text{ind}}$ is proportional to **E**, so that

$$\boldsymbol{\mu}_{\text{ind}} = \alpha \mathbf{E} \tag{1.27}$$

The proportionality constant α is defined as the atomic (or molecular) *polarizability*. The total dipole moment of the polarized molecule in the electric field is

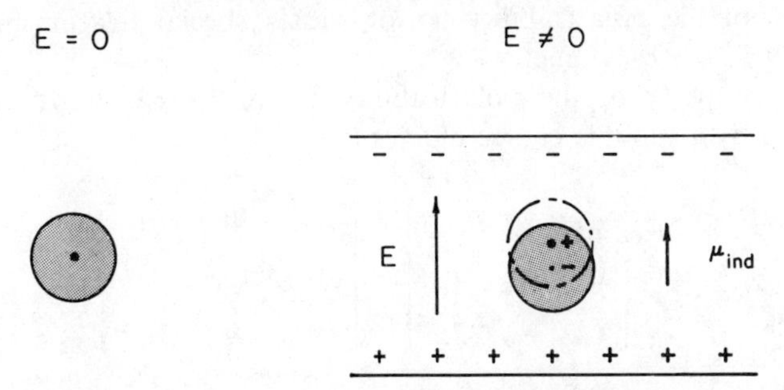

Figure 1.3 Polarizable atom in the absence of an external electric field (left) and in the presence of a uniform electric field **E** (right).

then

$$\boldsymbol{\mu} = \boldsymbol{\mu}_0 + \boldsymbol{\mu}_{\text{ind}} = \boldsymbol{\mu}_0 + \alpha\mathbf{E} \tag{1.28}$$

and the classical energy becomes [4]

$$\begin{aligned} E &= E_0 - \int_0^{\mathbf{E}} \boldsymbol{\mu} \cdot d\mathbf{E}' \\ &= E_0 - \boldsymbol{\mu}_0 \cdot \mathbf{E} - \tfrac{1}{2}\alpha\mathbf{E}\cdot\mathbf{E} \end{aligned} \tag{1.29}$$

instead of Eq. 1.25. In molecules lacking special symmetry (e.g., CO_2, CH_3OH) the induced moment $\boldsymbol{\mu}_{\text{ind}}$ does not generally point parallel to the external field $\mathbf{E}$, because the electron cloud in such molecules is more easily distorted in certain directions than in others (Fig. 1.4). In such cases, the polarizability is a tensor rather than a scalar quantity. The induced dipole moment is then given by

$$\boldsymbol{\mu}_{\text{ind}} = \boldsymbol{\alpha} \cdot \mathbf{E} \tag{1.30}$$

with

$$\boldsymbol{\alpha} = \begin{bmatrix} \alpha_{xx} & \alpha_{xy} & \alpha_{xz} \\ \alpha_{yx} & \alpha_{yy} & \alpha_{yz} \\ \alpha_{zx} & \alpha_{zy} & \alpha_{zz} \end{bmatrix} \tag{1.31}$$

$$\boldsymbol{\mu}_{\text{ind}} = \begin{bmatrix} \mu_{\text{ind},x} \\ \mu_{\text{ind},y} \\ \mu_{\text{ind},z} \end{bmatrix} = \begin{bmatrix} \alpha_{xx}E_x + \alpha_{xy}E_y + \alpha_{xz}E_z \\ \alpha_{yx}E_x + \alpha_{yy}E_y + \alpha_{yz}E_z \\ \alpha_{zx}E_x + \alpha_{zy}E_y + \alpha_{zz}E_z \end{bmatrix} \tag{1.32}$$

Thus, in general, $\boldsymbol{\mu}_{\text{ind}}$ is not parallel to $\mathbf{E}$ (i.e., $\mu_{\text{ind},y}/\mu_{\text{ind},x} \neq E_y/E_x$, etc.) unless the elements α_{ij} of the polarizability tensor satisfy special relationships—as in molecules of T_d or O_h symmetry.

Another property of the polarizability tensor is that it can always be diagonalized by a suitable choice of axes $(x'y'z')$:

$$\boldsymbol{\alpha} = \begin{bmatrix} \alpha_{xx} & \alpha_{xy} & \alpha_{xz} \\ \alpha_{yx} & \alpha_{yy} & \alpha_{yz} \\ \alpha_{zx} & \alpha_{zy} & \alpha_{zz} \end{bmatrix} \rightarrow \begin{bmatrix} \alpha_{x'x'} & 0 & 0 \\ 0 & \alpha_{y'y'} & 0 \\ 0 & 0 & \alpha_{z'z'} \end{bmatrix} \tag{1.33}$$

This is analogous to the choice of principal axes $(x'y'z')$ in calculating the three principal moments of inertia of a rigid rotor (Chapter 5), and it shows that only

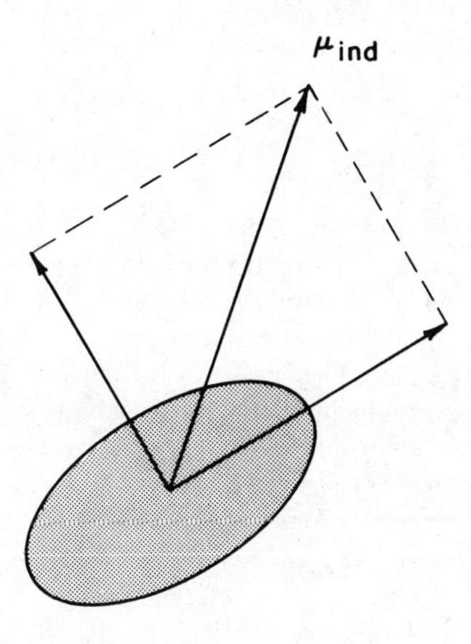

Figure 1.4 Polarization of an anisotropic molecule by an electric field. Since the electronic charge distribution is more polarizable along the long axis than along the short axis, the induced dipole moment $\boldsymbol{\mu}_{ind}$ is not parallel to **E**.

three of the components of a polarizability tensor are independent. In molecules with sufficiently high symmetry, the principal polarizability axes coincide with symmetry axes of the molecule. In CO_2, one of these principal axes is the C_∞ molecular axis, while the other two may be any choice of orthogonal C_2 axes perpendicular to the molecular axis. In a less symmetric molecule like CH_3OH, the directions of the principal polarizability axes must be evaluated numerically.

When the polarizability is a tensor rather than a scalar, Eq. 1.29 for the classical energy becomes

$$E = E_0 - \boldsymbol{\mu}_0 \cdot \mathbf{E} - \tfrac{1}{2}\mathbf{E} \cdot \boldsymbol{\alpha} \cdot \mathbf{E} \tag{1.34}$$

Comparison of the terms quadratic in **E** in Eqs. 1.19, 1.22, and 1.34 then reveals that the quantum mechanical expression for the molecular polarizability is

$$\boldsymbol{\alpha} = 2 \sum_{l \neq n} \frac{\langle \psi_n^{(0)} | \boldsymbol{\mu} | \psi_l^{(0)} \rangle \langle \psi_l^{(0)} | \boldsymbol{\mu} | \psi_n^{(0)} \rangle}{E_l^{(0)} - E_n^{(0)}} \tag{1.35}$$

If the spatial part $|\psi_n^{(0)}\rangle$ of the unperturbed molecular wave function is accurately known in state n, the permanent dipole moment $\boldsymbol{\mu}_0$ can be evaluated for that state using Eq. 1.26. However, Eq. 1.35 shows that an accurate knowledge of all the molecular eigenstates $|\psi_l^{(0)}\rangle$ with $l \neq n$ are normally required in addition to $|\psi_n^{(0)}\rangle$ to calculate the molecular polarizability in state n. In particular, the xy component of the polarizability tensor (which will be nonvanishing only if the coordinate axes (xyz) are not chosen to coincide with the principal molecular axes) will be

$$\alpha_{xy} = 2 \sum_{l \neq n} \frac{\langle \psi_n^{(0)} | \Sigma e_i x_i | \psi_l^{(0)} \rangle \langle \psi_l^{(0)} | \Sigma e_i y_i | \psi_n^{(0)} \rangle}{E_l^{(0)} - E_n^{(0)}} \tag{1.36}$$

where the index i is summed over all of the molecular charges. Inspection of Eq. 1.35 or 1.36 shows that large values of the polarizability tensor components are favored by large $\langle \psi_n^{(0)} | \mathbf{r}_i | \psi_l^{(0)} \rangle$ and by small energy denominators $(E_l^{(0)} - E_n^{(0)})$. For this reason, the alkali atoms, with their voluminous valence orbitals and closely spaced energy levels, exhibit the largest atomic polarizabilities in the periodic table (Table 1.2), and such atoms have figured prominently in the development of nonlinear optics. These expressions for molecular polarizabilities become useful in Chapter 10, where Raman transition probabilities are discussed.

Table 1.2 Polarizabilities of several atoms[a]

Atom	α
H	0.666793
He	0.204956
Li	24.3
N	1.10
O	0.802
F	0.557
Ne	0.3946
Na	23.6
Ar	1.64
K	43.4
Cs	59.6

[a]In units of Å^3. Data taken from T. M. Miller and B. Bederson, *Adv. At. Mol. Phys.* **13**: 1 (1977).

The stationary-state perturbation theory used in this section is applicable only to nondegenerate states $|\psi_n^{(0)}\rangle$; degenerate perturbation theory must otherwise be used. The polarizability expressions developed here are only good for time-independent (static) external fields. The polarizability turns out to depend on the frequency ω of the applied field, since the electronic motion cannot respond instantaneously to changes in **E**. Finally, since light contains time-dependent electromagnetic rather than static electric fields, the results of this section are not directly applicable to radiation–molecule interactions.

1.3 CLASSICAL DESCRIPTION OF MOLECULES IN TIME-DEPENDENT FIELDS

In the classical electromagnetic theory of light, light in vacuum consists of transverse electromagnetic waves that obey Maxwell's equations [1, 2]

$$\nabla \cdot \mathbf{E} = 0 \tag{1.37a}$$

$$\nabla \cdot \mathbf{B} = 0 \tag{1.37b}$$

$$\nabla \times \mathbf{E} = -\frac{\partial \mathbf{B}}{\partial t} \tag{1.37c}$$

$$\nabla \times \mathbf{B} = \mu_0 \varepsilon_0 \frac{\partial \mathbf{E}}{\partial t} \tag{1.37d}$$

where $\varepsilon_0 = 8.854 \times 10^{-12}\ \mathrm{C^2/J \cdot m}$ and $\mu_0 = 1.257 \times 10^{-6}$ H/m are the electric permeability and magnetic susceptibility of free space.* Examples of electric and magnetic fields satisfying Eqs. 1.37 are given by

$$\mathbf{E}(\mathbf{r}, t) = \mathbf{E}_0 e^{i(\mathbf{k}\cdot\mathbf{r} - \omega t)} \tag{1.38a}$$

$$\mathbf{B}(\mathbf{r}, t) = \mathbf{B}_0 e^{i(\mathbf{k}\cdot\mathbf{r} - \omega t)} \tag{1.38b}$$

It is easy to show from Maxwell's equations that $\mathbf{E}_0 \cdot \mathbf{k} = \mathbf{B}_0 \cdot \mathbf{k} = 0$ (i.e., both fields point normal to the direction $\hat{k}$ of propagation as required in a transverse electromagnetic wave), that $\mathbf{E}_0 \cdot \mathbf{B}_0 = 0$, and that Eqs. 1.38 describe linearly polarized light with its electric polarization parallel to $\mathbf{E}_0$ as shown in Fig. 1.5.

Since $\nabla \cdot \mathbf{B} = 0$ (as magnetic monopoles do not exist), **B** can be expressed in terms of a vector potential $\mathbf{A}(\mathbf{r}, t)$ as [1, 2]

$$\mathbf{B} = \nabla \times \mathbf{A} \tag{1.39}$$

*This book uses the International System of Units (abbreviated SI), an extension of the mks system. The units for length, mass, time, and current are meters, kilograms, seconds, and amperes, respectively. The equations in this and the following sections differ from the corresponding equations written in cgs units in that the factors of the speed of light $c = (\varepsilon_0 \mu_0)^{-1/2} = 2.998 \times 10^8$ m/s, which appear in many of the cgs equations, are absent in the SI versions.

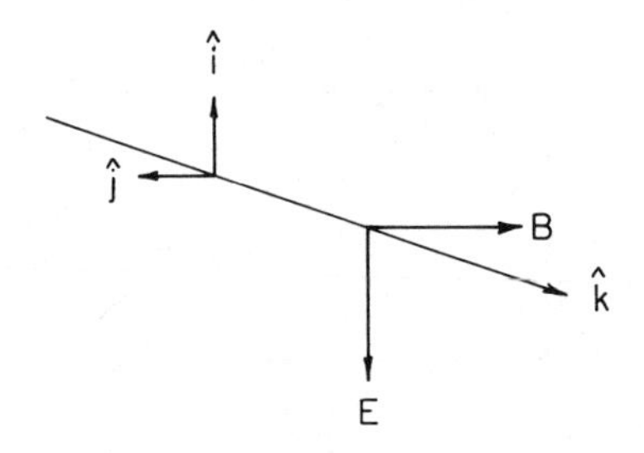

Figure 1.5 Orientations of the vectors $\mathbf{E}_0$, $\mathbf{B}_0$, and $\mathbf{k}$ for the light wave described by Eqs. 1.38.

This automatically satisfies $\nabla \cdot \mathbf{B} = 0$ in consequence of the vector identity

$$\nabla \cdot (\nabla \times \mathbf{A}) \equiv 0 \tag{1.40}$$

(The latter identity is apparent because $\nabla \times \mathbf{A}$ formally yields another vector that is normal to both ∇ and $\mathbf{A}$; hence a scalar product between ∇ and $(\nabla \times \mathbf{A})$ invariably vanishes.) From the third of Maxwell's equations, we have $\nabla \times \mathbf{E} = -\partial \mathbf{B}/\partial t$, which implies that

$$\nabla \times \mathbf{E} = -\nabla \times \dot{\mathbf{A}}$$

and so

$$\nabla \times (\mathbf{E} + \dot{\mathbf{A}}) = 0 \tag{1.41}$$

In the absence of magnetic fields, $\mathbf{B}$ and the vector potential $\mathbf{A}$ vanish, and the electric field $\mathbf{E}$ is related to the scalar potential ϕ by $\mathbf{E} = -\nabla\phi$. Hence Eq. 1.41 and the third Maxwell equation are consistent with setting

$$\mathbf{E} = -\nabla\phi - \frac{\partial \mathbf{A}}{\partial t} \tag{1.42}$$

because then

$$\begin{aligned} \nabla \times \mathbf{E} &= -\nabla \times (\nabla\phi) - \frac{\partial}{\partial t}(\nabla \times \mathbf{A}) \\ &\equiv -\partial \mathbf{B}/\partial t \end{aligned} \tag{1.43}$$

in view of the vector identity $\nabla \times (\nabla\phi) \equiv 0$ for arbitrary scalar fields ϕ. This enables us to express the measurable $\mathbf{E}$ and $\mathbf{B}$ fields in terms of a scalar potential $\phi(\mathbf{r}, t)$ and a vector potential $\mathbf{A}(\mathbf{r}, t)$ using Eqs. 1.39 and 1.42.

The vector and scalar potentials are not directly measurable themselves, since their definition has an arbitrariness analogous to the setting of standard states in thermodynamic functions [2, 5]. Suppose ϕ and $\mathbf{A}$ are altered using some scalar function $X(\mathbf{r}, t)$ via

$$\mathbf{A}' = \mathbf{A} + \nabla X \tag{1.44a}$$

$$\phi' = \phi - \frac{\partial X}{\partial t} \tag{1.44b}$$

Under this transformation, the magnetic field $\mathbf{B} = \nabla \times \mathbf{A}$ becomes

$$\begin{aligned} \mathbf{B}' &= \nabla \times (\mathbf{A} + \nabla X) = \nabla \times \mathbf{A} + \nabla \times (\nabla X) \\ &= \mathbf{B} \end{aligned} \tag{1.45}$$

since $\nabla \times (\nabla X) \equiv 0$. Similarly $\mathbf{E} = -\nabla\phi - \partial\mathbf{A}/\partial t$ becomes

$$\begin{aligned} \mathbf{E}' &= -\nabla(\phi - \dot{X}) - \frac{\partial}{\partial t}(\mathbf{A} + \nabla X) \\ &= -\nabla\phi + \nabla\dot{X} - \frac{\partial \mathbf{A}}{\partial t} - \nabla\dot{X} \\ &= \mathbf{E} \end{aligned} \tag{1.46}$$

Thus, the physical **E** and **B** fields are both unaffected by this so-called *gauge transformation*, and this gives us latitude to select algebraically convenient expressions for ϕ and **A** without affecting the measurable electromagnetic fields. It may be shown [5] that it is possible to choose the *Coulomb gauge* in which $\nabla \cdot \mathbf{A} = 0$. This gauge is often used to describe electromagnetic waves in free space, where the scalar potential $\phi = 0$. The fields are then given by

$$\mathbf{E} = -\frac{\partial \mathbf{A}}{\partial t} \tag{1.47a}$$

$$\mathbf{B} = \nabla \times \mathbf{A} \tag{1.47b}$$

It now remains to formulate the classical Hamiltonian for a charged particle subjected to potentials ϕ and **A** (or, equivalently, their associated fields **E** and **B**). Nonrelativistic classical mechanics is based on Newton's law of motion

$$\mathbf{F} = m\ddot{\mathbf{r}} \tag{1.48}$$

for a particle of mass m subjected to an external force **F**. The kinetic energy of the particle is

$$T = \tfrac{1}{2}m\mathbf{v}^2 \tag{1.49}$$

and (for conservative forces) the force is derivable from a potential V by

$$\mathbf{F} = -\nabla V \tag{1.50}$$

Since for each Cartesian component i of the particle velocity $\mathbf{v}$

$$\frac{\partial T}{\partial v_i} = mv_i \tag{1.51}$$

and

$$\frac{d}{dt}\left(\frac{\partial T}{\partial v_i}\right) = \frac{d}{dt}(mv_i) = m\ddot{r}_i = -\frac{\partial V}{\partial x_i} \tag{1.52}$$

it follows that

$$\frac{d}{dt}\left(\frac{\partial T}{\partial v_i}\right) + \frac{\partial V}{\partial x_i} = 0 \tag{1.53}$$

If one forms the *Lagrangian function* [6]

$$L \equiv T - V \tag{1.54}$$

Eq. 1.53 assumes the form of the *Lagrangian equations*

$$\frac{d}{dt}\left(\frac{\partial L}{\partial v_i}\right) - \frac{\partial L}{\partial x_i} = 0 \tag{1.55}$$

provided the external force $\mathbf{F}$ is conservative (i.e., the potential function V is independent of the particle velocity $\mathbf{v}$). It may be shown that if one chooses any convenient set of $3N$ generalized coordinates q_i for an N-particle system, so that for each particle n the Cartesian components of its position are expressible in the form

$$\begin{aligned} x_n &= x_n(q_1, q_2, \ldots, q_{3N}) \\ y_n &= y_n(q_1, q_2, \ldots, q_{3N}) \\ z_n &= z_n(q_1, q_2, \ldots, q_{3N}) \end{aligned} \tag{1.56}$$

the transformation from Cartesian to generalized coordinates yields the more powerful *generalized* Lagrangian equations

$$\frac{d}{dt}\left(\frac{\partial L}{\partial \dot{q}_i}\right) - \frac{\partial L}{\partial q_i} = 0 \tag{1.57}$$

In an N-particle system with r constraints, $(3N - r)$ of the generalized coordinates will be independent, and there will be $(3N - r)$ independent equations of the form (1.57). The Lagrangian is treated as a function of the conjugate variables q_i and $\dot{q}_i$.

Differentiating the Lagrangian function with respect to v_i in Cartesian coordinates gives (for conservative forces)

$$\frac{\partial L}{\partial v_i} = \frac{\partial T}{\partial v_i} = mv_i = p_i \tag{1.58}$$

which is the ith component of linear momentum. In generalized coordinates, the same procedure

$$\frac{\partial L}{\partial \dot{q}_i} = p_i \tag{1.59}$$

yields the component of generalized momentum conjugate to the generalized coordinate q_i. Differentiation of L with respect to x_i in Cartesian coordinates yields

$$\frac{\partial L}{\partial x_i} = -\frac{\partial V}{\partial x_i} = F_i = \dot{p}_i \tag{1.60}$$

which expresses the conservation law that if the Lagrangian is independent of x_i, the linear momentum component p_i is a constant of the motion. The analogous equation

$$\frac{\partial L}{\partial q_i} = \dot{p}_i \tag{1.61}$$

can also be shown to hold for generalized momenta.

The classical Hamiltonian is now defined as [6]

$$\begin{aligned} H &= \sum_{i=1}^{3N-r} p_i \dot{q}_i - L \\ &= H(p_i, q_i) \end{aligned} \tag{1.62}$$

and is handled as a function of the conjugate variables q_i, p_i. It is readily shown that the classical Hamiltonian gives the total energy for a single particle experiencing conservative forces in Cartesian coordinates:

$$\begin{aligned} H &= \sum_{i=1}^{3} p_i \dot{q}_i - L = \sum_{i=1}^{3} \left(\frac{\partial T}{\partial \dot{q}_i}\right) \dot{q}_i - L \\ &= \sum_{i=1}^{3} (m\dot{x}_i)\dot{x}_i - \sum_{i=1}^{3} \tfrac{1}{2} m\dot{x}_i^2 + V = T + V \end{aligned} \tag{1.63}$$

A similar result can be obtained for the Hamiltonian (1.62) in generalized coordinates.

The Lorentz force $\mathbf{F}$ experienced by a charge e of mass m in the presence of electric and magnetic fields $\mathbf{E}$ and $\mathbf{B}$ is given in SI units by [1, 2, 5]

$$\mathbf{F} = e[\mathbf{E} + \mathbf{v} \times \mathbf{B}] = \frac{d}{dt}(m\mathbf{v}) \tag{1.64}$$

This nonconservative ($\mathbf{v}$ – dependent) force is not derivable from a potential energy V in the manner of Eq. 1.50. If a Lagrangian function can nevertheless be found that obeys Eq. 1.57, the formulation of the Hamiltonian using Eq. 1.62 will still be valid. Recasting Eq. 1.64 in terms of the vector and scalar potentials (Eqs. 1.39 and 1.42), we obtain

$$\mathbf{F} = e\left[-\nabla\phi - \frac{\partial \mathbf{A}}{\partial t} + \mathbf{v} \times (\nabla \times \mathbf{A})\right] \tag{1.65}$$

This expression for the force can be simplified with the identities

$$\frac{d\mathbf{A}}{dt} \equiv \frac{\partial \mathbf{A}}{\partial t} + (\mathbf{v}\cdot\nabla)\mathbf{A} \tag{1.66}$$

$$\mathbf{v} \times (\nabla \times \mathbf{A}) \equiv \nabla(\mathbf{v}\cdot\mathbf{A}) - (\mathbf{v}\cdot\nabla)\mathbf{A} \tag{1.67}$$

to give

$$\mathbf{F} = e\left[-\nabla(\phi - \mathbf{v}\cdot\mathbf{A}) - \frac{d\mathbf{A}}{dt}\right] = \frac{d}{dt}(m\mathbf{v}) \tag{1.68}$$

We are now in a position to show that the Lagrangian for this system happens to be

$$L = \tfrac{1}{2}m\mathbf{v}^2 + e(\mathbf{v}\cdot\mathbf{A}) - e\phi \tag{1.69}$$

Substitution of (1.69) into the Lagrangian equations (1.57), using the particle Cartesian coordinates for the q_i, yields

$$\frac{d}{dt}(mv_i + eA_i) - e\frac{\partial}{\partial x_i}(\mathbf{v}\cdot\mathbf{A} - \phi) = 0 \tag{1.70}$$

since the external vector potential $\mathbf{A} = \mathbf{A}(\mathbf{r}, t)$ depends only on position and time. This result is in fact just Eq. 1.68 in component form, which confirms that the system Lagrangian is correctly given by Eq. 1.69. According to Eq. 1.62, the system Hamiltonian becomes

$$
\begin{aligned}
H &= \sum_{i=1}^{3} p_i \dot{q}_i - L \\
&= \sum_{i=1}^{3} \left(\frac{\partial L}{\partial \dot{x}_i}\right) \dot{x}_i - \tfrac{1}{2} m\mathbf{v}^2 - e(\mathbf{v} \cdot \mathbf{A}) + e\phi \\
&= m\mathbf{v}^2 + e(\mathbf{v} \cdot \mathbf{A}) - \tfrac{1}{2} m\mathbf{v}^2 - e(\mathbf{v} \cdot \mathbf{A}) + e\phi \\
&= \tfrac{1}{2} m\mathbf{v}^2 + e\phi = \frac{1}{2m} (m\mathbf{v})^2 + e\phi
\end{aligned}
\tag{1.71}
$$

The Hamiltonian is conventionally written in terms of the position coordinates q_i (x_i if Cartesian) and their conjugate momenta p_i. From Eq. 1.58, the latter are

$$
p_i = \frac{\partial L}{\partial \dot{x}_i} = m\dot{x}_i + eA_i \tag{1.72}
$$

or $\mathbf{p} = m\mathbf{v} + e\mathbf{A}$. Substitution of $(\mathbf{p} - e\mathbf{A})$ for $m\mathbf{v}$ in Eq. 1.71 then produces the classical Hamiltonian for a charged particle in an electromagnetic field as a function of the conjugate Cartesian variables $\mathbf{r}$ and $\mathbf{p}$,

$$
H = \frac{1}{2m} (\mathbf{p} - e\mathbf{A})^2 + e\phi \tag{1.73}
$$

(the $\mathbf{r}$ – dependence in H stems from the r – dependence in the scalar and vector potentials). Physically, the conjugate momentum $\mathbf{p}$ does not equal $m\mathbf{v}$ because the particle's linear momentum in the electromagnetic field is influenced by the vector potential $\mathbf{A}$. This Hamiltonian is straightforwardly modified if, in addition to the external fields we have just treated, the particle experiences a conservative potential $V(\mathbf{r})$ (e.g., that arising from electrostatic interactions with other charges in a molecule). The correct Hamiltonian in this case is given by [7, 8]

$$
H = \frac{1}{2m} (\mathbf{p} - e\mathbf{A})^2 + e\phi + V \tag{1.74}
$$

1.4 TIME-DEPENDENT PERTURBATION THEORY OF RADIATION–MATTER INTERACTIONS

The quantum mechanical Hamiltonian operator corresponding to the classical Hamiltonian (1.74) is

$$
\hat{H} = \frac{1}{2m} \left(\frac{\hbar}{i} \nabla - e\mathbf{A}\right)^2 + e\phi + V \tag{1.75}
$$

which may be expanded into

$$\hat{H} = \frac{1}{2m}\left[-\hbar^2\nabla^2 - \frac{\hbar e}{i}(\nabla\cdot\mathbf{A}) - \frac{2\hbar e}{i}(\mathbf{A}\cdot\nabla) + e^2\mathbf{A}^2\right]$$
$$+e\phi + V \tag{1.76}$$

The parentheses surrounding the quantity $\nabla\cdot\mathbf{A}$ in Eq. 1.76 indicate that ∇ operates only on the vector potential $\mathbf{A}$ immediately following it. The choice of Coulomb gauge ($\nabla\cdot\mathbf{A} = 0$) for electromagnetic waves propagating in free space ($\phi = 0$) reduces the Hamiltonian to

$$\hat{H} = \left[-\frac{\hbar^2}{2m}\nabla^2 + V(\mathbf{r})\right] + \left[\frac{e^2\mathbf{A}^2}{2m} - \frac{\hbar e}{im}(\mathbf{A}\cdot\nabla)\right] \equiv \hat{H}_0 + W \tag{1.77}$$

where the terms have now been grouped in the form of Eq. 1.16. $\hat{H}_0$ represents the zero-order Hamiltonian for the particle unperturbed by the external fields, and the terms arising from the radiation–matter interaction have been isolated in the perturbation W. For particles bound in atoms or molecules experiencing ordinary electromagnetic waves, the internal electric fields due to $V(\mathbf{r})$ are orders of magnitude larger than the external fields, with the consequence that $e\mathbf{A} \ll \mathbf{p}$. The quadratic term $e^2\mathbf{A}^2$ in W then becomes negligible next to $e\hbar\mathbf{A}\cdot\nabla$, with the result that the perturbation is well approximated by the linear term,

$$W = \frac{ie\hbar}{m}\mathbf{A}\cdot\nabla \tag{1.78}$$

As a concrete example, the vector potential for a linearly polarized monochromatic electromagnetic plane wave with wave vector $\mathbf{k}$ may be written

$$\mathbf{A}(\mathbf{r}, t) = \mathrm{Re}(\mathbf{A}_0 e^{i(\mathbf{k}\cdot\mathbf{r} - \omega t)})$$
$$= \mathbf{A}_0 \cos(\mathbf{k}\cdot\mathbf{r} - \omega t) \tag{1.79}$$

where $\mathbf{k}$ and the circular frequency ω are related to the wavelength λ and frequency ν by

$$|\mathbf{k}| = 2\pi/\lambda \tag{1.80a}$$

$$\omega = 2\pi\nu \tag{1.80b}$$

Since $\phi = 0$, the electric field is

$$\mathbf{E}(\mathbf{r}, t) = -\frac{\partial\mathbf{A}}{\partial t} = -\omega\mathbf{A}_0 \sin(\mathbf{k}\cdot\mathbf{r} - \omega t)$$
$$= -c|\mathbf{k}|\mathbf{A}_0 \sin(\mathbf{k}\cdot\mathbf{r} - \omega t) \tag{1.81}$$

so that the electric field points antiparallel to **A**. The magnetic field associated with the light wave is

$$\mathbf{B}(\mathbf{r}, t) = \nabla \times \mathbf{A} = -\mathbf{k} \times \mathbf{A}_0 \sin(\mathbf{k}\cdot\mathbf{r} - \omega t) \tag{1.82}$$

Another way of writing Eq. 1.82 is

$$\mathbf{B}(\mathbf{r}, t) = \begin{vmatrix} \hat{i} & \hat{j} & \hat{k} \\ \dfrac{\partial}{\partial x} & \dfrac{\partial}{\partial y} & \dfrac{\partial}{\partial z} \\ A_x & A_y & A_z \end{vmatrix} \tag{1.83}$$

Comparison of Eqs. 1.81 and 1.82 shows that **E** and **B** are mutually orthogonal, and the latter equation requires that **B** is orthogonal to the wave's propagation vector **k**. Hence the vector potential in Eq. 1.79 describes a linearly polarized transverse electromagnetic wave. The wave propagates at the speed of light c, because **E** and **B** are both functions of $(\mathbf{k}\cdot\mathbf{r} - \omega t) \equiv |\mathbf{k}|\ (\hat{k}\cdot\mathbf{r} - \nu\lambda t) \equiv |\mathbf{k}|$ $(\hat{k}\cdot\mathbf{r} - ct)$ according to Eqs. 1.80.

Since the vector potential **A** in Eq. 1.79 depends explicitly on time, the perturbation $W = (ie\hbar/m)\mathbf{A}\cdot\nabla$ is time-dependent as well. A perturbation theory based on the time-dependent Schrödinger equation (1.14) must therefore be used to describe the radiation–matter coupling. The Hamiltonian is assumed to have the form of Eq. 1.77, except that the perturbation $W(t)$ is now explicitly acknowledged to depend on time. The molecule has zero-order eigenstates $|\psi_n^{(0)}(\mathbf{r}, t)\rangle$ obeying Eqs. 1.13 through 1.15. It is assumed initially (at $t = -\infty$) that the molecule is in state $|k\rangle \equiv |\psi_k^{(0)}\rangle$. We then turn on the perturbation $W(t)$, which can cause the molecule to undergo a transition to some other state $|m\rangle \equiv |\psi_m^{(0)}\rangle$ because of its interaction with the radiation field. We wish to calculate the probability that the molecule ends up in state $|m\rangle$ by some later time t.

In general, the state of the interacting molecule–radiation system $|\Psi(\mathbf{r}, t)\rangle$ will not coincide with one of the zero-order states $|\Psi_n(\mathbf{r}, t)\rangle|\chi_{\text{rad}}\rangle$, because the Schrödinger equation is modified by the presence of the coupling term $W(t)$. (In what follows, we will drop $|\chi_{\text{rad}}\rangle$ from our discussion, since including it could only tell us how many photons of each type (energy, polarization, etc.) will be absorbed or emitted in a given transition, and we have other ways of obtaining this information. By focusing on the *molecular* states $|\Psi_n(\mathbf{r}, t)\rangle$, we gain the far more interesting information about what happens in the molecule.) If we have a complete orthonormal set of zeroth-order (i.e., isolated-molecule) eigenstates $|\Psi_n(\mathbf{r}, t)\rangle$ of $\hat{H}_0$, the mixed state $|\Psi(\mathbf{r}, t)\rangle$ can always be expressed as

$$|\Psi(\mathbf{r}, t)\rangle = \sum_n c_n(t)|\Psi_n(\mathbf{r}, t)\rangle \equiv \sum_n c_n(t) \exp(-iE_n^{(0)}t/\hbar)\,|n\rangle \tag{1.84}$$

by a suitable choice of coefficients $c_n(t)$. The $c_n(t)$ are assumed to be normalized and to obey the initial conditions

$$\left.\begin{aligned} c_k(-\infty) &= 1 \\ c_{n\neq k}(-\infty) &= 0 \end{aligned}\right\} \qquad c_n(-\infty) = \delta_{nk} \tag{1.85}$$

$$\sum_n |c_n(t)|^2 = 1 \tag{1.86}$$

Equation 1.85 states that $|\Psi(\mathbf{r}, -\infty)\rangle = \exp(-iE_k^{(0)}t/\hbar)\,|k\rangle$; i.e., the molecule is initially in state $|k\rangle$ with energy $E_k \equiv E_k^{(0)}$. The expansion (1.84) can be substituted into the time-dependent Schrödinger equation to give

$$[\hat{H}_0 + W(t)] \sum_n c_n(t)\, e^{-iE_nt/\hbar}|n\rangle = i\hbar \frac{\partial}{\partial t} \sum_n c_n(t)\, e^{-iE_nt/\hbar}|n\rangle \tag{1.87}$$

Using the fact that $\hat{H}_0|n\rangle = E_n|n\rangle$ and multiplying on the left by the bra $\langle m|$, we have

$$\begin{aligned} \langle m| \sum_n c_n(t) W(t) e^{-iE_nt/\hbar}|n\rangle &= i\hbar \langle m| \sum_n \frac{\partial c_n}{\partial t} e^{-iE_nt/\hbar}|n\rangle \\ &= i\hbar \sum \frac{\partial c_n}{\partial t} e^{-iE_nt/\hbar}\langle m|n\rangle \\ &= i\hbar\, e^{-iE_mt/\hbar} \frac{\partial c_m}{\partial t} \end{aligned} \tag{1.88}$$

The latter follows since $\langle m|n\rangle = \delta_{mn}$ in an orthonormal set, so that only the term with $n = m$ survives in the summation. Hence, the time-dependent coefficients obey the coupled equations

$$\frac{dc_m}{dt} = \frac{1}{i\hbar} \sum_n c_n(t)\, e^{-i(E_n - E_m)t/\hbar}\langle m|W(t)|n\rangle \tag{1.89}$$

This expression is *exact*. We can now introduce the spirit of the perturbation theory by assuming that the transition probability from the initial state $|k\rangle$ to some other state $|m\rangle$ is small. This would imply that $c_n(t) \simeq \delta_{kn}$ at all times t, not just at $t = -\infty$. Hence, as a *zeroth* approximation, we take [9]

$$c_k^{(0)}(t) = 1 \tag{1.90}$$

$$c_{n\neq k}^{(0)}(t) = 0 \tag{1.91}$$

and, using ω_{nm} to denote $(E_n - E_m)/\hbar$ and substituting Eq. 1.90 into the *right* side of Eq. 1.89, we can get an expression for the next order of approximation to $c_m(t)$:

$$\frac{dc_m^{(1)}}{dt} = \frac{1}{i\hbar} \sum_n \delta_{nk} e^{-i\omega_{nm}t} \langle m|W(t)|n\rangle$$
$$= \frac{1}{i\hbar} e^{-i\omega_{km}t} \langle m|W(t)|k\rangle \qquad (1.92)$$

From this, considering the initial condition that $c_m(t) \to c_m^{(0)}(t)$ as $t \to -\infty$ (and hence that $c_m^{(1)}(t) \to 0$ for $m \neq k$ in this limit), we can integrate Eq. 1.92 to obtain

$$c_m^{(1)}(t) = \frac{1}{i\hbar} \int_{-\infty}^{t} e^{-i\omega_{km}t_1} \langle m|W(t_1)|k\rangle dt_1 \qquad (1.93)$$

For a second approximation to $c_m(t)$, we can place $c_m^{(1)}(t)$ into the right side of Eq. 1.89 to obtain

$$\frac{dc_m^{(2)}}{dt} = \frac{1}{i\hbar} \sum_n c_n^{(1)}(t) e^{-i\omega_{nm}t} \langle m|W(t)|n\rangle$$
$$= \frac{1}{(i\hbar)^2} \sum_n e^{-i\omega_{nm}t} \langle m|W(t)|n\rangle \int_{-\infty}^{t} e^{-i\omega_{kn}t_1} \langle n|W(t_1)|k\rangle dt_1 \qquad (1.94)$$

Integrating this with the proper initial condition leads to

$$c_m^{(2)}(t) = \frac{1}{(i\hbar)^2} \sum_n \int_{-\infty}^{t} e^{-i\omega_{nm}t_1} \langle m|W(t_1)|n\rangle dt_1$$
$$\times \int_{-\infty}^{t_1} e^{-i\omega_{kn}t_2} \langle n|W(t_2)|k\rangle dt_2 \qquad (1.95)$$

Iteration of this process will show that [9]

$$c_m(t) = c_m^{(0)}(t) + c_m^{(1)}(t) + c_m^{(2)}(t) + \cdots$$
$$= \delta_{km} + \frac{1}{i\hbar} \int_{-\infty}^{t} e^{-i\omega_{km}t_1} \langle m|W(t_1)|k\rangle dt_1$$
$$+ \frac{1}{(i\hbar)^2} \sum_n \int_{-\infty}^{t} e^{-i\omega_{nm}t_1} \langle m|W(t_1)|n\rangle dt_1 \int_{-\infty}^{t_1} e^{-i\omega_{kn}t_2} \langle n|W(t_2)|k\rangle dt_2$$
$$+ \frac{1}{(i\hbar)^3} \sum_{nn'} \int_{-\infty}^{t} e^{-i\omega_{nm}t_1} \langle m|W|n\rangle dt_1 \int_{-\infty}^{t_1} e^{-i\omega_{n'n}t_2} \langle n|W|n'\rangle dt_2$$
$$\times \int_{-\infty}^{t_2} e^{-i\omega_{kn'}t_3} \langle n'|W|k\rangle dt_3$$
$$+ \cdots \qquad (1.96)$$

In the absence of any perturbation $W(t)$, $c_m(t)$ is given by $c_m^{(0)} = \delta_{km}$, and no transitions can occur from the initial state $|k\rangle$. The next term $c_m^{(1)}(t)$ corresponds to one-photon processes (absorption and emission of single photons), and covers most of classical spectroscopy. The two-photon processes (two-photon absorption and Raman spectroscopy) are contained in the second-order term $c_m^{(2)}(t)$, the three-photon processes (e.g., second-harmonic generation and three-photon absorption) correspond to $c_m^{(3)}(t)$, and so on. We will concentrate on the consequences of the first-order (one-photon) term $c_m^{(1)}(t)$ in the next few chapters. Higher order terms like $c_m^{(2)}(t)$ and $c_m^{(3)}(t)$ require intense electromagnetic fields (i.e., lasers) to gain importance, and indeed the practicality of Raman spectroscopy bloomed dramatically with the advent of lasers.

Under the normalization and initial conditions (1.85) and (1.86), the probability that the molecule has reached state $|m\rangle$ at time t is equal to $|c_m(t)|^2$. In first order, $c_m(t)$ is given by

$$c_m^{(1)}(t) = \frac{1}{i\hbar} \int_{-\infty}^{t} e^{-i\omega_{km}t_1} \langle m|W(t_1)|k\rangle dt_1 \tag{1.97}$$

and so we must have $\langle m|W(t)|k\rangle \neq 0$ for an *allowed* $k \to m$ one-photon transition. The transition is otherwise said to be *forbidden.* To calculate molecular transition probabilities more concretely and to derive general selection rules for allowed transitions, we need only to substitute specific expressions for $W(t)$.

1.5 SELECTION RULES FOR ONE-PHOTON TRANSITIONS

Heuristic selection rules for one-photon transitions may be obtained by using Eq. 1.9 or 1.10 for the perturbation W in the expression for $c_m^{(1)}(t)$, Eq. 1.97. This procedure yields the matrix element

$$\begin{aligned}\langle m|W|k\rangle &= \langle m|e\phi|k\rangle - \langle m|\boldsymbol{\mu}\cdot\mathbf{E}|k\rangle \\ &\quad - \langle m|\frac{1}{2}e\sum_{ij} x_i x_j \frac{\partial E_j}{\partial x_i}|k\rangle + \cdots \\ &= 0 - \mathbf{E}\cdot\langle m|\boldsymbol{\mu}|k\rangle - \frac{1}{2}\sum_{ij} e\frac{\partial E_j}{\partial x_i}\langle m|x_i x_j|k\rangle + \cdots \end{aligned} \tag{1.98}$$

which controls the probability of transitions from state k to state m. The first term vanishes ($\langle m|k\rangle = 0$) due to orthogonality between eigenstates of $\hat{H}_0$ having different energy eigenvalues. The second term results from interaction of the instantaneous molecular dipole moment with the external electric field $\mathbf{E}$, and leads to *electric dipole* (E1) *transitions* from state k to state m. The third term arises from interaction of the instantaneous molecular quadrupole moment tensor with the electric field gradients $\partial E_j/\partial x_i$; it is responsible for *electric quadrupole* (E2) transitions from state k to state m. Our qualification that it is the instantaneous (rather than permanent) moments that are critical here is

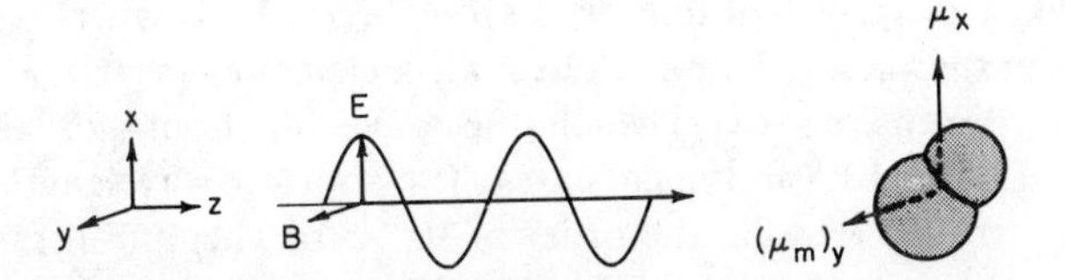

Figure 1.6 Orientations of the **E** and **B** fields associated with the linearly polarized light wave described by Eqs. 1.99. The **E** field, directed along the *x* axis, interacts with the *x* component of the molecule's instantaneous electric dipole moment; the **B** field, directed along the *y* axis, interacts with the *y* component of the molecule's instantaneous magnetic dipole moment.

important, since, for example, E1 transitions can occur in atoms (e.g., in Na and Hg lamps) even though no atom has any nonvanishing permanent dipole moment $\boldsymbol{\mu}_0$. The foregoing discussion can be summarized in the following *selection rules*:

For allowed E1 transitions, $\langle m|\boldsymbol{\mu}|k\rangle \neq 0$
For allowed E2 transitions, $\langle m|x_i x_j|k\rangle \neq 0$ for some (i, j)

While this discussion based on electrostatics ignores the time dependence in $W(t)$ and omits the effects of magnetic fields associated with the light wave, it does anticipate some of our final results in this section regarding electric dipole and electric quadrupole contributions to the matrix elements $\langle m|W(t)|k\rangle$. It yields no insight into magnetic multipole transitions or into the nature of the time-ordered integrals in the Dyson series expansion of Eq. 1.96.

Next we calculate the matrix elements using the correct time-dependent perturbation $W = (ie\hbar/m)(\mathbf{A}\cdot\nabla)$, Eq. 1.78. We assume for clarity that the vector potential is that for a linearly polarized plane wave (Eq. 1.79) with $\mathbf{A}_0 = A_0\hat{i}$ and $\mathbf{k} = |\mathbf{k}|\hat{k}$. This vector potential points along the x axis and propagates along the z axis (Fig. 1.6); results for the more general case are given at the end of this discussion. Following Eqs. 1.81 and 1.82, the electric and magnetic fields corresponding to this vector potential are

$$\mathbf{E}(\mathbf{r}, t) = -c|\mathbf{k}|A_0\hat{i}\sin(kz - \omega t) \tag{1.99a}$$

$$\mathbf{B}(\mathbf{r}, t) = -\hat{j}|\mathbf{k}|A_0\sin(kz - \omega t) \tag{1.99b}$$

so that the **E** and **B** fields point along the negative x and y axes, respectively. The matrix element $\langle m|W(t)|k\rangle$ becomes

$$\begin{aligned}\langle m|W(t)|k\rangle &= \frac{i\hbar e}{m}\langle m|\mathbf{A}_0 e^{i(\mathbf{k}\cdot\mathbf{r}-\omega t)}\cdot\nabla|k\rangle \\ &= \frac{i\hbar e}{m}\mathbf{A}_0 e^{-i\omega t}\cdot\langle m|e^{i\mathbf{k}\cdot\mathbf{r}}\nabla|k\rangle \\ &= \frac{i\hbar e}{m}\mathbf{A}_0 e^{-i\omega t}\cdot\langle m|\left(1 + i\mathbf{k}\cdot\mathbf{r} + \frac{(i\mathbf{k}\cdot\mathbf{r})^2}{2} + \cdots\right)\nabla|k\rangle\end{aligned} \tag{1.100}$$

The quantity $\mathbf{k}\cdot\mathbf{r}$ is equivalent to $|\mathbf{k}|\,|\mathbf{r}|\cos\theta = 2\pi|\mathbf{r}|\cos\theta/\lambda$, where θ is the angle formed between the vectors $\mathbf{k}$ and $\mathbf{r}$. The matrix elements $\langle m|W(t)|k\rangle$ limit $|\mathbf{r}|$ to the molecular dimensions over which the wave functions $|k\rangle$ and $\langle m|$ are appreciable, i.e., $|\mathbf{r}| \leqslant 10\,\text{Å}$ in typical cases. The shortest wavelengths λ used in molecular spectroscopy are on the order of 10^3 Å for vacuum-ultraviolet light, and are of course much longer for visible, IR, and microwave spectroscopy. Hence $\mathbf{k}\cdot\mathbf{r}$ is typically much less than 1, and the series expansion of $\exp(i\mathbf{k}\cdot\mathbf{r})$ converges rapidly. In the special geometry we have assumed for our vector potential,

$$\begin{aligned}\langle m|W(t)|k\rangle &= \frac{i\hbar e}{m} A_0 e^{-i\omega t}\langle m|1 \frac{\partial}{\partial x}|k\rangle \\ &+ \frac{i\hbar e}{m} A_0 e^{-i\omega t}\langle m|(ikz)\frac{\partial}{\partial x}|k\rangle \\ &+ \frac{i\hbar e}{m} A_0 e^{-i\omega t}\langle m|\frac{(ikz)^2}{2}\frac{\partial}{\partial x}|k\rangle + \cdots \end{aligned} \tag{1.101}$$

The first term in $\langle m|W(t)|k\rangle$ requires the matrix element $\langle m|\partial/\partial x|k\rangle$. This can be obtained by evaluating the commutator

$$\begin{aligned}[\hat{H}_0, x] &= [\hat{p}^2/2m + V(x), x] = \frac{1}{2m}[\hat{p}^2, x] \\ &= \frac{1}{2m}[\hat{p}_x^2, x] = \frac{1}{2m}[\hat{p}_x, x]\hat{p}_x + \frac{\hat{p}_x}{2m}[\hat{p}_x, x] \\ &\equiv -\frac{i\hbar}{m}\hat{p}_x \equiv \hat{H}_0 x - x\hat{H}_0 \end{aligned} \tag{1.102}$$

where we have used the commutator identity $[AB, C] = A[B, C] + [A, C]B$ [3]. Then

$$\frac{\partial}{\partial x} = \frac{i}{\hbar}\hat{p}_x = -\frac{m}{\hbar^2}\left(-\frac{i\hbar}{m}\hat{p}_x\right) = -\frac{m}{\hbar^2}(\hat{H}_0 x - x\hat{H}_0) \tag{1.103}$$

and so

$$\begin{aligned}\langle m|\frac{\partial}{\partial x}|k\rangle &= -\frac{m}{\hbar^2}\langle m|\hat{H}_0 x - x\hat{H}_0|k\rangle \\ &= -\frac{m}{\hbar^2}\langle m|E_m x - xE_k|k\rangle \\ &= -\frac{m}{\hbar^2}(E_m - E_k)\langle m|x|k\rangle \\ &= -\frac{m\omega_{mk}}{\hbar}\langle m|x|k\rangle \end{aligned} \tag{1.04}$$

The second term in $\langle m|W(t)|k\rangle$ requires

$$\begin{aligned}\langle m|z\frac{\partial}{\partial x}|k\rangle &= \tfrac{1}{2}\langle m|z\frac{\partial}{\partial x} - x\frac{\partial}{\partial z}|k\rangle + \tfrac{1}{2}\langle m|z\frac{\partial}{\partial x} + x\frac{\partial}{\partial z}|k\rangle \\ &= \frac{i}{2\hbar}\langle m|z\frac{\hbar}{i}\frac{\partial}{\partial x} - x\frac{\hbar}{i}\frac{\partial}{\partial z}|k\rangle + \tfrac{1}{2}\langle m|z\frac{\partial}{\partial x} + x\frac{\partial}{\partial z}|k\rangle \\ &= \frac{i}{2\hbar}\langle m|\hat{L}_y|k\rangle + \tfrac{1}{2}\langle m|z\frac{\partial}{\partial x} + x\frac{\partial}{\partial z}|k\rangle \end{aligned} \tag{1.105}$$

since the y component of the *orbital* (not spin or total) angular momentum is $\hat{L}_y = z\hat{p}_x - x\hat{p}_z$. The last matrix element on the right side above can be obtained using the commutator

$$\begin{aligned}[\hat{H}_0, xz] &= \frac{1}{2m}[\hat{p}^2, xz] \\ &= \frac{1}{2m}[\hat{p}_x^2 + \hat{p}_z^2, xz] \\ &= \frac{1}{2m}2\hat{p}_x[\hat{p}_x, x]z + \frac{1}{2m}2\hat{p}_z x[\hat{p}_z, z] \\ &= -\frac{i\hbar}{m}(\hat{p}_x z + x\hat{p}_z) \\ &= -\frac{\hbar^2}{m}\left(z\frac{\partial}{\partial x} + x\frac{\partial}{\partial z}\right)\end{aligned} \tag{1.106}$$

Then

$$\begin{aligned}\langle m|z\frac{\partial}{\partial x} + x\frac{\partial}{\partial z}|k\rangle &= -\frac{m}{\hbar^2}\langle m|\hat{H}_0 xz - xz\hat{H}_0|k\rangle \\ &= \frac{m}{\hbar^2}(E_k - E_m)\langle m|xz|k\rangle \\ &= \frac{m\omega_{km}}{\hbar}\langle m|xz|k\rangle\end{aligned} \tag{1.107}$$

Collecting these results for the first and second terms in $\langle m|W(t)|k\rangle$, we summarize that

$$\begin{aligned}\langle m|W(t)|k\rangle = &+ie\omega_{km}A_0 e^{-i\omega t}\langle m|x|k\rangle \\ &-\frac{ike}{2m}A_0 e^{-i\omega t}\langle m|\hat{L}_y|k\rangle + \frac{ke\omega_{km}}{2}A_0 e^{-i\omega t}\langle m|xz|k\rangle \\ &-\frac{i\hbar k^2}{2m}eA_0 e^{-i\omega t}\langle m|z^2\frac{\partial}{\partial x}|k\rangle + \cdots\end{aligned} \tag{1.108}$$

For a single particle, the electric dipole operator is $\boldsymbol{\mu} = e\mathbf{r}$; the first term on the right side of Eq. 1.108 is therefore

$$i\omega_{km}A_0e^{-i\omega t}\langle m|\mu_x|k\rangle$$

and it represents the electric dipole (E1) contribution to the total transition probability. Only the x component of $\boldsymbol{\mu}$ appears here, because in our example the $\mathbf{E}$ field of the light wave has only an x component (Fig. 1.6), and the electric dipole interaction behaves as $\boldsymbol{\mu}\cdot\mathbf{E}$. The second term in Eq. 1.108 can be recast in terms of the *magnetic dipole moment operator* in SI units

$$\boldsymbol{\mu}_m = e\mathbf{L}/2m \tag{1.109}$$

(the orbital angular moment $\mathbf{L}$ of a moving charged particle physically gives rise to a proportional magnetic dipole moment $\boldsymbol{\mu}_m$ in the same direction as $\mathbf{L}$ for $e > 0$), and so it becomes

$$-ikA_0e^{-i\omega t}\langle m|(\boldsymbol{\mu}_m)_y|k\rangle$$

This corresponds to the magnetic dipole (M1) contribution. The energy of a magnetic dipole moment $\boldsymbol{\mu}_m$ in a uniform magnetic field $\mathbf{B}$ is $-\boldsymbol{\mu}_m\cdot\mathbf{B}$, and the magnetic field of our current problem is directed along the y axis (Fig. 1.6)—which is why only the y component of $\boldsymbol{\mu}_m$ appears in this term. The third term in Eq. 1.108 embodies the electric quadrupole (E2) xz component, which is the only contributing electric quadrupole tensor component since $\mathbf{E}$ in our example has only an x component that spatially depends only on z (all of the other $\partial E_j/\partial x_i$ are zero). The succeeding terms not shown in Eq. 1.108 describe higher order (electric octupole, magnetic quadrupole, etc.) transitions; their importance decreases rapidly with increasing order because of the increasingly high powers in (kz). Since the E2 and M1 transition amplitudes contain a factor of kz that is absent in the E1 term of $\langle m|W(t)|k\rangle$, they are inherently much weaker transitions than electric dipole transitions. The vast majority of one-photon spectroscopic transitions that are exploited in practice are E1 transitions.

We now give without proof the matrix element of $W(t)$ for the more general case of an incident plane wave $\mathbf{A}(\mathbf{r}, t) = \mathbf{A}_0 \exp(i\mathbf{k}\cdot\mathbf{r} - \omega t)$ in which $\mathbf{k}$ and $\mathbf{A}_0$ point in arbitrary directions ($\mathbf{A}_0$ must be normal to $\mathbf{k}$ to give a physically real light wave in vacuum, however):

$$\begin{aligned}\langle m|W(t)|k\rangle &= \frac{i\hbar e}{m}\, e^{-i\omega t}\left[\frac{m\omega_{km}}{\hbar}\,\langle m|(\mathbf{A}_0\cdot\mathbf{r})|k\rangle\right.\\ &\qquad \left. -\frac{1}{2\hbar}\,\langle m|L_\perp|\mathrm{k}\rangle + \frac{im\omega_{km}}{\hbar}\,\langle m|(\mathbf{k}\cdot\mathbf{r})(\mathbf{A}_0\cdot\mathbf{r})|k\rangle + \cdots\right]\\ &\equiv e^{-i\omega t}C\end{aligned} \tag{1.110}$$

in which $L_\perp$ denotes the component of orbital angular momentum about an axis normal to both $\mathbf{A}_0$ and $\mathbf{k}$. These terms in order correspond to the E1, M1, and E2 $k \to m$ transition amplitudes.

From Eq. 1.96, the $k \to m$ *one-photon* transition probability becomes

$$
\begin{aligned}
|c_m^{(1)}(t)|^2 &= \frac{1}{\hbar^2}\left|\int_{-\infty}^{t} e^{-i\omega_{km}t_1}\langle m|W(t_1)|k\rangle dt_1\right|^2 \\
&= \frac{|C|^2}{\hbar^2}\left|\int_{-\infty}^{t} e^{-i(\omega_{km}+\omega)t_1} dt_1\right|^2
\end{aligned} \tag{1.111}
$$

meaning that the transition probability is proportional to the absolute value squared of the weighted sum of matrix elements in Eq. 1.110. To see the significance of the time integral, we may take the limit as $t \to +\infty$ (the continuous-wave limit) to get

$$
|c_m^{(1)}(t)|^2 = \frac{4\pi^2|C|^2}{\hbar^2}[\delta(\omega_{km} + \omega)]^2 \tag{1.112}
$$

since the integral representation of the Dirac delta function is

$$
\delta(x) = \frac{1}{2\pi}\int_{-\infty}^{\infty} e^{ixt}\, dt \tag{1.113}
$$

Hence, the $k \to m$ transition in this limit cannot occur unless $\omega = -\omega_{km} = +\omega_{mk} = (E_m - E_k)/\hbar$. The frequency in the external radiation field must exactly match the energy level difference between the initial and final states, in accordance with the Ritz combination principle. Thus, the time integral leads to an "energy-conserving" delta function. This energy-matching condition should not be taken too seriously at this point, because in fact the energies E_m and E_k themselves are not sharply defined in general owing to lifetime broadening [10] (sometimes referred to as the "*time-energy uncertainty principle*"). Rather, the time integral in Eq. 1.111 expresses the ω-dependence of the transition probability in the idealized case when the energies of the two levels are infinitely sharp. It is interesting to note that if the upper integration limit t is set to some finite number rather than $+\infty$, the function $\delta(\omega_{km} + \omega)$ will be replaced by some function $g(\omega)$ with a finite, rather than zero, width. This corresponds to the fact that a light wave with less than infinite length has some uncertainty in its frequency ω, so that its center (or most probable) frequency can be detuned from $(E_m - E_k)/\hbar$ and still have some finite probability of effecting the molecular transition.

The selection rules we have derived in this section form the basis for all of the one-photon spectroscopies treated in Chapters 2 through 7 of this book. They may be succinctly summarized as follows for general one-photon transitions from state k to state m:

$$\text{Electric dipole (E1):} \quad \langle m|\boldsymbol{\mu}|k\rangle \neq 0 \tag{1.114a}$$

$$\text{Magnetic dipole (M1):} \quad \langle m|\mathbf{L}|k\rangle \neq 0 \tag{1.114b}$$

$$\text{Electric quadrupole (E2):} \quad \langle m|x_i x_j|k\rangle \neq 0 \quad \text{for some } (i, j) \tag{1.114c}$$

Obtaining the selection rules for two-photon and higher order multiphoton processes requires analysis of the expansion coefficients $c_m^{(2)}(t)$, $c_m^{(3)}(t)$, ... in Eq. 1.96. This is done explicitly for two-photon processes in Chapter 10, where two-photon absorption and Raman spectroscopy are discussed. This formalism becomes increasingly unwieldy when applied to three- and four-photon processes, and diagrammatic techniques then become useful for organizing the calculation of the pertinent transition probabilities (Chapter 11).

REFERENCES

1. M. H. Nayfeh and M. K. Brussel, *Electricity and Magnetism*, Wiley, New York, 1985.
2. J. D. Jackson, *Classical Electrodynamics*, Wiley, New York, 1962.
3. E. Merzbacher, *Quantum Mechanics*, Wiley, New York, 1961.
4. K. S. Pitzer, *Quantum Chemistry*, Prentice-Hall, Englewood Cliffs, NJ, 1953.
5. W. K. H. Panofsky and M. Phillips, *Classical Electricity and Magnetism*, 2d ed., Addison-Wesley, Reading, MA, 1962.
6. J. B. Marion, *Classical Dynamics of Particles and Systems*, Academic, New York, 1965.
7. R. H. Dicke and J. P. Wittke, *Introduction to Quantum Mechanics*, Addison-Wesley, Reading, MA, 1960.
8. W. H. Flygare, *Molecular Structure and Dynamics*, Prentice-Hall, Englewood Cliffs, NJ, 1978.
9. A. S. Davydov, *Quantum Mechanics*, NEO Press, Peaks Island, ME, 1966.
10. P. W. Atkins, *Molecular Quantum Mechanics*, 2d ed., Oxford Univ. Press, London, 1983.

PROBLEMS

1. For the electric and magnetic fields given in Eqs. 1.38, show that Maxwell's equations in vacuum (Eqs. 1.37) require that $\mathbf{E}_0 \cdot \mathbf{B}_0 = \mathbf{E}_0 \cdot \mathbf{k} = \mathbf{B}_0 \cdot \mathbf{k} = 0$.

2. A vector potential is given by $\mathbf{A}(\mathbf{r}, t) = A_0(\hat{j} + \hat{k})\cos(\mathbf{k} \cdot \mathbf{r} - \omega t)$, in which the wave vector $\mathbf{k} = \hat{i}|\mathbf{k}|$. Compute $\mathbf{E}(\mathbf{r}, t)$ and $B(\mathbf{r}, t)$ in the Coulomb gauge, and show that these fields obey Maxwell's equations in vacuum.

3. The evaluation of ground-state atomic or molecular polarizabilities using Eq. 1.35 requires accurate knowledge of all of the molecular eigenstates in principle. This proves to be unnecessary in the hydrogen atom (A. Dalgarno and J. T. Lewis, *Proc. R. Soc. London, Ser. A* **233**: 70 (1955); E. Merzbacher, *Quantum Mechanics*, Wiley, New York, 1961), where the second-order perturbation sum (1.35) can be evaluated exactly. In this problem, we evaluate α_{zz} for the 1s ground state $|0\rangle$

$$\alpha_{zz} = 2e^2 \sum_{l \neq 0} \frac{\langle 0|z|l\rangle\langle l|z|0\rangle}{E_l - E_0}$$

in which $|l\rangle$ denotes an excited state in hydrogen and the summation is evaluated over all such states.

(a) Verify by substitution that the function

$$F = -\frac{\mu a_0}{\hbar^2}\left(\frac{r}{2} + a_0\right) z$$

satisfies the commutation relation

$$z|0\rangle = (F\hat{H}_0 - \hat{H}_0 F)|0\rangle$$

Here μ and a_0 are the hydrogen atom reduced mass and Bohr radius, and $\hat{H}_0$ is the hydrogen atom Hamiltonian.

(b) Show that this result leads to

$$\alpha_{zz} = \frac{\mu a_0 e^2}{\hbar^2} \langle 0| \left(\frac{r}{2} + a_0\right) z^2|0\rangle$$

so that no information about excited states $|l\rangle$ with $l \neq 0$ is required to compute the polarizability in the hydrogen atom.

(c) Compute α_{zz} in Å^3. Compare this value with the polarizabilities of He (0.205 Å^3) and Li (24.3 Å^2) and discuss the differences.

4. An electromagnetic wave with vector potential

$$\mathbf{A}(\mathbf{r}, t) = A_0(\hat{j} + \hat{k}) \cos(kx - \omega t)$$

is incident on a 1s hydrogen atom.

(a) Calculate the **E** and **B** fields, assuming $\phi(\mathbf{r}) = 0$.

(b) Write down all of the *nonvanishing* terms in the matrix elements $\langle 1s|W|2s\rangle$, $\langle 1s|W|2p_x\rangle$, $\langle 1s|W|2p_y\rangle$, and $\langle 1s|W|3d_{xz}\rangle$ for this vector potential up to first order in $(\mathbf{k}\cdot\mathbf{r})$ in Eq. 1.100. [In terms of the hydrogen atom stationary states $|\psi_{nlm}(\mathbf{r})\rangle$, $|1s\rangle$ is $|\psi_{100}(\mathbf{r})\rangle$, $|2s\rangle$ is $|\psi_{200}(\mathbf{r})\rangle$, $|2p_x\rangle$ is the linear combination $2^{-1/2}(-|\psi_{211}\rangle + |\psi_{21,-1}\rangle)$, etc. Use elementary symmetry arguments to determine which of the matrix elements will vanish.]

(c) For this particular vector potential, which of the transitions $1s \to 2s$, $1s \to 2p_x$, $1s \to 2p_y$, and $1s \to 3d_{xz}$ are E1-allowed by symmetry? E2-allowed? M1-allowed?

5. Combine Maxwell's equations in vacuum with Eqs. 1.39 and 1.42 to generate the homogeneous wave equation for the vector potential in the Coulomb gauge,

$$\left(\nabla^2 - \mu_0\varepsilon_0 \frac{\partial^2}{\partial t^2}\right)\mathbf{A}(\mathbf{r}, t) = 0$$

Show that the most general solution to this wave equation is

$$\mathbf{A}(\mathbf{r}, t) = f(\mathbf{k}\cdot\mathbf{r} - \omega t)$$

where f is *any* function of the argument $(\mathbf{k}\cdot\mathbf{r} - \omega t)$ having first and second derivatives with respect to $\mathbf{r}$ and t. What physical significance does this function have in general?

6. After expanding the exponential in the matrix element $\langle m|\exp(ikz)(\partial/\partial x)|k\rangle$ in Eq. 1.101, we demonstrated that the first-order term $\langle m|(ikz)(\partial/\partial x)|k\rangle$ breaks down into a sum of contributions proportional to $\langle m|\hat{L}_y|k\rangle$ and $\langle m|xz|k\rangle$. These account for the M1 and E2 transition probabilities, respectively. Reduce the second-order term $\langle m|(ikz)^2(\partial/\partial x)|k\rangle$ into a similar pair of physically interpretable matrix elements, using the identity

$$z^2\frac{\partial}{\partial x} \equiv \frac{1}{3}\left[2z\left(z\frac{\partial}{\partial x} - x\frac{\partial}{\partial z}\right) + \left(z^2\frac{\partial}{\partial x} + 2xz\frac{\partial}{\partial z}\right)\right]$$

Determine which types of multipole transitions are embodied in this second-order term. What kinds of electric and magnetic field gradients are generally required to effect these types of transitions? By what factor do these transition probabilities differ from those of M1 and E2 transitions?

7. The time-dependent perturbation theory developed in Section 1.4 is useful for small perturbations, and is widely applied in spectroscopy. The contrasting situation in which the perturbation is *not* small compared to the energy separations between unperturbed levels is often more difficult to treat. A simplification occurs when the Hamiltonian changes *suddenly* at $t = 0$ from $\hat{H}_i$

to $\hat{H}_f$, where $\hat{H}_i$ and $\hat{H}_f$ are time-independent Hamiltonians satisfying

$$H_i|s; i\rangle = E_s|s; i\rangle$$
$$H_f|k; f\rangle = E_f|k; f\rangle$$

It can be shown that in the *sudden approximation* (which is applicable when the time τ during which the Hamiltonian changes satisfies $\tau(E_k - E_s) \ll h$), a system initially in state s of $\hat{H}_i$ will evolve into state k of $\hat{H}_f$ after $t = 0$ with probability

$$P_{s \to k} = |\langle s; i|k; f\rangle|^2$$

A 1s tritium atom undergoes 18 keV β decay to form He^+. With what probabilities is He^+ formed in the 1s, 2s, and 3d states?

2

ATOMIC SPECTROSCOPY

Atomic spectra accompany electronic transitions in neutral atoms and in atomic ions. One-photon transitions involving outer-shell (valence) electrons in neutral atoms yield spectral lines in the vacuum ultraviolet to the far infrared regions of the electromagnetic spectrum (Fig. 2.1), corresponding to wavelengths between several hundred angstroms and several meters. Transitions involving the more tightly bound inner-shell electrons give rise to spectra in the X-ray region at wavelengths below ~ 100 Å; we will not be concerned with X-ray spectra in this chapter.

Atomic *emission spectra* are commonly obtained by generating atoms in their electronic excited states in a vapor and analyzing the resulting emission with a spectrometer. Electric discharges produce excited atoms by allowing ground-state atoms to collide with electrons or ions that have been accelerated in an electric field. Such collisions convert part of the ion's translational kinetic energy into electronic excitation in the atom. Low-pressure mercury (Hg) calibration lamps operate by this mechanism. Atomic excited states may also be produced by excitation with lasers (Chapter 9), which are intense, highly monochromatic light sources. This monochromaticity permits selective laser excitation of single atomic states, a feature that is not possible in ordinary electric discharges. A less common method of generating excited atoms is by chemical reactions, and the resulting emission is called *chemiluminescence.* An important example is the bimolecular reaction between sodium dimers and chlorine atoms, $Na_2 + Cl \rightarrow Na^* + NaCl$, which creates electronically excited sodium atoms Na^* in a large number of different excited states. The photodissociation process $CH_3I + h\nu \rightarrow CH_3 + I^*$ initiated by ultraviolet light is an efficient method of producing iodine atoms I^* in their lowest excited electronic state, which cannot be reached by E1 one-photon transitions from ground-state I.

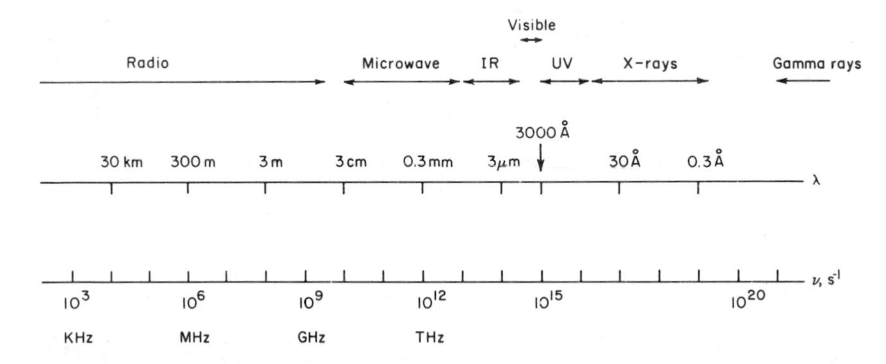

Figure 2.1 The electromagnetic spectrum. IR and UV are acronyms for infrared and ultraviolet, respectively; the abbreviations kHz, MHz, GHz, and THz stand for kilohertz, megahertz, gigahertz, and terahertz.

Light emitted at different wavelengths is spatially dispersed in grating spectrometers, and emission spectra may be recorded on photographic film. Alternatively, the grating instrument may be operated as a scanning monochromator that transmits a single wavelength (more precisely, a narrow bandwidth of wavelengths) at a time. Emission spectra may then be recorded using a sensitive photomultiplier tube to detect the emission transmitted by the monochromator while the latter is scanned through a range of wavelengths.

Atomic *absorption spectra* may be obtained by passing light from a source that emits a continuous spectrum (e.g., a tungsten filament lamp, whose output spectrum approximates that of a blackbody emitting at the filament temperature) through a cell containing the atomic vapor. The transmitted continuum is then dispersed in a grating spectrometer, and may be recorded either photographically or electronically using a vidicon (television camera tube) or linear photodiode array. Characteristic absorption wavelengths are associated with optical density minima in developed photographic negatives, or with transmitted light intensity minima detected on a vidicon or photodiode array grid. Emission spectroscopy is preferable to absorption spectroscopy for detection of atoms in trace amounts, since emitted photons are readily monitored photoelectrically with useful signal-to-noise ratios at atom concentrations at which absorption lines would be barely detectable in samples of reasonable size.

Representative emission spectra are shown schematically in Fig. 2.2 for hydrogen, potassium, and mercury on a common wavelength scale from the near infrared to the ultraviolet. Under the coarse wavelength resolution of this figure, the emitted light intensities are concentrated at single, well-defined emission lines. In H, the displayed emission consists of four convergent series of lines, the so-called Ritz-Paschen and Pfund series in the near infrared, the Lyman series in the vacuum ultraviolet, and the Balmer series in the visible. Johann Balmer, a schoolteacher in Basel in the late nineteenth century,

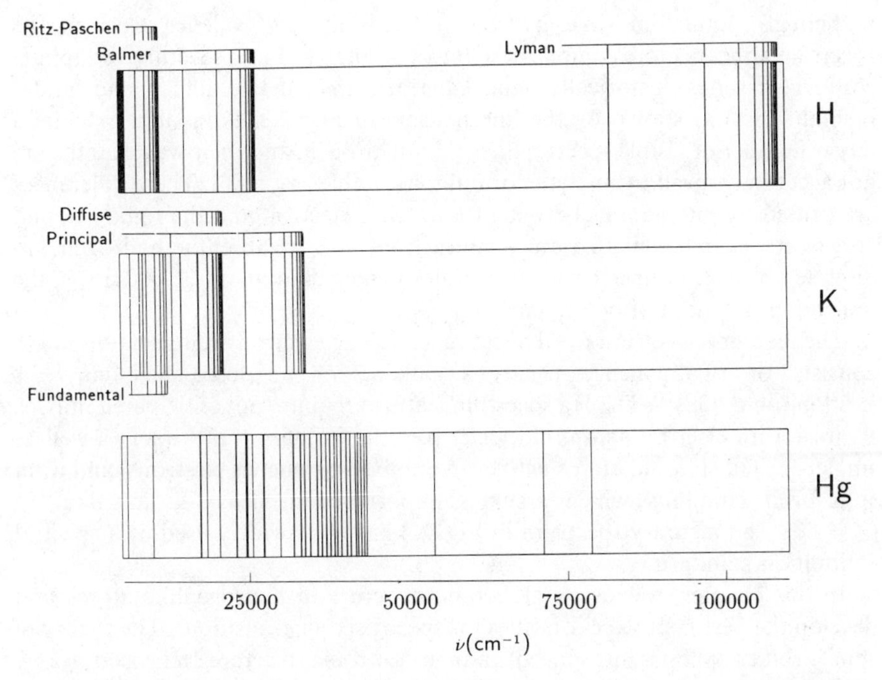

Figure 2.2 Schematic emission spectra of H, K, and Hg atoms. These are plotted versus the line frequencies $\bar{\nu} \equiv 1/\lambda$ in units of cm^{-1}, where the λ are the emission wavelengths in vacuum. Only the strongest emission lines are included, and relative line intensities are not shown. Line headers for H and K denote series of lines resulting from transitions terminating at a common lower level. Line headers are omitted for the Pfund series in H (which appears at extreme left) and for the sharp series in K, which closely overlaps the diffuse series.

discovered that the wavelengths in angstroms of lines in the latter series closely obey the remarkably simple formula

$$\lambda = 3645.6 \frac{m^2}{(m^2 - 2^2)} \tag{2.1}$$

with $m = 3, 4, 5, \ldots$. In the limit of large m, this expression converges to a *series limit* at $\lambda = 3645.6$ Å. (This limit is not directly observable in spectra like that in Fig. 2.2, because the line intensities become weak for large m.) The wavelengths of lines in the Ritz-Paschen series are similarly well approximated by $\lambda = 8202.6\, m^2/(m^2 - 3^2)$ with $m = 4, 5, 6, \ldots$. Hydrogenlike ions with atomic number $Z \neq 1$ (He^+, Li^{2+}, etc.) exhibit analogous series in which the emission wavelengths are scaled by the factor $1/Z^2$ relative to those in H. The compactness of the analytic expressions (cf. Eq. 2.1) for spectral line positions in hydrogenlike atoms is, of course, a consequence of their simple electronic structure.

Though potassium (like hydrogen) has only one valence electron, its spectrum appears more complicated. It can be analyzed into several overlapping convergent series (historically named the principal, sharp, diffuse, and fundamental series) as shown by the line headers in Fig. 2.2. Potassium exhibits a larger number of visible spectral lines than hydrogen, and their wavelengths are not accurately given by analytic formulas as simple as Eq. 2.1. These differences are caused by interactions between the valence electron and the tightly bound core electrons in the alkali atom—interactions that are absent in hydrogen. No discrete absorption lines occur at energies higher than about 35,000 cm^{-1}, the ionization potential of potassium.

The mercury spectrum is even less regular. The electron configuration in Hg consists of two valence electrons outside of a closed-shell core ... $(5s)^2(5p)^6(4f)^{14}(5d)^{10}$. The Hg spectrum features that are not anticipated in H or K arise from electron spin multiplicity (i.e., the formation of triplet as well as singlet excited states in atoms with even numbers of valence electrons) and from spin–orbit coupling, which assumes importance in heavy atoms like Hg ($Z = 80$). The mercury spectrum in Fig. 2.2 has been widely used as a spectral calibration standard.

In this chapter, we review electronic structure in hydrogenlike atoms and develop the pertinent selection rules for spectroscopic transitions. The theory of spin–orbit coupling is introduced, and the electronic structure and spectroscopy of many-electron atoms is greated. These discussions enable us to explain details of the spectra in Fig. 2.2. Finally, we deal with atomic perturbations in static external magnetic fields, which lead to the normal and anomalous Zeeman effects. The latter furnishes a useful tool for the assignment of atomic spectral lines.

2.1 HYDROGENLIKE SPECTRA

The unperturbed Hamiltonian for an electron in a hydrogenlike atom with nuclear charge $+Ze$ is

$$\hat{H}_0 = -\frac{\hbar^2}{2\mu}\nabla^2 - \frac{Ze^2}{4\pi\varepsilon_0 r} \tag{2.2}$$

The atomic reduced mass μ is related to the nuclear mass m_N and electron mass m_e by $\mu = m_e m_N/(m_e + m_N)$, ∇^2 operates on the electronic coordinates, and r is the electron–nuclear separation. The eigenfunctions $|\psi_{nlm}(r, \theta, \phi)\rangle$ and eigenvalues E_n of this Hamiltonian exhibit the properties

$$\hat{H}_0|\psi_{nlm}(r, \theta, \phi)\rangle = E_n|\psi_{nlm}(r, \theta, \phi)\rangle \tag{2.3}$$

$$n = 1, 2, \ldots$$

$$l = 0, 1, \ldots, (n-1)$$

$$m = 0, \pm 1, \ldots, \pm l$$

$$|\psi_{nlm}(r, \theta, \phi)\rangle = R_{nl}(r)Y_{lm}(\theta, \phi) \tag{2.4}$$

$$E_n = -\frac{\mu Z^2(e^2/4\pi\varepsilon_0)^2}{2n^2\hbar^2} \tag{2.5}$$

The eigenfunctions factor into a radial part $R_{nl}(r)$ and the well-known spherical harmonics $Y_{lm}(\theta, \phi)$; n, l, and m are the principal, azimuthal, and magnetic quantum numbers, respectively. The energy eigenvalues E_n depend only on the principal quantum number n. Since the total number of independent spherical harmonics $Y_{lm}(\theta, \phi)$ for $l \leqslant (n-1)$ is equal to n^2 for a given n, each energy eigenvalue E_n is n^2-fold degenerate. While n controls the hydrogenlike orbital energy as well as size, the quantum numbers l and m govern the orbital anisotropy (shape) and angular momentum. Since the orbital angular momentum operators $\hat{L}^2$ and $\hat{L}_z$ commute with the Hamiltonian $\hat{H}_0$, the eigenfunctions $|\psi_{nlm}\rangle$ of $\hat{H}_0$ are also eigenfunctions of $\hat{L}^2$ and $\hat{L}_z$,

$$\hat{L}^2 Y_{lm}(\theta, \phi) = l(l+1)\hbar^2 Y_{lm}(\theta, \phi) \tag{2.6}$$

$$\hat{L}_z Y_{lm}(\theta, \phi) = m\hbar Y_{lm}(\theta, \phi) \tag{2.7}$$

(The radial part $R_{nl}(r)$ of $|\psi_{nlm}\rangle$ cancels out in Eqs. 2.6 and 2.7, because $\hat{L}^2$ and $\hat{L}_z$ operate only on θ and ϕ.) This implies that the orbital angular momentum quantities $\hat{L}^2$ and $\hat{L}_z$ are constants of the motion in stationary state $|\psi_{nlm}\rangle$ with values $l(l+1)\hbar^2$ and $m\hbar$, respectively. A common notation for one-electron orbitals combines the principal quantum number n with the letter s, p, d, or f for orbitals with $l = 0$, 1, 2, and 3, respectively. (This notation is a vestige of the nomenclature sharp, principal, diffuse, and fundamental for the emission series observed in alkali atoms, as shown for K in Fig. 2.2.) An orbital with $n = 2$, $l = 0$ is called a $2s$ orbital, one with $n = 4$, $l = 3$ a $4f$ orbital, and so on. Numerical subscripts are occasionally added to indicate the pertinent m value: the $2p_0$ orbital exhibits $n = 2$, $l = 1$, and $m = 0$. Chemists frequently work with real (rather than complex) orbitals which transform as Cartesian vector (or tensor) components. A normalized $2p_x$ orbital is the linear combination $(-|2p_1\rangle + |2p_{-1}\rangle)/\sqrt{2} \equiv (-|211\rangle + |21, -1\rangle)/\sqrt{2}$, because the spherical harmonics Y_{11} and $Y_{1,-1}$ are given in Cartesian coordinates by

$$Y_{1,1} = -\sqrt{\frac{3}{8\pi}}\frac{x+iy}{r}$$

$$Y_{1,-1} = \sqrt{\frac{3}{8\pi}}\frac{x-iy}{r} \tag{2.8}$$

with $r = (x^2 + y^2 + z^2)^{1/2}$. Contours of some of the lower-energy hydrogenlike orbitals are shown in Fig. 2.3.

In accordance with the selection rules developed in Chapter 1, one must have

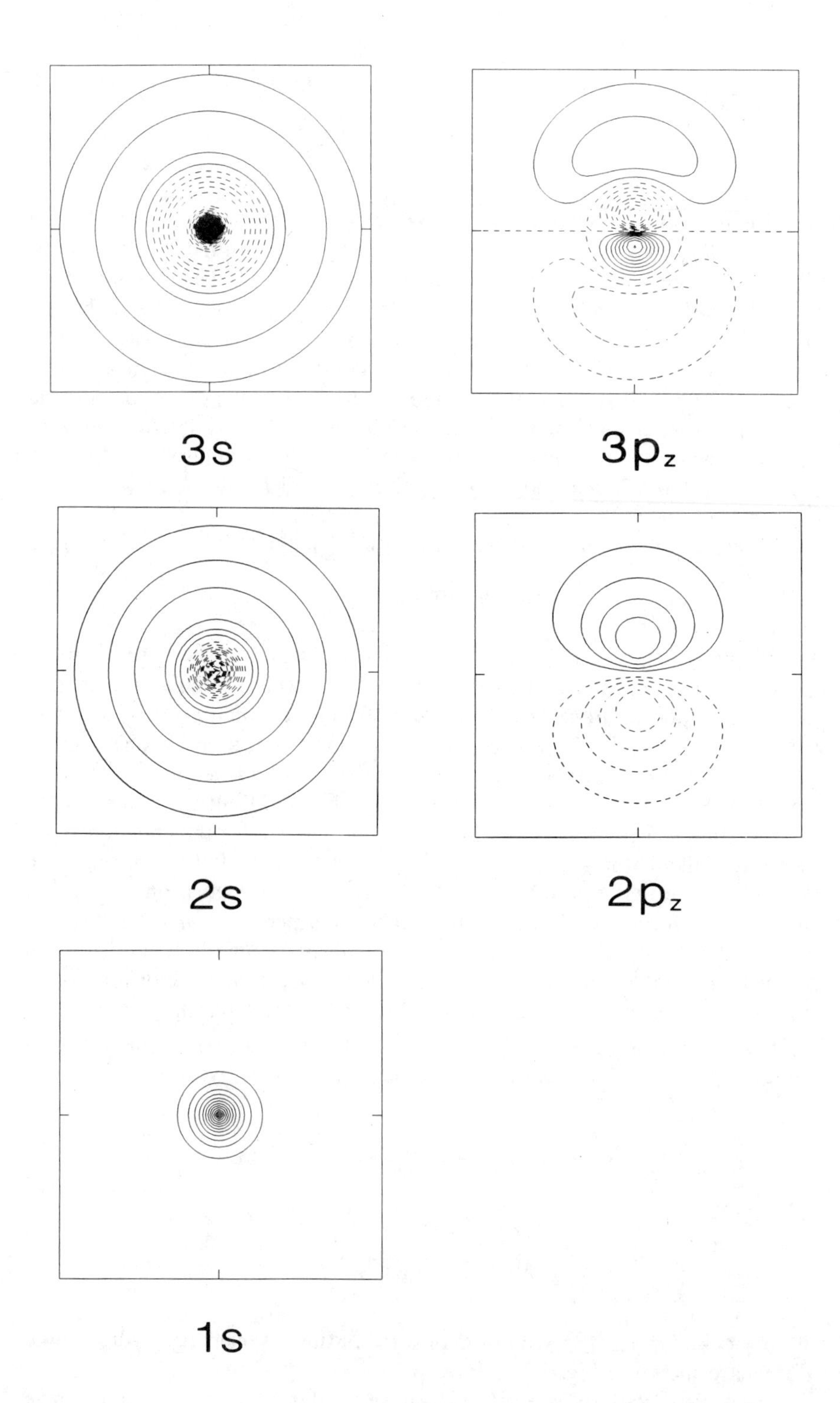
3s
3p$_z$
2s
2p$_z$
1s

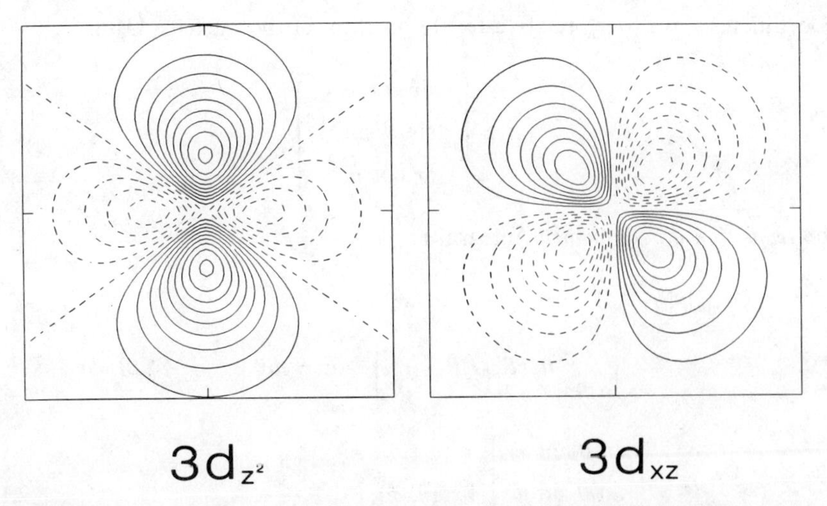

Figure 2.3 Contour plots of several low-lying H atom orbitals. Curves are surfaces on which the wavefunction exhibits constant values; solid and dashed curves correspond to positive and negative values, respectively. The outermost contour in all cases defines a surface containing ~90% of the electron probability density. The incremental change in wavefunction value between adjacent contours is 0.04, 0.008, 0.015, 0.003, 0.005, and 0.003 bohr$^{-3/2}$ respectively for the $1s$, $2s$, $2p$, $3p$, and $3d$ orbitals. Boxes exhibit side lengths of 20 bohrs ($1s$, $2s$, $2p$) and 40 bohrs ($3s$, $3p$, $3d$), so that orbital sizes can be compared. Straight dashed lines in $3p_z$ and $3d_{z^2}$ plots show locations of nodes.

for allowed one-photon transitions from state $|\psi_{nlm}\rangle$ to state $|\psi_{n'l'm'}\rangle$

$$\begin{aligned} \langle\psi_{nlm}|\boldsymbol{\mu}|\psi_{n'l'm'}\rangle &\neq 0, &&\text{E1 transitions} \\ \langle\psi_{nlm}|\mathbf{L}|\psi_{n'l'm'}\rangle &\neq 0, &&\text{M1 transitions} \\ \langle\psi_{nlm}|\mathbf{Q}|\psi_{n'l'm'}\rangle &\neq 0, &&\text{E2 transitions} \end{aligned} \tag{2.9}$$

To consider the specific selection rules on electric dipole (E1) transitions in one-electron atoms, we evaluate the matrix elements of the pertinent electric dipole moment operator $\boldsymbol{\mu} = \Sigma_i e_i \mathbf{r}_i = Ze\mathbf{r}_N - e\mathbf{r}'$, where $\mathbf{r}_N$ and $\mathbf{r}'$ are the positions of the nucleus and electron referenced to an arbitrary origin in space. Then

$$\begin{aligned} \langle\psi_{nlm}|\boldsymbol{\mu}|\psi_{n'l'm'}\rangle &= \langle\psi_{nlm}|Ze\mathbf{r}_N - e\mathbf{r}'|\psi_{n'l'm'}\rangle \\ &= (Z-1)e\mathbf{r}_N\langle\psi_{nlm}|\psi_{n'l'm'}\rangle - e\langle\psi_{nlm}|\mathbf{r}|\psi_{n'l'm'}\rangle \\ &= 0 - e\langle\psi_{nlm}|\mathbf{r}|\psi_{n'l'm'}\rangle \end{aligned} \tag{2.10}$$

since the hydrogenlike states are orthogonal and depend only on the electron's coordinates $\mathbf{r} = \mathbf{r}' - \mathbf{r}_N$ relative to the position of the nucleus. Using

$$\mathbf{r} = \begin{bmatrix} r \sin\theta \cos\phi \\ r \sin\theta \sin\phi \\ r \cos\theta \end{bmatrix} \tag{2.11}$$

the *transition dipole moment* integral is

$$\begin{aligned} &\langle \psi_{nlm} | \boldsymbol{\mu} | \psi_{n'l'm'} \rangle \\ &\quad = -e \int_0^\infty r^2 dr\, r R_{nl}(r) R_{n'l'}(r) \int_0^\pi \sin\theta\, d\theta \int_0^{2\pi} d\phi\, Y^*_{lm}(\theta, \phi) \\ &\quad \times \begin{bmatrix} \sin\theta \cos\phi \\ \sin\theta \sin\phi \\ \cos\theta \end{bmatrix} Y_{l'm'}(\theta, \phi) \end{aligned} \tag{2.12}$$

It can be shown that the angular part of this integral vanishes [1] unless $\Delta l \equiv l' - l = \pm 1$ (*not* zero), and unless $\Delta m \equiv m' - m = 0$ or ± 1. Evaluation of the radial part is difficult; this factor is nonzero regardless of $\Delta n \equiv n' - n$, provided $\Delta l = \pm 1$. Hence we can summarize the E1 *selection rules* for one-photon transitions in *hydrogenlike atoms*,

$$\Delta l = \pm 1 \tag{2.13a}$$

$$\Delta m = 0, \pm 1 \tag{2.13b}$$

Since the hydrogenlike energy levels are given by Eq. 2.5, the transition energy accompanying emission of a photon of frequency ω will be

$$\begin{aligned} -\hbar\omega = E_{n'} - E_n &= \frac{-\mu Z^2 (e^2/4\pi\varepsilon_0)^2}{2\hbar^2} \left[\frac{1}{(n')^2} - \frac{1}{n^2} \right] \\ &= -hc/\lambda \end{aligned} \tag{2.14}$$

in going from state $|\psi_{nlm}\rangle$ to state $|\psi_{n'l'm'}\rangle$. The corresponding photon wavelength is then

$$\lambda = \frac{8h^3 c}{\mu Z^2 (e^2/\varepsilon_0)^2} \left[\frac{n^2 (n')^2}{n^2 - (n')^2} \right] \tag{2.15}$$

which has a form identical to Eq. 2.1 if $n' = 2$. The visible H atom lines in the Balmer series thus result from transitions from $n = 3, 4, 5, \ldots$ down to $n' = 2$.

Other H atom series arise from transitions terminating in different values of n' (Fig. 2.4). Since the hydrogenlike energy levels E_n are independent of l and m, the selection rules (2.12) do not preclude the appearance of a spectral line for any energy separation $(E_{n'} - E_n)$, and lines appear for all combinations (n', n). These facts quantitatively account for the wavelengths of all of the H atom spectral lines in the low-resolution spectrum of Fig. 2.2.

Another consequence of the selection rules (2.13) is that a hydrogenlike atom

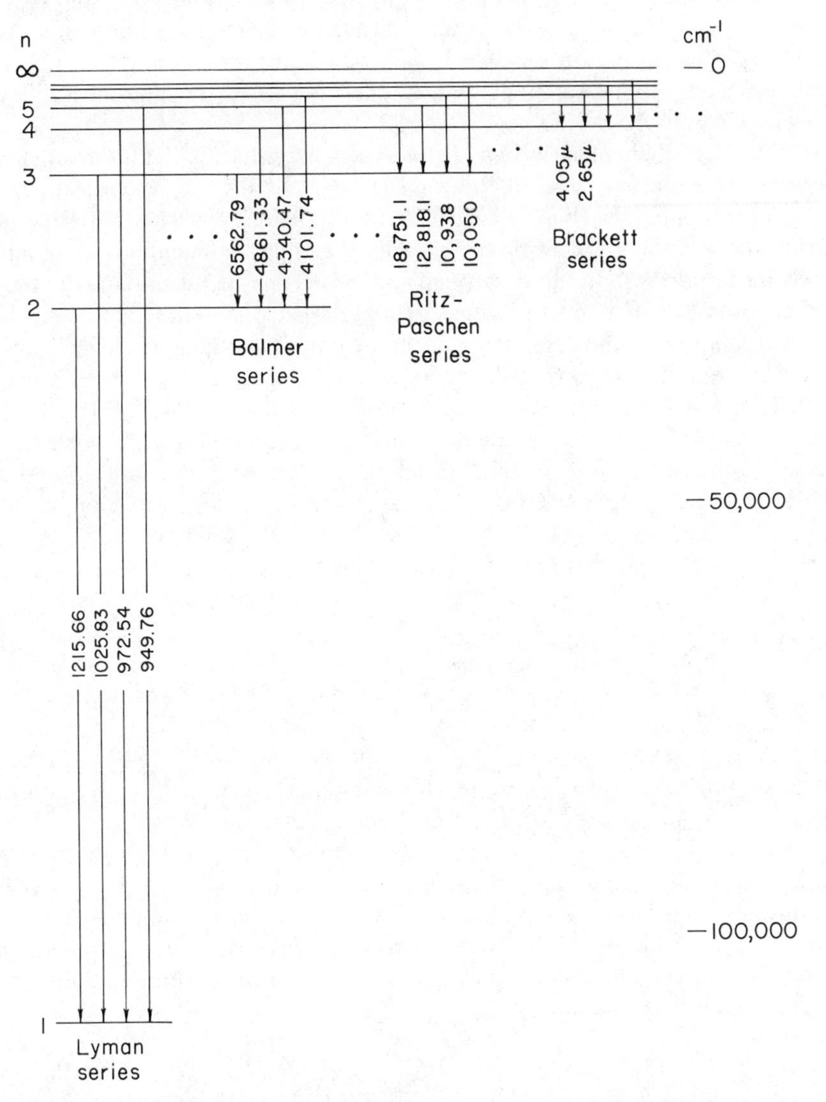

Figure 2.4 Hydrogen atom energy levels and transitions. The Lyman, Balmer, Ritz-Paschen, and Brackett series occur in the vacuum ultraviolet, visible, near-infrared, and infrared regions of the electromagnetic spectrum, respectively.

in *any* excited electronic state except the $2s$ state can emit a photon spontaneously by electric dipole radiation, and thereby relax to some lower energy state. For any such excited state $|\psi_{n'l'm'}\rangle$, there exists a lower state $|\psi_{nlm}\rangle$ for which $\Delta l = \pm 1$ and $\Delta m = 0, \pm 1$. The $2s$ state is the exception, because the only state with lower energy than the $2s$ state is the $1s$ state, and the $2s \rightarrow 1s$ fluorescence transition ($\Delta l = 0$) is E1-forbidden. The $2s$ state is therefore called *metastable* since it will exhibit an unusually long lifetime, lacking an electric dipole-allowed radiative transition to any lower state. Metastable states are not simply states which have no E1-allowed transition to the ground state: to be metastable, a state must have *no* E1-allowed transitions to *any* states of lower energy. Such atomic states play an important role in the efficiency of He/Ne lasers (Chapter 9).

Since the spherical harmonics $Y_{lm}(\theta, \phi)$ also describe angular momenta for the single valence electron in alkali atoms (Li, Na, K, Rb, Cs), the selection rules (2.13) apply equally well to valence electron transitions in such atoms (but not to transitions involving core electrons, where angular momentum coupling can become important). In this case, (nlm) and $(n'l'm')$ denote the initial and final sets of quantum numbers in the valence orbital. Valence–core interactions in alkali atoms split the n^2-fold degeneracy of energy levels belonging to a given principal quantum number n, so that the energies now depend on l as well as n. This is illustrated in the energy level diagram for potassium in Fig. 2.5. (Each of the levels is still $(2l + 1)$-fold degenerate, since the m sublevels for given n and l are isoenergetic in the absence of external magnetic fields.) The larger number of distinct levels resulting from this degeneracy-breaking in K does complicate the emission spectrum. However, the selection rules (2.13) still limit the observed E1 transitions to a small subset of the total number that could conceivably occur. The allowed transitions ($\Delta l = \pm 1$) are restricted to ones that connect levels in *adjacent* vertical groups of levels in Fig. 2.5, where the levels are organized in columns according to their l values. The principal series in K arises from $np \rightarrow 4s$ transitions with $n \geqslant 4$; the sharp series results from $ns \rightarrow 4p$ transitions with $n \geqslant 5$; the diffuse series occurs in $nd \rightarrow 4p$ transitions with $n \geqslant 4$; and the fundamental series is produced by $nf \rightarrow 4d$ transitions with $n \geqslant 4$. The origin of each of the lines in the low-resolution potassium spectrum (Fig. 2.2) can now be qualitatively understood with reference to Fig. 2.5.

Each of the series in the H and K spectra converges in principle to a series limit, as the discrete atomic energy levels must converge when the onset of the ionization continuum is approached (Figs. 2.4 and 2.5). The lines are very weak in the neighborhood of the series limits, because their intensities are proportional to the absolute value square of the transition dipole moment. The latter contains the factor

$$\left| \int_0^\infty r^3 dr R_{nl}(r) R_{n'l'}(r) \right|^2$$

which falls off rapidly with $(n - n')$ for given (l, l') [1]. For example, this quantity

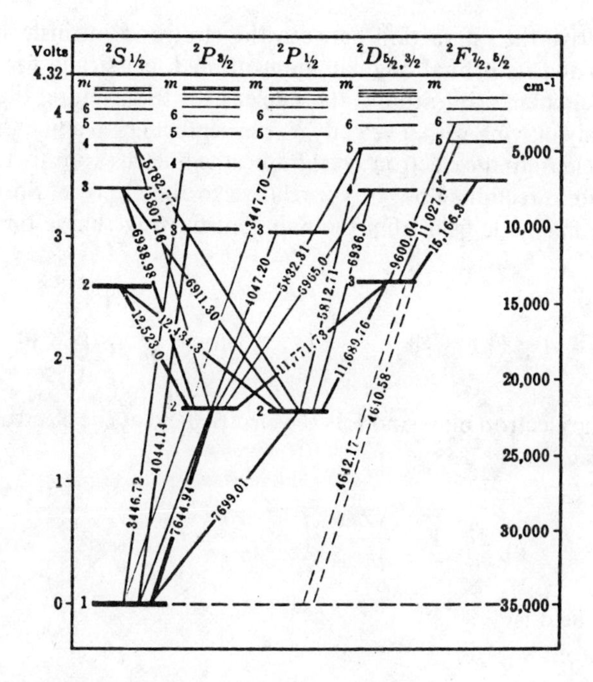

Figure 2.5 Energy levels and observed transitions in K. This type of diagram is commonly referred to as a *Grotrian diagram*. All of these low-lying energy levels arise from electron configurations of the type $(1s)^2(2s)^2(2p)^6(3s)^2(3p)^6(nl)^1 \equiv \cdots (nl)^1$. The $^2S_{1/2}$ level labeled "1", the $^2P_{1/2}$ and $^2P_{3/2}$ levels labeled "2", the $^2D_{3/2}$ and $^2D_{5/2}$ levels labeled "3", and the $^2F_{5/2}$ and $^2F_{7/2}$ levels labeled "4" correspond to the electron configurations $\cdots (4s)^1$, $\cdots (4p)^1$, $\cdots (4d)^1$, and $\cdots (4f)^1$ respectively. Reproduced, by permission, from G. Herzberg, *Atomic Spectra and Atomic Structure*, Dover Publications, Inc., New York, 1944.

equals 0.464, 0.075, and 0.026 Å^2, respectively, for the $1s \to 2p$, $1s \to 3p$, and $1s \to 4p$ transitions in hydrogen, and so good sensitivity is required to observe lines near the series limit.

2.2 SPIN–ORBIT COUPLING

If the low-resolution potassium spectrum in Fig. 2.2 is reexamined using a scanning monochromator that can distinguish between wavelengths that are 5 Å apart, each of the lines becomes split into closely spaced *multiplets* or groups of lines. Lines in the principal and sharp series become doublets, and lines in the diffuse series appear as triplets. This *fine structure* arises from the interaction between the orbital angular momentum **L** and the spin angular momentum **s** of the valence electron. Analogous splittings occur in the hydrogen spectrum, but much higher resolution is required to observe them in this atom, and other relativistic effects have comparable importance in hydrogen.

We can derive the interaction between the electron's intrinsic spin angular momentum $\mathbf{s}$ and its orbital angular momentum $\mathbf{L}$ classically for an electron moving in a circular orbit around the nucleus. In this picture, the electron is instantaneously moving with a velocity $\mathbf{V}_e$ perpendicular to a line that connects it with the nucleus. In the electron's rest frame, the nucleus appears to be moving in the opposite direction, $\mathbf{V}_N = -\mathbf{V}_e$, relative to the electron. So the electron experiences a magnetic field from the apparent moving charge on the nucleus [2],

$$\mathbf{B} = \frac{1}{c^2}(\mathbf{V}_N \times \mathbf{E}) = -\frac{1}{c^2}(\mathbf{V}_e \times \mathbf{E}) = +\frac{1}{m_e c^2}(\mathbf{E} \times \mathbf{p}) \tag{2.16}$$

where m_e is the electron mass and $\mathbf{E}$ is the electric field at the electron due to the nucleus. Since

$$\mathbf{E} = \frac{Ze\mathbf{r}}{4\pi\varepsilon_0 r^3}\left(= \frac{Ze}{4\pi\varepsilon_0 r^2}\hat{r}\right) \tag{2.17}$$

the magnetic field is

$$\mathbf{B} = \frac{Ze}{4\pi\varepsilon_0 m_e c^2 r^3}(\mathbf{r} \times \mathbf{p}) = \frac{Ze}{4\pi\varepsilon_0 m_e c^2 r^3}\mathbf{L} \tag{2.18}$$

where $\mathbf{L}$ is the electron's orbital angular momentum. The electron's intrinsic spin $\mathbf{s}$ carries an associated magnetic dipole moment [3]

$$\boldsymbol{\mu}_s = \frac{-g_s e\mathbf{s}}{2m_e} \tag{2.19}$$

where g_s is the electronic g factor ($g_s = 2$ according to Dirac, 2.0023 according to Schwinger [4]). The negative sign in Eq. 2.19 is due to the negative charge on the electron. The energy of interaction of the electron spin magnetic moment with the magnetic field $\mathbf{B}$ due to the moving nuclear charge is [2]

$$\begin{aligned}\hat{H}_{\text{so}} &= -\boldsymbol{\mu}_s \cdot \mathbf{B} \\ &= -\left(\frac{-g_s e\mathbf{s}}{2m_e}\right)\left(\frac{Ze\mathbf{L}}{4\pi\varepsilon_0 m_e c^2 r^3}\right) \\ &= \frac{Zg_s e^2}{8\pi\varepsilon_0 m_e^2 c^2 r^3}(\mathbf{L}\cdot\mathbf{s})\end{aligned} \tag{2.20}$$

Since the potential energy of attraction between the electron and nucleus is

$$V(r) = -Ze^2/4\pi\varepsilon_0 r$$

the spin–orbital energy is

$$\hat{H}_{so} = \frac{g_s}{2m_e^2c^2r}\frac{\partial V}{\partial r}(\mathbf{L}\cdot\mathbf{s}) \tag{2.21}$$

This actually overestimates the spin–orbital energy by a factor of 2, because we have neglected the fact that an electron in a circular or elliptical orbit does not travel at a uniform velocity $\mathbf{V}_e$, but experiences acceleration. The effect of correcting for this is to cancel [5] (or nearly cancel, according to Schwinger) the g factor g_s, and we write

$$\hat{H}_{so} = \frac{1}{2\mu^2c^2r}\frac{\partial V}{\partial r}(\mathbf{L}\cdot\mathbf{s}) \tag{2.22}$$

where m_e has been replaced by the atomic reduced mass μ. We note that since $V(\mathrm{r}) = -Ze^2/4\pi\varepsilon_0 r$, the magnitude of the spin–orbital Hamiltonian increases with atomic number Z.

In hydrogenlike atoms, the total electronic Hamiltonian now becomes

$$\begin{aligned}\hat{H} &= \hat{H}_0 + \hat{H}_{so} \\ &= -\frac{\hbar^2}{2\mu}\nabla^2 - \frac{Ze^2}{4\pi\varepsilon_0 r} + \hat{H}_{so}\end{aligned} \tag{2.23}$$

In the limit where $\hat{H}_{so}$ can be treated as a stationary perturbation, the energy corrected to first order becomes

$$E_{nlm} \simeq -\frac{\mu Z^2(e^2/4\pi\varepsilon_0)^2}{2n^2\hbar^2} + \langle\psi_{nlm}|\hat{H}_{so}|\psi_{nlm}\rangle \tag{2.24}$$

The latter matrix element requires an expression for $\mathbf{L}\cdot\mathbf{s}$ according to Eq. 2.22. It also requires knowledge of the total angular momentum states that can arise in an atom with orbital and spin angular momenta $\mathbf{L}$ and $\mathbf{s}$ (Appendix E). The spherical harmonics in the atomic states ψ_{nlm} are eigenfunctions of $\hat{L}^2$ and $\hat{L}_z$ (Eqs. 2.6, 2.7). The electron spin states $|sm_s\rangle$ obey

$$\hat{s}^2|sm_s\rangle = s(s+1)\hbar^2|sm_s\rangle \tag{2.25a}$$

$$\hat{s}_z|sm_s\rangle = m_s\hbar|sm_s\rangle \tag{2.25b}$$

with $s = \frac{1}{2}$ and $m_s = \pm\frac{1}{2}$ for a single electron [3]. Two alternative commuting sets of angular momentum operators are then $\hat{L}^2$, $\hat{L}_z$, $\hat{s}^2$, $\hat{s}_z$ and $\hat{L}^2$, $\hat{s}^2$, $\hat{J}^2$, $\hat{J}_z$, where the total angular momentum $\mathbf{J}$ is defined as $\mathbf{J} = \mathbf{L} + \mathbf{s}$. Eigenfunctions of the first commuting set form the *uncoupled representation* $|lm_l sm_s\rangle \equiv |lm_l\rangle|sm_s\rangle$. Since $\hat{J}^2$ does not commute with $\hat{L}_z$ or $\hat{s}_z$ (Appendix E), these uncoupled states are not eigenfunctions of $\hat{J}^2$. Eigenfunctions of the second commuting set form the *coupled representation* $|lsjm\rangle$. The total angular momentum squared $\mathbf{J}^2$ must be a constant of the motion in an isolated atom, which is then appropriately described using the coupled representation. In this representation, the states $|lsjm\rangle$ are eigenfunctions of $\hat{J}^2$ with eigenvalue $j(j+1)\hbar^2$, where the possible values of j are

$$j = l + s,\ l + s - 1, \ldots, |l - s| \tag{2.26}$$

Since

$$\mathbf{L}\cdot\mathbf{s} = \tfrac{1}{2}(\mathbf{J}^2 - \mathbf{L}^2 - \mathbf{s}^2) \tag{2.27}$$

the energy 2.24 corrected for spin–orbit coupling becomes

$$\begin{aligned} E_{nlm} &\simeq \frac{-\mu Z^2(e^2/4\pi\varepsilon_0)^2}{2n^2\hbar^2} + \frac{1}{2\mu^2c^2}\langle\psi_{nlm}|\frac{1}{r}\frac{\partial V}{\partial r}\frac{1}{2}(\mathbf{J}^2 - \mathbf{L}^2 - \mathbf{s}^2)|\psi_{nlm}\rangle \\ &= \frac{-\mu Z^2(e^2/4\pi\varepsilon_0)^2}{2n^2\hbar^2} + \frac{[j(j+1) - l(l+1) - s(s+1)]\hbar^2}{4\mu^2c^2}\langle\psi_{nlm}|\frac{1}{r}\frac{\partial V}{\partial r}|\psi_{nlm}\rangle \end{aligned} \tag{2.28}$$

Letting

$$A_{nl} \equiv \frac{\hbar^2}{2\mu^2c^2}\int_0^\infty r^2dr\,\frac{1}{r}\frac{\partial V}{\partial r}\,R_{nl}^2(r) \tag{2.29}$$

the corrected energies are

$$E_{nlm} \simeq \frac{-\mu Z^2(e^2/4\pi\varepsilon_0)^2}{2n^2\hbar^2} + A_{nl}[j(j+1) - l(l+1) - s(s+1)]/2 \tag{2.30}$$

The possible values of the quantum number $j = l + s, \ldots, |l - s|$ reduce to $j = l \pm \frac{1}{2}$ for $l \neq 0$ in hydrogenlike atoms. As an example, the possible j values for $2p$ states in hydrogenlike atoms are $j = 1 + \frac{1}{2}, 1 - \frac{1}{2} = \frac{3}{2}, \frac{1}{2}$. The corresponding energies of the j sublevels would then be*

$$E_{3/2} = \frac{-\mu Z^2(e^2/4\pi\varepsilon_0)^2}{8\hbar^2} + A_{21}[\tfrac{3}{2}(\tfrac{3}{2} + 1) - 2 - \tfrac{3}{4}]/2 \qquad \text{for } j = \tfrac{3}{2}$$

*It can be shown that for hydrogenlike atoms, the spin–orbit coupling constant A_{nl} is given by

$$A_{nl} = \frac{Z^4e^2\hbar^2}{2\mu^2c^2a_0^3}\cdot\frac{1}{n^3l(l+\frac{1}{2})(l+1)}$$

The radial wave functions $R_{nl}(r)$ do not have closed-form expressions in many-electron atoms (Section 2.3), and so A_{nl} is not given by simple formulas in such atoms. Note the sensitivity of A_{nl} to the atomic number Z; this gives rise to large spin–orbit coupling in heavy atoms.

$$E_{1/2} = \frac{-\mu Z^2(e^2/4\pi\varepsilon_0)^2}{8\hbar^2} + A_{21}[\tfrac{1}{2}(\tfrac{1}{2}+1) - 2 - \tfrac{3}{4}]/2 \qquad \text{for } j = \tfrac{1}{2} \tag{2.31}$$

and the energy splitting between these spin–orbital states is

$$E_{3/2} - E_{1/2} = 3A_{21}/2 \tag{2.32}$$

This is an example of the Landé interval rule (1933) for the energy separation of spin–orbital states with successive j values:

$$E_j - E_{j-1} = jA_{nl} \tag{2.33}$$

To keep track of the different angular momenta in atomic states, *term symbols* are used to specify the values of l, m, and j:

$$^{2s+1}L_j \tag{2.34}$$

When $l = 0$, L is denoted with an S; when $l = 1$, L is denoted with a P, and so on. Similar term symbols are used to notate the angular momenta in many-electron atoms. For the hydrogenlike $2p$ sublevels with $j = \frac{3}{2}$ and $\frac{1}{2}$, the term symbols are $2^2\mathrm{P}_{3/2}$ and $2^2\mathrm{P}_{1/2}$ respectively, where the first digit indicating $n = 2$ is useful for specifying the principal quantum number of the valence electron in hydrogenlike and alkali atom states.

Since each of the hydrogenlike levels with $l \neq 0$ is now split into doublets with $j = l \pm \frac{1}{2}$, it becomes necessary to augment the E1 selection rules on Δn, Δl, and Δm with E1 selection rules on Δj. It turns out that these are $\Delta j = 0, \pm 1$. We will demonstrate this selection rule in the case of $n^2S_{1/2} \to n'^2\mathrm{P}_j$ transitions, where j can be $\frac{1}{2}$ or $\frac{3}{2}$. The coupled (total) angular momentum states $|lsjm\rangle \equiv |jm\rangle$ in atoms can be expressed as a superposition of uncoupled states $|lm_l sm_s\rangle$ weighted by Clebsch-Gordan coefficients [1, 3],

$$|jm\rangle = \sum_{\substack{m_l \\ m_s}} |lm_l sm_s\rangle\langle lm_l sm_s|jm\rangle \tag{2.35}$$

For the n'^2P_j states, it is understood that $l = 1$, $s = \frac{1}{2}$; the possible m_l values are $0, \pm 1$ and the possible m_s values are $\pm\frac{1}{2}$. The four components of the $^2\mathrm{P}_{3/2}$ state ($m = -\frac{3}{2}$ through $+\frac{3}{2}$) are then

$$\underset{|\frac{3}{2}, \frac{3}{2}\rangle}{jm} = \underset{|1, \frac{1}{2}\rangle}{m_l m_s} \tag{2.36a}$$

$$|\tfrac{3}{2}, \tfrac{1}{2}\rangle = (\tfrac{2}{3})^{1/2}|0, \tfrac{1}{2}\rangle + (\tfrac{1}{3})^{1/2}|1, -\tfrac{1}{2}\rangle \tag{2.36b}$$

$$|\tfrac{3}{2}, -\tfrac{1}{2}\rangle = (\tfrac{1}{3})^{1/2}|-1, \tfrac{1}{2}\rangle + (\tfrac{2}{3})^{1/2}|0, -\tfrac{1}{2}\rangle \tag{2.36c}$$

$$|\tfrac{3}{2}, -\tfrac{3}{2}\rangle = |-1, -\tfrac{1}{2}\rangle \tag{2.36d}$$

The first of Eqs. 2.36 arises because there is only one combination of m_l and m_s in a $^2P_{3/2}$ state that can give a total m of $\frac{3}{2}$ ($m_l = 1$, $m_s = \frac{1}{2}$). The second can be shown using the fact that the raising/lowering operator $J_\pm$ has the effect

$$J_\pm|jm\rangle = \sqrt{j(j+1) - m(m \pm 1)}|j(m \pm 1)\rangle \tag{2.37}$$

when applied to state $|jm\rangle$, and that

$$J_\pm = L_\pm + S_\pm \tag{2.38}$$

with

$$L_\pm|lm_l\rangle = \sqrt{l(l+1) - m_l(m_l \pm 1)}|l(m_l \pm 1)\rangle \tag{2.39a}$$

$$S_\pm|sm_s\rangle = \sqrt{s(s+1) - m_s(m_s \pm 1)}|s(m_s \pm 1)\rangle \tag{2.39b}$$

The $|jm\rangle$ state $|\frac{3}{2}, -\frac{1}{2}\rangle$ can thus be obtained by applying the J_- operator to $|\frac{3}{2}, \frac{1}{2}\rangle$ and using Eqs. 2.37 and 2.39.

The two components of the $^2P_{1/2}$ state are given by

$$\underbrace{|\tfrac{1}{2}, \tfrac{1}{2}\rangle}_{jm} = (\tfrac{1}{3})^{1/2}\underbrace{|0, \tfrac{1}{2}\rangle}_{m_l m_s} - (\tfrac{2}{3})^{1/2}\underbrace{|1, -\tfrac{1}{2}\rangle}_{m_l m_s} \tag{2.40a}$$

$$|\tfrac{1}{2}, -\tfrac{1}{2}\rangle = (\tfrac{2}{3})^{1/2}|-1, \tfrac{1}{2}\rangle - (\tfrac{1}{3})^{1/2}|0, -\tfrac{1}{2}\rangle \tag{2.40b}$$

These follow because the $|jm\rangle$ state $|\frac{1}{2}, \frac{1}{2}\rangle$ must be normalized and orthogonal to the $|jm\rangle$ state $|\frac{3}{2}, \frac{1}{2}\rangle$ in Eq. 2.36, and because $|\frac{1}{2}, -\frac{1}{2}\rangle$ can be obtained from $|\frac{1}{2}, \frac{1}{2}\rangle$ by application of the J_- operator and use of Eqs. 2.37 and 2.39. Finally, the $n^2S_{1/2}$ state ($l = 0$) has the two components ($m = m_s = \pm\frac{1}{2}$)

$$|0, \tfrac{1}{2}\rangle \quad \text{and} \quad |0, -\tfrac{1}{2}\rangle \tag{2.41}$$

The electric dipole transition moments for the various fine structure transitions between the $n^2S_{1/2}$ and $n'^2P_{1/2,3/2}$ manifolds (i.e., groups) of levels can now be evaluated:

<u>($\Delta j = +1$, $\Delta m = +1$)</u>

$$\langle ^2S_{1/2,1/2}|\boldsymbol{\mu}|^2P_{3/2,3/2}\rangle = \langle 0, \tfrac{1}{2}|\boldsymbol{\mu}|1, \tfrac{1}{2}\rangle \tag{2.42a}$$

<u>($\Delta j = +1$, $\Delta m = 0$)</u>

$$\langle ^2S_{1/2,1/2}|\boldsymbol{\mu}|^2P_{3/2,1/2}\rangle = (\tfrac{2}{3})^{1/2}\langle 0, \tfrac{1}{2}|\boldsymbol{\mu}|0, \tfrac{1}{2}\rangle + (\tfrac{1}{3})^{1/2} \times \langle 0, \tfrac{1}{2}|\boldsymbol{\mu}|1, -\tfrac{1}{2}\rangle \tag{2.42b}$$

$(\Delta j = +1, \Delta m = -1)$

$$\langle {}^2S_{1/2,1/2}|\boldsymbol{\mu}|{}^2P_{3/2,-1/2}\rangle = (\tfrac{1}{3})^{1/2}\langle 0, \tfrac{1}{2}|\boldsymbol{\mu}| -1, \tfrac{1}{2}\rangle + (\tfrac{2}{3})^{1/2}\langle 0, \tfrac{1}{2}|\boldsymbol{\mu}|0, -\tfrac{1}{2}\rangle \tag{2.42c}$$

$(\Delta j = +1, \Delta m = -2)$

$$\langle {}^2S_{1/2,1/2}|\boldsymbol{\mu}|{}^2P_{3/2,-3/2}\rangle = \langle 0, \tfrac{1}{2}|\boldsymbol{\mu}| -1, -\tfrac{1}{2}\rangle \tag{2.42d}$$

In evaluating these, we note that

$$\langle m_l m_s|\boldsymbol{\mu}|m_l' m_s'\rangle = \langle m_l|\boldsymbol{\mu}|m_l'\rangle \delta_{m_s, m_s'}$$

since $\boldsymbol{\mu}$ does not depend on the spin coordinates. Further, this matrix element vanishes unless $\Delta m_l = m_l' - m_l = 0$ or ± 1. Hence, the second terms on the right sides of Eqs. 2.42b and 2.42c both vanish, but none of the total transition moments in Eqs. 2.38a–2.38c vanish. The only E1-forbidden transition is the one whose (zero) transition moment is given by Eq. 2.42d.

For the $n^2S_{1/2} \rightarrow n^2P_{1/2}$ transitions, we have

$(\Delta j = 0, \Delta m = 0)$

$$\langle {}^2S_{1/2,1/2}|\boldsymbol{\mu}|{}^2P_{1/2,1/2}\rangle = (\tfrac{1}{3})^{1/2}\langle 0, \tfrac{1}{2}|\boldsymbol{\mu}|0, \tfrac{1}{2}\rangle - (\tfrac{2}{3})^{1/2}\langle 0, \tfrac{1}{2}|\boldsymbol{\mu}|1, -\tfrac{1}{2}\rangle \tag{2.43a}$$

$(\Delta j = 0, \Delta m = -1)$

$$\langle {}^2S_{1/2,1/2}|\boldsymbol{\mu}|{}^2P_{1/2,-1/2}\rangle = (\tfrac{2}{3})^{1/2}\langle 0, \tfrac{1}{2}|\boldsymbol{\mu}| -1, \tfrac{1}{2}\rangle - (\tfrac{1}{3})^{1/2}\langle 0, \tfrac{1}{2}|\boldsymbol{\mu}|0, -\tfrac{1}{2}\rangle \tag{2.43b}$$

The first terms on both right sides are nonzero, and so these are both allowed transitions. A set of equations analogous to (2.42) and (2.43) can be obtained for transitions from $|{}^2S_{1/2,-1/2}\rangle$, and will not be included here. Equations 2.42 and 2.43 typify the E1 selection rules $\Delta l = \pm 1$; $\Delta j = 0, \pm 1$; and $\Delta m = 0, \pm 1$ for electronic transitions in hydrogenlike atoms.

These angular momentum selection rules figure prominently in the fine structure of alkali atom spectra. The filled-shell core electrons have zero net orbital and spin angular momentum, so the term symbols $2^2S_{1/2}$, $3^2S_{1/2}$, $4^2S_{1/2}$, $5^2S_{1/2}$, and $6^2S_{1/2}$ of ground states Li, Na, K, Rb, and Cs respectively are composed from the angular momenta of the single valence electron. The *principal series* of alkali atomic lines arises from $n^2S_{1/2} \rightarrow n'^2P_{1/2,3/2}$ transitions;

since $\Delta j = 0, \pm 1$, there will always be two fine structure components (e.g., $3^2S_{1/2} \rightarrow 3^2P_{1/2}$ and $3^2S_{1/2} \rightarrow 3^2P_{3/2}$ in Na) in this series. (Bear in mind that the various m sublevels of any $|jm\rangle$ state are degenerate, so that only transitions from a given level to final states with *different* j values will give rise to more than one spectral line—unless an applied magnetic field splits the m sublevels.) The *diffuse series* $n^2P_{1/2,3/2} \rightarrow n'^2D_{3/2,5/2}$ always yields triplets (e.g., $3^2P_{1/2} \rightarrow 3^2D_{3/2}$, $3^2P_{3/2} \rightarrow 3^2D_{3/2}$, and $3^2P_{3/2} \rightarrow 3^2D_{5/2}$; but not $3^2P_{1/2} \rightarrow 3^2D_{5/2}$, for which $\Delta j = +2$.) Doublets occur in the *sharp series* $n^2P_{1/2,3/2} \rightarrow n'^2S_{1/2}$. These E1-allowed fine structure transitions are all summarized in Fig. 2.6.

In the alkali atoms, the spin–orbit coupling is a small perturbation to the zero-order electronic Hamiltonian. In Na, the energy separation between the $3^2P_{1/2}$ and $3^2P_{3/2}$ spin–orbit sublevels of the lowest excited 3^2P state is only $\sim 17\,\text{cm}^{-1}$, versus $\sim 17{,}000\,\text{cm}^{-1}$ for the difference between the 3^2P and ground state (3^2S) levels. At the other extreme, the 5^2P ground state of the I atom is split by spin–orbit coupling into $5^2P_{3/2}$ and $5^2P_{1/2}$ sublevels which are about $8000\,\text{cm}^{-1}$ apart! In this limit, the spin–orbit sublevels behave much like different electronic states—which they are, because the spin–orbit coupling is no longer a small perturbation to the electronic structure.

When $\hat{H}_{so}$ is *not* a small perturbation, it becomes important to know which dynamical observables are still conserved in the atom. In the absence of spin–orbit coupling, the electronic Hamiltonian $\hat{H}_0$ and the angular momenta obey the commutation relationships

$$
\begin{aligned}
&[\hat{H}_0, \hat{L}^2] = 0 \qquad [\hat{H}_0, \hat{J}^2] = 0 \qquad [\hat{H}_0, \hat{S}_z] = 0 \\
&[\hat{H}_0, \hat{S}^2] = 0 \qquad [\hat{H}_0, \hat{L}_z] = 0 \qquad [\hat{H}_0, \hat{J}_z] = 0 \qquad (2.44)
\end{aligned}
$$

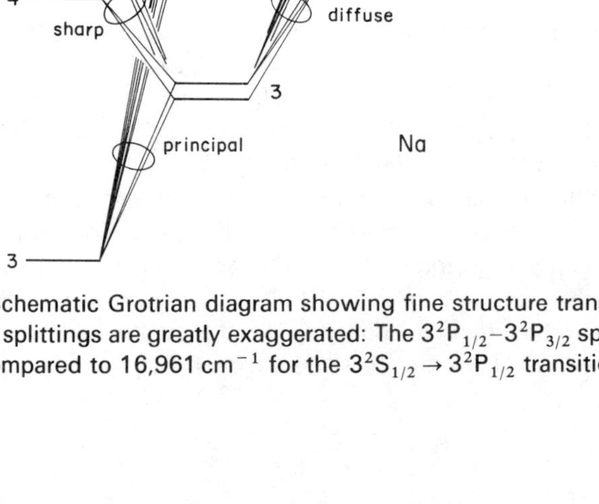

Figure 2.6 Schematic Grotrian diagram showing fine structure transitions in Na. The spin–orbit splittings are greatly exaggerated: The $3^2P_{1/2}$–$3^2P_{3/2}$ splitting is only $17\,\text{cm}^{-1}$, as compared to $16{,}961\,\text{cm}^{-1}$ for the $3^2S_{1/2} \rightarrow 3^2P_{1/2}$ transition.

When $\hat{H}_{so}$ is turned on, the total Hamiltonian $\hat{H}$ becomes $\hat{H}_0 + \hat{H}_{so}$, gaining a term proportional to $\hat{L}\cdot\hat{S}$. In this case, it can be shown (Problem 2.4) that

$$\begin{aligned} &[\hat{H}, \hat{L}^2] = 0 \qquad [\hat{H}, \hat{J}^2] = 0 \qquad [\hat{H}, \hat{S}_z] \neq 0 \\ &[\hat{H}, \hat{S}^2] = 0 \qquad [\hat{H}, \hat{L}_z] \neq 0 \qquad [\hat{H}, \hat{J}_z] = 0 \end{aligned} \tag{2.45}$$

so that m_l and m_s are not good quantum numbers (and L_z and S_z are not conserved) in the presence of spin–orbit coupling.

We now briefly consider the magnetic dipole (M1) selection rules for transitions in hydrogenlike and alkali atoms. The relevant matrix element is $\langle\psi_{nlm}|\mathbf{L}|\psi_{n'l'm'}\rangle$ (not $\langle\psi_{nlm}|\mathbf{J}|\psi_{n'l'm'}\rangle$ as has sometimes been implied, because the derivation of the M1 selection rules in Chapter 1 makes it clear that only the orbital part of the angular momentum enters in this matrix element). Using [3]

$$\hat{L}_x = \tfrac{1}{2}(\hat{L}_+ + \hat{L}_-) \tag{2.46a}$$

$$\hat{L}_y = \frac{1}{2i}(\hat{L}_+ - \hat{L}_-) \tag{2.46b}$$

$$\hat{L}_z = \hat{L}_z \tag{2.46c}$$

and using Eq. 2.39 immediately shows that since all matrix elements of $\hat{L}$ are diagonal in l (i.e., proportional to $\delta_{ll'}$), the M1 selection rule on Δl is $\Delta l = 0$. To obtain the M1 selection rule on Δj, one must again expand the coupled $|jm\rangle$ states in terms of the uncoupled states (e.g., Eqs. 2.36 and 2.40) and then get expressions analogous to Eqs. 2.42 and 2.43, with $\mathbf{L}$ replacing $\boldsymbol{\mu}$. An example of an M1 (but not E1) allowed atomic transition is the $^2P_{1/2} \rightarrow {}^2P_{3/2}$ spin–orbital transition between the lowest two levels in the I atom, which forms the basis of the 1.2 μm CH_3I dissociation laser. That such a laser works at all is somewhat startling, because M1 transitions are inherently weak, and the overwhelming majority of laser transitions (e.g., in the He/Ne laser) operate on strong E1 transitions.

2.3 STRUCTURE OF MANY-ELECTRON ATOMS

We now extend our discussion of hydrogenlike atoms to complex atoms with a total of p electrons. The nonrelativistic Hamiltonian operator for such atoms in the absence of external fields is

$$\hat{H}_0 = \sum_{i=1}^{p}\left(-\frac{\hbar^2}{2m_e}\nabla_i^2 - \frac{Ze^2}{4\pi\varepsilon_0 r_i}\right) + \sum_{i<j}^{p}\frac{e^2}{4\pi\varepsilon_0 r_{ij}} \tag{2.47}$$

where r_i is the distance of electron i from the nucleus and r_{ij} is the separation between electrons i and j. This Hamiltonian consists of a sum of p one-electron hydrogenlike Hamiltonians

$$\hat{H}_i \equiv -\frac{\hbar^2}{2m_e}\nabla_i^2 - \frac{Ze^2}{4\pi\varepsilon_0 r_i} \tag{2.48}$$

combined with a sum of electron–electron repulsion terms of the form $e^2/4\pi\varepsilon_0 r_{ij}$. Were it not for these pairwise repulsion terms, the many-electron Hamiltonian $\hat{H}_0$ would reduce to

$$\hat{H}_0 = \sum_{i=1}^{p} \hat{H}_i \tag{2.49}$$

whose eigenfunctions are simply products of p hydrogenlike states $|\psi_{nlm}\rangle$,

$$|\psi(1, 2, \ldots, p)\rangle = |\psi_{n_1 l_1 m_1}(1)\psi_{n_2 l_2 m_2}(2)\cdots\psi_{n_p l_p m_p}(p)\rangle \tag{2.50}$$

Such expressions incorporating hydrogenlike states do not in fact provide useful approximations to electronic wave functions in many-electron atoms: the electron repulsions have a large effect on the total energy, and the wave function (2.50) is not properly antisymmetrized (see below). The concept of writing many-electron wave functions as products of generalized one-electron orbitals nonetheless provides a viable starting point for developing accurate approximations to the true nonrelativistic wave functions. As a prototype example, we consider the neutral He atom, for which the Hamiltonian operator is

$$\hat{H}_0 = \hat{H}_1 + \hat{H}_2 + e^2/4\pi\varepsilon_0 r_{12} \tag{2.51}$$

Since the Schrödinger equation using this two-electron Hamiltonian cannot be exactly solved, we use as a trial wave function for ground-state He the product of one-electron orbitals $|\phi_1(1)\rangle$ and $|\phi_2(2)\rangle$,

$$|\psi(1, 2)\rangle = |\phi_1(1)\phi_2(2)\rangle \tag{2.52}$$

According to the variational theorem [3], the trial energy

$$W_0 = \frac{\langle\psi(1, 2)|\hat{H}_1 + \hat{H}_2 + e^2/4\pi\varepsilon_0 r_{12}|\psi(1, 2)\rangle}{\langle\psi(1, 2)|\psi(1, 2)\rangle} \tag{2.53}$$

is bounded from below by the true ground-state energy E_0,

$$E_0 \leqslant W_0 \tag{2.54}$$

As a first approximation to $|\psi(1, 2)\rangle$, we may start with

$$|\phi_1(1)\phi_2(2)\rangle = |\psi_{100}(1)\psi_{100}(2)\rangle \tag{2.55}$$

where $|\psi_{100}(i)\rangle = N\exp(-Zr_i/a_0)$ is the normalized hydrogenlike 1*s* orbital with $Z = 2$ for electron *i*. This arbitrarily places both electrons in identical orbitals which are undistorted by electron repulsion. Substitution of this zeroth-order wave function into Eq. 2.53 for the trial energy in He yields

$$\begin{aligned} W_0 &= \langle\phi_1(1)|\hat{H}_1|\phi_1(1)\rangle + \langle\phi_2(2)|\hat{H}_2|\phi_2(2)\rangle \\ &\quad + \langle\phi_1(1)\phi_2(2)|e^2/4\pi\varepsilon_0 r_{12}|\phi_1(1)\phi_2(2)\rangle \\ &= 2\langle\psi_{100}(1)|\hat{H}_1|\psi_{100}(1)\rangle \\ &\quad + \langle\psi_{100}(1)\psi_{100}(2)|e^2/4\pi\varepsilon_0 r_{12}|\psi_{100}(1)\psi_{100}(2)\rangle \\ &= 2E_1 + (e^2/4\pi\varepsilon_0)\langle\psi_{100}(1)\psi_{100}(2)|1/r_{12}|\psi_{100}(1)\psi_{100}(2)\rangle \end{aligned} \tag{2.56}$$

where

$$E_1 = -\frac{\mu Z^2(e^2/4\pi\varepsilon_0)^2}{2\hbar^2} = -4(13.6058 \text{ eV}) \tag{2.57}$$

is the exact nonrelativistic ground-state energy of the hydrogenlike ion He^+ ($Z = 2$). The matrix element of $1/r_{12}$ can be evaluated [6] to yield $5\mu Z(e^2/4\pi\varepsilon_0)^2/8\hbar^2 = 34.0145$ eV. Then W_0 becomes -74.832 eV, as compared to the experimental energy 79.014 eV required to remove both electrons from a He atom. While W_0 so computed is obviously a large fraction of the true electronic energy, its error of 4.18 eV is of the same order as excited-state energy separations in He (cf. Fig. 2.11); a more sophisticated treatment is clearly necessary to obtain results of spectroscopic accuracy.

An improved wave function can be obtained by replacing the fixed atomic number $Z = 2$ in $|\psi_{100}(i)\rangle$ with a single variational parameter ζ. The trial energy W_0 is then calculated in a manner analogous to Eq. 2.56, and is minimized with respect to ζ by setting $\partial W_0/\partial\zeta = 0$. This procedure yields $\zeta = 27/16 = 1.688$ in He; this is physically smaller than $Z = 2$, because each electron screens part of the nuclear charge from the other electron. The corresponding trial energy $W_0 = -77.490$ eV is a closer approximation to true energy, but its error is still large. It is then logical to consider trial wave functions with more flexibility than $N\exp(-\zeta r_i/a_0)$, which has only one variational parameter. An example of such a wave function is the *Slater-type orbital* (STO), which has the general form

$$|S_{nlm}\rangle = Nr_i^{n-1}e^{-\zeta r_i}Y_{lm}(\theta_i, \phi_i) \tag{2.58}$$

STOs exhibit no radial nodes (unlike hydrogenlike orbitals for 2*s* and higher energy states), but both *n* and ζ can simultaneously be varied to simulate the behavior of the outer (largest-*r*) lobes of orbitals in many-electron atoms. Optimization of *n* and ζ to minimize W_0, again using identical STOs for both electrons in ground-state He (with $l = m = 0$), yields $n = 0.955$, $\zeta = 1.612$, and $W_0 = -77.667$ eV. This is still closer to the true energy, but the error has been reduced by a factor of only 0.88 over that in the previous approximation. The use of arbitrarily flexible functions in variational calculations that restrict both electrons to occupying identical orbitals in He yields no trial energies lower than $-77.8714\,\text{eV}$ (Fig. 2.7). The remaining error in the energy is $(79.014 - 77.871)\,\text{eV} = 1.143\,\text{eV}$, and is called the *correlation error*. It arises physically from the electrons' tendency to avoid each other in order to minimize their average Coulomb repulsion energy—a tendency that is ignored when identical hydrogenlike or Slater-type orbitals are used for both electrons.

To reduce the trial energy W_0 below $-77.8714\,\text{eV}$, the electrons must be placed in functionally distinct orbitals, or trial functions more general than single products of the form (2.52) must be introduced. In pursuing the first of these alternatives, the electrons could be placed in hydrogenlike orbitals

$$\begin{aligned} |\psi_{100}^{(1)}(r_i)\rangle &= N_1 e^{-\zeta_1 r_i} \equiv \phi_1(i) \\ |\psi_{100}^{(2)}(r_i)\rangle &= N_2 e^{-\zeta_2 r_i} \equiv \phi_2(i) \end{aligned} \tag{2.59}$$

with the two independently variable parameters ζ_1 and ζ_2. Such a calculation requires explicit construction of two-electron wave functions that are antisymmetric [6] with respect to exchange of the electrons (which are fermions). Since the Pauli principle demands that no two electrons with the same spatial quantum numbers (n, l, m) can have the same spin, the electrons in ground-state He $(1s)^2$ must have opposite spin, $m_s = +\frac{1}{2}\,(\alpha)$ and $m_s = -\frac{1}{2}\,(\beta)$. An acceptable antisymmetrized trial function for He using the individualized orbitals (2.59) is then

$$\psi(1, 2) = \tfrac{1}{2}[\phi_1(1)\phi_2(2) + \phi_1(2)\phi_2(1)]\,[\alpha(1)\beta(2) - \alpha(2)\beta(1)] \tag{2.60}$$

(Antisymmetrization of trial functions placing the two electrons into identical spatial orbitals $\phi_1(i)$ was unnecessary, since the use of the correctly antisym-

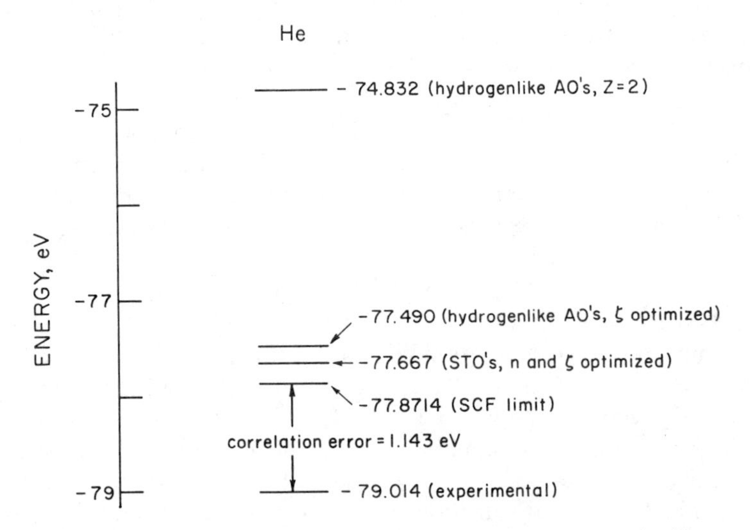

Figure 2.7 Energies obtained from variational calculations on the He atom.

metrized function $\phi_1(1)\phi_1(2)[\alpha(1)\beta(2) - \alpha(2)\beta(1)]$ in place of $\phi_1(1)\phi_1(2)$ in Eq. 2.53 does not influence the value of the trial energy.) The simultaneous optimization of the orbital exponents ζ_1 and ζ_2 for ϕ_1 and ϕ_2 in Eq. 2.60 yields $\zeta_1 = 1.189$, $\zeta_2 = 2.173$, and $W_0 = -78.252$ eV. It is clear that this simple calculation removes a substantial part of the correlation error, but the residual error of 0.762 eV still does not begin to approach spectroscopic accuracy. Better accuracy can be achieved by using trial functions that are linear combinations of many antisymmetrized functions like (2.60), in which a set of linearly independent basis functions of spherical symmetry (e.g., 2*s*, 3*s*, 4*s*, ... hydrogenlike orbitals in addition to 1*s*) is used for ϕ_1 and ϕ_2. This procedure corrects the so-called *radial* correlation error. Incorporation of a sufficient number of suitable nonspherical orbitals (i.e., with higher order spherical harmonics Y_{lm} in the angular part) as basis functions in such calculations removes the *angular* correlation error, and spectroscopic precision has been achieved in this manner for many atoms.

Electronic structure calculations in atoms with more than two electrons require properly antisymmetrized trial functions. For a closed-shell atom with *p* electrons having paired spins, such a function can be written in the form of a normalized *Slater determinant*:

$$\psi(1, 2, \ldots, p) = \frac{1}{\sqrt{p!}} \begin{vmatrix} \phi_1(1)\alpha(1) & \phi_1(2)\alpha(2) & \cdots & \phi_1(p)\alpha(p) \\ \phi_1(1)\beta(1) & \phi_1(2)\beta(2) & & \phi_1(p)\beta(p) \\ \phi_2(1)\alpha(1) & \phi_2(2)\alpha(2) & & \phi_2(p)\alpha(p) \\ \vdots & & & \\ \phi_{p/2}(1)\alpha(1) & & & \phi_{p/2}(p)\alpha(p) \\ \phi_{p/2}(1)\beta(1) & \cdots & & \phi_{p/2}(p)\beta(p) \end{vmatrix} \quad (2.61)$$

This determinant vanishes if any two rows are identical, which occurs if any two electrons are assigned the same spatial and spin states. The determinant changes sign if any two columns are interchanged, corresponding to exchange of two electrons. Hence, both the Pauli principle and antisymmetrization are built into the Slater determinant. Note that the determinant (2.61) assigns the same spatial function ϕ_i to both electrons in each orbital. Since the many-electron Hamiltonian has the form

$$\hat{H}_0 = \sum_{i=1}^{p} \hat{H}_i + \sum_{i<j}^{p} e^2/4\pi\varepsilon_0 r_{ij} \quad (2.62)$$

it can be shown that that trial energy becomes [6]

$$\begin{aligned} W_0 &= \frac{\langle \psi(1, 2, \ldots, p) | H_0 | \psi(1, 2, \ldots, p) \rangle}{\langle \psi(1, 2, \ldots, p) | \psi(1, 2, \ldots, p) \rangle} \\ &= \sum_{i=1}^{p/2} \left[2H_{ii} + \sum_{j=1}^{p/2} (2\langle ij|ij \rangle - \langle ij|ji \rangle) \right] \end{aligned} \quad (2.63)$$

with

$$H_{ii} = \langle \phi_i | H_i | \phi_i \rangle \tag{2.64}$$

$$\langle ij|ij \rangle = \langle \phi_i(1)\phi_j(2)|e^2/4\pi\varepsilon_0 r_{12}|\phi_i(1)\phi_j(2)\rangle \tag{2.65}$$

$$\langle ij|ji \rangle = \langle \phi_i(1)\phi_j(2)|e^2/4\pi\varepsilon_0 r_{12}|\phi_j(1)\phi_i(2)\rangle \tag{2.66}$$

The matrix elements in Eqs. 2.64 through 2.66 are referred to as the core integral, the Coulomb integral, and the exchange integral, respectively. Minimization of the trial energy by varying the ϕ_i under the constraint that the basis functions remain orthonormal leads to the *Hartree-Fock equation* [6]

$$\left\{H_{ii} + \sum_{j=1}^{p/2} [2\langle \phi_j(2)|e^2/4\pi\varepsilon_0 r_{12}|\phi_j(2)\rangle - \langle \phi_j(2)|e^2/4\pi\varepsilon_0 r_{12}|\phi_i(2)\rangle]\right\}\phi_i(1)$$

$$\equiv \left\{H_{ii} + \sum_{j=1}^{p/2} [2J_j(1) - K_j(1)]\right\}\phi_i(1) = \sum_{j=1}^{p/2} \phi_j(1)\varepsilon_{ji} \tag{2.67}$$

where the ε_{ji} are elements of a $(p/2) \times (p/2)$ matrix having units of energy. This equation is frequently abbreviated as

$$F\phi_i = \sum \phi_j \varepsilon_{ji} \tag{2.68}$$

where F, called the Hartree-Fock operator, depends on the basis functions ϕ_i through Eq. 2.67. The Hartree-Fock equation can be solved numerically by setting the ε_{ji} equal to $\varepsilon_{ji}\delta_{ij}$, computing the F operator from an assumed set of basis function ϕ_i, and using Eq. 2.68 to compute a new set of functions ϕ_i. These new functions are used in turn to compute a new F operator. This cycle is repeated until the ϕ_i used for calculating the Hartree-Fock operator converge to the final solutions ϕ_i to within desired precision. When this *self-consistent field* (SCF) limit has been reached, the orbital energies ε_i may be evaluated from Eq. 2.63,

$$\varepsilon_i = H_{ii} + \sum_j (2\langle ij|ij\rangle - \langle ij|ji\rangle) \tag{2.69}$$

and the optimized trial energy (Eq. 2.63) becomes

$$W_0 = \sum_{i=1}^{p/2} (H_{ii} + \varepsilon_i) \tag{2.70}$$

The spherical harmonics are ordinarily used for the angular part of the orbitals, and the Hartree-Fock equations are solved to obtain the radial wave functions numerically. The Hartree-Fock wave functions are the best radial wave functions that can be obtained in the form of the Slater determinant 2.61. For

He, the Hartree-Fock wave function is equivalent to the infinite-parameter variational wave function (2.52) with both electrons in identical spatial orbitals; the SCF energy of He is -77.871 eV (Fig. 2.7). Slater determinants like (2.61) and the derivation of the Hartree-Fock equations given here are specific to closed-shell atoms. For open-shell atoms (e.g., K, F) and for electronically excited atoms, different procedures must be followed.

The difference between the SCF energy and the true nonrelativistic energy is the correlation error. For first-row atoms, the correlation error is less than 2% of the true energy. This implies that the physical picture offered by the Hartree-Fock treatment—in which each electron experiences a centrosymmetric field due to the averaged interactions $\langle 1/r_{ij} \rangle$ with other electrons—accounts for the major portion of the electronic energy. However, the absolute correlation error is so large (1.14 eV for He) that the differences between SCF energies computed in atoms do not agree well with spectroscopically measured energy separations. A commonly used method of recouping part of the correlation error is *configurational interaction* (CI). In this technique, the ground-state wave function is expanded as a linear combination of determinants, rather than a single determinant as in closed-shell Hartree-Fock theory. One of these is the Hartree-Fock determinant wave function (2.61), and the remainder are determinants for excited electron configurations. For He, a CI wave function may be expanded as

$$\psi(1, 2) = C_0\Delta(1s^2) + C_1\Delta(1s2s) + C_2\Delta(2s^2) + C_3\Delta(1s3s) + \cdots \tag{2.71}$$

where $\Delta(1s^2)$ is the determinant describing two electrons in identical $1s$ orbitals,

Table 2.1.

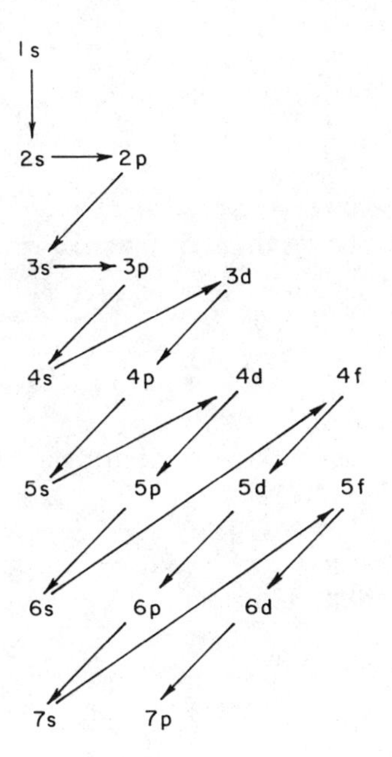

etc. The expansion coefficients C_0, C_1, etc., are optimized in a variational calculation. Inclusion of many excited configurations in CI determinant expansions like (2.71) reduces the radial correlation error to an arbitrary level, as the excitations permit the electrons to become more separated and decrease the average electron repulsion energy.

This section barely scratches the surface of many-electron atomic structure calculations. They have mushroomed in complexity to multiconfigurational SCF calculations in which linear combinations of atomic orbitals (LCAOs) are used for each of the spatial orbitals ϕ_i. With atomic orbital basis sets of sufficient size, close agreement is obtained with experiment.

The *Aufbau* or building-up principle by which electrons are placed in successive atomic orbitals in many-electron atoms has been experimentally established to follow the pattern shown in Table 2.1.

2.4 ANGULAR MOMENTUM COUPLING IN MANY-ELECTRON ATOMS

Since filled shells do not contribute to the net orbital or spin angular momentum in atoms, one needs to consider only the electrons in unfilled orbitals when calculating the possible angular momenta for a given electron configuration. The simplest case is an atom with two electrons in unfilled orbitals. We use $\mathbf{l}_1$ and $\mathbf{l}_2$ to denote the orbital angular momenta of the individual electrons, and likewise use $\mathbf{s}_1$ and $\mathbf{s}_2$ for their spin angular momenta. In the limit of weak spin–orbital coupling, the total atomic angular momentum is composed by first coupling the above vectors to obtain resultants for the total orbital and total spin angular momenta,

$$\mathbf{L} = \mathbf{l}_1 + \mathbf{l}_2$$

$$\mathbf{S} = \mathbf{s}_1 + \mathbf{s}_2 \tag{2.72}$$

According to the rules for composition of angular momenta, the possible quantum numbers L and S for the resultant orbital and spin angular momenta are

$$L = l_1 + l_2, \ldots, |l_1 - l_2|$$

$$S = s_1 + s_2, \ldots, |s_1 - s_2| \tag{2.73}$$

The total angular momentum $\mathbf{J}$ is then the vector sum

$$\mathbf{J} = \mathbf{L} + \mathbf{S} \tag{2.74}$$

and exhibits the possible quantum numbers

$$J = L + S, \ldots, |L - S| \tag{2.75}$$

This coupling scheme is known as *Russell-Saunders* coupling. As an example, we treat the *equivalent* p^2 configuration in which the two valence electrons have the same principal quantum number n (e.g., the ground state of a carbon atom that has the configuration $(1s)^2(2s)^2(2p)^2$). In this case $l_1 = l_2 = 1$, so that

$$\begin{aligned} L &= l_1 + l_2, \ldots, |l_1 - l_2| = 2, 1, 0 \\ S &= s_1 + s_2, \ldots, |s_1 - s_2| = 1, 0 \end{aligned} \tag{2.76}$$

In an equivalent p^2 atom we thus expect to find S, P, and D states (i.e., $L = 0, 1, 2$); some will be singlet states ($S = 0$) and some will be triplets ($S = 1$). Schematic vector diagrams illustrating these resultant angular momenta are shown in Fig. 2.8. At first sight, one might predict that all possible combinations of L and S could appear (^{1}S, ^{3}S, ^{1}P, ^{3}P, ^{1}D, ^{3}D)—but some of these combinations will violate the Pauli principle, and the possible states must be considered more directly. One may count the ways in which two electrons can legally be distributed among the equivalent (degenerate) p states; for example, the allowed configuration

$$\begin{array}{ccc} m_l = -1 & m_l = 0 & m_l = +1 \\ \underline{\quad\uparrow\quad} & \underline{\qquad\quad} & \underline{\quad\downarrow\quad} \end{array}$$

contributes one state with $M_L = m_{l1} + m_{l2} = 0$, $M_S = m_{s1} + m_{s2} = 0$. The total numbers of states counted for all combinations of M_L and M_S may be tabulated

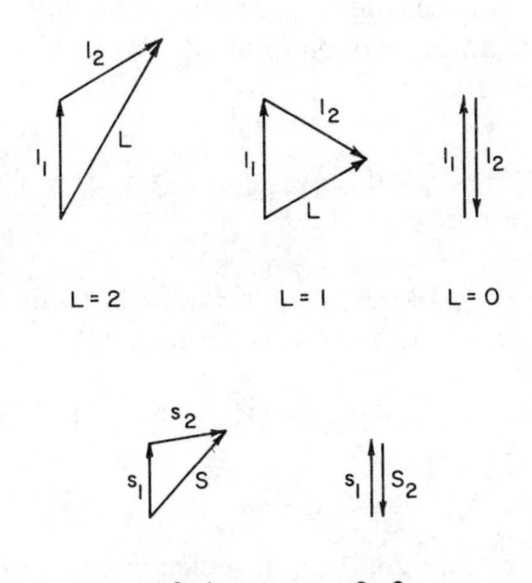

Figure 2.8 Vector addition of orbital and spin angular momenta for two valence electrons with $l_1 = 1$, $l_2 = 1$ in the Russell-Saunders coupling scheme.

in the *Slater diagram*

$\uparrow M_L$				
$+2$		1		
$+1$	1	2	1	
0	1	3	1	
-1	1	2	1	
-2		1		
	-1	0	$+1$	$M_S \rightarrow$

which can be viewed as the sum of the diagrams

$$
\begin{matrix} \\ \\ 1 \\ \\ \\ \\ (^1\mathrm{S}) \end{matrix}
\quad + \quad
\begin{matrix} \\ 1 & 1 & 1 \\ 1 & 1 & 1 \\ 1 & 1 & 1 \\ \\ & (^3\mathrm{P}) & \end{matrix}
\quad + \quad
\begin{matrix} 1 \\ 1 \\ 1 \\ 1 \\ 1 \\ (^1\mathrm{D}) \end{matrix}
$$

The last diagram, for example, represents a ^{1}D state, because it restricts M_S to zero (i.e., it is a singlet state) but allows M_L to range from $+2$ to -2. Thus, the possible states in an equivalent p^2 configuration are ^{1}S, ^{3}P, and ^{1}D. In each of these, the possible J values range between $L + S$ and $|L - S|$; i.e., the allowed term symbols are $^1\mathrm{S}_0$, $^1\mathrm{D}_2$, $^3\mathrm{P}_2$, $^3\mathrm{P}_1$, and $^3\mathrm{P}_0$. The J subscript is superflous in singlet term symbols (since only one J value is possible for a given L in singlet states, namely $J = L$) and it is often omitted. The total number of states in an equivalent p^2 configuration is

$$\sum_{L,S} (2L + 1)(2S + 1) = \underset{^1\mathrm{S}}{1(1)} + \underset{^3\mathrm{P}}{3(3)} + \underset{^1\mathrm{D}}{5(1)} = 15$$

which must equal the sum of the integers in the Slater diagram. Another way of summing the states is to count $(2J + 1)$ for each $^{2S+1}L_J$ term symbol,

$$\sum_J (2J + 1) = \underset{^1\mathrm{S}_0}{1} + \underset{^1\mathrm{D}_2}{5} + \underset{^3\mathrm{P}_2}{5} + \underset{^3\mathrm{P}_1}{3} + \underset{^3\mathrm{P}_0}{1} = 15$$

For *nonequivalent* p^2 configurations in which the two electrons are in different shells (e.g., the $(2p)^1(3p)^1$ excited state of carbon), states exist for *all* combinations of the allowed L and S values generated in Eqs. 2.76. Slater diagram tabulations of the allowed (M_L, M_S) combinations are then superfluous; the nonequivalent p^2 configuration gives rise to ^{1}S, ^{3}S, ^{1}P, ^{3}P, ^{1}D, and ^{3}D term symbols. Slater diagram tabulations are readily extended to open-shell configurations with three or more electrons (although they can become tedious). A simplification occurs for nearly filled (n, l) shells: The possible term symbols are identical to those for configurations consisting of the absent electrons. For example, equivalent p^5 and p^1 configurations both exhibit only $^2P_{1/2}$ and $^2P_{3/2}$ term symbols, and equivalent d^2 and d^8 configurations both yield the term symbols $^1S_0$, $^3P_{0,1,2}$, $^1D_2$, $^3F_{2,3,4}$, and $^1G_4$.

The energy level scheme for a p^2 atom can now be described for the case of small spin–orbit coupling (Fig. 2.9), where Russell-Saunders coupling applies. In the absence of electron–electron interactions, *all* of the configurations counted in the Slater diagram would have the same energy, because the one-electron p orbitals with $m_l = 0, \pm 1$ are degenerate. When the electron–electron interactions are turned on, the states with term symbols $^{2S+1}L_J$ have different energies for different L, S. By *Hund's rule* (the states with the highest multiplicity from a given configuration—in this case p^2—will be lowest in energy), the ^{3}P states empirically lie below the others. When the spin–orbit coupling is turned on, the energy levels assume a J-dependence of the form $E(L, S) + A[J(J+1) - L(L+1) - S(S+1)]/2$; in this case A is not analytic, because the radial wave functions in a many-electron atom are of course no longer hydrogenic.

Russell-Saunders coupling applies only when the spin–orbital coupling is small enough to be treated as a perturbation to the electronic Hamiltonian. In the limit of large $\hat{H}_{so}$, an alternative scheme called *jj coupling* applies. In this case, the total angular momenta of each electron are added to form the resultant total angular momentum $\mathbf{J}$. For a two-electron configuration, we have

$$\begin{aligned} \mathbf{l}_1 + \mathbf{s}_1 &= \mathbf{j}_1 \\ \mathbf{l}_2 + \mathbf{s}_2 &= \mathbf{j}_2 \\ \mathbf{J} &= \mathbf{j}_1 + \mathbf{j}_2 \end{aligned} \tag{2.77}$$

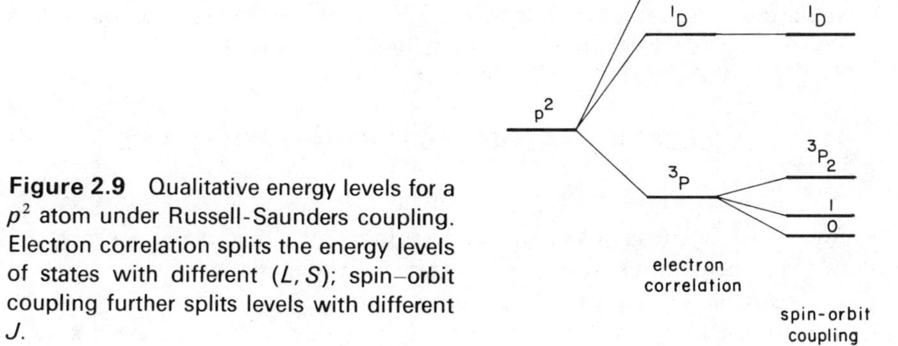

Figure 2.9 Qualitative energy levels for a p^2 atom under Russell-Saunders coupling. Electron correlation splits the energy levels of states with different (L, S); spin–orbit coupling further splits levels with different J.

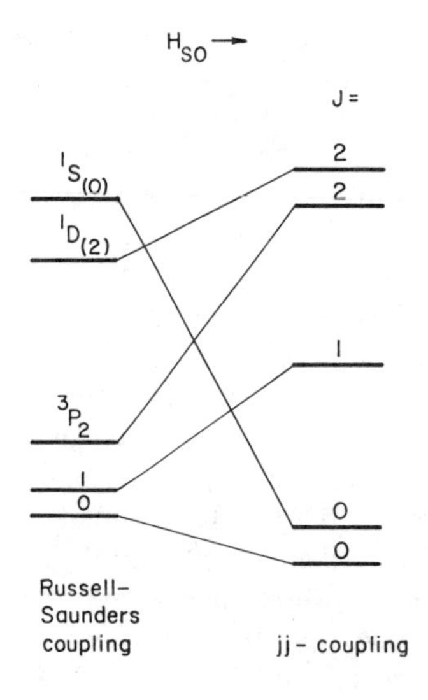

Figure 2.10 Correlation between levels under Russell-Saunders coupling (left) and *jj* coupling (right).

In a p^2 configuration, each electron i can have $j_i = l_i + s_i, \ldots, |l_i - s_i|$, or $j_i = \frac{3}{2}, \frac{1}{2}$. Then the possible J values are $j_1 + j_2, \ldots, |j_1 - j_2|$; this leads to $J = 0, 1, 2, 3$. However, the last of these values violates the Pauli principle. (Allowing $J = 3$ would require $j_1 = j_2 = \frac{3}{2}$; this means that $l_1 = l_2 = 1$ and $s_1 = s_2 = \frac{1}{2}$, which would require that one of the possible states has $m_{l_1} = m_{l_2} = 1$ and $m_{s_1} = m_{s_2} = \frac{1}{2}$. This would place two electrons with the same spin into the same orbital). Hence, the only allowed J values in an equivalent p^2 configuration are 0, 1, and 2. This is reasonable, since $\mathbf{J}^2$ is conserved even for large $\hat{H}_{so}$ according to the commutation relationships in Eqs. 2.45. Hence, the J values that are possible for a given electron configuration in the limit of Russell-Saunders coupling will be the same as the ones accessible in the limit of *jj* coupling, and a conceptual correlation diagram can be drawn between J states in the Russell-Saunders and *jj* limits (Fig. 2.10). In the latter limit, the splittings between term energies depend primarily on J, rather than on L and S. The *noncrossing rule* states that no two energy curves representing states with the same symmetry (which is indicated by J, M in the full rotation group of spherical atoms) can intersect, and this rule is observed in Fig. 2.10. Most atoms are Russell-Saunders or intermediate coupling cases; few atoms are *jj*-coupled.

2.5 MANY-ELECTRON ATOMS: SELECTION RULES AND SPECTRA

An interesting example of a many-electron spectrum is that of He, in which the shown low-energy transitions involve orbital jumps of one of the two electrons. For this case our one-electron atomic selection rules ($\Delta l = \pm 1, \Delta j = 0, \pm 1$) hold for the electron involved in the transition. The He electronic spectrum resembles a superposition of *two independent* alkali spectra [7]: It exhibits two independent principal series, two independent sharp series, and so on (Fig. 2.11). This occurs because in He, where the spin–orbit coupling is very small ($Z = 2$), the electronic states have nearly pure singlet or triplet character. Since the electric dipole operator $\boldsymbol{\mu}$ does not contain any spin coordinates, the spin selection rule $\Delta S = 0$ is strict in He. Hence there are two families of E1 transitions, one among the singlet levels ($S = 0$) and the other among the triplet levels ($S = 1$). The separations between lines within fine structure multiplets in this light atom are too small to depict in Fig. 2.11, and the J subscripts in the level term symbols are omitted in this figure.

The spin–orbit coupling is much larger in Hg ($Z = 80$), whose strongest transitions are shown in Fig. 2.12. This complicates the emission spectrum in two ways. The fine structure levels arising from each triplet multiplet (e.g., the $^3P_0$, $^3P_1$, and $^3P_2$ levels arising from the $(6s)^1(6p)^1$ configuration) are now well separated in energy, increasing the number of spectral lines observed under low

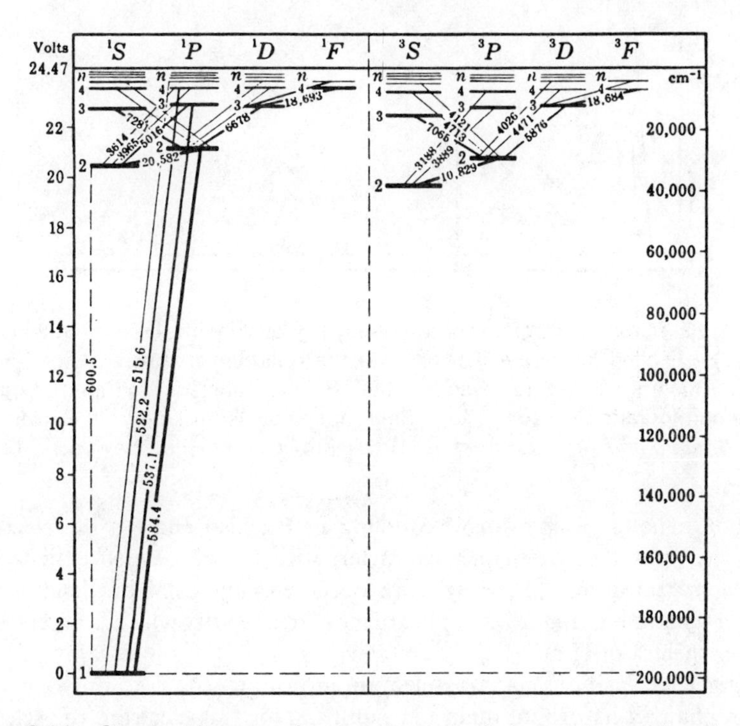

Figure 2.11 Grotrian diagram of energy levels and observed transitions in He. All shown levels arise from electron configurations of the type $(1s)^1(nl)^1$. The ^{1}S level labeled "1" is the $(1s)^2$ ground state. The ^{1}S, ^{3}S levels labeled "2" arise from $(1s)^1(2s)^1$ configurations, the ^{1}P, ^{3}P levels labeled "2" arise from $(1s)^1(2p)^1$ configurations, and so on. Reproduced by permission from G. Herzberg, *Atomic Spectra and Atomic Structure*, Dover Publications, Inc., New York, 1944.

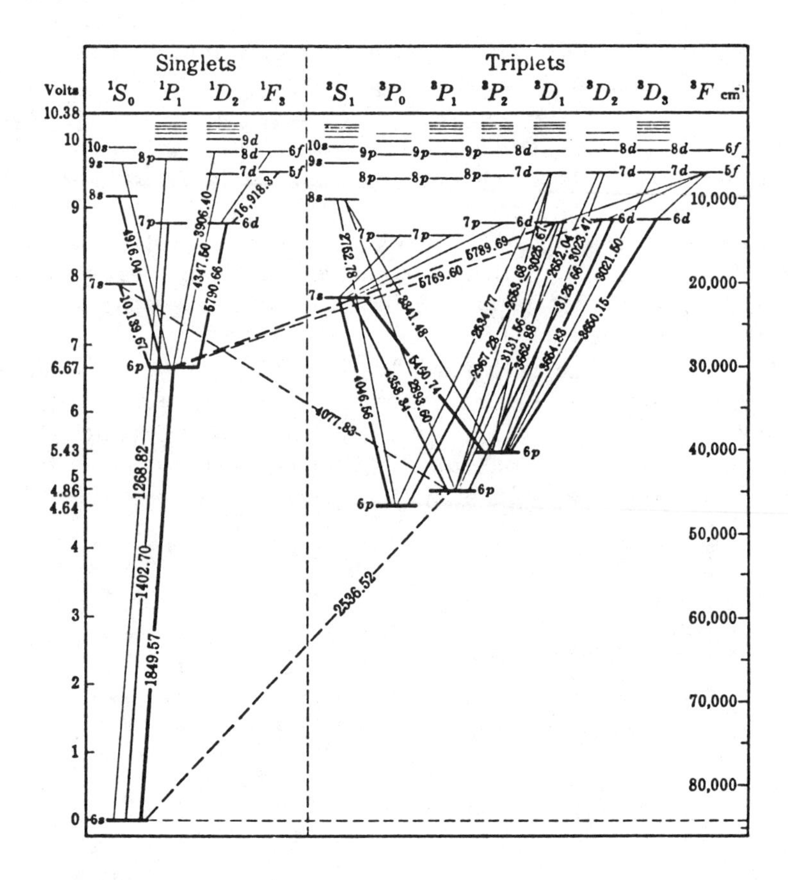

Figure 2.12 Grotrian diagram for low-lying Hg levels with the configuration $\cdots (6s)^1(n1)^1$. Excited levels are labeled with the quantum numbers of the valence electron which is excited; for example, the $^1P_1$ level labeled "7p" arises from the electron configuration $\cdots (6s)^1(7p)^1$. Reproduced by permission from G. Herzberg, *Atomic Spectra and Atomic Structure*, Dover Publications, Inc., New York, 1944.

resolution. The large spin–orbit coupling in Hg also endows the electronic states with mixed singlet/triplet character, with the consequence that many intersystem transitions ($\Delta S = \pm 1$) are observed. These factors lend the Hg emission spectrum in Fig. 2.2 an appearance of irregularity which is absent in the emission spectra of H and K.

Strong E1 transitions in many-electron atoms are observed only when *one* electron changes its orbital quantum numbers; for this electron, the selection rule $\Delta l = \pm 1$ must be obeyed (cf. our discussion following Eq. 2.12). To appreciate this, we recall that spatial wavefunctions in many-electron atoms may be expressed (Section 2.3) in terms of products $\psi(1, 2, \ldots, p) = \phi_1(1)\phi_2(2) \ldots \phi_p(p)$ of one-electron orbitals $\phi_1(1), \phi_2(2), \ldots, \phi_p(p)$. Since the pertinent electric dipole operator is $\boldsymbol{\mu} = -e\sum \mathbf{r}_i$, the E1 transition moment from electronic state $\psi(1, 2, \ldots, p)$ to state $\psi'(1, 2, \ldots, p) = \phi'_1(1)\phi'_2(2) \ldots \phi'_p(p)$

behaves as

$$
\begin{aligned}
&-e\langle \phi_1(1)\phi_2(2) \cdots \phi_p(\mathrm{p}) \left| \sum_{i=1}^{p} \mathbf{r}_i \right| \phi_i'(1)\phi_2'(2) \cdots \phi_p'(p)\rangle \\
&\quad = -e\langle \phi_1(1)|\mathbf{r}_1|\phi_1'(1)\rangle\langle \phi_2(2)|\phi_2'(2)\rangle \cdots \langle \phi_p(p)|\phi_p'(p)\rangle \\
&\quad\quad - e\langle \phi_2(2)|\mathbf{r}_2|\phi_2'(2)\rangle\langle \phi_1(1)|\phi_1'(1)\rangle \cdots \langle \phi_p(p)|\phi_p'(p)\rangle \\
&\quad\quad - e\langle \phi_p(p)|\mathbf{r}_p|\phi_p'(p)\rangle\langle \phi_1(1)|\phi_1'(1)\rangle \cdots \langle \phi_{p-1}(p-1)|\, \phi_{p-1}'(p-1)\rangle
\end{aligned}
$$

Each of these terms contains a factor $\langle \phi_i(i)|\mathbf{r}_i|\phi_i'(i)\rangle$ analogous to the hydrogen-like transition moment (2.10), multiplied by orthogonality integrals $\langle \phi_j(j)|\phi_j'(j)\rangle$ for all other electrons $j \neq i$. Hence, if electron i jumps to a new orbital, i.e., $\phi_i'(i) \neq \phi_i(i)$, orthogonality requires that all other electrons remain in their original orbitals because $\langle \phi_j(j)|\phi_j'(j)\rangle = 0$ unless $\phi_j' = \phi_j$.

The E1 selection rules may also be formulated in terms of the quantum numbers L, S, and J for the many-electron orbital, spin, and total angular momentum. These prove to be

E1

$$
\begin{aligned}
&\Delta L = 0, \pm 1 \qquad \text{except } L = 0 \not\leftrightarrow L' = 0 \\
&\Delta S = 0 \\
&\Delta J = 0, \pm 1 \qquad \text{except } J = 0 \not\leftrightarrow J' = 0
\end{aligned}
\tag{2.78}
$$

Not all transitions consistent with these selection rules are E1-allowed, since $\Delta l = \pm 1$ must be simultaneously obeyed by the electron that jumps. For example, a transition from a state with $L = 3$, $l_1 = 3$, $l_2 = 1$ to a state with $L = 2$, $l_1 = 1$, $l_2 = 1$ is forbidden because $\Delta l_1 = -2$ and $\Delta l_2 = 0$, even though the $\Delta L = 0, \pm 1$ selection rule is obeyed.

The corresponding M1 and E2 selection rules are

M1

$$
\begin{aligned}
&\Delta J = 0, \pm 1 \qquad \text{except } J = 0 \not\leftrightarrow J' = 0 \\
&\Delta L = 0, \pm 1 \\
&\Delta S = 0
\end{aligned}
\tag{2.79}
$$

E2

$$
\begin{aligned}
&\Delta J = 0, \pm 1, \pm 2 \quad \text{and} \quad (J + J') \geqslant 2 \\
&\Delta L = 0, \pm 1, \pm 2 \quad \text{except } L = 0 \not\leftrightarrow L' = 0 \\
&\Delta S = 0
\end{aligned}
\tag{2.80}
$$

In all cases, the $\Delta S = 0$ selection rule becomes relaxed when the spin–orbit coupling is large.

To illustrate one final point, we return to the example of the equivalent p^2 configuration, for which the possible atomic term symbols are ^1S, $^3P_{0,1,2}$, and ^1D. The E1 selection rules in Eq. 2.78 indicate that there are no E1-allowed transitions connecting any two of these levels with different L if the spin–orbit coupling is small. This is an example of a general rule that no E1 transitions connect two term symbols arising from the *same* electron configuration (in this case, p^2). E1-allowed transitions can, however, occur between states from different electron configurations, say $s^1p^1 \to p^2$. This proves to be a useful principle that carries over to the electronic spectroscopy of diatomic and polyatomic molecules.

2.6 THE ZEEMAN EFFECT

When atoms are subjected to external magnetic fields, their spectral lines become split into several components whose separations depend on the field strength **B**. In atoms with zero net electronic spin ($S = 0$), the splittings turn out to be identical for all spectral lines. For weak, static magnetic fields **B**, the interaction energy for a spinless atom of angular momentum $\mathbf{J} = \mathbf{L}$ with the field is

$$W = -\boldsymbol{\mu}_m \cdot \mathbf{B} = \frac{e}{2m_e}\,\mathbf{L}\cdot\mathbf{B} \tag{2.81}$$

and it may be treated as a stationary perturbation to the atomic Hamiltonian $\hat{H}_0$. The first-order correction to the electronic energy of any level characterized by angular momentum quantum numbers (L, M_L) in the presence of a magnetic field $\mathbf{B} = B_z\hat{z}$ is

$$\begin{aligned}\frac{e}{2m_e}\langle LM_L|\mathbf{L}\cdot\mathbf{B}|LM_L\rangle &= \frac{e}{2m_e}\,B_z\langle LM_L|\hat{L}_z|LM_L\rangle \\ &= \frac{e\hbar}{2m_e}\,B_z M_L\end{aligned} \tag{2.82}$$

Since M_L ranges from $+L$ to $-L$, a level with orbital angular momentum L becomes split into $(2L + 1)$ components separated in energy by the amount $e\hbar B_z/2m_e$. A spectral line that appears at energy ΔE in zero magnetic field then appears at energy

$$\hbar\omega = \Delta E + \frac{e\hbar}{2m_e}\,B_z\Delta M_L \tag{2.83}$$

in the presence of the field, where ΔM_L is the change in M_L accompanying the electronic transition. The E1 selection rules on ΔM_L (Eq. 2.78) then imply that two new lines will appear for $\Delta M_L = \pm 1$, in addition to an unshifted line corresponding to $\Delta M_L = 0$ (Fig. 2.13). The only exception to this nominally occurs when both electronic levels in the transition are S states with $L = 0$, but such a transition would be E1-forbidden to begin with. Such splittings were predicted classically by Larmor many years before they were observed in atoms with $S = 0$. This phenomenon is historically known as the *normal Zeeman effect*.

In atoms with nonvanishing spin as well as orbital angular momentum, the situation is far more interesting. The Hamiltonian for such atoms in external magnetic fields assumes the form

$$\hat{H} = \hat{H}_0 + f(r)\mathbf{L}\cdot\mathbf{S} - \boldsymbol{\mu}_m\cdot\mathbf{B} \tag{2.84}$$

where the second and third terms represent the spin–orbit and Zeeman Hamiltonians, respectively. (The function $f(r)$ in the spin–orbit Hamiltonian

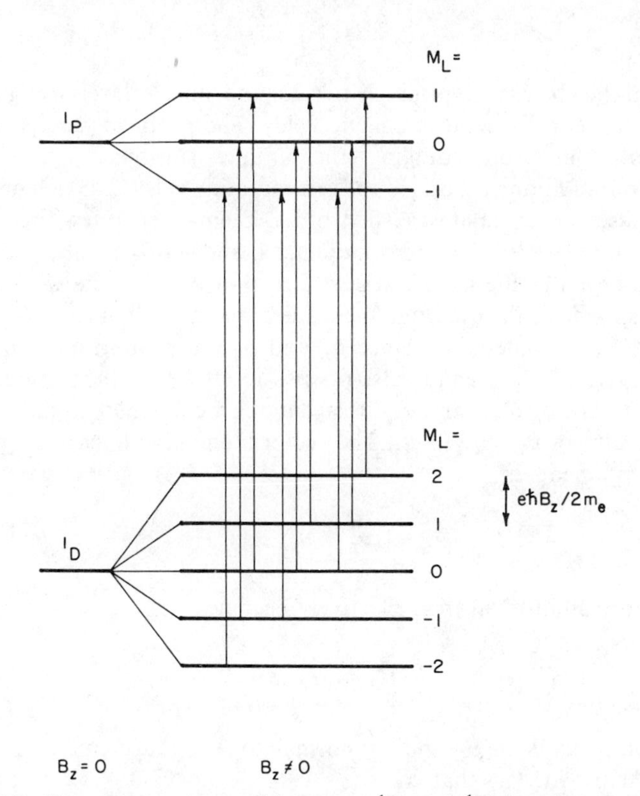

Figure 2.13 Transitions observed between ^{1}D and ^{1}P states under the normal Zeeman effect. Since the separation between adjacent sublevels is invariably $e\hbar B_z/2m_e$ and the selection rule restricts ΔM_L to 0, ± 1, spectral lines appear at only two new frequencies when the magnetic field is turned on.

depends only on the electronic radial coordinates.) The atomic magnetic moment $\boldsymbol{\mu}_m$ is now given by

$$\boldsymbol{\mu}_m = -\frac{e\mathbf{L}}{2m_e} - \frac{g_s e\mathbf{S}}{2m_e}$$

$$\equiv \boldsymbol{\mu}_L + \boldsymbol{\mu}_S \tag{2.85}$$

where the electron spin g factor is approximately given by $g_s \simeq 2.0$. The Zeeman Hamiltonian

$$\hat{H}_Z = -\boldsymbol{\mu}_m \cdot \mathbf{B} \tag{2.86}$$

for $\mathbf{B} = B_z\hat{z}$ could be taken to be

$$\hat{H}_Z = \frac{e}{2m_e}(\hat{L}_z + g_s\hat{S}_z)B_z \tag{2.87}$$

which could then be used in principle to calculate the new levels using first-order perturbation theory in weak magnetic fields. The problem here is that in the presence of spin–orbit coupling the atomic stationary states are not eigenfunctions of L_z and S_z, due to the commutation rules (2.45). Hence, Eq. 2.87 cannot be used to calculate the first-order Zeeman energies directly. In the classical view of Russell-Saunders coupling, **L** and **S** precess about **J**, which in turn precesses about the magnetic field direction $\hat{B} \equiv \hat{z}$ as shown in Fig. 2.14. (This corresponds to the quantum mechanical picture in which J_z is stationary, but L_z and S_z are indefinite.) Since $\boldsymbol{\mu}_L$ and $\boldsymbol{\mu}_S$ are proportional to **L** and **S**, respectively (Eq. 2.85), $\boldsymbol{\mu}_L$ and $\boldsymbol{\mu}_S$ also precess about **J**, with the consequence that only the components of $\boldsymbol{\mu}_L$ and $\boldsymbol{\mu}_S$ directed along **J** contribute to the expectation (averaged) value of $\boldsymbol{\mu}_M = \boldsymbol{\mu}_L + \boldsymbol{\mu}_S$. The vector along **J** with length equal to sum of these components is

$$\boldsymbol{\mu}_J = [(\boldsymbol{\mu}_L + \boldsymbol{\mu}_S)\cdot\mathbf{J}]\mathbf{J}/\mathbf{J}^2 \tag{2.88}$$

The Zeeman Hamiltonian then effectively becomes

$$\hat{H}_Z = -\boldsymbol{\mu}_J \cdot \mathbf{B}$$

because components of $\boldsymbol{\mu}_L$ and $\boldsymbol{\mu}_S$ normal to **J** average out to zero during precession. Now the fact that

$$\frac{(\boldsymbol{\mu}_L + \boldsymbol{\mu}_S)\cdot\mathbf{J}}{\mathbf{J}^2} = -\frac{e}{2m_e}(\mathbf{L} + g_s\mathbf{S})\cdot\mathbf{J}\cdot\frac{1}{\mathbf{J}^2} \tag{2.89}$$

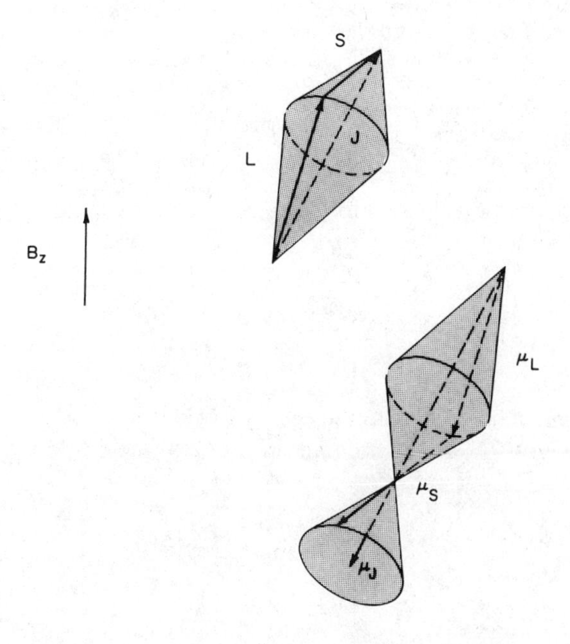

Figure 2.14 Angular momentum vectors **L**, **S**, **J** (left) and their associated magnetic moments $\boldsymbol{\mu}_L$, $\boldsymbol{\mu}_S$, $\boldsymbol{\mu}_J$ (right). Under Russell-Saunders coupling, the orbital and spin angular momenta **L** and **S** precess about the total angular momentum **J**. The magnetic moments $\boldsymbol{\mu}_L$, $\boldsymbol{\mu}_S$ (which are antiparallel to **L** and **S**) undergo similar precession. Since $g_s \simeq 2$, we have $|\boldsymbol{\mu}_S|/|\mathbf{S}| \simeq 2|\boldsymbol{\mu}_L|/|\mathbf{L}|$, with the consequence that the vector sum $(\boldsymbol{\mu}_L + \boldsymbol{\mu}_S)$ is not parallel to **J**. The *effective g* factor g_J is the projection of $(\boldsymbol{\mu}_L + \boldsymbol{\mu}_S)$ upon the direction of **J**, since precession rapidly averages the other components of $(\boldsymbol{\mu}_L + \boldsymbol{\mu}_S)$ to zero.

combined with the identities

$$\mathbf{S}^2 = (\mathbf{J} - \mathbf{L})^2 = \mathbf{J}^2 + \mathbf{L}^2 - 2\mathbf{L}\cdot\mathbf{J}$$
$$\mathbf{L}^2 = (\mathbf{J} - \mathbf{S})^2 = \mathbf{J}^2 + \mathbf{S}^2 - 2\mathbf{S}\cdot\mathbf{J} \tag{2.90}$$

allows us to write

$$\boldsymbol{\mu}_J = -\frac{e}{2m_e}\frac{\mathbf{J}^2 + \mathbf{L}^2 - \mathbf{S}^2 + g_s(\mathbf{J}^2 + \mathbf{S}^2 - \mathbf{L}^2)}{2\mathbf{J}^2}\mathbf{J} \tag{2.91}$$

Replacing $\mathbf{J}^2$, $\mathbf{L}^2$, and $\mathbf{S}^2$ with $\hbar^2 J(J+1)$, $\hbar^2 L(L+1)$, and $\hbar^2 S(S+1)$, respectively, yields

$$\boldsymbol{\mu}_J = -\frac{g_J e\mathbf{J}}{2m_e} \tag{2.92}$$

with the effective Landé g factor

$$g_J = \frac{(1 + g_s)J(J + 1) + (1 - g_s)[L(L + 1) - S(S + 1)]}{2J(J + 1)} \tag{2.93}$$

Using the approximate value 2.0 for the electron spin g factor g_s results in the common expression [1]

$$g_J = 1 + \frac{J(J + 1) + S(S + 1) - L(L + 1)}{2J(J + 1)} \tag{2.94}$$

Finally, the Zeeman correction to the energy in a state with angular momentum quantum numbers L, S, J, M_J becomes in the weak-field limit

$$\begin{aligned} -\langle LSJM_J|\boldsymbol{\mu}_m \cdot \mathbf{B}|LSJM_J\rangle &= \frac{eg_J}{2m_e} \langle LSJM_J|\mathbf{J} \cdot \mathbf{B}|LSJM_J\rangle \\ &= \frac{g_J e\hbar}{2m_e} B_z M_J \end{aligned} \tag{2.95}$$

The remarkable property of the effective g factor g_J is that it depends on L and S as well as on J (Eqs. 2.93 and 2.94). In the limits where $\mathbf{L} = 0$ and $\mathbf{S} = 0$, g_J reasonably approaches g_s and unity, respectively. If g_s were unity rather than approximately 2.0, g_J would also be unity according to Eq. 2.93; $\boldsymbol{\mu}_m$ would then be parallel to $\mathbf{J}$. Since the orbital and spin angular momenta have different g factors, $\boldsymbol{\mu}_m$ does not point along $\mathbf{J}$ (Fig. 2.14), and the Zeeman energies depend on the orbital and spin as well as total angular momentum quantum numbers. They are *independent* of the radial quantum numbers. Since atomic spectral lines in the absence of external fields do not appear with little flags exclaiming "I came from $J = 4$, $L = 3$, $S = 1$," this *anomalous Zeeman effect* observed in atoms with nonvanishing $\mathbf{L}$, $\mathbf{S}$ has proved to be an invaluable experimental tool for assigning them. The fact that g_J depends only on L, S, and J is known as *Preston's Law*.

Figure 2.15 illustrates the anomalous Zeeman effect in Na atom transitions from the $3^2S_{1/2}$ ground state to the lowest lying excited fine structure levels $3^2P_{1/2}$ and $3^2P_{3/2}$. (These are the well-known D line transitions which occur at 5895.93 and 5889.96 Å in zero field.) According to Eq. 2.94, the respective g_J values for the $3^2S_{1/2}$, $3^2P_{1/2}$, and $3^2P_{3/2}$ terms symbols are 2, $\frac{2}{3}$, and $\frac{4}{3}$; a weak magnetic field splits the fine structure levels into $2J + 1 = 2$, 2, and 4 components, respectively. The E1 transitions displayed in Fig. 2.15 obey the selection rules $\Delta J = 0, \pm 1$ and $\Delta M_J = 0, \pm 1$. Their wavelengths may be analyzed using Eqs. 2.94 and 2.95 to deduce the angular momentum quantum numbers a priori.

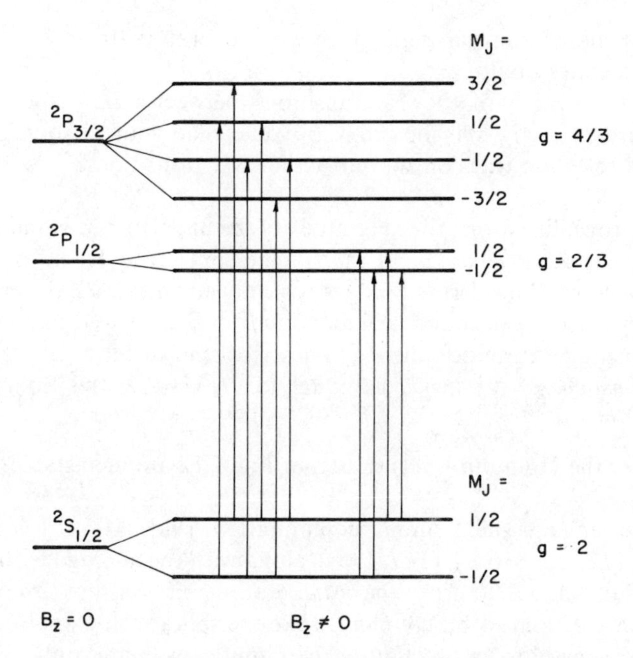

Figure 2.15 Transitions observed between a $^2S_{1/2}$ state and $^2P_{1/2}$ and $^2P_{3/2}$ states under the anomalous Zeeman effect. Since the Lande g factor $g_J = 1 + [J(J+1) + S(S+1) - L(L+1)]/2J(J+1)$ has a different value in each of these electronic states, a unique frequency is associated with each of the shown transitions. These frequencies can be analyzed to infer the angular momentum quantum numbers for each electronic state.

REFERENCES

1. E. U. Condon and G. H. Shortley, *The Theory of Atomic Spectra*, Cambridge University Press, London, 1970.
2. M. H. Nayfeh and M. K. Brussell, *Electricity and Magnetism*, Wiley, New York, 1985.
3. E. Merzbacher, *Quantum Mechanics*, Wiley, New York, 1961.
4. K. Gottfried, *Quantum Mechanics*, Vol. 1, W. A. Benjamin, New York, 1966.
5. W. H. Furry, *Am. J. Phys.* **23**: 517 (1955).
6. W. H. Flygare, *Molecular Structure and Dynamics*, Prentice-Hall, Englewood Cliffs, NJ, 1978.
7. G. Herzberg, *Atomic Spectra and Atomic Structure*, Dover, New York, 1944.

PROBLEMS

1. Consider the $^2D_{3/2}$ and $^2D_{5/2}$ fine structure levels in a one-electron atom.

(a) Express the coupled states $|lsjm\rangle$ in terms of the uncoupled states $|lm_l sm_s\rangle$ for these levels by working out the Clebsch-Gordan coefficients using angular momentum raising and lowering operators (there are 10 of these coupled states in all).

(b) Determine which of the 24 transitions between $^2D_{3/2}$ and $^2D_{5/2}$ fine structure levels are M1-allowed. Can you extend your reasoning to obtain general selection rules on Δj and Δm for M1 transitions?

2. In a hydrogenlike atom, the magnitude of the unperturbed Hamiltonian $\hat{H}_0$ is on the order of $Ze^2/4\pi\varepsilon_0 r$, and that of the spin–orbit Hamiltonian $\hat{H}_{so}$ is $\hbar^2 Ze^2/8\pi\varepsilon_0 m_e^2 c^2 r^3$. Considering that r is typically equal to a_0/Z (where a_0 is the Bohr radius), calculate the numerical magnitude of $\hat{H}_{so}/\hat{H}_0$, and comment on the validity of using perturbation theory to describe spin–orbit coupling. (The *fine structure constant* $\alpha = e^2/4\pi\varepsilon_0\hbar c$ is equal to 1/137.0372; the Bohr radius is $a_0 = 4\pi\varepsilon_0\hbar^2/m_e e^2$.)

3. Which of the Hg atom levels shown in Fig. 2.12 are metastable?

4. Use the angular momentum commutation rules $[L_x, L_y] = i\hbar L_z$, $[L_y, L_z] = i\hbar L_x$, $[L_z, L_x] = i\hbar L_y$, $[\mathbf{L}, L^2] = 0$, along with the analogous rules for the components of $\mathbf{S}$ and $\mathbf{J}$, to prove the commutation relationships (2.45) that hold in the presence of spin–orbit coupling. Take the spin–orbit Hamiltonian to be $\hat{H}_{so} = f(r)\mathbf{L}\cdot\mathbf{S}$, with $f(r)$ a function of the radial coordinates only.

5. Determine the atomic term symbols $^{2S+1}L_J$ that may be obtained from the equivalent p^3 configuration using a Slater diagram. Ascertain explicitly whether there are any E1-allowed transitions among sublevels of this p^3 multiplet. According to Hund's rule, what should be the term symbol for ground state N?

6. Which levels in K could be reached by E2 transitions from the $4^2S_{1/2}$ ground state?

7. An E1 transition at 3125.66 Å connects the $6p\ ^3P_1$ and $6d\ ^3D_2$ fine structure levels in Hg. A weak magnetic field splits these levels via the anomalous Zeeman effect. Evaluate the g_J factors for both levels, indicate all of the E1 transitions that will occur using a schematic energy level diagram, and draw the resulting "stick spectrum" on a wavelength scale.

3

ROTATION AND VIBRATION IN DIATOMICS

Many of the features that are prominent in atomic spectroscopy have close analogies in electronic spectroscopy of diatomic molecules. Like atoms, diatomics exhibit electron correlation, spin–orbit coupling with its associated fine structure and angular momentum coupling schemes, and the Zeeman effect. In diatomics as in atoms, the symmetry of the electronic Hamiltonian plays a major role in the electronic state degeneracies and selection rules. The distinctions encountered here in going from atoms to diatomics are rather trivial ones that accompany the reduction of spherical symmetry in atoms into the $C_{\infty v}$ and $D_{\infty h}$ point group symmetries in diatomics.

These analogies fail to explain the coarse structure observed in diatomic electronic spectra under low resolution (Chapter 4), because the additional nuclear coordinates introduced by the second atom in a diatomic molecule AB create new internal modes that have no counterpart in atoms. The nuclear positions relative to an arbitrary origin fixed in space can be specified using the Cartesian vectors $\mathbf{R}_A$ and $\mathbf{R}_B$ (Fig. 3.1). They may be equivalently described in terms of the coordinates

$$\mathbf{R}_{cm} \equiv (M_A\mathbf{R}_A + M_B\mathbf{R}_B)/(M_A + M_B) \tag{3.1}$$

and

$$\mathbf{R} \equiv \mathbf{R}_B - \mathbf{R}_A \tag{3.2}$$

where M_A and M_B are the nuclear masses. The vector $\mathbf{R}_{cm}$ locates the nuclear center of mass; $\mathbf{R}$ is a vector whose length R gives the internuclear separation

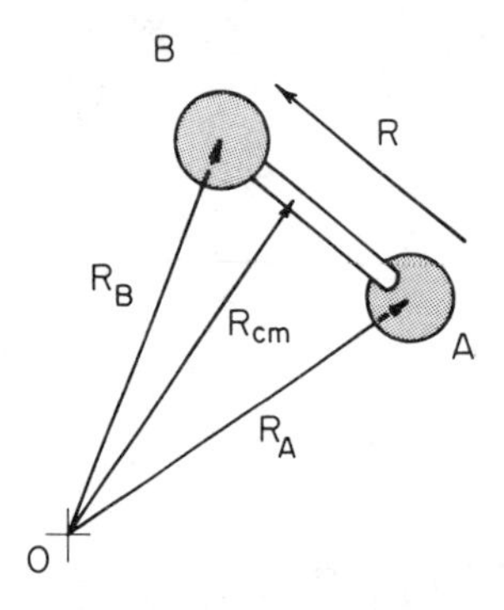

Figure 3.1 The nuclear positions in a diatomic molecule may be specified in terms of either the laboratory coordinates $\mathbf{R}_A$, $\mathbf{R}_B$, or the center-of-mass coordinates $\mathbf{R}_{cm}$, $\mathbf{R}$.

and whose direction $\hat{R}$ gives the orientation of the internuclear axis in space. It can readily be shown [1] (see also Appendix G) that the kinetic energy

$$T = \tfrac{1}{2}M_A\dot{\mathbf{R}}_A^2 + \tfrac{1}{2}M_B\dot{\mathbf{R}}_B^2 \tag{3.3}$$

of nuclear motion in Cartesian coordinates becomes transformed using Eqs. 3.1 and 3.2 into

$$T = \tfrac{1}{2}M\dot{\mathbf{R}}_{cm}^2 + \tfrac{1}{2}\mu_N\dot{\mathbf{R}}^2 = \tfrac{1}{2}M\dot{\mathbf{R}}_{cm}^2 + \tfrac{1}{2}\mu\dot{R}^2 + \tfrac{1}{2}\mu R^2\dot{\theta}^2 \tag{3.4}$$

in terms of the center-of-mass coordinates $\mathbf{R}_{cm}$ and $\mathbf{R}$. The masses appearing in Eq. 3.4 are the total nuclear mass

$$M = M_A + M_B \tag{3.5}$$

and the nuclear *reduced mass*

$$\mu_N = M_A M_B/(M_A + M_B) \tag{3.6}$$

The first term in the kinetic energy (3.4), associated with center of mass translation, does not affect the internal energy levels in the molecule. The $\tfrac{1}{2}\mu_N\dot{R}^2$ term is the *vibrational* kinetic energy arising from changes in the diatomic's bond length R. The last term, $\tfrac{1}{2}\mu_N R^2\dot{\theta}^2$, is the nuclear *rotational* kinetic energy connected with molecular rotation through an angle θ about an axis normal to the molecular axis. The usefulness of the center-of-mass coordinates in Eq. 3.4 is that they separate the total nuclear kinetic energy into contributions from the overall molecular translation and the internal modes (vibration and rotation). These internal modes in diatomics are responsible for spectroscopic transitions occurring in the microwave to near-infrared regions of the electromagnetic spectrum.

An example of a *far-infrared* absorption spectrum of a diatomic gas is shown in Fig. 3.2. The vertical coordinate is light intensity transmitted by an HCl

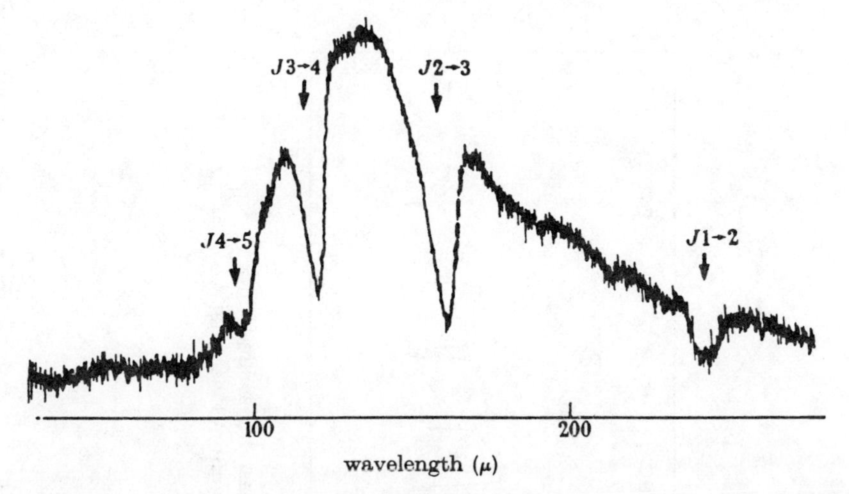

Figure 3.2 Far-infrared absorption spectrum of HCl gas. The vertical coordinate is light intensity transmitted by the gas sample; the horizontal coordinate is wavelength. Reproduced by permission from D. Bloor et al., *Proc. London Ser. A* **260**: 510 (1961).

sample after passing through a scanning monochromator operating at wavelengths λ betwen $\sim$90 and 300 μm. The intensity envelope peaking near 130 μm is determined primarily by wavelength-dependent light transmission (technically, the *grating blaze*) in the monochromator. The characteristic HCl absorption lines appear as minima at $\lambda = 240, 160, 120$, and 96 μm. The regularity in this sequence of lines is more striking when these wavelengths are converted into frequencies $\bar{\nu} = 1/\lambda = 41.7, 62.5, 83.3$, and 104 cm^{-1}, respectively: These frequencies are uniformly spaced by 20.8 cm^{-1}. The energy changes associated with these absorption lines, which arise from transitions between different *rotational levels* in HCl, are smaller than those observed in the atomic line spectra in Fig. 2.2 by factors of typically 10^3.

The far-infrared spectrum in Fig. 3.2 exhibits poor signal-to-noise ratios (S/N) and low resolution in comparison to other forms of molecular spectroscopy that will be studied later. As far-infrared spectra go, it represented nearly state-of-the-art technology in 1961. The experimental difficulties of far-infrared spectroscopy stem from the low intensities available from continuum light sources at such wavelengths (a 10^4 K blackbody emits only 1 part in 10^8 of its total output between 150 and 350 μm) and from the low efficiency of specialized light detectors (e.g., Golay thermal detectors, superconducting bolometers) that can be used in this regime. The linewidths displayed in Fig. 3.2 (defined as full width at half maximum, or fwhm) are on the order of 5 cm^{-1}. They are not intrinsically a property of the HCl gas, but arise from the necessity of relaxing the monochromator resolution (using wider entrance and exit slits) to bring the signal to a detectable level. Far higher resolution is now routinely available in this wavelength regime using Fourier transform techniques.

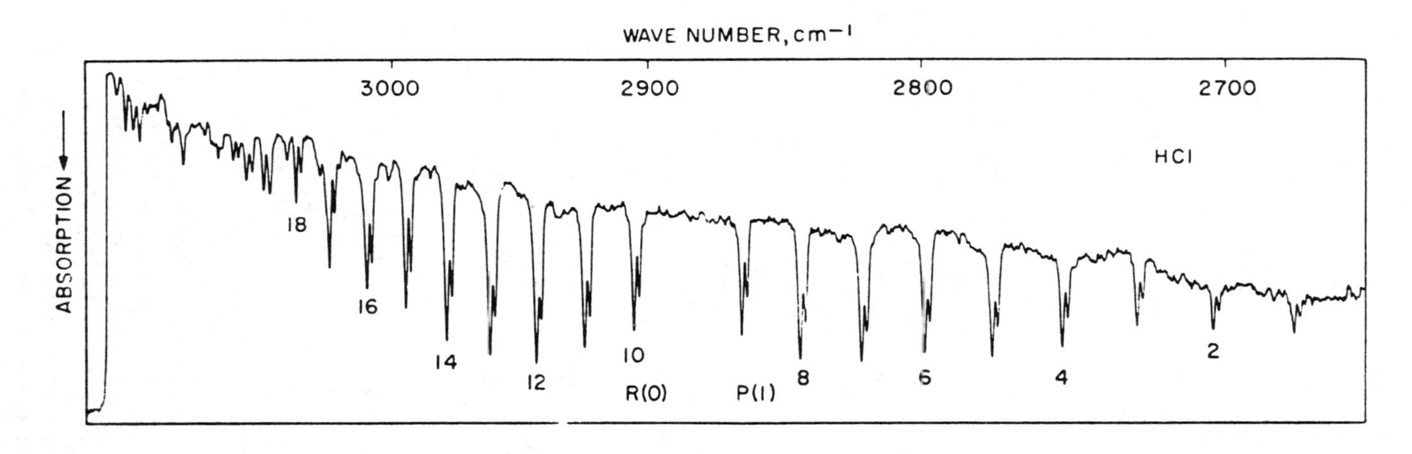

Figure 3.3 Near-infrared absorption spectrum of HCl gas: transmitted light intensity versus frequency in cm^{-1}. Reproduced by permission from Pyler et al., *J. Res. Nat. Bur. Stand. A* **64**: 29 (1960).

HCl gas also exhibits a series of absorption lines in the *near-infrared*, as shown in Fig. 3.3. The S/N in this spectrum is considerably higher than in the far-infrared spectrum; the fwhm linewidths in Fig. 3.3 are approximately 1 cm^{-1}. This reflects on the relative facility of generating and detecting light with near-infrared wavelengths; the energy changes associated with the transitions in Fig. 3.3 are larger than those in the far-infrared spectrum by factors on the order of 10^2. The absorption lines in the near-infrared HCl spectrum are grouped into closely spaced doublets separated by $\sim$2 cm^{-1}. It will be shown that these doublets appear because HCl naturally occurs in two common isotopes, $H^{35}Cl$ and $H^{37}Cl$, with 75.5 and 24.5% abundance, respectively. (Under sufficient resolution, the far-infrared absorption lines in Fig. 3.2 would similarly be split into isotopic doublets.) The near-infrared spectrum results from simultaneous rotational and vibrational level changes in HCl, and is called a *vibration–rotation* spectrum. The frequency spacings of $\sim$20 cm^{-1} between successive absorption lines are similar to those occurring in the far-infrared spectrum, although they are not nearly as constant as in the *pure rotation* spectrum of Fig. 3.2. The vibration–rotation spectrum shown in Fig. 3.3 is centered near 2900 cm^{-1}; similar (but weaker) families of vibration–rotation lines can also be observed at frequencies near multiples of 2900 cm^{-1}.

These spectra can be analyzed to yield detailed information about the molecular bond length, about the potential energy function that governs the vibrational motion, and about the interactions between vibrational and rotational motions which result in small corrections to the total molecular energy. A remarkable property of pure rotational spectra is that *no* knowledge of electronic structure is required to interpret them. This situation contrasts strikingly with that in atomic line spectra, where extensive configurational interaction calculations must be performed to correctly predict spectral line positions. The implication here is that electronic and nuclear rotational motions are in some way essentially decoupled. Vibrational and electronic motions often prove to be separable in a similar sense, although details of electronic structure do influence vibrational motion via their effect on the vibrational potential energy function. The concept of separability between electronic and nuclear motions forms the basis of the *Born-Oppenheimer approximation*, which we discuss in Section 3.1. In the remainder of this chapter, we build on the foundation provided by the Born-Oppenheimer principle to develop the eigenfunctions, energy levels, and spectroscopy of diatomic rotations and vibrations.

3.1 THE BORN-OPPENHEIMER PRINCIPLE

The Born-Oppenheimer principle is a cornerstone of molecular spectroscopy, an organizing principle that vastly simplifies the assignment of different spectral features to different types of molecular motion. Without it, electronic and nuclear motions would be scrambled in complicated molecular Hamiltonians, and extensive numerical calculations would be necessary to extract even the most qualitative features of vibrational and rotational structure—as must now be done to compute energy levels that are influenced by correlation in many-electron atoms. The well-known approximations to diatomic vibrational and rotational energy levels [$E_v \simeq \hbar\omega(v + \frac{1}{2})$ and $E_J \simeq hcBJ(J + 1)$] would have no reality. The Born-Oppenheimer principle sometimes breaks down, and the cases in which it does so have attracted considerable research interest from people interested in molecular dynamics and energy transfer. Examples of such breakdowns are vibronic coupling effects in electronic spectroscopy of polynuclear aromatic hydrocarbons, and nonadiabatic transitions in reactive molecular collisions.

The total Hamiltonian for a diatomic molecule with n electrons is

$$\hat{H} = \sum_{N=1}^{2} \frac{-\hbar^2}{2m_N} \nabla_N^2 + \sum_{i=1}^{n} \frac{-\hbar^2}{2m_e} \nabla_i^2 + \frac{1}{4\pi\varepsilon_0} \sum_{i<j}^{n} \frac{e^2}{|\mathbf{r}_i - \mathbf{r}_j|}$$

$$- \frac{1}{4\pi\varepsilon_0} \sum_{i=1}^{n} \sum_{N=1}^{2} \frac{Z_N e^2}{|\mathbf{r}_i - \mathbf{R}_N|} + \frac{1}{4\pi\varepsilon_0} \frac{Z_A Z_B e^2}{|\mathbf{R}_A - \mathbf{R}_B|} + \hat{H}_{so} \tag{3.7}$$

These contributions to $\hat{H}$ include the nuclear kinetic energy, the electronic kinetic energy, the electron–electron repulsions, the electron–nuclear attractions, the nuclear–nuclear repulsion, and the spin–orbit coupling. A priori, the electronic coordinates $\mathbf{r}$ and nuclear coordinates $\mathbf{R}$ appear to be inseparably mixed in $\hat{H}$, and the electronic and nuclear coordinates are strongly coupled. The Schrödinger equation for the diatomic becomes

$$\hat{H}(\mathbf{r}, \mathbf{R})|\Psi(\mathbf{r}, \mathbf{R})\rangle = E|\Psi(\mathbf{r}, \mathbf{R})\rangle \tag{3.8}$$

where $\mathbf{r}$ and $\mathbf{R}$ represent the sets of electronic and nuclear coordinates, respectively.

We now define the *electronic* Hamiltonian $\hat{H}_{el}$ as

$$\begin{aligned} \hat{H}_{el} &\equiv \hat{H} - \sum_{N=1}^{2} \frac{-\hbar^2}{2m_N} \nabla_N^2 - \hat{H}_{so} \\ &\equiv \hat{H} - \hat{T}_N - \hat{H}_{so} \end{aligned} \tag{3.9}$$

$\hat{H}_{el}$ differs from $\hat{H}$ by lacking the nuclear kinetic energy and spin–orbit operators $\hat{T}_N$ and $\hat{H}_{so}$. It is possible to find eigenfunctions of $\hat{H}_{el}$ for fixed $\mathbf{R}$ (i.e., for nuclei motionlessly clamped in position), such that

$$\hat{H}_{el}(\mathbf{r}, \mathbf{R})|\psi_k(\mathbf{r}; \mathbf{R})\rangle = \varepsilon_k(R)|\psi_k(\mathbf{r}; \mathbf{R})\rangle \tag{3.10}$$

The resulting electronic states $|\psi_k(\mathbf{r}; \mathbf{R})\rangle$ will depend parametrically on $\mathbf{R}$ in that the choice of fixed nuclear positions influences the electronic states (physically, pulling the nuclei apart will naturally distort the molecule's electron cloud). The $\varepsilon_k(R)$ are the diatomic potential energy curves—the electronic energies of the various states k as functions of the internuclear separation R. Numerous potential energy curves are shown for several diatomic molecules in Figs. 3.4, 3.10, 4.2, 4.6, 4.7, 4.8, 4.9, 4.14, 4.29, 4.30, and 4.31 in this book.

For the solutions $|\Psi(\mathbf{r}, \mathbf{R})\rangle$ to the full diatomic Hamiltonian, we now try expansions of the form

$$|\Psi(\mathbf{r}, \mathbf{R})\rangle = \sum_k a_k |\psi_k(\mathbf{r}; \mathbf{R})\rangle |\chi_k(\mathbf{R})\rangle \tag{3.11}$$

where the $|\chi_k(\mathbf{R})\rangle$ are the diatomic nuclear (vibrational–rotational) wave functions in electronic state k and the a_k are expansion coefficients. If we have *complete sets* $|\psi_k\rangle$ and $|\chi_k\rangle$, $|\Psi(\mathbf{r}, \mathbf{R})\rangle$ can *always* be expanded as in Eq. 3.11. It is only when $|\Psi(\mathbf{r}, \mathbf{R})\rangle$ can be well approximated by a *single* term of the form $|\psi_k\rangle|\chi_k\rangle$ that one can be said to have achieved separation between electronic and nuclear motion, by factorizing the total wave function into electronic and nuclear parts. Such a factorized state is called a Born-Oppenheimer state.

If we substitute from Eq. 3.11 for $|\Psi(\mathbf{r}, \mathbf{R})\rangle$ into the diatomic Schrödinger equation using the Hamiltonian of Eq. 3.7, we obtain [2]

$$a_k \left[\sum_{N=1}^{2} \frac{-\hbar^2}{2m_N} \nabla_N^2 + T''_{kk} + U_{kk} - E \right] |\chi_k(\mathbf{R})\rangle$$

$$= -\sum_{k' \neq k} a_{k'} [T'_{kk'} + T''_{kk'} + U_{kk'}] |\chi_{k'}(\mathbf{R})\rangle \tag{3.12}$$

According to Eq. 3.4, the nuclear kinetic energy operator may be recast in terms of center-of-mass coordinates,

$$\sum_{N=1}^{2} -\frac{\hbar^2}{2m_N} \nabla_N^2 \equiv -\frac{\hbar^2}{2M_\mathrm{A}} \nabla_\mathrm{A}^2 - \frac{\hbar^2}{2M_\mathrm{B}} \nabla_\mathrm{B}^2$$

$$= -\frac{\hbar^2}{2M} \nabla_\mathrm{cm}^2 - \frac{\hbar^2}{2\mu_N} \nabla_R^2 \tag{3.13}$$

Since the electronic and nuclear wave functions are independent of the center-of-mass position $\mathbf{R}_\mathrm{cm}$, the term in $\hat{T}_N$ involving ∇_cm^2 drops out in Eq. 3.12. The new quantities introduced in Eq. 3.12 are defined as

$$U_{kk'} = \langle \psi_k | \hat{H}_\mathrm{el} + \hat{H}_\mathrm{so} | \psi_{k'} \rangle = \varepsilon_k(R)\delta_{kk'} + \langle \psi_k | \hat{H}_\mathrm{so} | \psi_{k'} \rangle \tag{3.14a}$$

$$T'_{kk'} = \frac{-\hbar^2}{2\mu_N} \mathbf{d}_{kk'} \cdot \nabla_R \tag{3.14b}$$

$$T''_{kk'} = \frac{-\hbar^2}{2\mu_N} D_{kk'} \tag{3.14c}$$

$$d_{kk'} = \langle \psi_k | \nabla_R | \psi_{k'} \rangle \tag{3.14d}$$

$$D_{kk'} = \langle \psi_k | \nabla_R^2 | \psi_{k'} \rangle \tag{3.14e}$$

The operators ∇_R and ∇_R^2 are given by

$$\nabla_R = \hat{\imath}\frac{\partial}{\partial R} + \hat{\jmath}\frac{1}{R}\frac{\partial}{\partial \theta} + \hat{k}\frac{1}{R \sin\theta}\frac{\partial}{\partial \phi}$$

$$\nabla_R^2 = \frac{1}{R^2}\frac{\partial}{\partial R}\left(R^2\frac{\partial}{\partial R}\right) + \frac{1}{R^2 \sin\theta}\frac{\partial}{\partial \theta}\left(\sin\theta\frac{\partial}{\partial \theta}\right) + \frac{1}{R^2 \sin^2\theta}\frac{\partial^2}{\partial \phi^2} \tag{3.15}$$

where (θ, ϕ) define the orientation of the internuclear axis in space, R is the

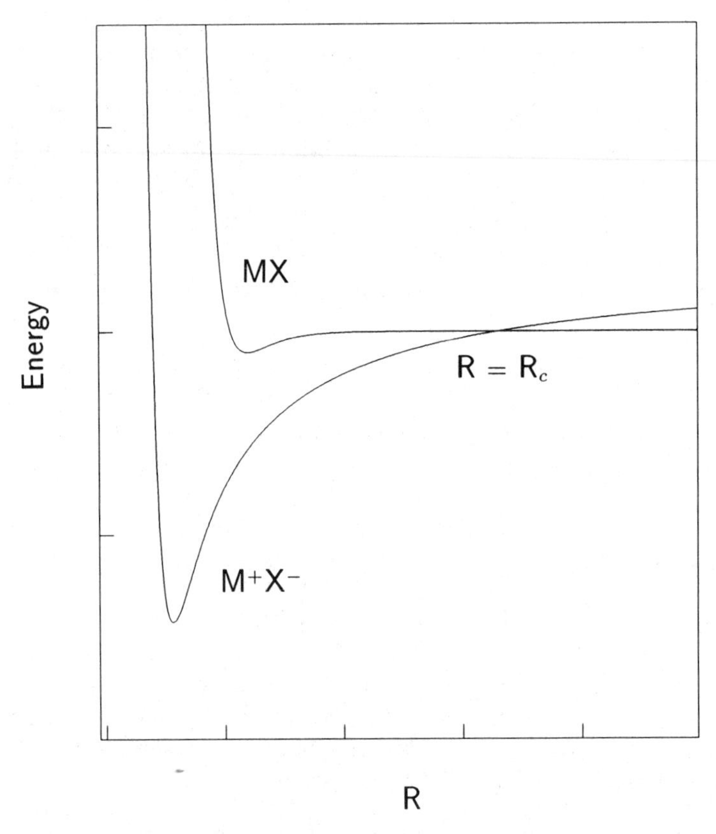

Figure 3.4 Lowest two bound potential energy curves for an alkali halide molecule MX. The zeroth-order ionic and covalent curves, which intersect at the *crossing radius* $R = R_c$, are shown at left. The adiabatic curves $\varepsilon_1(R)$ and $\varepsilon_2(R)$ are shown at right; the avoided crossing at $R = R_c$ is caused by mixing of the degenerate ionic and covalent states, which have the same symmetry. The lower adiabatic state $|\psi_1(\mathbf{r}; \mathbf{R})\rangle$ describes ionic M^+X^- for $R < R_c$, but correlates with uncharged separated atoms M + X for large R. The higher adiabatic state $|\psi_2(\mathbf{r}; \mathbf{R})\rangle$ describes covalent MX (which is far more weakly bound that ionic M^+X^-) for $R < R_c$, and correlates with $M^+ + X^-$ at large separations.

internuclear separation, and $(\hat{i}, \hat{j}, \hat{k})$ in this context are unit vectors associated with the coordinates (R, θ, ϕ).

Equation 3.12 is actually an infinite set of coupled equations that tell us that nuclear motion generally takes place in *all* electronic states simultaneously (since motion in a given state k is coupled by the right side to motion in all other electronic states k'). When this is true, the electronic and nuclear motions cannot be separated.

We now briefly consider the special case in which the right side of Eq. 3.12 is negligible. The quantity T''_{kk} on the left side of Eq. 3.12 is usually small (i.e., the electronic states $|\psi_k\rangle$ do not normally oscillate violently with internuclear separation R!) and so in this case we obtain

$$\left[\frac{-\hbar^2}{2\mu_N}\nabla_R^2 + U_{kk}(R) - E\right]|\chi_k(\mathbf{R})\rangle = 0 \tag{3.16}$$

When the spin–orbit coupling is small, $U_{kk}(R) = \varepsilon_k(R)$ according to Eq. 3.14a: $U_{kk}(R) = \langle\psi_k|H_{\mathrm{el}}|\psi_k\rangle = \varepsilon_k(R)$. Then Eq. 3.16 becomes a Schrödinger equation

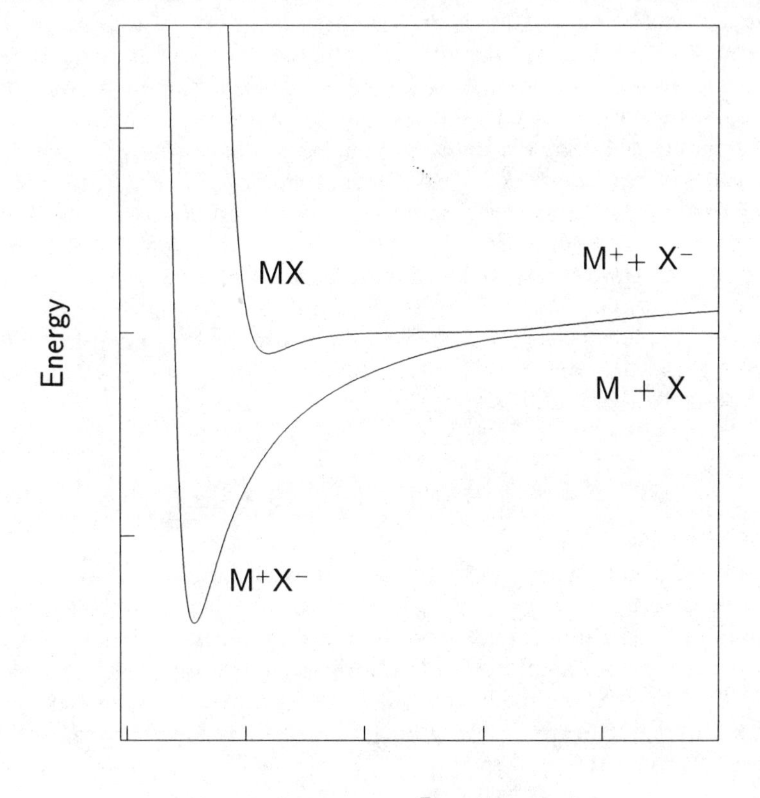

Figure 3.4 ***(Continued)***

for nuclear motion (vibration and rotation) in the potential $U_{kk}(R)$ specified by the potential energy curve $\varepsilon_k(R)$ of electronic state k. This eigenvalue equation then leads to the diatomic rotational and vibrational states as they are commonly understood in molecular spectroscopy. When Eq. 3.16 is approximately valid, the Born-Oppenheimer principle applies.

We now consider reasons why the Born-Oppenheimer approximation can break down, i.e., why the *right* side of Eq. 3.12 can become appreciable. Since $U_{kk'} = \langle\psi_k|\hat{H}_{so}|\psi_{k'}\rangle$ for $k \neq k'$, large spin–orbital Hamiltonians $\hat{H}_{so}$ can be a culprit—and Born-Oppenheimer violations can indeed result from large spin–orbit coupling. The term $T''_{kk'}$ in Eq. 3.14 is generally small, for the same reason that T''_{kk} is usually negligible. So we now focus on T'_{kk}. This term can become large when the potential energy curves for two electronic states k and k' approach closely, causing nonadiabatic transitions from nuclear motion in state k to nuclear motion in state k'.

As an example, we discuss the approach of an alkali atom M and a halogen atom X. The lowest two potential energy curves $\varepsilon_1(R)$ and $\varepsilon_2(R)$ for the alkali halide MX are schematically shown in Fig. 3.4. The lower of the two states $|\psi_1(\mathbf{r}; \mathbf{R})\rangle$ correlates with the ground-state neutral alkali atoms M + X, and has $^1\Sigma^+$ symmetry. The upper state $|\psi_2(\mathbf{r}; \mathbf{R})\rangle$ correlates with the ions $M^+ + X^-$, and also has $^1\Sigma^+$ symmetry. The two potential energy curves $\varepsilon_1(R)$ and $\varepsilon_2(R)$ exhibit an *avoided crossing* at $R = R_c$, in consequence of the noncrossing rule [3] for curves corresponding to states of the same symmetry. Because of this noncrossing rule, the adiabatic states $|\psi_1\rangle$ and $|\psi_2\rangle$ rapidly change in character near $R = R_c$. In particular, $|\psi_1\rangle$ describes ionic M^+X^- for $R \lesssim R_c$—since the deep well in $\varepsilon_1(R)$ results primarily from Coulomb attraction between the ions at smaller separations—but corresponds to covalent MX for $R \gtrsim R_c$, where covalent MX has lower energy than ionic M^+X^-. Conversely, $|\psi_2\rangle$ switches from covalent to ionic character as R_c is passed going outward; $\varepsilon_2(R)$ has a shallow well since covalent MX is considerably more weakly bound than ionic M^+X^-. *In the neighborhood of* R_c where the potential curves $\varepsilon_1(R)$ and $\varepsilon_2(R)$ approach closely, $T'_{kk'}$ becomes large for $k = 1$, $k' = 2$, because the quantity

$$\mathbf{d}_{12} = \langle\psi_1|\nabla_R|\psi_2\rangle = \hat{i}\langle\psi_1|\frac{\partial}{\partial R}|\psi_2\rangle \tag{3.17}$$

is large in this region where $|\psi_1\rangle$ and $|\psi_2\rangle$ are switching between ionic and covalent behavior.

The implications of this for collisions between the neutral atoms M + X are as follows. The collision starts out at large R in state $|\psi_1\rangle$ (corresponding to M + X). As the atoms approach at $R \gg R_c$, the system remains in state $|\psi_1\rangle$, because in this region the right side of Eq. 3.12 is small. But as R approaches R_c, $\mathbf{d}_{12}$ becomes appreciable, and then

$$T'_{12} = \frac{-\hbar^2}{2\mu_N}\,\mathbf{d}_{12}\cdot\nabla_R \tag{3.18}$$

can become important if the atoms approach with sufficient relative velocity (which is proportional to $\mathbf{V}_R$). At this point, Eq. 3.12 informs us that the system has a finite probability of hopping to state $|\psi_2\rangle$ near $R = R_c$. If it does, the atoms will momentarily form covalent MX (following curve $\varepsilon_2(R)$) on closer approach, $R \lesssim R_c$. If the curve-hopping fails to occur, the system remains adiabatically in state $|\psi_1\rangle$, and forms ionic M^+X^- when R becomes smaller than R_c. The probability of this curve-hopping (which amounts to a breakdown of the Born-Oppenheimer principle) depends on the relative nuclear velocity: the larger $\mathbf{V}_R$ is, the more probable the nonadiabatic jump. The widespread use of Eq. 3.16 in molecular spectroscopy rests on the assumption that such nonadiabatic transitions never occur, and that nuclear motion is restricted to motion on one potential energy curve (or surface in polyatomic molecules) corresponding to a single electronic state. This is valid in the limit of small spin–orbit coupling, and when the electronic potential energy surface in question is not closely approached by other surfaces at the total (electronic plus nuclear) energies of interest.

Interested readers are referred to the chapter by J. C. Tully in *Dynamics of Molecular Collisions, Part B*, edited by W. H. Miller (Plenum, New York, 1976) for additional information on this subject.

3.2 DIATOMIC ROTATIONAL ENERGY LEVELS AND SPECTROSCOPY

In the remainder of this chapter, we assume that the Born-Oppenheimer approximation is good, and that Eq. 3.16 holds. In this section we consider the rotational motion of a idealized rigid diatomic rotor, in which R is fixed at R_0. Then $U_{kk}(R) = U_{kk}(R_0)$ is a constant that we may set to zero for convenience. Using Eq. 3.15 for ∇_R^2 in the Schrödinger equation (3.16), we immediately have

$$\frac{-\hbar^2}{2\mu_N}\left[\frac{1}{R^2}\frac{\partial}{\partial R}\left(R^2\frac{\partial}{\partial R}\right) - \frac{\hat{J}^2}{\hbar^2 R^2}\right]|\chi_{\text{rot}}\rangle$$
$$= \frac{\hat{J}^2}{2\mu_N R_0^2}|\chi_{\text{rot}}(\theta,\phi)\rangle = E_{\text{rot}}|\chi_{\text{rot}}(\theta,\phi)\rangle \tag{3.19}$$

where we have used the fact that rotor's rotational angular momentum operator $\hat{J}^2$ is

$$\hat{J}^2 = -\hbar^2\left[\frac{1}{\sin\theta}\frac{\partial}{\partial\theta}\left(\sin\theta\frac{\partial}{\partial\theta}\right) + \frac{1}{\sin^2\theta}\frac{\partial^2}{\partial\phi^2}\right] \tag{3.20}$$

and the fact that terms in $\partial/\partial R$ drop out when R is held constant. Hence the *rigid rotor Hamiltonian* is

$$\hat{H}_{\text{rot}} = \hat{J}^2/2\mu_N R_0^2 = \hat{J}^2/2I \tag{3.21}$$

where I is the rotational moment of inertia, $I = \mu_N R_0^2$. The eigenfunctions of $\hat{H}_{\text{rot}}$ are the spherical harmonics

$$|\chi_{\text{rot}}(\theta, \phi)\rangle = Y_{JM}(\theta, \phi)$$
$$J = 0, 1, 2, \ldots$$
$$M = J, \ldots, -J \tag{3.22}$$

and its eigenvalues are

$$E_{\text{rot}} = \frac{J(J+1)\hbar^2}{2I} \qquad J = 0, 1, 2, \ldots \tag{3.23}$$

According to Eq. 3.22, the rotational levels are $(2J + 1)$-fold degenerate. Equation 3.23 gives the rotational levels in joules (J) when h is in J · s and I is in kg · m^2. They are more commonly expressed in cm^{-1} (wave numbers) via

$$E_{\text{rot}}(\text{cm}^{-1}) = BJ(J+1) \tag{3.24}$$

where B (the *rotational constant* in cm^{-1}) is

$$B = \hbar^2/2hcI = h/8\pi^2 cI \tag{3.25}$$

The rotational energy spacing in cm^{-1} between level J and level $(J - 1)$ is $BJ(J + 1) - B(J - 1)J = 2BJ$, which increases linearly with J (Fig. 3.5).

We now derive the *pure rotational* selection rules (i.e., in the absence of vibrational or electronic transitions) for the rigid rotor. If a rotor of fixed length R_0 has charges $\pm\delta$ glued to its ends, it has a permanent dipole moment

$$\boldsymbol{\mu}_0 = \delta\mathbf{R} = \delta R_0 \hat{R} \tag{3.26}$$

along the molecular axis. The E1 transition moment for the $JM \to J'M'$ rotational transition is then $\delta R_0 \langle Y_{JM}(\theta, \phi)|\hat{R}|Y_{J'M'}(\theta, \phi)\rangle$, which vanishes unless $\Delta J \equiv J' - J = \pm 1$, $\Delta M \equiv M' - M = 0, \pm 1$, and $\delta \neq 0$. Accordingly, E1 transitions occur only between adjacent rotational levels for which $\Delta E_J = E(J) - E(J - 1) = 2BJ$. Note that this E1 transition moment integral is formally identical to the angular part of the E1 transition moment for hydrogenlike atoms—which is why it yields identical selection rules. The factor of δ means that only heteronuclear diatomics with *permanent* dipole moments $(\delta \neq 0)$—such as HBr and CO—can exhibit a pure E1 rotational spectrum.

Since the energy changes in cm^{-1} for transitions from state $(J - 1)$ to state J are equal to $2BJ$ for a rigid rotor, the observed rotational absorption lines are predicted to be uniformly separated by the spacing $2B$ (Fig. 3.5). This is borne out to within experimental error in the far-infrared spectrum of HCl, Fig. 3.2. If the nuclear masses (and hence the nuclear reduced mass μ_N) are independently

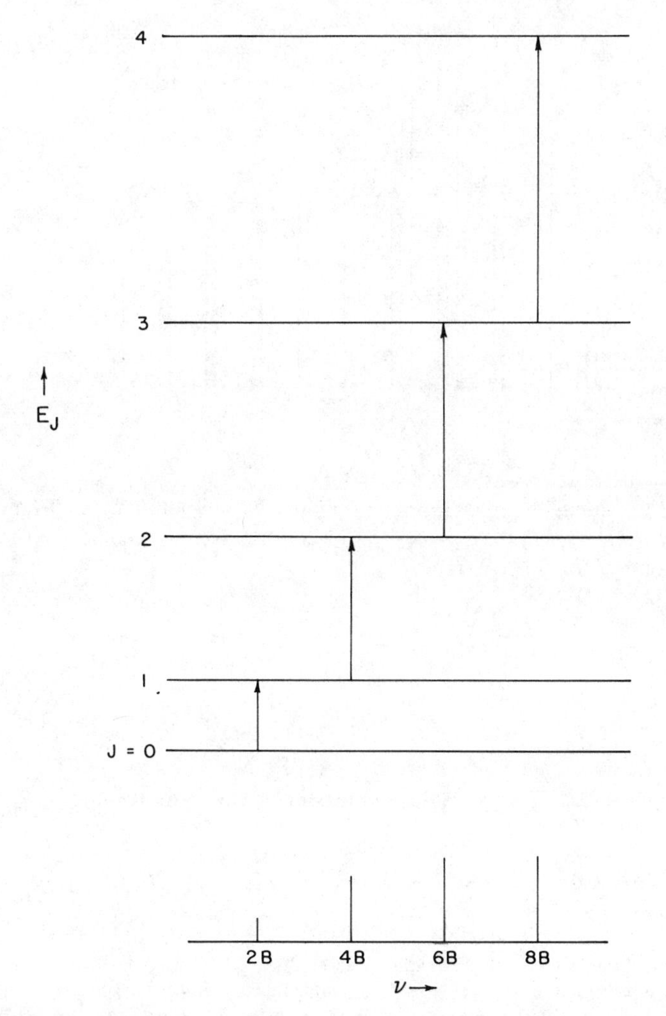

Figure 3.5 Rotational levels in a diatomic molecule. Allowed absorptive transitions ($\Delta J = +1$) are shown by vertical arrows. Schematic absorption spectrum is shown at bottom.

known, the rotational constant B inferred from such a spectrum can be used to calculate the apparent internuclear separation

$$R_0 = (\hbar^2/2hc\mu_N B)^{1/2} \tag{3.27}$$

Indeed, a principal application of rotational (and vibrational–rotational) spectroscopy is determination of molecular bond lengths (and bond angles in polyatomics, Chapter 5).

The intensities of the absorption lines in a far-infrared spectrum like Fig. 3.2

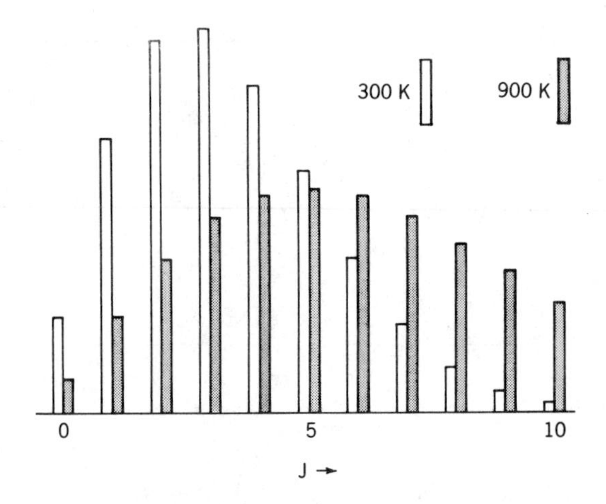

Figure 3.6 Relative rotational state populations, given by $g_J \exp[-hcBJ(J+1)/kT]/\Sigma_J g_J \exp[-hcBJ(J+1)/kT]$, versus rotational quantum number J for $B = 10.6\,\text{cm}^{-1}$.

Table 3.1 Rotational constants of diatomic molecules, cm^{-1}

Molecule	B_e	α_e	D_e
$^{40}Ar_2$	0.05975	0.00375	11.3E-7
$^{138}Ba^{16}O$	0.3126140	0.0013921	2.724E-7
$^{12}C^{16}O$	1.93128087	0.01750441	6.1214E-6
$^{1}H_2$	60.853	3.062	4.71E-2
$^{1}H^{35}Cl$	10.59341	0.30718	5.3194E-4
$^{127}I_2$	0.03737	0.000113	4.2E-9
$^{127}I^{35}Cl$	0.1141587	0.0005354	4.03E-8
$^{23}Na^{35}Cl$	0.2180630	0.0016248	3.12E-7
$^{23}Na^{1}H$	4.9012	0.1353	3.32E-4
$^{23}Na_2$	0.154707	0.0008736	5.81E-7
$^{14}N_2$	1.99824	0.01731	5.76E-6
$^{14}N^{16}O$	1.67195	0.0171	0.5E-6
$^{16}O_2$	1.44563	0.0159	4.839E-6
$^{16}O^{1}H$	18.910	0.7242	19.38E-4

Data taken from K. P. Huber and G. Herzberg, *Molecular Spectra and Molecular Structure: Constants of Diatomic Molecules*, Van Nostrand-Reinhold, New York, 1979.

are also of interest. The optical densities of absorption lines in a sample at thermal equilibrium will be weighted by the Boltzmann factors

$$q_{J_l} \exp(-E_{J_l}/kT) = (2J_l + 1)\exp[-hcBJ_l(J_l + 1)/kT] \tag{3.28}$$

where J_l is the rotational quantum number of the lower (absorbing) level. (Optical densities are defined in Appendix D.) This causes the pure rotational spectrum to peak its intensity at some J_l determined by the sample temperature and rotational moment. In Fig. 3.6, relative rotational state populations are shown for $J = 1$ to 10 in HCl for two contrasting temperatures.

Most values of heteronuclear diatomic rotational constants B fall between the extremes of HF (20.9 cm^{-1}) and ICl (0.114 cm^{-1}). For these species, the frequencies of the $J = 0 \to 1$ rotational transitions are 41.8 and 0.228 cm^{-1}; both are in the far-infrared to microwave region of the electromagnetic spectrum. Some representative rotational constants are listed in Table 3.1.

3.3 VIBRATIONAL SPECTROSCOPY IN DIATOMICS

With the aid of Eqs. 3.15 and 3.20, the Schrödinger equation (3.16) for nuclear motion in the Born-Oppenheimer approximation becomes

$$\left[U_{kk}(R) - \frac{\hbar^2}{2\mu_N}\left(\frac{1}{R^2}\frac{\partial}{\partial R}\left(R^2\frac{\partial}{\partial R}\right) - \frac{\hat{J}^2}{\hbar^2 R^2}\right)\right]|\chi_k(\mathbf{R})\rangle = E|\chi_k(\mathbf{R})\rangle \tag{3.29}$$

where $|\chi_k(\mathbf{R})\rangle$ is the wave function for nuclear motion in electronic state k. If we make the substitution $|\chi_k(\mathbf{R})\rangle = S_k(R)Y_{JM}(\theta, \phi)/R$, and recall that $\hat{J}^2 Y_{JM}(\theta, \phi) = J(J + 1)\hbar^2 Y_{JM}(\theta, \phi)$, we obtain

$$\frac{d^2S_k(R)}{dR^2} + \frac{2\mu_N}{\hbar^2}\left[E - U_{kk}(R) - \frac{J(J+1)\hbar^2}{2\mu_N R^2}\right]S_k(R) = 0 \tag{3.30}$$

This is the Schrödinger equation for vibrational motion, with eigenstates $S_k(R)$ representing vibrational wave functions in electronic state k, under the *effective vibrational potential*

$$U_{\text{vib}}(R) = U_{kk}(R) + \frac{J(J+1)\hbar^2}{2\mu_N R^2} \tag{3.31}$$

We recall that $U_{kk}(R) = \langle\psi_k|\hat{H}_{\text{el}}|\psi_k\rangle + \langle\psi_k|\hat{H}_{\text{so}}|\psi_k\rangle$; it is the electronic potential energy curve $\varepsilon_k(R)$, corrected by the expectation value of the spin–orbit coupling Hamiltonian. The second term in Eq. 3.31 is the *centrifugal potential* which is occasioned by the rotational motion in state $|\chi_{\text{rot}}\rangle = Y_{JM}(\theta, \phi)$; it reminds us that the rotational and vibrational motions cannot be truly independent, since the rotational state influences the effective vibrational potential.

Needless to say, the R-dependence of $U_{kk}(R) = \varepsilon_k(R)$ is not analytic. For bound electronic states that exhibit local minima in $\varepsilon_k(R)$, it can be usefully expanded in a Taylor series about the equilibrium separation $R = R_e$,

$$U_{kk}(R) = U_{kk}(R_e) + \left(\frac{dU_{kk}}{dR}\right)_{R_e}(R - R_e) + \frac{1}{2}\left(\frac{d^2U_{kk}}{dR^2}\right)_{R_e}(R - R_e)^2 + \cdots \tag{3.32}$$

If we set $U_{kk}(R_e) = 0$ and recognize that dU_{kk}/dR vanishes at the local minimum $R = R_e$, then

$$U_{kk}(R) \simeq \tfrac{1}{2}k(R - R_e)^2 \tag{3.33}$$

where the vibrational *force constant* k is defined as

$$k = \left(\frac{d^2U_{kk}}{dR^2}\right)_{R_e} \tag{3.34}$$

This is the familiar potential for a one-dimensional harmonic oscillator (Fig.

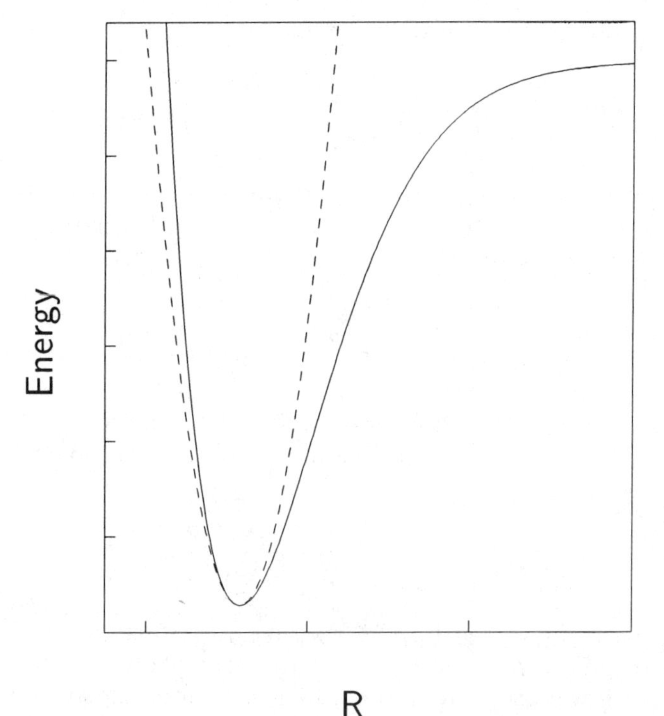

Figure 3.7 Harmonic oscillator approximation (dashed curve) to true potential energy curve $\varepsilon(R)$ in a diatomic molecule.

3.7). If we restrict ourselves to pure vibrational motion ($J = 0$), Eq. 3.30 then becomes

$$\left(\frac{-\hbar^2}{2\mu_N}\frac{d^2}{dR^2} + \tfrac{1}{2}k(R - R_e)^2\right)S_k(R) = ES_k(R) \tag{3.35}$$

From prior experience with the one-dimensional quantum mechanical harmonic oscillator [4], we know that its eigenfunctions and eigenvalues are

$$S_k(R) = N_v e^{-x^2/2} H_v(x) \equiv |v\rangle \tag{3.36a}$$

$$E_{\text{vib}} = \hbar\omega(v + \tfrac{1}{2}) \tag{3.36b}$$

$$v = 0, 1, 2, \ldots$$

where $x = (R - R_e)\sqrt{\mu_N \omega/\hbar}$, $\omega = \sqrt{k/\mu_N}$, and the H_v are the Hermite polynomials in x. To keep things in perspective, we recall that these are the diatomic vibrational states only in the harmonic approximation (ignoring higher than second-order terms in the Taylor series expansion (3.32)) in the absence of rotation ($J = 0$).

In the E1 approximation, absorption or emission of radiation can accompany a one-photon transition between vibrational states $|v\rangle$ and $|v'\rangle$ only if $\langle v|\boldsymbol{\mu}|v'\rangle \neq 0$. Since the vibrational states depend on R, the value of this transition moment is affected by the R-dependence of $\boldsymbol{\mu} = \boldsymbol{\mu}(R)$, the permanent dipole moment. In homonuclear diatomics, $\boldsymbol{\mu}(R) \equiv 0$ for all R. In heteronuclear diatomics, the R-dependence of $\boldsymbol{\mu}$ must resemble the schematic behavior shown in Fig. 3.8. This is so because $\boldsymbol{\mu}(R) \to 0$ both in the united-atom limit ($R = 0$) and in the separated-atom limit ($R = \infty$) where the diatomic dissociates into neutral atoms. Thus $\boldsymbol{\mu}(R)$ generally is not given by a closed-form expression, but can be approximated by a Taylor series about $R = R_e$:

$$\boldsymbol{\mu}(R) = \boldsymbol{\mu}(R_e) + \left(\frac{\partial \boldsymbol{\mu}}{\partial R}\right)_{R_e}(R - R_e) + \frac{1}{2}\left(\frac{\partial^2 \boldsymbol{\mu}}{\partial R^2}\right)_{R_e}(R - R_e)^2 + \cdots \tag{3.37}$$

This should not be confused with the Taylor series expansion of $U_{kk}(R)$, Eq. 3.32. The validity of terminating the expansion for $U_{kk}(R)$ with the second-order term (i.e., the harmonic approximation) is not connected with the validity of breaking off the expansion of $\boldsymbol{\mu}(R)$ with some low-order term—these two approximations are unrelated. Textbooks commonly base their derivation of E1 vibrational selection rules on the premise that the series for $\boldsymbol{\mu}(R)$ can be terminated with the term that is linear in $(R - R_e)$. This can be approximately valid over limited-R regions, but is (strictly speaking) unphysical for general R. The error of using this linear approximation is mitigated by the fact that the vibrational eigenstates tend to limit contributions of $\boldsymbol{\mu}(R)$ in the matrix element $\langle v|\boldsymbol{\mu}(R)|v'\rangle$ to small regions near $R = R_e$ (Fig. 3.8).

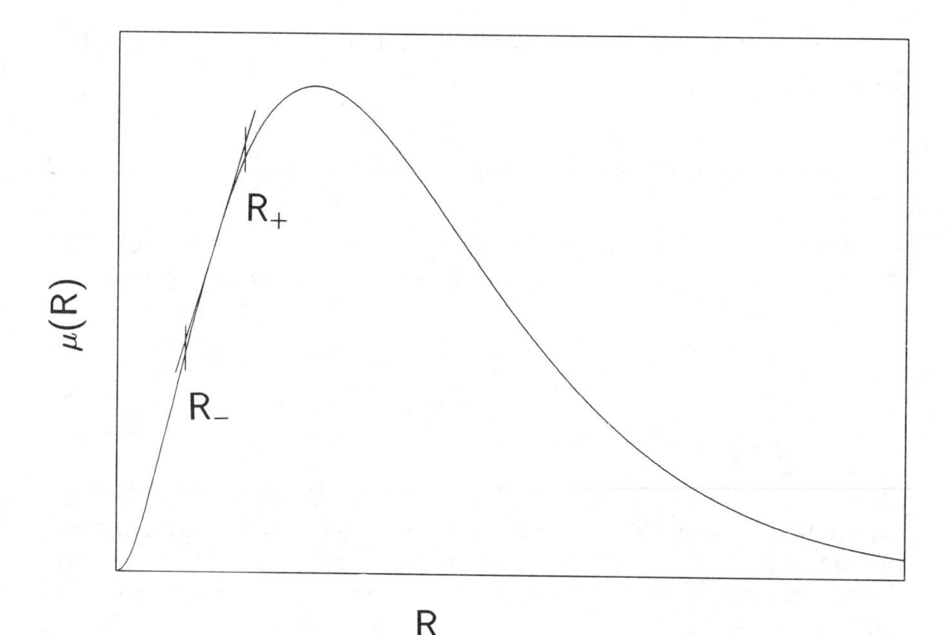

Figure 3.8 Qualitative behavior of the dipole moment function $\mu(R)$ for a heteronuclear diatomic molecule that dissociates into neutral atoms. The straight line tangent to $\mu(R)$ at $R = R_e$ represents the linear approximation to $\mu(R)$ which breaks off the Taylor series (3.39) after the term proportional to $(R - R_e)$. When the vibrational motion is limited to a small range of internuclear separations $R_- \lesssim R \lesssim R_+$, the deviations between the true dipole moment function and its linear approximation remain small.

Using the series expansion for $\boldsymbol{\mu}(R)$, the vibrational transition dipole moment becomes

$$\langle v|\boldsymbol{\mu}(R)|v'\rangle = \boldsymbol{\mu}(R_e)\langle v|v'\rangle + \left(\frac{\partial \boldsymbol{\mu}}{\partial R}\right)_{R_e} \langle v|(R - R_e)|v'\rangle + \frac{1}{2}\left(\frac{\partial^2 \boldsymbol{\mu}}{\partial R^2}\right)_{R_c} \langle v|(R - R_e)^2|v'\rangle + \cdots \tag{3.38}$$

We now need a systematic way to evaluate matrix elements like $\langle v|(R - R_e)^n|v'\rangle$. This is provided by the *second quantization* formulation [5] of the one-dimensional harmonic oscillator problem, which parallels in some ways the ladder operator treatment of angular momentum. The harmonic oscillator Hamiltonian is

$$\hat{H}_{\text{vib}} = -\frac{\hbar^2}{2\mu_N}\frac{d^2}{dR^2} + \tfrac{1}{2}k(R - R_e)^2 \equiv \frac{\hat{p}^2}{2\mu_N} + \tfrac{1}{2}kq^2$$

with $\hat{p} = (\hbar/i)(\partial/\partial R)$ and $q = (R - R_e)$. Since the harmonic oscillator frequency is $\omega = (k/\mu_N)^{1/2}$, this is the same as

$$\hat{H}_{\text{vib}} = \frac{\hat{p}^2}{2\mu_N} + \tfrac{1}{2}\mu_N\omega^2 q^2 \tag{3.39}$$

We now define the *destruction operator*

$$a \equiv (\hat{p} - i\mu_N\omega q)/\sqrt{2\mu_N\hbar\omega} \tag{3.40}$$

Its Hermitian conjugate, known as the *creation operator*, is

$$a^+ = (\hat{p} + i\mu_N\omega q)/\sqrt{2\mu_N\hbar\omega} \tag{3.41}$$

because both $\hat{p}$ and q are Hermitian. The inverses of Eqs. 3.40 and 3.41 are

$$\hat{p} = \left(\frac{\mu_N\hbar\omega}{2}\right)^{1/2}(a + a^+)$$

$$q = i\left(\frac{\hbar}{2\mu_N\omega}\right)^{1/2}(a - a^+) \tag{3.42}$$

and these allow us to rewrite the Hamiltonian (3.39) compactly in terms of the creation and destruction operators,

$$\hat{H}_{\text{vib}} = \frac{\hbar\omega}{2}(aa^+ + a^+a) \tag{3.43}$$

A useful commutation relationship for these operators is

$$[a, a^+] = \frac{1}{2\mu_N\hbar\omega}[\hat{p} - i\mu_N\omega q, \hat{p} + i\mu_N\omega q]$$

$$= \frac{-i}{2\hbar}[q, \hat{p}] + \frac{i}{2\hbar}[\hat{p}, q] = 1$$

so that

$$[a, a^+] = 1 \tag{3.44}$$

Using this, the Hamiltonian becomes

$$\hat{H}_{\text{vib}} = \frac{\hbar\omega}{2}(a^+a + 1 + a^+a) = \hbar\omega(a^+a + \tfrac{1}{2}) \tag{3.45}$$

because $aa^+ - a^+a = 1$ from Eq. 3.44. If we call the operator product $a^+a \equiv \hat{N}$ the *number operator*, the Hamiltonian (3.45) can be rewritten as

$$\hat{H}_{\text{vib}} = \hbar\omega(\hat{N} + \tfrac{1}{2}). \tag{3.46}$$

However, we know from prior experience that

$$\hat{H}_{\text{vib}}|v\rangle = \hbar\omega(v + \tfrac{1}{2})|v\rangle \tag{3.47}$$

Comparison of Eqs. (3.46) and (3.47) then shows that

$$\hat{N}|v\rangle = v|v\rangle \tag{3.48}$$

The state $|v\rangle$ is then an eigenstate of the number operator $\hat{N}$ with eigenvalue v, which shows the number of vibrational quanta in that state.

We now wish to determine what kind of state $a^+|v\rangle$ is. We can do this by testing the new state $a^+|v\rangle$ with the number operator,

$$\hat{N}(a^+|v\rangle) \equiv (a^+\hat{N} + [\hat{N}, a^+])|v\rangle \tag{3.49}$$

But

$$[\hat{N}, a^+] = [a^+a, a^+] = a^+[a, a^+] = a^+ \tag{3.50}$$

using the commutator in Eq. 3.44. Then Eq. 3.49 becomes

$$\hat{N}(a^+|v\rangle) = (a^+\hat{N} + a^+)|v\rangle = (v + 1)(a^+|v\rangle) \tag{3.51}$$

Hence the new state $(a^+|v\rangle)$ is an eigenstate of $\hat{N}$ with eigenvalue $(v + 1)$. Since the eigenstates of a one-dimensional harmonic oscillator are nondegenerate, this implies that

$$a^+|v\rangle \propto |v + 1\rangle \tag{3.52}$$

We can similarly show that

$$a|v\rangle \propto |v - 1\rangle \tag{3.53}$$

so that the operators a^+ and a have the effect of creating and annihilating one quantum respectively in the harmonic oscillator. It can further be demonstrated that

$$\begin{aligned}\langle v|aa^+|v\rangle &\equiv (a^+|v\rangle)^+(a^+|v\rangle) \\ &= (v + 1)\langle v + 1|v + 1\rangle = (v + 1)\end{aligned} \tag{3.54}$$

and that

$$\langle v|a^+a|v\rangle \equiv (a|v\rangle)^+(a|v\rangle) = v \tag{3.55}$$

so that to within a phase factor

$$\begin{aligned} a^+|v\rangle &= \sqrt{v+1}\,|v+1\rangle \\ a|v\rangle &= \sqrt{v}\,|v-1\rangle \end{aligned} \tag{3.56}$$

Then we finally have the matrix elements useful for evaluating the contributions to the transition moment in Eq. 3.38,

$$\begin{aligned} \langle v|a^+|v'\rangle &= \sqrt{v}\,\delta_{v,v'+1} \\ \langle v|a|v'\rangle &= \sqrt{v+1}\,\delta_{v,v'-1} \end{aligned} \tag{3.57}$$

We now calculate the first few terms in the transition moment (3.38) explicitly. The leading term $\boldsymbol{\mu}(R_e)\langle v|v'\rangle$ vanishes by orthogonality of the harmonic oscillator states. For the next (first-order) term, the substitution for q using Eq. 3.42 leads to

$$\langle v|(R - R_e)|v'\rangle = i\left(\frac{\hbar}{2\mu_N\omega}\right)^{1/2}[\sqrt{v+1}\,\delta_{v,v'-1} - \sqrt{v}\,\delta_{v,v'+1}] \tag{3.58}$$

so that the linear term in the expansion (3.37) of $\boldsymbol{\mu}(R)$ yields the selection rule $\Delta v = v' - v = \pm 1$ for vibrational transitions. The second-order term is proportional to

$$\langle v|(R - R_e)^2|v'\rangle = -\frac{\hbar}{2\mu_N\omega}\langle v|aa - a^+a - aa^+ + a^+a^+|v'\rangle \tag{3.59}$$

The respective integrands resulting from expansion of $(a - a^+)^2$ can be easily be shown to give $\Delta v = +2$, 0, 0, and -2. Hence the second-order term allows $\Delta v = \pm 2$ *overtone* vibrational transitions to occur, and higher overtones can result from the succeeding terms in the expansion of $\boldsymbol{\mu}(R)$. In practice, the fundamental transitions $\Delta v = \pm 1$ are usually the ones with the largest probabilities, and the overtone transition probabilities for $\Delta v = \pm n$ fall off rapidly with increasing n, because the transition moment integral samples a comparatively limited region of R near R_e.

Typical vibrational frequencies in heteronuclear diatomics tend to be bracketed between those of ICl and those in hydrides (in which low reduced masses are accompanied by high vibrational frequencies). In wave numbers, the fundamental vibrational frequencies $\omega_e = \omega/2\pi c$ are $\sim 384\,\text{cm}^{-1}$ and $\sim 2990\,\text{cm}^{-1}$ for ICl and HCl, respectively. Vibrational frequencies are listed for

Table 3.2 Vibrational constants of diatomic molecules, cm^{-1}

Molecule	ω_e	$\omega_e x_e$	$\omega_e y_e$
$^{40}Ar_2$	25.74		
$^{138}Ba^{16}O$	669.76	2.02	−0.003
$^{12}C^{16}O$	2169.81358	13.28831	0.010511
$^{1}H_2$	4401.21	121.33	0.812
$^{1}H^{35}Cl$	2990.946	52.8186	0.2243
$^{127}I_2$	214.50	0.614	
$^{127}I^{35}Cl$	384.29	1.501	
$^{23}Na^{35}Cl$	366.0	2.0	
$^{23}Na^{1}H$	1172.2	19.72	0.160
$^{23}Na_2$	159.12	0.7254	−0.00109
$^{14}N_2$	2358.57	14.32	−0.00226
$^{14}N^{16}O$	1904.20	14.075	0.011
$^{16}O_2$	1580.19	11.98	0.0474
$^{16}O^{1}H$	3737.76	84.881	0.540

Data taken from K. P. Huber and G. Herzberg, *Molecular and Molecular Structure: Constants of Diatomic Molecules*, Van Nostrand-Reinhold, New York, 1979.

several other diatomic molecules in Table 3.2. These frequencies lie in the infrared to near-infrared region of the electromagnetic spectrum.

3.4 VIBRATION–ROTATION SPECTRA IN DIATOMICS

Molecular vibrational spectra exhibit fine structure in gases, because rotational transitions can occur simultaneously with vibrational transitions. In diatomics with small vibration–rotation coupling, the selection rules on Δv and ΔJ are exactly as in the cases of pure vibrational and pure rotational spectroscopy, respectively:

$$\Delta J = \pm 1$$
$$\Delta v = \pm 1, (\pm 2, \pm 3, \ldots) \tag{3.60}$$

These rules hold provided the electronic state in question has no component of orbital angular momentum along the molecular axis (Chapter 4). When this component is *nonvanishing*, the selection rules are influenced by interaction between electronic and rotational angular momenta, and the $\Delta J = 0$ transition becomes allowed:

$$\Delta J = 0, \pm 1$$
$$\Delta v = \pm 1, (\pm 2, \pm 3, \ldots) \tag{3.61}$$

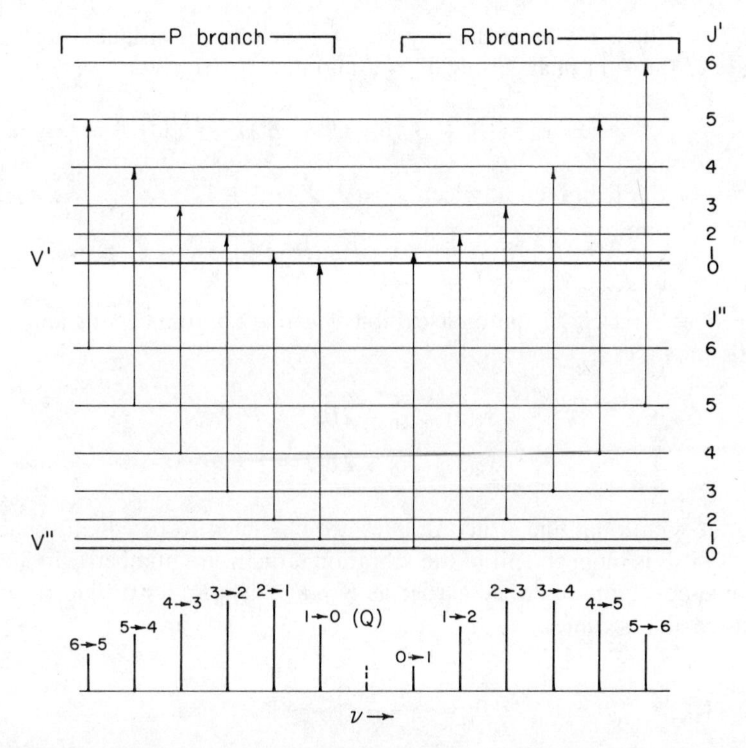

Figure 3.9 Energy levels and transitions giving rise to vibration–rotation spectra in a diatomic molecule. The upper state and lower state quantum numbers are (v', J') and (v'', J''), respectively. The schematic spectrum at bottom shows line intensities weighted by rotational state populations, which are proportional to $(2J''+1)\exp[-hcBJ''(J''+1)/kT]$. The rotational constants $B_{v''}$ and $B_{v'}$ are assumed to be equal, resulting in equally spaced rotational lines. This assumption is clearly not valid in the HCl vibration–rotation spectrum in Fig. 3.3.

The vibration–rotation spectrum can be understood on the basis of a partial energy level diagram for the lowest vibrational and rotational levels (Fig. 3.9). The upper and lower level quantum numbers are denoted by ($v'J'$) and ($v''J''$), respectively. Transitions for which $\Delta J = +1$, $\Delta J = -1$, and $\Delta J = 0$ (the latter occurring only in states with nonzero orbital angular momentum along the molecular axis) are called R-, P-, and Q-branch transitions. The transition frequencies are derived from

$$h\nu = \Delta E_{\text{vib}} + \Delta E_{\text{rot}} = \hbar\omega(v' + \tfrac{1}{2}) - \hbar\omega(v'' + \tfrac{1}{2}) + hcB'J'(J'+1) - hcB''J''(J''+1) \tag{3.62}$$

or

$$\bar{\nu}(\text{cm}^{-1}) = \bar{\nu}_0 + B'J'(J'+1) - B''J''(J''+1) \tag{3.63}$$

where $\bar{\nu}_0 = \omega(v' - v'')/2\pi c$ is the frequency for the pure vibrational transition. In the case of the P branch, letting $J'' = J$ and $J' = J - 1$ gives

$$\bar{\nu} = \bar{\nu}_0 - (B' + B'')J + (B' - B'')J^2 \equiv \bar{\nu}_\mathrm{P}(J) \tag{3.64}$$

whereas for the R branch in which $J'' = J$, $J' = J + 1$,

$$\bar{\nu} = \bar{\nu}_0 + (3B' - B'')J + (B' - B'')J^2 + 2B' \equiv \bar{\nu}_\mathrm{R}(J) \tag{3.65}$$

If $B' = B'' \equiv B$ (i.e., if the diatomic exhibits *identical* rotational constants in both vibrational states $|v'\rangle$ and $|v''\rangle$),

$$\begin{aligned} \bar{\nu}_\mathrm{P}(J) &= \bar{\nu}_0 - 2BJ \\ \bar{\nu}_\mathrm{R}(J) &= \bar{\nu}_0 + 2B(J + 1) \end{aligned} \tag{3.66}$$

Hence the rotational fine structure lines are predicted to be equally spaced in frequency if B is independent of the vibrational quantum number v. In fact, the rotational constant, described earlier as $B = \hbar^2/2hc\mu_N R_0^2$ for a rigid rotor with separation R_0, becomes

$$B_v = \frac{\hbar^2}{2hc\mu_N} \langle v| \frac{1}{R^2} |v\rangle \tag{3.67}$$

in a vibrating diatomic. B_v then acquires a v-dependence, largely because the harmonic oscillator potential is asymmetric about $R = R_e$. Since $U_{kk}(R)$ levels off to the separated atom asymptote for large R (Fig. 3.10) but falls rapidly for small R, $\langle v|1/R^2|v\rangle$ (and therefore B_v) decreases as v increases. This fact is accommodated experimentally by fitting measured B_v values to the expression [6]

$$B_v = B_e - \alpha_e(v + \tfrac{1}{2}) + \gamma_e(v + \tfrac{1}{2})^2 + \delta_e(v + \tfrac{1}{2})^3 + \cdots \tag{3.68}$$

The v-dependence in the rotational constant B_v is clearly visible in the HCl near-infrared spectrum shown in Fig. 3.3. This vibration–rotation spectrum consists of the $v' = 0$ to $v' = 1$ absorptive transition with rotational fine structure in a P branch (whose absorption lines at successive J values appear at ever lower frequencies according to Eq. 3.64) and an R branch (whose absorption lines run to higher frequencies as J is increased, Eq. 3.65). No Q branch line occurs, because HCl in its closed-shell electronic ground state has no electronic angular momentum. The absorption lines are labeled according to the value of J'' in the lower vibrational state: R(0) is the R-branch line from $J'' = 0$, P(1) is the P-branch line from $J'' = 1$, etc. The rotational line spacings decrease at higher frequencies in both branches, in consequence of the quadratic

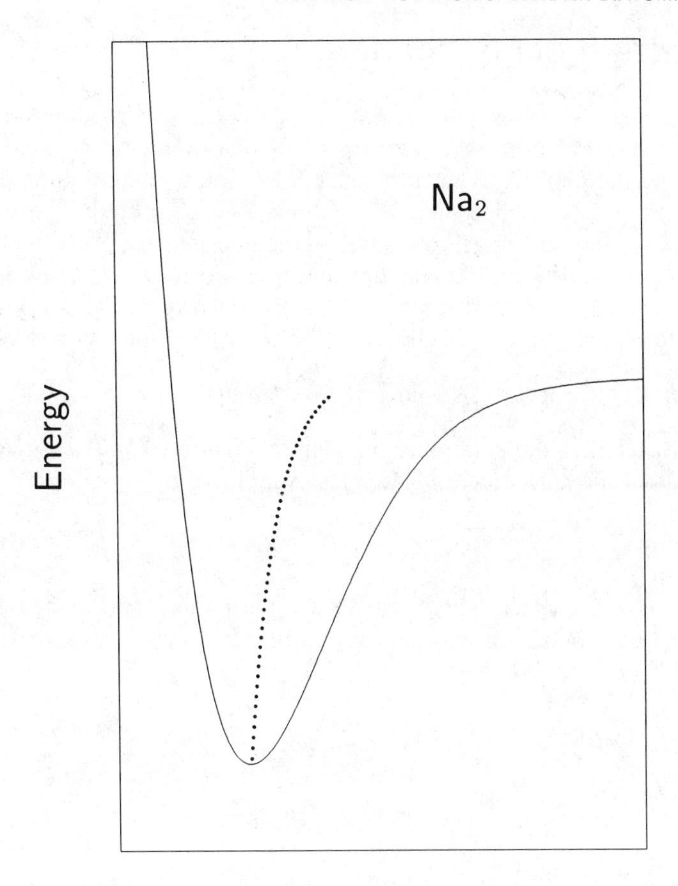

Figure 3.10 Potential energy curve for the electronic ground state of Na_2, with effective internuclear separations R_v indicated by dots for $v = 0$ through 45. These R_v values are computed from the experimental rotational constants B_v via $hcB_v = \hbar^2/2\mu_N R_v^2$. R_v is close to R_e in the lowest vibrational states, and increases with v.

$(B' - B'')J^2$ terms in $\nu_P(J)$ and $\nu_R(J)$. This implies that $B'' > B'$, or that the rotational constant is larger in vibrational state $v = 0$ than in $v = 1$.

Each of the rotational lines in Fig. 3.3 is split into doublets spaced by $\sim 2\ \mathrm{cm}^{-1}$ because the isotopes $H^{35}Cl$ and $H^{37}Cl$ have slightly different reduced masses, and therefore different rotational constants according to Eq. 3.25. Vibration–rotation spectra like the one in Fig. 3.3 can be analyzed to obtain accurate values of the rotational constant in the upper and lower vibrational levels for both isotopes.

3.5 CENTRIFUGAL DISTORTION

Since chemical bonds can be stretched, the centrifugal force accompanying diatomic rotation pushes the nuclei farther apart than they would be in a nonrotating diatomic. This in turn reduces the rotational constant B_v, which then depends on J as well as on v. To treat this effect quantitatively, we begin by defining F_c as the centrifugal force pulling the nuclei apart, F_v as the harmonic oscillator restoring force pulling the nuclei together, R_e as the equilibrium separation when $J = 0$, and R_c as the separation when $J \neq 0$. Clearly $R_c > R_e$, due to the centrifugal force. Classically, the centrifugal force is given by

$$F_c = \mu_N \omega^2 R_c = J^2/\mu_N R_c^3 \tag{3.69}$$

using the fact that the rotational angular momentum $J = I\omega = \mu_N R_c^2 \omega$. The magnitude of the harmonic oscillator restoring force is

$$|F_r| = k(R_c - R_e) \tag{3.70}$$

because $U_{kk}(R) = \frac{1}{2}k(R_c - R_e)^2$ in the harmonic approximation, and $F_r = -dU/dR_c$. Balancing the centrifugal and restoring forces gives

$$J^2/\mu_N R_c^3 = k(R_c - R_e) \tag{3.71}$$

or

$$R_c - R_e = J^2/\mu_N R_c^3 k. \tag{3.72}$$

Then the rotational energy is

$$E_{\text{rot}} = \frac{J^2}{2\mu_N R_c^2} + \frac{1}{2}k(R_c - R_e)^2 \tag{3.73}$$

where the latter term is included in the rotational energy because it reflects the change in the diatomic potential due to the centrifugal displacement of R from R_e to R_c. Using twice the expression for R_c from Eq. 3.72, the rotational energy then becomes

$$\begin{aligned}
E_{\text{rot}} &= \frac{J^2}{2\mu_N} \frac{1}{(R_e + J^2/\mu_N R_c^3 k)^2} + \frac{1}{2}k\left(\frac{J^2}{\mu_N R_c^3 k}\right)^2 \\
&\simeq \frac{J^2}{2\mu_N} \frac{1}{(R_e + J^2/\mu_N R_e^3 k)^2} + \frac{1}{2}k\left(\frac{J^2}{\mu_N R_e^3 k}\right)^2 \\
&\simeq \frac{J^2}{2\mu_N R_e^2}\left(1 - \frac{2J^2}{\mu_N R_e^4 k} + \cdots\right) + \frac{1}{2}\frac{J^4}{\mu_N^2 R_e^6 k} \\
&= \frac{J^2}{2\mu_N R_e^2} - \frac{J^4}{2\mu_N^2 R_e^6 k} + \cdots
\end{aligned} \tag{3.74}$$

where we have used the identity $(1 + x)^{-2} \simeq 1 - 2x + 0(x^2)$ for small x. Quantum mechanically, this expression for the rotational energy becomes

$$E_{\text{rot}} = J(J + 1)\hbar^2/2\mu_N R_e^2 - J^2(J + 1)^2\hbar^4/2\mu_N^2 R_e^6 k \tag{3.75}$$

and the rotational energy in wave numbers (with v-dependence of the rotational constants included) is now

$$E_{\text{rot}}(\text{cm}^{-1}) = B_v J(J + 1) - D_v J^2(J + 1)^2 \tag{3.76}$$

with

$$D_v = \frac{\hbar^4}{2hc\mu_N^2 k} \langle v| \frac{1}{R^6} |v\rangle \tag{3.77}$$

Like B_v, D_v decreases as v increases; its v-dependence has often been fitted for diatomics using expressions of the form [6]

$$D_v = D_e - \beta_e(v + \tfrac{1}{2}) + \cdots \tag{3.78}$$

If the v-dependence in B_v and D_v is ignored, the fact that

$$B \simeq \frac{\hbar^2}{2hc\mu_N R_e^2} \tag{3.79}$$

and

$$D \simeq \frac{\hbar^4}{2hc\mu_N R_e^6}$$

implies that the centrifugal distortion constant is

$$D \simeq \frac{4B^3}{\omega_e^2} \tag{3.80}$$

where ω_e is the widely used notation for the vibrational fundamental frequency in cm^{-1} (which we have been calling $\bar{\nu}_0$). The physical significance of Eq. 3.80 is that bonds with stronger force constants (and hence larger ω_e) experience less centrifugal distortion. It is interesting that knowledge of the rotational and vibrational constants B and ω_e permits one to predict (with reasonable accuracy) the more subtle quantity D.

When $D_v \neq 0$, the rotational lines in a pure rotational spectrum (or in a vibration–rotation spectrum) are no longer equally spaced, but become more closely spaced at higher J. However, D_v is usually not particularly large. In I_2, which has one of the weaker vibrational force constants among ordinary diatomics (excluding van der Waals molecules!), $B_e = .0374\,\mathrm{cm}^{-1}$ and $\omega_e = 214.6\,\mathrm{cm}^{-1}$—yielding $D_e \sim 4.55 \times 10^{-9}\,\mathrm{cm}^{-1}$, which is seven orders of magnitude smaller than B_e. In HCl ($B_e = 10.6\,\mathrm{cm}^{-1}$), the distortion constant D_e is $5.3 \times 10^{-4}\,\mathrm{cm}^{-1}$; centrifugal distortion is far too small to observe in the HCl far-infrared spectrum in Fig. 3.2.

3.6 THE ANHARMONIC OSCILLATOR

Most of our discussion of vibrational eigenstates and selection rules has been centered on the harmonic approximation. Of course, the effective vibrational potential $U_{kk}(R)$ is not well approximated by a parabola for energies corresponding to large vibrational quantum numbers v (cf. Fig. 3.7), and considerable work has been expended to find alternative expressions for either $U_{kk}(R)$ or the vibrational energy levels which are both compact and accurate. It has become conventional to fit experimentally determined vibrational levels $G_v(\mathrm{cm}^{-1})$ to expressions of the form [6]

$$G_v = \omega_e(v + \tfrac{1}{2}) - \omega_e x_e(v + \tfrac{1}{2})^2 + \omega_e y_e(v + \tfrac{1}{2})^3 + \cdots \tag{3.81}$$

The first term in G_v is the harmonic approximation to the vibrational energy, and the remaining terms are the *anharmonic* corrections. The negative sign in the leading anharmonic term corresponds to the fact that real vibrational levels become more closely spaced with higher v. A typical value of $\omega_e x_e/\omega_e$ is 0.0028 for I_2, which indicates that its ground-state potential energy curve is very nearly harmonic near the bottom of its well. For the $A^1\Sigma^+$ excited state in NaH (Chapter 4), $\omega_e x_e/\omega_e$ is -0.0174, which is anomalous because it is negative; this occurs because the $A^1\Sigma^+$ potential curve in NaH is pathologically misshapen due to perturbation by other nearby excited states.

If the vibrational energy level expression (3.81) is cut off after its leading anharmonic term

$$G_v = \omega_e(v + \tfrac{1}{2}) - \omega_e x_e(v + \tfrac{1}{2})^2 \tag{3.82}$$

the resulting levels are the *exact* eigenvalues of an approximate vibrational Hamiltonian in which the potential is given by the analytic function

$$U_{kk}(R) = D_e[1 - e^{-a(R-R_e)}]^2 \tag{3.83}$$

This is known as the *Morse potential.* At $R = R_e$, the first and second derivatives of the Morse potential are [7]

$$\frac{dU_{kk}(R)}{dR} = 0$$

$$\frac{d^2U_{kk}(R)}{dR^2} = 2a^2D_e \equiv k \tag{3.84}$$

with the consequence that

$$\omega_e = a\left(\frac{\hbar D_e}{\pi c \mu_N}\right)^{1/2} = \frac{1}{2\pi c}\sqrt{\frac{k}{\mu_N}} \tag{3.85}$$

with D_e in cm^{-1}. It can also be shown that

$$\omega_e x_e = \frac{\hbar a^2}{4\pi c \mu_N} \tag{3.86}$$

The Morse potential expression for $U_{kk}(R)$ has the limits D_e and 0 as R approaches ∞ and R_e, respectively. When $R \to 0$, $U_{kk}(R) \to D_e \exp(2aR_e)$, which is typically large. Hence the Morse potential is capable of modeling the qualitative features of a realistic diatomic potential energy curve. If the latter is well described by a Morse potential, then spectroscopically determined values of ω_e and $\omega_e x_e$ can be combined using Eqs. 3.85 and 3.86 to estimate the molecular *dissociation energy* [7]

$$D_e = \omega_e^2/4\omega_e x_e \tag{3.87}$$

measured between the bottom of the potential well and the asymptote of $U_{kk}(R)$ at large R. This can yield useful approximations to D_e if the dissociation energy is not available by other means. In more general cases where the vibrational energy levels are given by Eq. 3.81 rather than 3.82, the dissociation energy D_0 measured from level $v = 0$ is rigorously given in terms of the level differences

$$\begin{aligned}\Delta G(v) &\equiv G(v+1) - G(v)\\ &= \omega_e - 2\omega_e x_e(v+1) + \omega_e y_e(3v^2 + 6v + \tfrac{13}{4})\\ &\quad + \cdots\end{aligned} \tag{3.88}$$

as

$$D_0 = \sum_{v=0}^{v_{max}} \Delta G(v) \tag{3.89}$$

if the vibrational level spacing vanishes between level v_{max} and level $(v_{max} + 1)$ at the dissociation limit. The area under an experimental plot of $\Delta G(v)$ versus v then equals D_0 in principle (Fig. 3.11). Many of the terms $G(v)$ for larger v will not be known in practice, so that the experimental $\Delta G(v)$ curve is frequently extrapolated to $\Delta G = 0$ to estimate the dissociation energy. The use of Eq. 3.87 to estimate D_e amounts to making a linear extrapolation of $\Delta G(v)$, since for the Morse potential only the two leading terms in Eq. 3.88 contribute to $\Delta G(v)$. Such *Birge-Sponer* extrapolations frequently yield dissociation energies that are

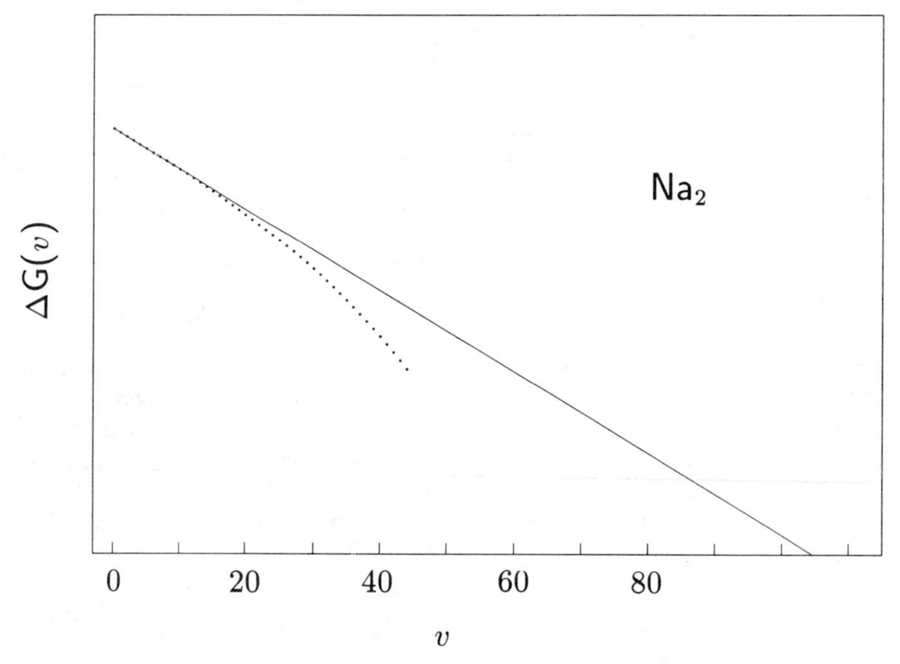

Figure 3.11 Birge-Sponer plot of $\Delta G(v)$ versus v for the electronic ground state of Na_2. The area under the $\Delta G(v)$ curve yields the ground state dissociation energy D_0. Linear extrapolation of the points for low v (straight line) would clearly yield a gross overestimate of D_0. Nonlinear extrapolation of the points exhibited for $0 \leqslant v \leqslant 45$ (P. Kusch and M. M. Hessel, J. Chem. Phys. *68*: 2591 (1978)) leads to an improved estimate of D_0. A still better approximation to D_0 can be made by analyzing vibrational levels for v up to 55 from the more recent Na_2 fluorescence spectrum shown in Fig. 4.1.

~30% too large for diatomics that dissociate into neutral atoms [6]. Diatomics that dissociate into ions exhibit potentials that behave as R^{-1} at long range. Such potentials exhibit infinite numbers of vibrational levels; plots of $G(v)$ versus v then run asymptotically along the v axis for large v, and the Birge-Sponer extrapolation becomes inapplicable.

REFERENCES

1. J. B. Marion, *Classical Dynamics of Particles and Systems*, Academic, New York, 1965.
2. J. C. Tully, Nonadiabatic processes in molecular collisions. In W. H. Miller (Ed.), *Dynamics of Molecular Collisions, Part B*, Plenum, New York, 1976.
3. L. D. Landau and E. M. Lifshitz, *Quantum Mechanics: Nonrelativistic Theory*, 2d ed., Addison-Wesley, Reading, MA, 1958.

4. E. Merzbacher, *Quantum Mechanics*, Wiley, New York, 1961.
5. K. Gottfried, *Quantum Mechanics*, Vol. 1, W. A. Benjamin, New York, 1966.
6. G. Herzberg, *Molecular Spectra and Molecular Structure, I. Spectra of Diatomic Molecules*, 2d ed., Van Nostrand, Princeton, NJ, 1950.
7. G. W. King, *Spectroscopy and Molecular Structure*, Holt, Rinehart, & Winston, New York, 1964.

PROBLEMS

1. In the pure rotational spectrum of $^1H^{35}Cl$, lines from the initial states $J = 2$ and $J = 3$ are observed with equal intensity. Calculate the temperature of the sample. (For $^1H^{35}Cl$, the rotational constant is $10.6\,\text{cm}^{-1}$.)

2. Use the second quantization formalism to develop the selection rule on $\langle v'|(R - R_e)^n|v\rangle$ for arbitrary n.

3. For a heteronuclear diatomic molecule AB, the dipole moment function in the neighborhood of $R = R_e$ is given by

$$\mu(R) = a + b(R - R_e) + c(R - R_e)^2 + d(R - R_e)^3$$

in which a, b, c, and d are constants. Treating this molecule as a harmonic oscillator, calculate the relative intensities of the $v = 0 \rightarrow 1$ fundamental and $v = 0 \rightarrow 2$ and $0 \rightarrow 3$ overtone transitions in the E1 approximation in terms of these constants and the harmonic oscillator constants μ and ω.

4. Some of the frequencies and assignments of the $^1H^{35}Cl$ vibration–rotation lines in Fig. 3.3 are given below.

$\bar{\nu}(\text{cm}^{-1})$	Assignment
2963	R(3)
2944	R(2)
2906	R(0)
2865	P(1)
2843	P(2)
2821	P(3)

Determine the values of the rotational constants (in cm^{-1}) and the associated bond lengths (in Å) of $^1H^{35}Cl$ in vibrational states $v = 0$ and $v = 1$. The nuclear masses of ^1H and ^{35}Cl are 1.007825 and 34.96885 amu, respectively.

5. From an analysis of the $B^1\Pi_u \rightarrow X^1\Sigma_g^+$ fluorescence bands of $^{23}Na_2$ (see Chapter 4), the vibrational energy levels in the electronic ground state can be

represented by

$$G(v) = 159.12\,(v + \tfrac{1}{2}) - 0.725(v + \tfrac{1}{2})^2 - 0.0011(v + \tfrac{1}{2})^3$$

in cm^{-1}. Determine the vibrational quantum number v_{max} at which the vibrational level spacing vanishes, and estimate the dissociation energy D_0. Compare this dissociation energy with that estimated using a linear Birge-Sponer extrapolation, and with the directly measured value $D_0 = 0.73\,\text{eV}$.

4

ELECTRONIC STRUCTURE AND SPECTRA IN DIATOMICS

Like the atomic spectra discussed in Chapter 2, electronic band spectra in diatomic molecules arise from transitions between different electronic states. Both types of spectra occur at wavelengths ranging from the vacuum ultraviolet to the infrared regions of the electromagnetic spectrum. A complication in diatomic band spectra is that changes in vibrational and/or rotational state generally accompany electronic transitions in molecules, endowing band spectra with rich *rovibrational structure*. Analysis of this structure (which often lends a bandlike appearance to diatomic spectra, in contrast to the discrete line spectra characteristic of atoms) can yield a wealth of information about ground and excited electronic state symmetries, detailed potential energy curves, and vibrational wave functions. An anthology of such information is presented for several diatomic molecules in this chapter.

An uncommonly lucid example of an electronic band spectrum is the Na_2 fluorescence spectrum shown in Fig. 4.1, which was obtained by exciting Na_2 vapor in a 453°C oven using nearly monochromatic 5682 Å light from a Kr^+ ion laser. At this temperature, Na_2 exists in its electronic ground state (the $X^1\Sigma_g^+$ state, in notation to be developed later) with appreciable populations in several vibrational and many rotational levels. However, the 5862-Å excitation wavelength is uniquely matched in Na_2 by the energy level difference between the $X^1\Sigma_g^+$ state ($v'' = 3$, $J'' = 51$) and an electronically excited state ($A^1\Sigma_u^+$) with $v' = 34$, $J' = 50$. The latter level then becomes selectively pumped by the laser. It subsequently relaxes by fluorescence transitions to $X^1\Sigma_g^+$ Na_2 in vibrational levels v'' between 0 and 56, producing the exhibited spectrum. Only transitions to $v'' \geqslant 4$ are shown; a schematic energy level diagram showing some of these transitions is given in Fig. 4.2.

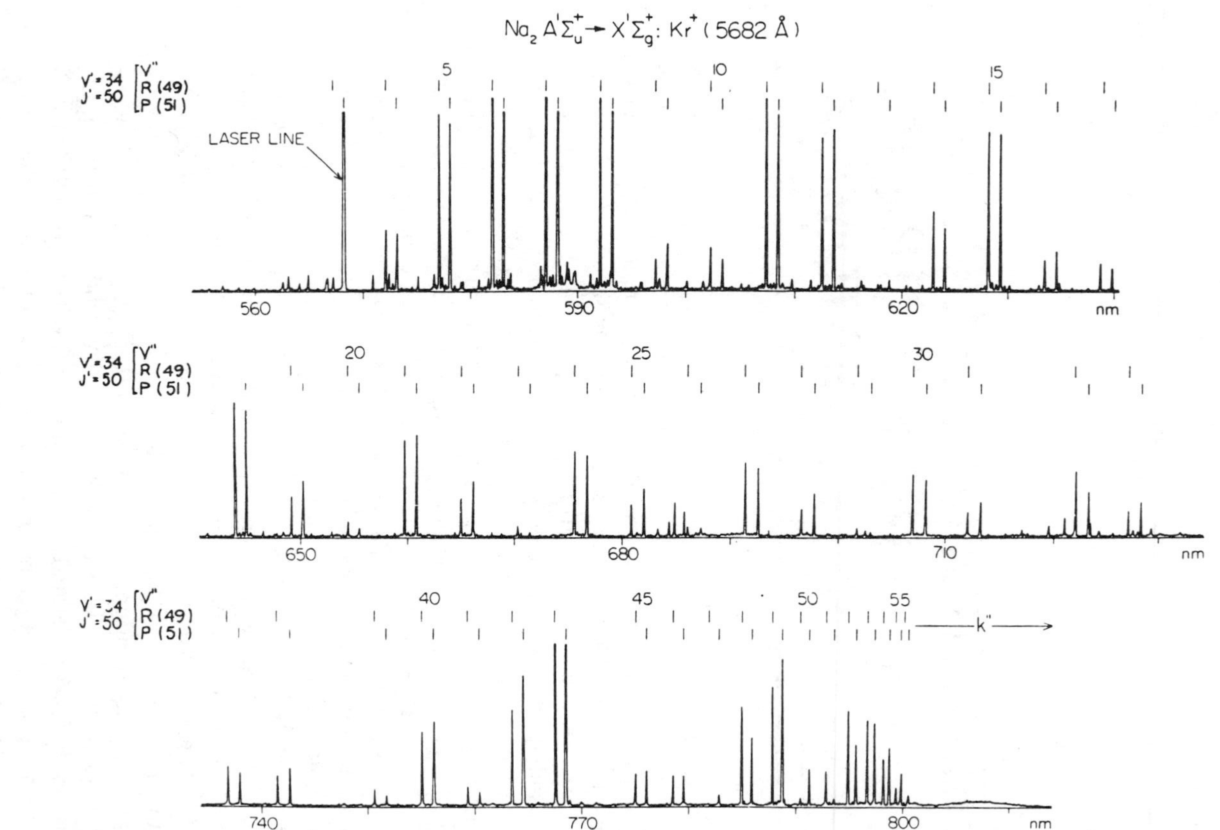

Figure 4.1 Fluorescence spectrum (fluorescence intensity as a function of fluorescence wavelength) for Na_2 vapor pumped by a 5682-A4 krypton ion laser. This wavelength excites Na_2 molecules from $v'' = 3$, $J'' = 51$ in the electronic ground state to $v' = 34$, $J' = 50$ in the $A^1\Sigma_u^+$ excited electronic state. The shown fluorescence lines result from transitions from the laser-excited level down to $v'' = 4$ through 56 in the electronic ground state. Reproduced by permission from K. K. Verma, A. R. Rajaei-Rizi, W. C. Stwalley, and W. T. Zemke, *J. Chem. Phys.* **78**: 3601 (1983).

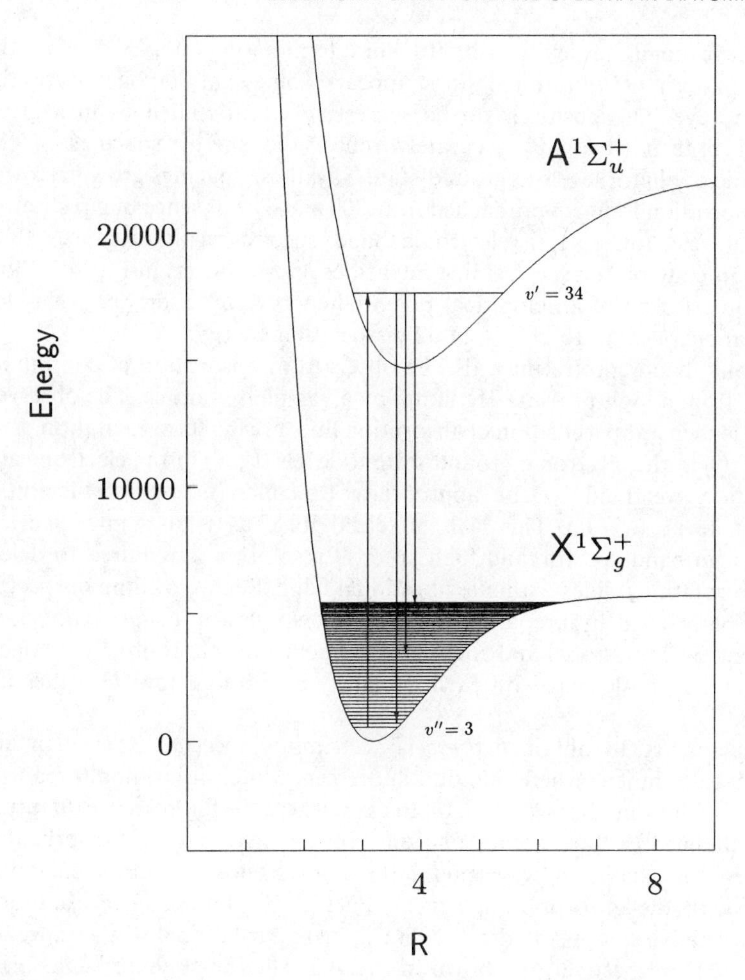

Figure 4.2 Energy level diagram for the fluorescence spectrum in Fig. 4.1. Absorption of the 5682-Å laser photon pumps Na_2 molecules from $v' = 3$, $J' = 51$ in the $X^1\Sigma_g^+$ electronic state to $v'' = 34$, $J' = 50$ in the $A^1\Sigma_u^+$ electronic state. Subsequent fluorescence transitions connect $v' = 34$ in the $A^1\Sigma_u^+$ state with $v'' = 0$ through 56 in the electronic ground state. Only three of these fluorescence transitions are shown for clarity. Rotational levels are not shown. Internuclear separations and energies are in Å and cm^{-1}, respectively.

Figure 4.1 illustrates some of the E1 selection rules on rotational and vibrational structure in electronic band spectra. The observed rotational selection rule, $\Delta J = \pm 1$ (the laser-excited level with $J' = 50$ fluorescences only to lower levels with $J'' = 49$ or 51) is reminiscent of that in vibration–rotation spectra of molecules with no component of electronic angular momentum along the molecular axis (Chapter 3). The R- and P-branch notations in Fig. 4.1 are identical in meaning to those used in vibration–rotation spectra. In marked contrast, there is *no* apparent selection rule on $\Delta v = v' - v''$: lines terminating in

$v'' = 43$ have comparable intensity to lines terminating in $v'' = 5$, and the intensity pattern of vibrational lines appears somewhat haphazard to the uneducated eye. The positions of the successive vibrational lines in a given rotational branch (R or P) accurately reflect the energy spacings of the anharmonic oscillator levels in ground-state Na_2; these spacings grow narrower as the dissociation limit is approached near 800 nm. The absence of a restrictive vibrational selection rule in electronic band spectra vastly enhances their information content; the spectral line intensities and positions in Fig. 4.1 allow precise construction of an empirical potential-energy curve for $X^1\Sigma_g^+$ Na_2 for vibrational energies up to >99% of its dissociation energy.

Electronic band spectra may also be observed in absorption of continuum light (e.g., from a high-pressure Hg lamp) by a gas-phase sample. The observed spectrum is then a superposition of absorption lines arising from excitation of *all* levels (v'', J'') in the electronic ground state to levels (v', J') in the electronically excited state, weighted by the appropriate Boltzman factors of the initial rovibronic levels (v'', J''). This lack of selectivity causes far greater spectral congestion in band spectra, and high spectral resolution is required to detect successive rotational lines within a vibrational band. Such crowding of spectral lines can be relieved by preparing the diatomic species in a *supersonic jet*, in which very low vibrational and rotational temperatures are routinely attained. In this manner, molecules with predominantly $v'' = 0$ and low J'' values are produced.

It is customary to obtain *fluorescence excitation spectra* rather than absorption spectra in jets, where the total fluorescence intensity is monitored as a function of excitation laser wavelength. In cases where the fluorescence quantum efficiency (defined as fluorescence photons emitted/laser photons absorbed) is independent of excitation wavelength, the fluorescence excitation spectrum coincides with the absorption spectrum. Part of the fluorescence excitation spectrum of an Na_2 jet operated with vibrational and rotational temperatures of ~50 and 30 K, respectively, is shown in Fig. 4.3. The four intense bands arise from electronic transitions from $v'' = 0$ in the $X^1\Sigma_g^+$ ground state to $v' = 25$ through 28 in the $A^1\Sigma_u^+$ excited state. The rotational fine structure in each band consists of a barely resolved series of narrow lines, creating an envelope that peaks asymmetrically toward the blue edge at the *bandhead.* Such rotational envelopes (or *contours*) are said to be *shaded to the red.* Electronic transitions are occasionally characterized by rotational contours that are shaded to the blue, i.e., by contours in which the bandhead lies at the long-wavelength edge. Such contours are absent in the fluorescence spectrum in Fig. 4.1, where the selective preparation of $J' = 50$ in the $A^1\Sigma_u^+$ state simplifies the rotational structure in the fluorescence spectrum. Analysis of absorption or fluorescence excitation spectra yields information about the vibrational structure and potential energy curve of the upper (as opposed to lower) electronic state.

This chapter begins with a treatment of symmetry and electronic structure in diatomic molecules. The symmetry selection rules for electronic transitions are derived, and vibrational band intensities (cf. Fig. 4.1) are described in terms of

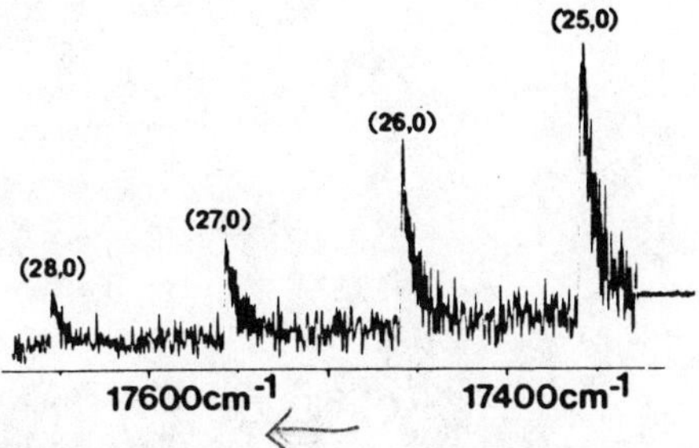

Figure 4.3 Fluorescence excitation spectrum (total fluorescence intensity versus excitation wavelength) of Na_2 molecules in a supersonic jet. The four intense groups of barely resolved lines are due to electronic transitions from $v'' = 0$ in the electronic ground state to $v' = 25$ through 28 in the $A^1\Sigma_u^+$ excited state. The individual lines are due to rotational fine structure, which is discussed in Section 4.6. Reproduced by permission from J. L. Gole, G. J. Green, S. A. Pace, and D. R. Preuss, *J. Chem. Phys.* **76**: 2251 (1982).

Franck-Condon factors. The most common angular momentum coupling cases are discussed, and rotational fine structure in electronic transitions (cf. Fig. 4.3) is rationalized for heteronuclear and homonuclear diatomics using Herzberg diagrams.

4.1 SYMMETRY AND ELECTRONIC STRUCTURE IN DIATOMICS

All diatomic molecules belong to either the $C_{\infty v}$ or $D_{\infty h}$ point group, and so much of their electronic structure and nomenclature is derived from the properties of these two groups. In what follows, the Cartesian z axis is always taken to be along the molecular axis (the line connecting the two nuclei). The x and y axes are both normal to the internuclear axis, as shown in Fig. 4.4.

The partial character table for the heteronuclear point group $C_{\infty v}$ (which has an infinite number of classes and irreducible representations) is

$C_{\infty v}$

	E	$2C_\varphi$		$\infty\sigma_v$	
Σ^+	1	1	$\cdots$	1	z
Σ^-	1	1		-1	R_z
Π	2	$2\cos\phi$		0	$(x, y), (R_x, R_y)$
Δ	2	$2\cos 2\phi$		0	
Φ	2	$2\cos 3\phi$		0	
$\vdots$	$\vdots$	$\vdots$		$\vdots$	

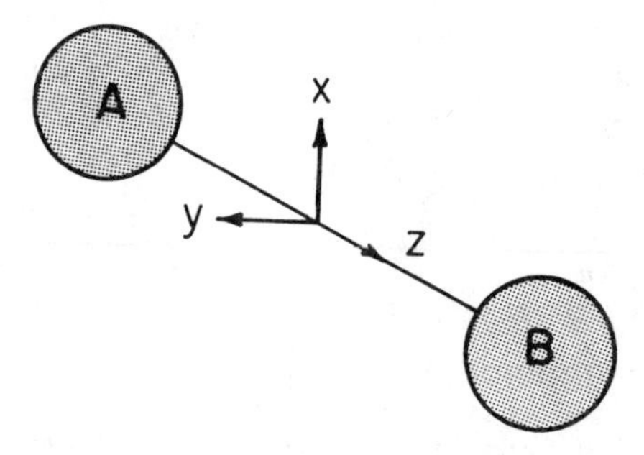

Figure 4.4 Orientation of Cartesian axes in a diatomic molecule AB.

The symmetry elements C_ϕ and σ_v are the rotation by an arbitrary angle ϕ about the principal (z) axis and a reflection plane containing the principal axis, respectively. The superscript in the notations for the Σ^+ and Σ^- irreducible representations (IRs) indicates the behavior of the IRs under the σ_v operation. Since the characters of the E operation in all IRs of $C_{\infty v}$ are either 1 or 2, this means that all diatomic electronic states are spatially either nondegenerate or doubly degenerate.

The behavior of the vector components (x, y, z) and the rotations (R_x, R_y, R_z) under the group operations proves to be important for determining the E1 and M1 selection rules for electronic transitions. In particular, the vector z is obviously unaffected by all of the $C_{\infty v}$ operations, so it transforms as the Σ^+ (totally symmetric) IR. The rotation R_z is not changed by E or C_ϕ, but changes sign (direction) under any σ_v operation—so that it belongs to the Σ^- IR. Some of the group operations transform the vectors (x, y) into linear combinations of x and y, so that (x, y) form a basis for a two-dimensional IR of $C_{\infty v}$. If the (x, y) basis vectors are rotated by an angle ϕ about the z axis, the resulting new basis (x', y') is related to the former basis by

$$\begin{bmatrix} x' \\ y' \end{bmatrix} = \begin{bmatrix} \cos\phi & -\sin\phi \\ \sin\phi & \cos\phi \end{bmatrix} \begin{bmatrix} x \\ y \end{bmatrix}$$

$$= \mathbf{C}_\phi \begin{bmatrix} x \\ y \end{bmatrix} \tag{4.1}$$

The character $\chi(\mathbf{C}_\phi)$ in this basis is then $2\cos\phi$. Similarly, suppose a reflection plane σ_v contains the x axis. Then under this σ_v operation,

$$\begin{bmatrix} x' \\ y' \end{bmatrix} = \begin{bmatrix} 1 & 0 \\ 0 & -1 \end{bmatrix} \begin{bmatrix} x \\ y \end{bmatrix}$$

$$= \boldsymbol{\sigma}_v \begin{bmatrix} x \\ y \end{bmatrix} \tag{4.2}$$

Then $\chi(\boldsymbol{\sigma}_v) = 0$; it can easily be shown that this result is independent of the orientation of the σ_v plane with respect to the x and y axes [1]. Finally, by the definition of the identity operation

$$\mathbf{E}\begin{bmatrix} x \\ y \end{bmatrix} \equiv \begin{bmatrix} x \\ y \end{bmatrix} \tag{4.3}$$

and $\chi(\mathbf{E}) = 2$. We conclude that the (x, y) vectors together form a basis for the Π IR of $C_{\infty v}$, and this fact is reflected in the character table. It may similarly be shown that the rotations (R_x, R_y) about the x and y axes also form a basis for the Π IR.

The homonuclear point group $D_{\infty h}$ can be generated from $C_{\infty v}$ by adding the inversion operation i about a center of symmetry, $(x, y, z) \rightarrow (-x, -y, -z)$. This creates two additional classes of group operations: the improper rotations $S_\phi = iC_\phi$, and the two fold rotations $C_2 = i\sigma_v$. The homonuclear point group character table is

$D_{\infty h}$

	E	$2C_\phi \cdots$	$\infty\sigma_v$	i	$2S_\phi \cdots$	∞C_2	
Σ_g^+	1	1	1	1	1	1	
Σ_g^-	1	1	-1	1	1	-1	R_z
Π_g	2	$2\cos\phi$	0	2	$-2\cos\phi$	0	(R_x, R_y)
Δ_g	2	$2\cos 2\phi$	0	2	$2\cos 2\phi$	0	
Σ_u^+	1	1	1	-1	-1	-1	z
Σ_u^-	1	1	-1	-1	-1	1	
Π_u	2	$2\cos\phi$	0	-2	$2\cos\phi$	0	(x, y)
Δ_u	2	$2\cos 2\phi$	0	-2	$-2\cos 2\phi$	0	
$\vdots$	$\vdots$	$\vdots$	$\vdots$	$\vdots$	$\vdots$	$\vdots$	

From our discussion of the $C_{\infty v}$ point group, z transforms as either Σ_g^+ or Σ_u^+; since $i \cdot z = -z$, it must belong to Σ_u^+. Since

$$\mathbf{i}\begin{bmatrix} x \\ y \end{bmatrix} = \begin{bmatrix} -x \\ -y \end{bmatrix} \tag{4.4}$$

the character $\chi(\mathbf{i})$ in the (x, y) basis is -2, and so (x, y) forms a basis for the Π_u rather than the Π_g IR. In general, the vector components always transform as u-type IRs in $D_{\infty h}$. The reverse is true of the rotations R_x, R_y, R_z, which always transform as g-type IRs.

We are now prepared to discuss the relationships between symmetry and electronic structure in diatomics. In the absence of spin–orbit coupling in *atoms*, one has the commutation relationships involving the electronic Hamiltonian and angular momenta:

$$
\begin{aligned}
[\hat{H}, \hat{L}^2) &= 0 \qquad [\hat{H}, \hat{L}_z] = 0 \\
[\hat{H}, \hat{S}^2] &= 0 \qquad [\hat{H}, \hat{S}_z] = 0 \\
[\hat{H}, \hat{J}^2] &= 0 \qquad [\hat{H}, \hat{J}_z] = 0
\end{aligned}
\tag{4.5}
$$

This means that L, S, J, M_L, M_S, M_J can all be good quantum numbers, in addition to the principal quantum number, in spherically symmetric potentials. (All six of these cannot *simultaneously* be good quantum numbers, for reasons explained in section 2.2 and Appendix E.) In the reduced cylindrical symmetry of diatomic molecules, however, two of these commutation relationships in the absence of spin–orbit coupling become modified [2],

$$
\begin{aligned}
[\hat{H}, \hat{L}^2] &\neq 0 \\
[\hat{H}, \hat{J}^2] &\neq 0
\end{aligned}
\tag{4.6}
$$

so that L and J are no longer good quantum numbers. The projections $L_z = M_L\hbar$, $S_z = M_S\hbar$, and $J_z = M_J\hbar$ of $\mathbf{L}$, $\mathbf{S}$, and $\mathbf{J}$ along the molecular axis remain conserved in the cylindrically symmetric potential. Diatomic states which are eigenfunctions of $\hat{L}_z$ (i.e. states in which M_L is a good quantum number) must have the ϕ-dependence $\exp(\pm i\Lambda\phi)$ with $\Lambda \equiv |M_L|$, because such functions are the only physically acceptable eigenfunctions of $\hat{L}_z$,

$$
\hat{L}_z \exp(\pm i\Lambda\phi) = \pm\Lambda\hbar \exp(\pm i\Lambda\phi) \tag{4.7}
$$

For continuity of this function at $\phi = 0$ (2π), Λ must be integral,

$$
\Lambda = 0, 1, 2, \ldots \tag{4.8}
$$

In analogy to $L_z = \pm\Lambda\hbar$, one also has $S_z = \pm\Sigma\hbar$, where Σ can be half-integral for spin angular momentum,

$$
\Sigma = 0, \tfrac{1}{2}, 1, \tfrac{3}{2}, \ldots \tag{4.9}
$$

When $\Lambda = 0$, the electronic state is obviously ϕ-independent and unaffected by the C_ϕ operation. Hence, a state with $\Lambda = 0$ exhibits $\chi(\mathbf{C}_\phi) = 1$ and therefore is some type of Σ state, which, according to the $C_{\infty v}$ and $D_{\infty h}$ character tables, is spatially nondegenerate. For $\Lambda = 1$, we have two diatomic states behaving as $\exp(\pm i\phi)$. These are degenerate a priori (i.e., in the absence of spin–orbit coupling), since the electronic energy is independent of whether L_z points in the $+z$ or $-z$ direction. One may then take linear combinations of these two states'

ϕ-dependence to form

$$\cos \phi = \tfrac{1}{2}(e^{i\phi} + e^{-i\phi}) \equiv x/r$$

$$\sin \phi = \frac{1}{2i}(e^{i\phi} - e^{-i\phi}) \equiv y/r \qquad (4.10)$$

with $r = (x^2 + y^2)^{1/2}$.

Clearly, these linear combinations transform as the vectors (x, y) under the group operations of $C_{\infty v}$ and $D_{\infty h}$—so they form a basis for the $\Pi(C_{\infty v})$ and $\Pi_u(D_{\infty h})$ IRs. It can similarly be shown that the functions $\exp(\pm 2i\phi)$ for $\Lambda = 2$ form a basis for the $\Delta(C_{\infty v})$ and $\Delta_g(D_{\infty h})$ IRs. These are examples of the fact that the Greek letter notations for the diatomic point group IRs give the Λ values associated with the electronic states directly:

Λ	*Type of state*
0	Σ
1	Π
2	Δ
3	Φ
⋮	⋮

Each of the Σ states in molecules belonging to $C_{\infty v}$ or $D_{\infty h}$ exhibits a definite behavior (+ or −) under the σ_v reflection operation, and this is always indicated in the superscript that accompanies the IR notation. For the doubly degenerate states (Π, Δ, Φ, . . .) it is always possible to choose linear combinations analogous to those in Eq. 4.10 in order that each of the two combinations is either unaffected or changes sign with respect to a particular reflection plane. The reflection symmetries of the $\cos \phi$ and $\sin \phi$ combinations in Eq. 4.10, for example, are respectively (+) and (−) with respect to a σ_v reflection plane containing the x and z axes.

4.2 CORRELATION OF MOLECULAR STATES WITH SEPARATED-ATOM STATES

The question of which diatomic term symbols may be obtained by adiabatically bringing together atoms A and B, initially in electronic states with angular momentum quantum numbers (l_A, s_A) and (l_B, s_B), can be answered without recourse to electronic structure calculations. Since the electronic (orbital plus spin) degeneracies on the respective atoms are $(2l_A + 1)(2s_A + 1)$ and $(2l_B + 1)(2s_B + 1)$, a total of $(2l_A + 1)(2s_A + 1)(2l_B + 1)(2s_B + 1)$ diatomic states must correlate with the separated-atom states. According to one of the commutation relationships (4.6), M_L is a good quantum number for diatomics at all internuclear separations R (in the absence of spin–orbit coupling). M_L must then be conserved as the atoms are pulled apart, and so $\Lambda = |M_L|$ must equal the absolute sum of magnetic quantum numbers m_{lA} and m_{lB} for the separated atoms. A similar argument applies to the spin angular momenta: M_S in any diatomic state is necessarily the sum of the separated-atom quantum numbers m_{sA} and m_{sB}. The notation $\Sigma \equiv |M_S|$ is commonly used to specify the projection of spin angular momentum along the molecular axis; it should not be confused with the unrelated use of the notation Σ to denote diatomic electronic states with $\Lambda = 0$.

We now consider the heteronuclear correlation problem for several examples of increasing complexity. An example of the simplest case (both atoms in ^2S states) is the ground state of NaH, which dissociates into the ground-state atoms $Na(3^2S) + H(1^2S)$. Since $l_A = l_B = 0$ in the separated atom S states, the total z component of orbital angular momentum is $L_z = \hbar M_L = \hbar(m_{lA} + m_{lB}) = 0$. Hence only Σ diatomic states ($\Lambda = 0$) can correlate with the S-state atoms. They must furthermore be Σ^+ rather than Σ^- states, because the atomic S states are even with respect to reflection in any σ_v plane containing the molecular axis (Fig. 4.5). The number of diatomic states which correlate with the S-state atoms is $(2l_A + 1)(2l_B + 1)(2s_A + 1)(2s_B + 1) = 4$. Hence the pertinent diatomic states are $^1\Sigma^+$ (one state with $M_S = 0$) and $^3\Sigma^+$ (three states with $M_S = 0, \pm 1$), which will later be seen to be bound and repulsive states, respectively.

The next few excited states in NaH correlate with $Na(3^2P) + H(1^2S)$, for which $l_A = 1$ and $l_B = 0$. The allowed m_l values for the separated atoms are then $m_{lA} = 0, \ \pm 1$ and $m_{lB} = 0$; these atomic states must give rise to $(2l_A + 1)(2s_A + 1)(2l_B + 1)(2s_B + 1) = 12$ diatomic states. It helps to tabulate the 12 possible combinations of $m_{lA}, m_{lB}, m_{sA}, m_{sB}$ as shown in Table 4.1, which also lists the resultant $\Lambda = |m_L|$ and M_S values. In this table, an upward (downward) arrow denotes the value $+\frac{1}{2}$ $(-\frac{1}{2})$ for either m_{sA} or m_{sB}. We obtain one Σ state each with $M_S = \pm 1$, and two Σ states with $M_S = 0$; this is possible only if there is one $^1\Sigma$ state (with $M_S = 0$) and three $^3\Sigma$ states (with $M_S = 0, \pm 1$). These Σ

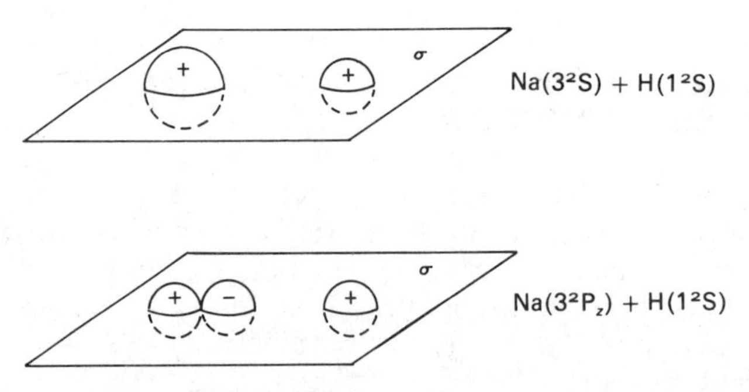

Figure 4.5 Reflection symmetry of NaH diatomic states correlating with (a) $H(1^2S) + Na(3^2S)$ and (b) $H(1^2S) + Na(3^2P_z)$.

Table 4.1 Diatomic states correlating with ^{2}P + ^{2}S atomic states

m_{lA}			m_{lB}		
−1	0	+1	0	Λ	M_S
↑			↑	1	1
↑			↓	1	0
	↑		↑	0	1
	↑		↓	0	0
		↑	↑	1	1
		↑	↓	1	0
↓			↑	1	0
↓			↓	1	−1
	↓		↑	0	0
	↓		↓	0	−1
		↓	↑	1	0
		↓	↓	1	−1

states must be Σ^+ states, since Table 4.1 shows they arise only from atomic states with $m_{lA} = m_{lB} = 0$. Such states (composed from the H s orbital and the Na p_z orbital oriented along the molecular axis) are even with respect to any σ_v reflection (Fig. 4.5). Table 4.1 similarly reveals two Π states each with $M_S = \pm 1$, and four Π states with $M_S = 0$. These are naturally grouped into one $^1\Pi$ and one $^3\Pi$ manifold of states; the inherent twofold spatial degeneracy of Π states (Section 4.1) is asserted by the appearance of two, four, and two (rather than one, two, and one) states, respectively, with $M_S = -1$, 0, and +1.

These correlations are summarized for NaH in Fig. 4.6, which shows theoretical potential energy curves for all diatomic states which dissociate into either Na(3^2S) + H(1^2S) or Na(3^2P) + H(1^2S). The ground atomic states are split into the X$^1\Sigma^+$ bound and a$^3\Sigma^+$ repulsive diatomic states as the atoms are brought together; the atomic states corresponding to Na(3^2P) + H(1^2S) are split into the A$^1\Sigma^+$, b$^3\Pi$, B$^1\Pi$, and c$^3\Sigma^+$ diatomic states. We will see in the discussion of Hund's coupling cases (Section 4.5) that when the spin–orbit coupling is small but nonnegligible, the orbital and spin angular momentum components Λ and Σ can couple to form a resultant electronic angular momentum Ω with possible values $\Omega = \Lambda + S, \ldots, |\Lambda - S|$. In the $^3\Pi$ states ($\Lambda = S = 1$), the possible Ω resultants are $(1 + 1), \ldots, (1 - 1) = 2, 1, 0$. The Ω values are notated as subscripts to the diatomic term symbols (and are analogous to the quantum number J in atomic term symbols); the $^3\Pi$ states are then said to be split into $^3\Pi_2$, $^3\Pi_1$, and $^3\Pi_0$ sublevels under spin–orbit coupling. The spin–orbit coupling is so small in NaH that these three sublevels have indistinguishable energies on the scale of Fig. 4.6, which shows only one potential energy curve for the b$^3\Pi$ state. In I_2 (Fig. 4.7), the reverse is true: the A$^3\Pi_{1u}$, B$^3\Pi_{0u}$, and $^3\Pi_{2u}$

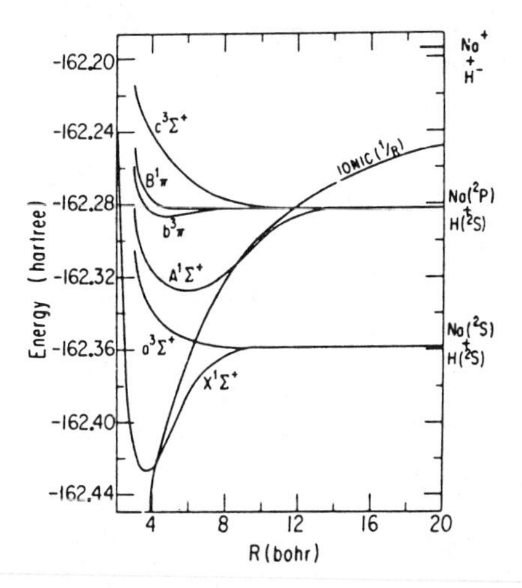

Figure 4.6 Potential energy curves for six electronic states of NaH. The curve labeled "ionic" is the function $e^2/4\pi\epsilon_0 R$, approaching the energy of $Na^+ + H^-$ at infinite separation. From multiconfigurational self-consistent field calculations by E. S. Sachs, J. Hinze, and N. H. Sabelli, *J. Chem. Phys.* **62**: 3367 (1975); used with permission.

states have such widely separated potential energy curves as a consequence of large spin–orbit coupling that they behave dynamically like separate electronic states. (The incorporation of the g and u labels in subscripts of I_2 state term symbols is necessary because I_2, unlike NaH, belongs to the $D_{\infty h}$ point group.)

A brief aside about diatomic electronic state notation is necessary here. A molecule can have many electronic states of a given symmetry and multiplicity (e.g., $^1\Pi$), and additional symbols are needed to tell them apart. The electronic *ground* state is always denoted with an X—as in $X^1\Sigma^+$ for NaH, $X^1\Sigma_g^+$ for I_2. The lowest excited state with the same multiplicity as the ground state is supposed to be denoted with A, the next one up is B, and so on. The lowest excited state with different multiplicity than the ground state should be labeled a, the next is called b, etc. These good intentions have not always been followed historically, partly because there are frequently "phantom states"—states that are difficult to observe spectroscopically because of selection rules, and are overlooked—so that labels that have become well established in the literature turn out to be incorrect when phantom states are flushed out by improved techniques. The way we have labeled the NaH potential energy curves in Fig. 4.5 is orthodox, because it is positively known that there are no other NaH states with comparable or lower energy than the ones pictured (our preceding discussion should convince you of that!). However, in I_2 the A and B state labeling in Fig. 4.6 is clearly wrong (they should not be capitalized for $^3\Pi_{0u}$ and $^3\Pi_{1u}$ states when the ground state is $X^1\Sigma_g^+$; they are not the lowest two states of their multiplicity, either). These A and B state notations in I_2 are well entrenched

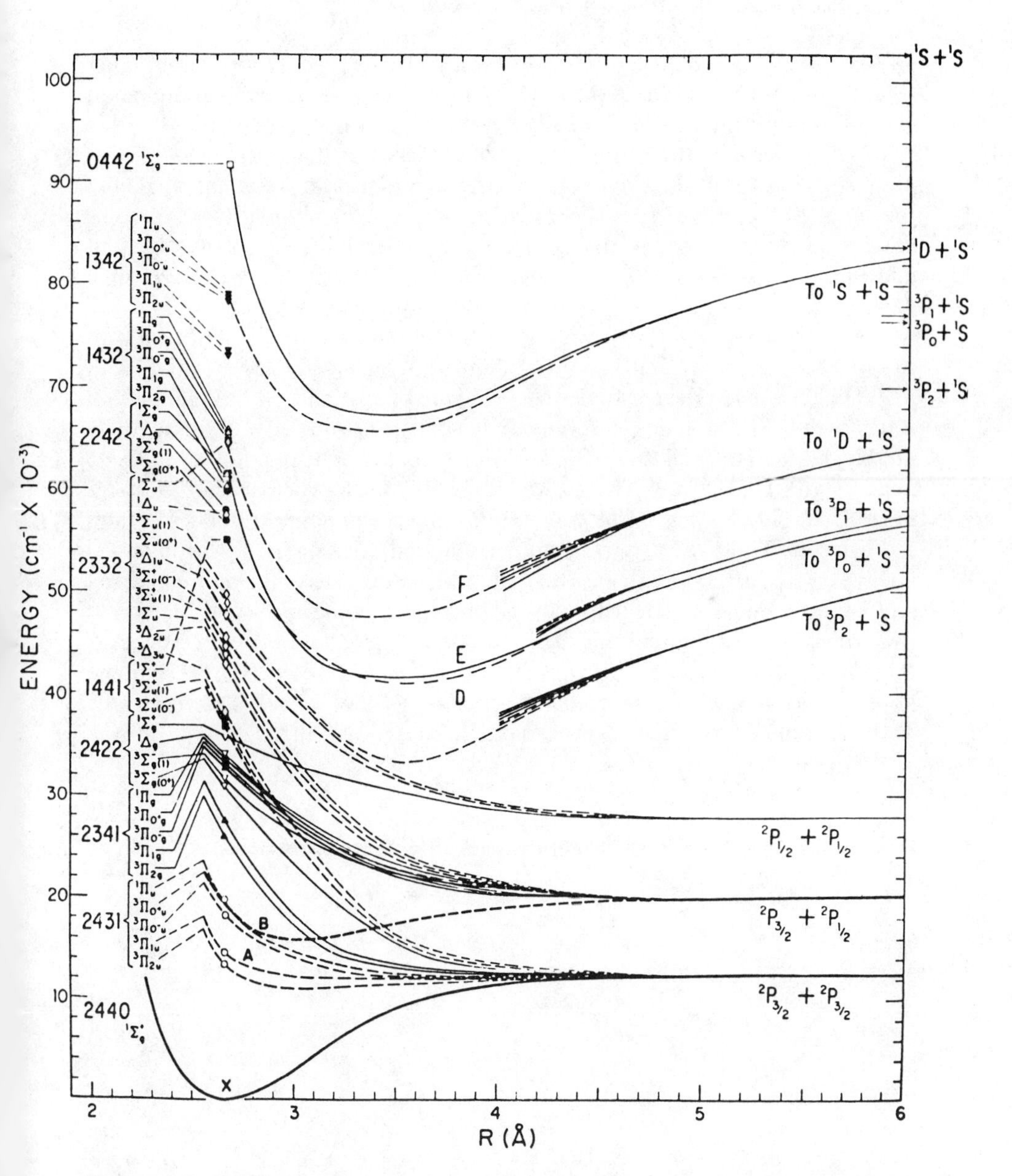

Figure 4.7 Potential energy curves for electronic states of I_2. The sets of numbers 2440, etc., give orbital occupancies of the $\sigma_g 5p$, $\pi_u 5p$, $\pi_g^* 5p$, and $\sigma_u^* 5p$ molecular orbitals, respectively (Section 4.3). The $X^1\Sigma_g^+$, $B^3\Pi_{ou}$, and $A^3\Pi_{1u}$ potential energy curves have been characterized spectroscopically. Used with permission from R. S. Mulliken, *J. Chem. Phys.* **55**: 288 (1971).

anyway. Such state prefixes should generally be regarded as interesting relics that draw attention to the most easily observed excited states; their supposed structural implications should not be taken at face value.

We consider next the heteronuclear diatomic states that correlate with two atoms in ^{2}P states, a situation that introduces complications not anticipated in the simpler cases. Since $l_A = l_B = 1$ and $s_A = s_B = \frac{1}{2}$, combining two ^{2}P atoms creates 36 diatomic states. By compiling a table similar to Table 4.1 (but allowing m_{lB} as well as m_{lA} to range from -1 to $+1$) for the 36 possible combinations of m_{lA}, m_{lB}, m_{sA}, and m_{sB}, one readily finds 8 Δ states with $\Lambda = 2$ (two each with $M_S = \pm 1$, four with $M_S = 0$), 16 π states with $\Lambda = 1$ (4 each with $M_S = \pm 1$, 8 with $M_S = 0$), and 12 Σ states (3 each with $M_S = \pm 1$, 6 with $M_S = 0$). One may thus conclude that there are one $^1\Delta$ and one $^3\Delta$ manifold of states with 2 and 6 states, respectively (since states with $\Lambda \neq 0$ are spatially doubly degenerate), and that there are two $^1\Pi$ and two $^3\Pi$ manifolds totaling 4 and 12 states, respectively. The question then arises as to whether the 12 Σ states have Σ^+ or Σ^- character. The configurations corresponding to these 12 states are listed in Table 4.2. The first four configurations (states ϕ_1 through ϕ_4) involve only p_z orbitals with $m_{lA} = m_{lB} = 0$, which point along the molecular axis and are even under σ_v. Hence states ϕ_1 through ϕ_4 (one each with $M_S = \pm 1$, two with $M_S = 0$) represent one $^1\Sigma^+$ and one $^3\Sigma^+$ manifold of states. States ϕ_5 through ϕ_{12} are composed of p orbitals with $m_{lA} = \pm 1$ and $m_{lB} = \pm 1$. Using the notations p_+ and p_- for p orbitals with $m_l = +1$ and $m_l = -1$, respectively, properly antisymmetrized expressions for states ϕ_5 through ϕ_{12} in the

Table 4.2 Diatomic Σ states correlating with ^{2}P + ^{2}P atomic states

m_{lA}			m_{lB}					
-1	0	$+1$	-1	0	$+1$	Λ	M_S	State
	↑			↑		0	1	ϕ_1
	↑			↓		0	0	ϕ_2
	↓			↑		0	0	ϕ_3
	↓			↓		0	-1	ϕ_4
↑					↑	0	1	ϕ_5
↑					↓	0	0	ϕ_6, ϕ_7
↓					↑	0	0	
↓					↓	0	-1	ϕ_8
		↑	↑			0	1	ϕ_9
		↑	↓			0	0	ϕ_{10}, ϕ_{11}
		↓	↑			0	0	
		↓	↓			0	-1	ϕ_{12}

separated-atom limit are

$$\phi_5 = [p_{A-}(1)p_{B+}(2) - p_{A-}(2)p_{B+}(1)]\alpha(1)\alpha(2)$$
$$\phi_6 = [p_{A-}(1)p_{B+}(2) - p_{A-}(2)p_{B+}(1)][\alpha(1)\beta(2) + \alpha(2)\beta(1)]$$
$$\phi_7 = [p_{A-}(1)p_{B+}(2) + p_{A-}(2)p_{B+}(1)][\alpha(1)\beta(2) - \alpha(2)\beta(1)]$$
$$\phi_8 = [p_{A-}(1)p_{B+}(2) - p_{A-}(2)p_{B+}(1)]\beta(1)\beta(2)$$
$$\phi_9 = [p_{A+}(1)p_{B-}(2) - p_{A+}(2)p_{B-}(1)]\alpha(1)\alpha(2)$$
$$\phi_{10} = [p_{A+}(1)p_{B-}(2) - p_{A+}(2)p_{B-}(1)][\alpha(1)\beta(2) + \alpha(2)\beta(1)]$$
$$\phi_{11} = [p_{A+}(1)p_{B-}(2) + p_{A+}(2)p_{B-}(1)][\alpha(1)\beta(2) - \alpha(2)\beta(1)]$$
$$\phi_{12} = [p_{A+}(1)p_{B-}(2) - p_{A+}(2)p_{B-}(1)]\beta(1)\beta(2)$$

Since the σ_v operation converts the function $\exp(\pm i\phi)$ into the function $\exp(\mp i\phi)$—σ_v reverses the sense of any rotation by an angle ϕ about the z axis—we have

$$\sigma_v p_{A\pm} = p_{A\mp}$$
$$\sigma_v p_{B\pm} = p_{B\mp} \tag{4.12}$$

with the result that

$$\begin{aligned} \sigma_v\phi_5 &= \phi_9 & \sigma_v\phi_9 &= \phi_5 \\ \sigma_v\phi_6 &= \phi_{10} & \sigma_v\phi_{10} &= \phi_6 \\ \sigma_v\phi_7 &= \phi_{11} & \sigma_v\phi_{11} &= \phi_7 \\ \sigma_v\phi_8 &= \phi_{12} & \sigma_v\phi_{12} &= \phi_8 \end{aligned} \tag{4.13}$$

One may thus form linear combinations of states ϕ_5 through ϕ_{12} exhibiting definite parity under σ_v:

$$\begin{aligned} \sigma_v(\phi_5 \pm \phi_9) &= \pm(\phi_5 \pm \phi_9) \\ \sigma_v(\phi_6 \pm \phi_{10}) &= \pm(\phi_6 \pm \phi_{10}) \\ \sigma_v(\phi_7 \pm \phi_{11}) &= \pm(\phi_7 \pm \phi_{11}) \\ \sigma_v(\phi_8 \pm \phi_{12}) &= \pm(\phi_8 \pm \phi_{12}) \end{aligned} \tag{4.14}$$

Hence, these linear combinations may be classified as shown in Table 4.3, using the properties of states ϕ_5 through ϕ_{12} taken from Table 4.2 and Eqs. 4.14. It is apparent that these states yield one manifold each of ${}^1\Sigma^+$, ${}^3\Sigma^+$, ${}^1\Sigma^-$, and ${}^3\Sigma^-$ character, with one, three, one, and three states, respectively.

Table 4.3

State	Symmetry	M_S
$\phi_5 + \phi_9$	Σ^+	$+1$
$\phi_5 - \phi_9$	Σ^-	$+1$
$\phi_6 + \phi_{10}$	Σ^+	0
$\phi_6 - \phi_{10}$	Σ^-	0
$\phi_7 + \phi_{11}$	Σ^+	0
$\phi_7 - \sigma_{11}$	Σ^-	0
$\phi_8 + \phi_{12}$	Σ^+	-1
$\phi_8 - \phi_{12}$	Σ^-	-1

In summary, the diatomic states that correlate with two ^{2}P atoms include

$$\begin{array}{ll} ^1\Delta,\ ^3\Delta & ^1\Sigma^+,\ ^3\Sigma^+ \\ ^1\Pi,\ ^3\Pi & ^1\Sigma^+,\ ^3\Sigma^+ \\ ^1\Pi,\ ^3\Pi & ^1\Sigma^-,\ ^3\Sigma^- \end{array} \tag{4.15}$$

In the presence of spin–orbit coupling, the triplet states with $\Lambda \neq 0$ become split into three components with $\Omega = \Lambda + S, \ldots, |\Lambda - S| = \Lambda + 1, \Lambda, \Lambda - 1$. The resulting diatomic states then become

$$\begin{array}{ll} ^1\Delta,\ ^3\Delta_3,\ ^3\Delta_2,\ ^3\Delta_1 & ^1\Sigma^+,\ ^3\Sigma^+ \\ ^1\Pi,\ ^3\Pi_2,\ ^3\Pi_1,\ ^3\Pi_0 & ^1\Sigma^+,\ ^3\Sigma^+ \\ ^1\Pi,\ ^3\Pi_2,\ ^3\Pi_1,\ ^3\Pi_0 & ^1\Sigma^+,\ ^3\Sigma^- \end{array} \tag{4.16}$$

Finally, it is instructive to discuss the homonuclear analogs to two of the simplest cases treated above. These are furnished by Na_2, a well-studied diatomic (cf. Figs. 4.1 and 4.3) whose ground state dissociates into 2Na(3^2S). The bound $X^1\Sigma_g^+$ and repulsive $a^3\Sigma_u^+$ diatomic Na_2 states correlate with the ground-state atoms, as shown in Fig. 4.8. Aside from the inclusion of the g and u labels appropriate to the $D_{\infty h}$ point group, these are entirely analogous to the $X^1\Sigma^+$ and $a^3\Sigma^+$ NaH states formed from ground-state Na and H atoms (Fig. 4.6). However, the next few Na_2 excited states correlate with the degenerate configurations

$$\begin{array}{c} \mathrm{Na}(3^2\mathrm{S}) + \mathrm{Na}(3^2\mathrm{P}) \\ \mathrm{Na}(3^2\mathrm{P}) + \mathrm{Na}(3^2\mathrm{S}) \end{array}$$

for which $l_A = 0$, $l_B = 1$ and $l_A = 1$, $l_B = 0$, respectively. Twelve diatomic states arise from the former configuration, and 12 additional states (of the same energy

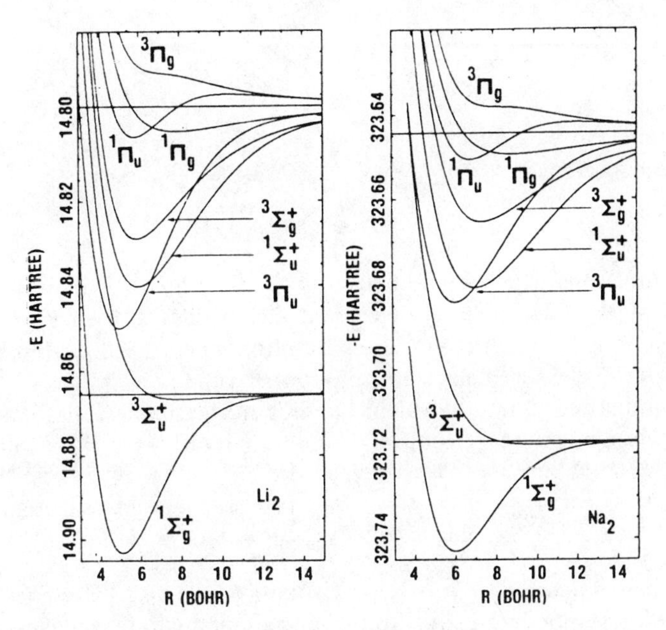

Figure 4.8 Multiconfigurational self-consistent field potential energy curves for low-lying electronic states of Li_2 and Na_2. Reproduced with permission from D. Konowalow, M. Rosenkrantz, and M. Olson, *J. Chem. Phys.* **72**: 2612 (1980).

in the separated-atom limit) arise from the latter. As a result, 24 Na_2 states correlate with one ^2S and one ^2P atom, in contrast to the 12 NaH states that dissociate into $Na(3^2P) + H(1^2S)$. Corresponding to each of the latter states in NaH (Fig. 4.6), namely

$$A^1\Sigma^+, b^3\Pi, B^1\Pi, c^3\Sigma^+$$

there is a *pair* of conjugate states with opposite inversion symmetry Na_2:

$$A^1\Sigma_u^+, b^3\Pi_u, B^1\Pi_u, {}^3\Sigma_u^+$$

$${}^1\Sigma_g^+, {}^3\Pi_g, {}^1\Pi_g, {}^3\Sigma_g^+$$

Theoretical potential energy curves are shown for six of these Na_2 states in Fig. 4.8.

4.3 LCAO–MO WAVE FUNCTIONS IN DIATOMICS

In our discussion of the Born-Oppenheimer principle (Section 3.1) we pointed out that eigenfunctions $|\psi_k(\mathbf{r}; \mathbf{R})\rangle$ of the electronic Hamiltonian

$$\hat{H}_{\text{el}} = \sum_{i=1}^{n} \frac{-\hbar^2}{2m_e} \nabla_i^2 - \frac{1}{4\pi\varepsilon_0} \sum_{i=1}^{n} \sum_{N=1}^{2} \frac{Z_N e^2}{|\mathbf{r}_i - \mathbf{R}_N|}$$

$$+ \frac{1}{4\pi\varepsilon_0} \sum_{i<j}^{n} \frac{e^2}{|\mathbf{r}_i - \mathbf{r}_j|} + \frac{Z_A Z_B e^2}{4\pi\varepsilon_0 |\mathbf{R}_A - \mathbf{R}_B|} \tag{4.17}$$

may be found for fixed nuclear positions $\mathbf{R}_A$, $\mathbf{R}_B$ by solving the clamped-nuclei eigenvalue equation

$$\hat{H}_{el}|\psi_k(\mathbf{r}; \mathbf{R})\rangle = \varepsilon_k(R)|\psi_k(\mathbf{r}; \mathbf{R})\rangle \tag{3.10}$$

The eigenfunctions depend parametrically on the choice of internuclear separation $R \equiv |\mathbf{R}_A - \mathbf{R}_B|$. The R-dependent eigenvalues $\varepsilon_k(R)$ act as potential energy functions for nuclear vibrational motion when the Born-Oppenheimer separation of nuclear and electronic motions is valid.

It is illuminating to treat covalent bonding in the simplest diatomic species, the H_2^+ molecule-ion. The electronic Hamiltonian (4.17) for H_2^+ reduces to

$$\hat{H}_{el} = \frac{-\hbar^2}{2m_e} \nabla^2 + \frac{e^2}{4\pi\varepsilon_0}\left(\frac{1}{R} - \frac{1}{r_A} - \frac{1}{r_B}\right) \tag{4.18}$$

with $r_A \equiv |\mathbf{r} - \mathbf{R}_A|$ and $r_B \equiv |\mathbf{r} - \mathbf{R}_B|$; it exhibits no electron–electron repulsion terms. The corresponding H_2^+ Schrödinger equation can be solved *exactly* for fixed R. The exact ground-state potential energy curve displays a minimum at $R_e = 2.00a_0$ (1.06 Å) with energy -2.79 eV relative to the energy of the separated proton and ground-state hydrogen atom (Fig. 4.9). The exact solutions for H_2^+ prove to be useful in assessing the accuracy of variational wave functions in this prototype diatomic.

A widely used approximation to true molecular wave functions employs linear combinations of atomic orbitals (LCAOs) to simulate the molecular orbitals (MOs). At the lowest level of approximation, an H_2^+ MO may be represented as a superposition of two H atom states centered on the respective nuclei,

$$|\psi\rangle = c_A|\phi_A\rangle + c_B|\phi_B\rangle \tag{4.19}$$

with expansion coefficients c_A, c_B to be determined by symmetry, normalization, and (in MOs with larger basis sets) the variational principle. Since H_2^+ belongs to the $D_{\infty h}$ point group, $|\psi|^2$ must be unaffected by the inversion i. This requires that $|c_A| = |c_B|$, and that the AOs $|\phi_A\rangle$ and $|\phi_B\rangle$ be identical apart from phase and the fact that they are centered on different nuclei. The MO (4.19) may then be rewritten (for real AOs and expansion coefficients)

$$|\psi_\pm\rangle = c(|\phi_A\rangle \pm |\phi_B\rangle) \tag{4.20}$$

Normalization then demands that

$$\begin{aligned} 1 &= c^2(\langle\phi_A|\phi_A\rangle \pm 2\langle\phi_A|\phi_B\rangle + \langle\phi_B|\phi_B\rangle \\ &\equiv c^2(2 \pm 2S_{AB}) \end{aligned} \tag{4.21}$$

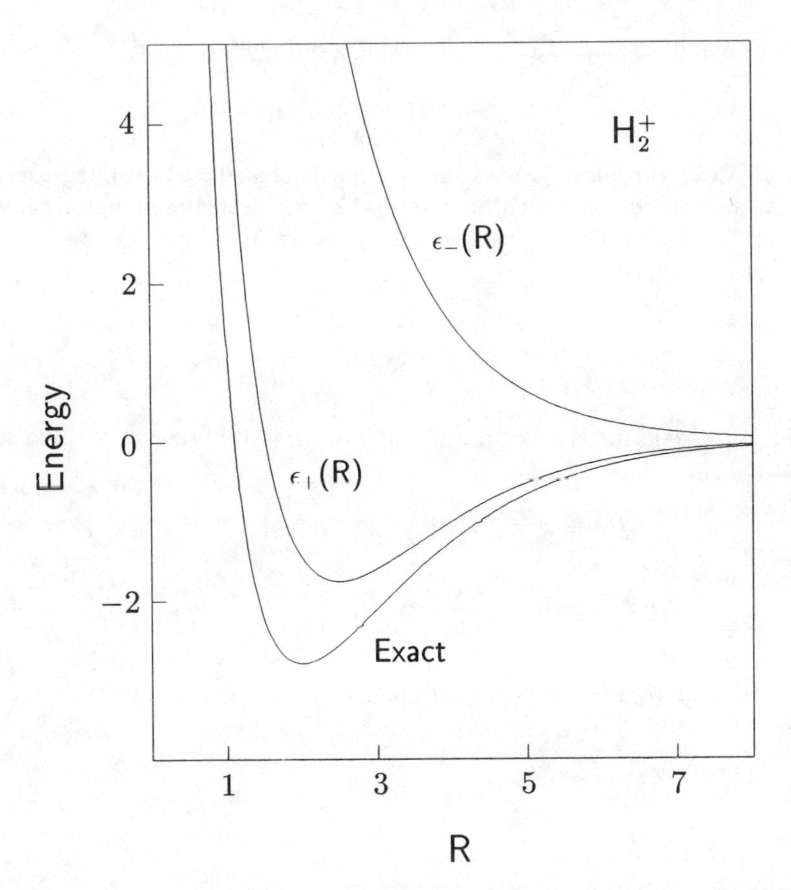

Figure 4.9 H_2^+ potential energy curves: trial energies $\epsilon_+(R)$ and $\epsilon_-(R)$ obtained from LCAO–MOs $|\psi_+\rangle$ and $|\psi_-\rangle$, respectively, and the exact ground-state potential energy curve. Energies and separations are in eV and in Bohr radii, respectively.

where $S_{AB} \equiv \langle\phi_A|\phi_B\rangle$ is the R-dependent *overlap integral* between the opposite AOs. The MO then assumes the explicit form

$$|\psi_\pm\rangle = \frac{1}{\sqrt{2(1 \pm S_{AB})}}(|\phi_A\rangle \pm |\phi_B\rangle) \tag{4.22}$$

In a trial LCAO–MO wave function for ground-state H_2^+, which correlates with proton and a 1s H atom as $R \to \infty$, we may use the hydrogen 1s states $|\phi_A\rangle = \exp(-r_A/a_0)/\sqrt{\pi a_0^3}$ and $|\phi_B\rangle = \exp(-r_B/a_0)/\sqrt{\pi a_0^3}$ for AOs. The overlap integral using these AOs becomes

$$S_{AB} = \frac{1}{\pi a_0^3}\int dV \exp[-(r_A + r_B)/a_0] \tag{4.23}$$

which is readily evaluated using ellipsoidal coordinates to yield

$$S_{AB} = e^{-\rho}(1 + \rho + \rho^2/3) \tag{4.24}$$

with $\rho = R/a_0$ (Problem 4.8). As the two nuclei coalesce ($R \to 0$), the overlap integral approaches unity according to Eq. 4.24. S_{AB} decreases monotonically to zero as the nuclei are pulled apart ($R \to \infty$). In this limit, the MO (4.22) consequently approaches

$$|\psi_\pm\rangle = \frac{1}{\sqrt{2}}(|\phi_A\rangle \pm |\phi_B\rangle) \tag{4.25}$$

In the same limit, the H_2^+ electronic Hamiltonian (4.18) becomes

$$\begin{aligned} \hat{H}_{el} &= \frac{-\hbar^2}{2m_e}\nabla^2 - \frac{e^2}{4\pi\varepsilon_0}\left(\frac{1}{r_A} + \frac{1}{r_B}\right) \\ &= \hat{H}_A - \frac{e^2}{4\pi\varepsilon_0 r_B} = \hat{H}_B - \frac{e^2}{4\pi\varepsilon_0 r_A} \end{aligned} \tag{4.26}$$

where $\hat{H}_A$ and $\hat{H}_B$ are the hydrogen atom Hamiltonians for atoms A and B. Since

$$\hat{H}_A|\phi_A\rangle = E_{1s}|\phi_A\rangle$$

and

$$\hat{H}_B|\phi_B\rangle = E_{1s}|\phi_B\rangle$$

the trial energies $\varepsilon_\pm$ may be evaluated in the limit $R \to \infty$ using the asymptotic Eqs. 4.25 and 4.26, respectively, for $|\psi_\pm\rangle$ and $\hat{H}_{el}$:

$$\begin{aligned} \langle\psi_\pm|\hat{H}_{el}|\psi_\pm\rangle &= 1/2(\langle\phi_A|\hat{H}_{el}|\phi_A\rangle \pm 2\langle\phi_A|\hat{H}_{el}|\phi_B\rangle + \langle\phi_B|\hat{H}_{el}|\phi_B\rangle) \\ &= 1/2\Bigg(\langle\phi_A|\hat{H}_A|\phi_A\rangle - \frac{e^2}{4\pi\varepsilon_0}\langle\phi_A|\frac{1}{r_B}|\phi_A\rangle \pm 2\langle\phi_A|\hat{H}_A|\phi_B\rangle \\ &\quad \mp \frac{e^2}{2\pi\varepsilon_0}\langle\phi_A|\frac{1}{r_B}|\phi_B\rangle + \langle\phi_B|\hat{H}_B|\phi_B\rangle - \frac{e^2}{4\pi\varepsilon_0}\langle\phi_B|\frac{1}{r_A}|\phi_B\rangle\Bigg) \\ &= \tfrac{1}{2}E_{1s}(\langle\phi_A|\phi_A\rangle \pm 2\langle\phi_A|\phi_B\rangle + \langle\phi_B|\phi_B\rangle) \\ &= E_{1s} \end{aligned} \tag{4.27}$$

Equation 4.27 follows because the integrals $\langle\phi_A|1/r_B|\phi_A\rangle \equiv \langle\phi_B|1/r_A|\phi_B\rangle$, $\langle\phi_A|\phi_B\rangle \equiv S_{AB}$, and $\langle\phi_A|1/r_B|\phi_B\rangle$ all tend to zero as $R \to \infty$. Hence both of the asymptotic LCAO–MOs (4.25) are exact eigenfunctions of the asymptotic Hamiltonian (4.26) with eigenvalue E_{1s} (the total electronic energy of the dissociated fragments H^+ + H(1s)). This correct asymptotic behavior provides a partial justification of the LCAO–MO method. The trial energies $\varepsilon_{\pm}(R)$ may now be obtained for general R using Eqs. 4.18 and 4.22, with the result that

$$\varepsilon_{\pm}(R) = \langle \psi_{\pm} | \hat{H}_{\mathrm{el}} | \psi_{\pm} \rangle = \frac{H_{\mathrm{AA}} \pm H_{\mathrm{AB}}}{1 \pm S_{\mathrm{AB}}} \tag{4.28}$$

where

$$\begin{aligned} H_{\mathrm{AA}} \equiv H_{\mathrm{BB}} &= \langle \phi_{\mathrm{A}} | \hat{H}_{\mathrm{el}} | \phi_{\mathrm{A}} \rangle \\ &= \langle \phi_{\mathrm{A}} | \hat{H}_{\mathrm{A}} | \phi_{\mathrm{A}} \rangle - \frac{e^2}{4\pi\varepsilon_0} \langle \phi_{\mathrm{A}} | \frac{1}{r_{\mathrm{B}}} | \phi_{\mathrm{A}} \rangle + \frac{e^2}{4\pi\varepsilon_0 R} \\ &= E_{1s} + J + \frac{e^2}{4\pi\varepsilon_0 R} \end{aligned} \tag{4.29}$$

and

$$\begin{aligned} H_{\mathrm{AB}} \equiv H_{\mathrm{BA}} &= \langle \phi_{\mathrm{A}} | \hat{H}_{\mathrm{el}} | \phi_{\mathrm{B}} \rangle \\ &= \langle \phi_{\mathrm{A}} | \hat{H}_{\mathrm{B}} | \phi_{\mathrm{B}} \rangle - \frac{e^2}{4\pi\varepsilon_0} \langle \phi_{\mathrm{A}} | \frac{1}{r_{\mathrm{A}}} | \phi_{\mathrm{B}} \rangle + \frac{e^2}{4\pi\varepsilon_0 R} \langle \phi_{\mathrm{A}} | \phi_{\mathrm{B}} \rangle \\ &= E_{1s} S_{\mathrm{AB}} + K + \frac{e^2 S_{\mathrm{AB}}}{4\pi\varepsilon_0 R} \end{aligned} \tag{4.30}$$

Here we have defined two new integrals, the *Coulomb integral*

$$J = \frac{-e^2}{4\pi\varepsilon_0} \langle \phi_{\mathrm{A}} | \frac{1}{r_{\mathrm{B}}} | \phi_{\mathrm{A}} \rangle \tag{4.31}$$

and the *exchange integral*

$$K = \frac{-e^2}{4\pi\varepsilon_0} \langle \phi_{\mathrm{A}} | \frac{1}{r_{\mathrm{A}}} | \phi_{\mathrm{B}} \rangle \tag{4.32}$$

As written in Eq. 4.31, the Coulomb integral is simply the expectation value of the energy due to electrostatic attraction between nucleus B and a 1s electron centered on nucleus A. The exchange integral is an intrinsically quantum phenomenon, and has no analogous classical electrostatic interpretation. Collating the results of Eqs. 4.28–4.30, we finally obtain the trial energies

$$\varepsilon_{\pm}(R) = E_{1s} + \frac{e^2}{4\pi\varepsilon_0 R} + \frac{J \pm K}{1 \pm S_{\mathrm{AB}}} \tag{4.33}$$

Since the Coulomb and exchange integrals (which are both negative-definite) approach zero as $R \to \infty$, the first term E_{1s} in $\varepsilon_{\pm}(R)$ is the dissociation limit of the trial energy, in agreement with Eq. 4.27. The second term is the electrostatic internuclear repulsion energy. It may be shown (Problem 4.8) that analytic expressions for the Coulomb and exchange integrals are

$$J = -\frac{e^2}{4\pi\varepsilon_0 a_0}\left[\frac{1}{\rho} - e^{-2\rho}\left(1 + \frac{1}{\rho}\right)\right]$$

$$K = -\frac{e^2}{4\pi\varepsilon_0 a_0} e^{-\rho}(1 + \rho) \tag{4.34}$$

with $\rho = R/a_0$. It is then straightforward to evaluate the LCAO–MO trial energies $\varepsilon_{\pm}(R)$, which are compared with the exact H_2^+ ground-state potential-energy curve in Fig. 4.9. The LCAO $|\psi_+\rangle$ generated from the positive superposition of AOs (Fig. 4.10) is a bound state with potential energy minimum at $R_e = 2.50a_0$, corresponding to an energy $-1.78\,\text{eV}$ relative to $H^+ + H(1s)$. This underestimates the true dissociation energy (2.79 eV) by 36%, and $\varepsilon_+(R)$ lies everywhere above the true ground state potential, in accordance with the variational principle. The LCAO $|\psi_-\rangle$, which has a nodal plane bisecting the internuclear axis (Fig. 4.10), is a purely repulsive state in which the nuclei experience a force pushing them apart at all finite R.

In the LCAO–MO perspective of Eq. 4.33, the chemical bonding (i.e., lowering of total energy relative to $H^+ + H\,(1s)$) in state $|\psi_+\rangle$ originates in part from the electrostatic consequences of concentrating electron density between the nuclei (the Coulomb term) and in part from the exchange term. It is easy to show that the presence of an electron in regions between the nuclei electrostatically tends to draw the nuclei together, whereas an electron in other regions exerts a net repulsive effect between the nuclei (Fig. 4.11). This suggests a tempting rationalization of the bonding and repulsive characters of MOs $|\psi_+\rangle$ and $|\psi_-\rangle$, respectively, in terms of the relative degrees to which charge density is concentrated between the nuclei [3,4]: the repulsive character of $|\psi_-\rangle$ could be attributed to the presence of the nodal plane, which diminishes the internuclear charge density. Such an interpretation overlooks kinetic energy effects (electron localization in a limited internuclear region increases the expectation value of kinetic energy), and analyses of the physical origin of chemical bonding are advisedly made on the basis of accurate rather than zeroth-order LCAO–MO wave functions. A detailed examination of contributions to the total energy using an exact ground-state H_2^+ wave function reveals that chemical bonding arises from a subtle balance between electrostatic and kinetic energy effects [5].

Zeroth-order approximations to higher excited states in H_2^+ may be obtained from linear combinations of higher-energy hydrogen atom AOs, subject to the symmetry and normalization constraints of Eqs. 4.20 and 4.22. Like the states $|\psi_{\pm}\rangle$ formed from the $1s$ AOs, the higher lying LCAOs yield inaccurate trial energies—but their nodal patterns do furnish useful illustrations of the

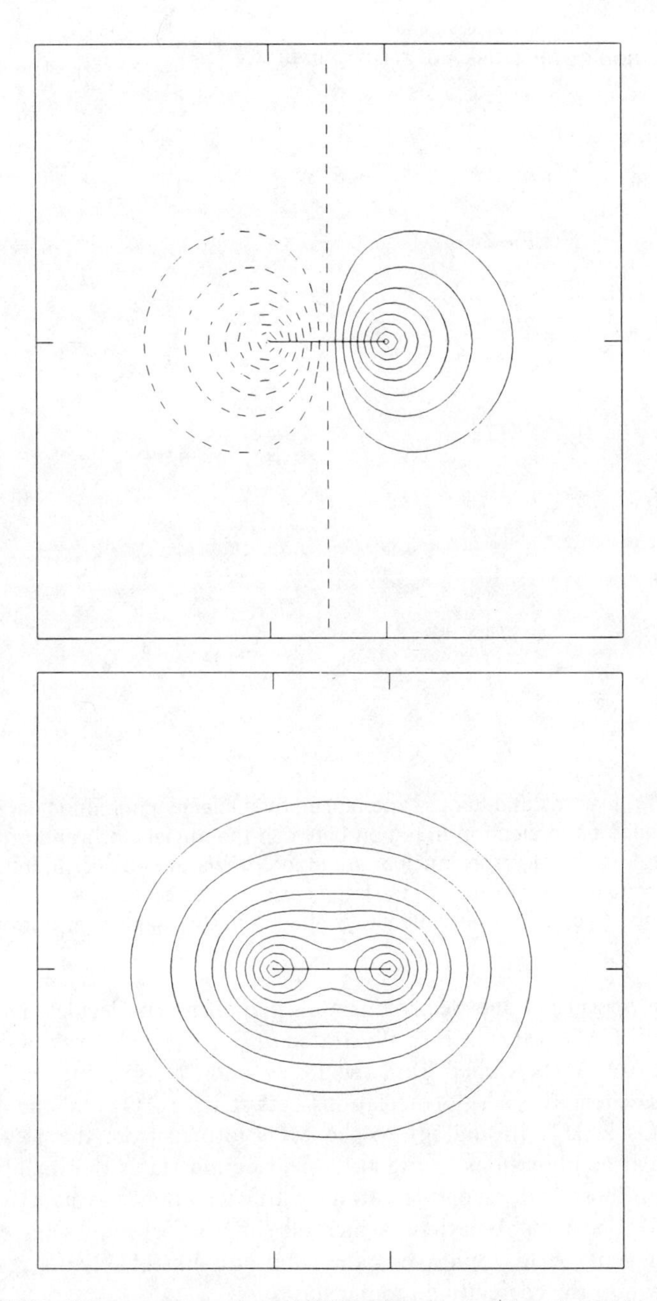

Figure 4.10 Contour plots of the LCAO-MOs $|\psi_-\rangle$ (above) and $|\psi_+\rangle$ (below). Curves are surfaces on which the wavefunction exhibits constant values; solid and dashed curves correspond to positive and negative values, respectively. The outermost contours in both cases define surfaces containing ~90% of the electron probability density. The incremental change in wavefunction value between adjacent contours is 0.04 bohr$^{-3/2}$. The border squares have sides 10 bohrs long; the internuclear separation is 2 bohrs, the equilibrium distance in ground-state H_2^+. Dashed straight line in $|\psi_-\rangle$ plot shows location of nodal plane.

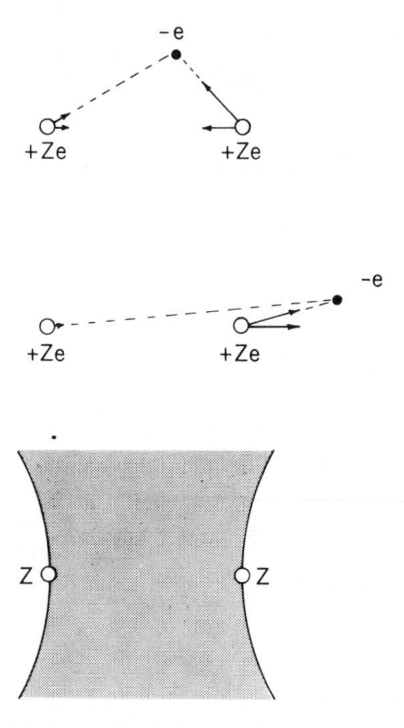

Figure 4.11 Electrostatic forces experienced by nuclei in a homonuclear molecule due to presence of an electron in region between the nuclei (top) and outside this region (middle). The electron–nuclear attraction draws the nuclear together in the former case, and pulls them apart in the latter case. The region in which the electron's presence tends to stabilize the molecule is shaded in the diagram at bottom.

symmetry properties of one-electron orbitals in diatomic molecules. The partial hierarchy of LCAO–MOs in F_2 is illustrated in Fig. 4.12, which shows contour plots of LCAO–MOs formed from the 1*s*, 2*s*, and 2*p* AOs, and in Fig. 4.13, which gives schematic energy correlations between the AOs and the diatomic LCAO–MOs in H_2^+. In analogy to the MOs formed from the 1*s* AOs, the positive linear combinations in Fig. 4.12 yield bound states that exhibit lower energies than those of the separated atoms with which they correlate (Fig. 4.13). The negative linear combinations, which all show nodal planes bisecting the internuclear axis, yield repulsive states that are unstable with respect to dissociation into the correlating atomic states.

Since L_z is conserved in each of these one-electron orbitals, the wave functions must exhibit a ϕ-dependence of the form $\exp(\pm i\lambda\phi)$ with $\lambda = 0, 1, 2, \ldots$ (We use the lower-case notation λ rather than Λ when discussing one-electron orbitals; Λ is reserved for characterizing the total component L_z of orbital angular momentum in many-electron diatomics.) One-electron orbitals with $\lambda = 0, 1, 2, \ldots$ are denoted σ, π, δ, ... orbitals, respectively; the subscripts g and u are appended to indicate the behavior of the homonuclear LCAO–MOs under inversion. To differentiate between MOs having the same point group

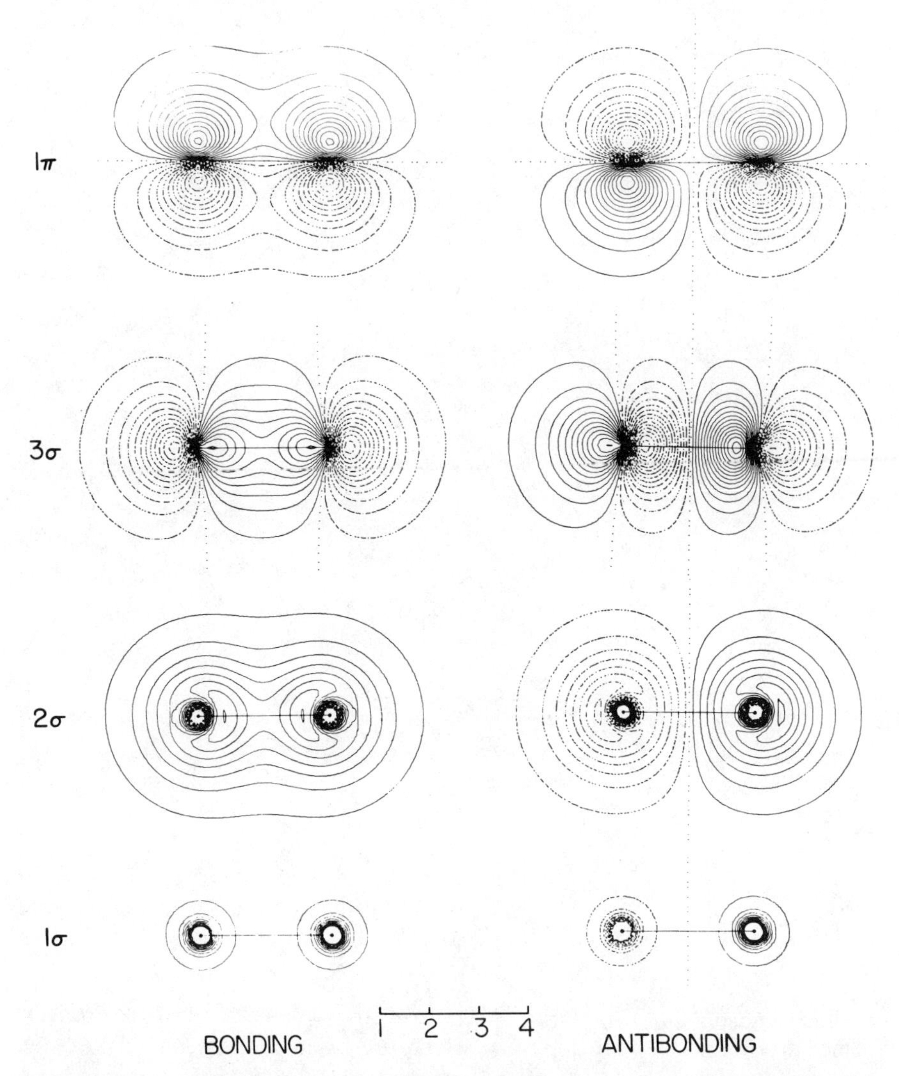

Figure 4.12 Contour plots of LCAO-MOs formed from $1s$, $2s$, $2p_z$, and $2p_x$ AOs (bottom to top) in the homonuclear diatomic molecule F_2. Distance scale is in bohrs. Solid and dashed contours correspond to positive and negative wavefunction values, respectively. Increments are 0.20 and 0.05 bohr$^{-3/2}$ for inner-shell and valence orbitals, respectively. Reproduced by permission from W. England, L. S. Salmon, and K. Ruedenberg, *Topics in Current Chemistry* **23**, 31 (1971).

symmetry, the quantum numbers nl of the AOs from which the LCAOs are formed are used as suffixes in Fig. 4.13. The lowest two H_2^+ states $|\psi_\pm\rangle$ are both $\sigma 1s$ states, since these LCAO–MOs have no ϕ-dependence and were formed from linear combinations of 1s AOs. Since $|\psi_+\rangle$ and $|\psi_-\rangle$ have g and u inversion symmetry, respectively, their orbital designations are $\sigma_g 1s$ and $\sigma_u 1s$. We similarly obtain bound $\sigma_g 2s$ and repulsive $\sigma_u 2s$ MOs from the 2s AOs, and

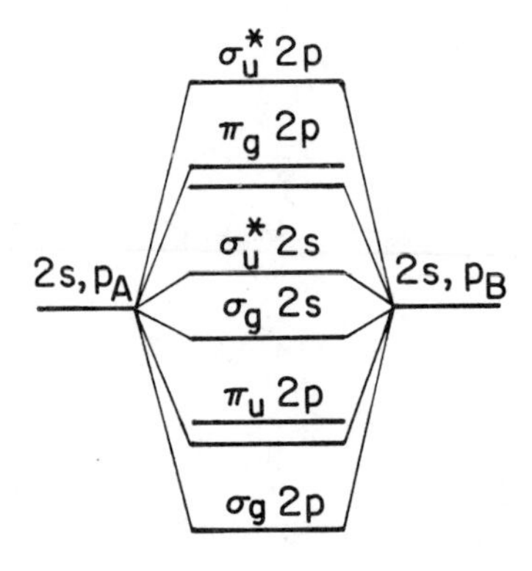

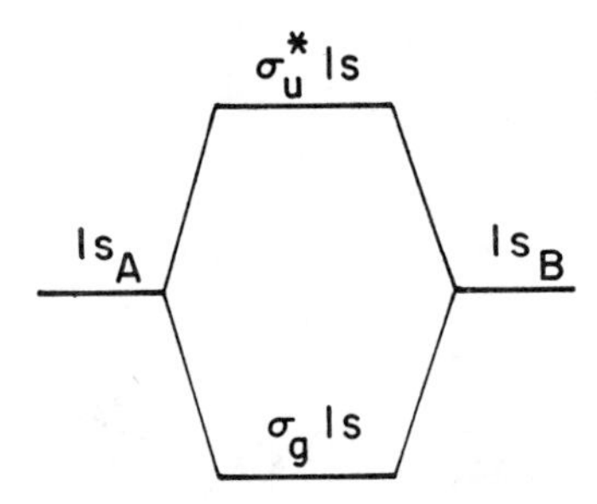

Figure 4.13 Qualitative energy ordering of MOs in H_2^+. This figure should be compared with Fig. 4.14.

bound $\sigma_g 2p_z$ and repulsive $\sigma_u 2p_z$ MOs from the $2p_z$ AOs. Since the LCAOs formed from the $2p_x$ and $2p_y$ AOs, which are oriented perpendicular to the molecular axis, behave as $\cos\phi$ $(=[\exp(i\phi)+\exp(-i\phi)]/2)$ and $\sin\phi$ $(=[\exp(i\phi)-\exp(-i\phi)]/2i)$, respectively, they are $\pi 2p$ orbitals. By inspection of the contour plots in Fig. 4.12, it is apparent that the bonding and repulsive π orbitals have u and g symmetry, respectively.

These one-electron orbitals may be used to conceptually build up many-electron configurations in heavier diatomics using the Aufbau prescription of placing electrons in orbitals according to the Pauli principle. The energy ordering of MOs varies with the diatomic [6]. For the diatomics with higher nuclear charge (e.g., O_2 and F_2 in the first row of the periodic table) the orbitals are ordered as shown in Fig. 4.14. In contrast to H_2^+, the atoms in these molecules exhibit large splittings between their 2s and 2p AOs as a consequence of configuration interaction (Chapter 2). There is thus comparatively little

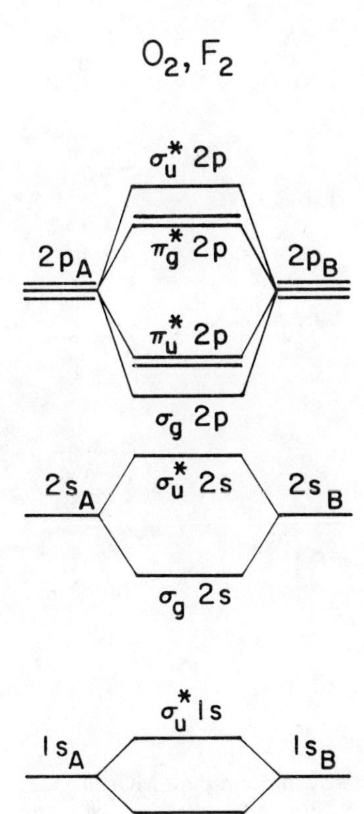

Figure 4.14 Energy ordering of MOs in O_2 and F_2.

mixing between the widely spearated $\sigma_g 2s$ and $\sigma_g 2p_z$ MOs in O_2 and F_2, so that partial hybridization endows them with relatively little $\sigma_g 2p_z$ and $\sigma_g 2s$ character, respectively. In N_2, C_2, B_2, Be_2, and Li_2 the $2s$–$2p$ atomic splitting is smaller (it vanishes in H_2), and thus the energy difference between the pure $\sigma_g 2s$ and $\sigma_g 2p_z$ MOs is smaller. Mutual mixing of the $\sigma_g 2s$ and $\sigma_g 2p_z$ MOs in these lighter molecules then increases the splitting between them, inverting the energy order of the $\sigma_g 2p$ and $\pi_u 2p$ levels (Fig. 4.15). The electron configuration in first-row diatomics can be read off from Figs. 4.14 and 4.15 by inspection. Ground-state N_2 (which has 14 electrons) has the configuration $(\sigma_g 1s)^2(\sigma_u 1s)^2(\sigma_g 2s)^2(\sigma_u 2s)^2(\pi_u 2p)^4(\sigma_g 2p)^2$. Since all of the MOs are fully occupied, ground-state N_2 is a totally symmetric state with zero net orbital and spin angular momentum. Hence $\Lambda = 0$ and $S = 0$, and the pertinent term symbol is $X^1\Sigma_g^+$. Ground-state F_2 (18 electrons) has the configuration $(\sigma_g 1s)^2(\sigma_u 1s)^2(\sigma_g 2s)^2(\sigma_u 2s)^2(\sigma_g 2p)^2(\pi_u 2p)^4(\pi_g 2p)^4$; it is also an $X^1\Sigma_g^+$ state, for the same reasons.

Evaluating the possible term symbols for diatomics with partially filled MOs is slightly more involved. A σ^1 configuration (e.g., H_2^+ $(\sigma_g 1s)^1$ or Li_2^+ $(\sigma_g 1s)^2(\sigma_u 1s)^2(\sigma_g 2s)^1$) has a single valence electron with $\lambda = 0$ and $s = \frac{1}{2}$. Since

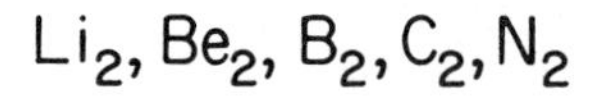

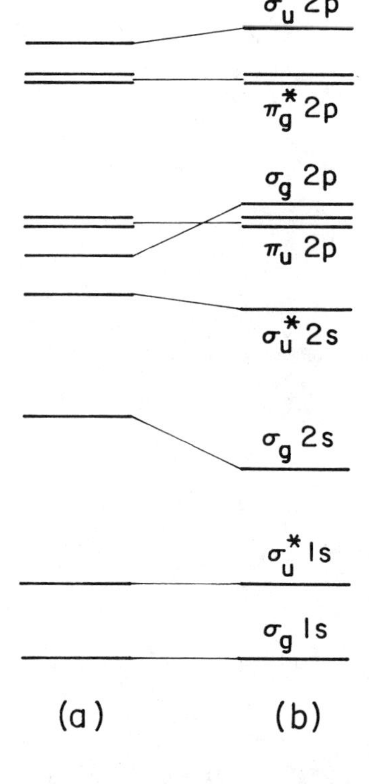

Figure 4.15 Energy ordering of MOs in Li_2, Be_2, B_2, C_2, and N_2: (a) prior to mixing of σ_g $2s$ with σ_g $2p$ and σ_u^* $2s$ with σ_u^* $2p$; and (b) after mixing.

filled orbitals do not contribute to orbital or spin angular momentum, one obtains $\Lambda = 0$ and $S = \frac{1}{2}$. Hence a σ^1 configuration gives rise to a ${}^2\Sigma$ state. By similar reasoning the $+/-$ and g/u classifications depend on the reflection and inversion symmetries of the partially occupied orbital. The H_2^+ and Li_2^+ ground states mentioned above are therefore both ${}^2\Sigma_g^+$ states; the Li_2^+ excited states with configurations $(\sigma_g 1s)^2(\sigma_u 1s)^2(\sigma_u 2s)^1$ and $(\sigma_g 1s)^2(\sigma_u 1s)^2(\pi_u 2p)^1$ are ${}^2\Sigma_u^+$ and ${}^2\Pi_u$ states, respectively.

In a *nonequivalent* σ^2 configuration (in which two valence electrons occupy different σ orbitals σ_a and σ_b), the four possible distributions of electron spins are shown in Table 4.4. We find that we have a singlet state ($M_S = 0$) and a triplet state ($M_S = 0, \pm 1$). The symmetry designation will be given by the direct product of point groups to which MOs σ_a and σ_b belong. For example, if σ_a and σ_b are σ_g and σ_u orbitals with positive reflection symmetry (e.g., LCAOs of s or p_z AOs), the resultant states are ${}^1\Sigma_u^+$, ${}^3\Sigma_u^+$ states because $\Sigma_g^+ \otimes \Sigma_u^+ = \Sigma_u^+$.

The interesting and important *equivalent* π^2 case is typified by ground-state O_2, which has the configuration $(\sigma_g 1s)^2(\sigma_u 1s)^2(\sigma_g 2s)^2(\sigma_u 2s)^2(\sigma_g 2p)^2(\pi_u 2p)^4$-$(\pi_g 2p)^2$ according to the level ordering in Fig. 4.14. Since the partially filled π_g

Table 4.4 Diatomic states from nonequivalent σ^2 configuration

σ_a	σ_b	Λ	M_S
↑	↑	0	+1
↓	↓	0	−1
↑	↓	0	0
↓	↑	0	0

Table 4.5 Diatomic states from equivalent π^2 configuration

$\lambda = -1$	$\lambda = +1$	Λ	M_S	State
	↑↓	2	0	ϕ_1
↑↓		2	0	ϕ_2
↑	↑	0	+1	ϕ_3
↑	↓	0	0	ϕ_4
↓	↑	0	0	ϕ_5
↓	↓	0	−1	ϕ_6

orbitals are doubly degenerate, we count the ways of placing two electrons in two π_g orbitals with $\lambda = \pm 1$ as shown in Table 4.5. We obviously obtain a $^1\Delta$ manifold of states ($\Lambda = 2$, $M_s = 0$), and $^3\Sigma$ and $^1\Sigma$ manifolds as well. We also know that since $\pi_g \otimes \pi_g = \Sigma_g^+ \oplus \Sigma_g^- \otimes \Delta_g$, the latter three states should evolve from an equivalent π_g^2 configuration. Then the possible states are either $^1\Sigma_g^+$, $^3\Sigma_g^-$, Δ_g or $^3\Sigma_g^+$, $^1\Sigma_g^-$, Δ_g. To choose between these alternatives, we inspect some of the wave functions for the states counted in Table 4.5. Let π_+ be the orbital for an electron with $\lambda = +1$, and π_- be that for an electron with $\lambda = -1$. Then the antisymmetrized state ϕ_3 is

$$\phi_3 = [\pi_+(1)\pi_-(2) - \pi_-(1)\pi_+(2)]\alpha(1)\alpha(2) \tag{4.35}$$

while the antisymmetrized state ϕ_6 is

$$\phi_6 = [\pi_+(1)\pi_-(2) - \pi_-(1)\pi_+(2)]\beta(1)\beta(2) \tag{4.36}$$

Under the σ_v operation, R_z changes sign and the z component of spatial angular momentum becomes reversed. This implies that

$$\begin{aligned}\sigma_v\pi_+ &= \pi_-\\ \sigma_v\pi_- &= \pi_+\end{aligned} \tag{4.37}$$

and so $\sigma_v\phi_3 = -\phi_3$, $\sigma_v\phi_6 = -\phi_6$. Then ϕ_3 and ϕ_6, which are clearly $M_S = +1$ and $M_S = -1$ components of a triplet state, are Σ^- rather than Σ^+ states. We may take linear combinations of ϕ_4 and ϕ_5 to form the $M_S = 0$ triplet component, $[\pi_+(1)\pi_-(2) - \pi_-(1)\pi_+(2)][\alpha(1)\beta(2) + \alpha(2)\beta(1)]$. The linear combination of ϕ_4 and ϕ_5 orthogonal to this forms the $^1\Sigma^+$ state, $[\pi_+(1)\pi_-(2) + \pi_-(1)\pi_+(2)][\alpha(1)\beta(2) - \alpha(2)\beta(1)]$. Note that the latter linear combination is even under σ_v as required for a Σ^+ state. We conclude that the possible term symbols in an equivalent π_g^2 configuration are $^1\Sigma_g^+$, $^3\Sigma_g^-$, and $^1\Delta_g$. The lowest three states in O_2 are in fact $X^3\Sigma_g^-$, $a^1\Delta_g$, and $b^1\Sigma_g^+$; the triplet state has the lowest energy according to Hund's rule.

The *nonequivalent* π^2 configuration can arise when one of two valence electrons is in a π_u orbital and the other is in a π_g orbital (e.g., in excited states of C_2). Finding the term symbols arising from this configuration is left as an exercise for the reader.

The atomic SCF calculations described in Section 2.3 may be extended in principle to diatomic molecules with closed-shell electron configurations. The diatomic electronic Hamiltonian in the clamped-nuclei approximation (Eqs. 3.7 and 3.9) may be broken down into a sum of one-electron operators $\hat{H}_i$ and electron repulsion terms $e^2/4\pi\varepsilon_0 r_{ij}$,

$$\hat{H}_{\text{el}} = \sum_{i=1}^{n}\left[-\frac{\hbar^2}{2m_e}\nabla_i^2 - \frac{1}{4\pi\varepsilon_0}\sum_{N=1}^{2}\frac{Z_N e^2}{|\mathbf{r}_i - \mathbf{R}_N|}\right] + \frac{1}{4\pi\varepsilon_0}\sum_{i<j}^{n}\frac{e^2}{|\mathbf{r}_i - \mathbf{r}_j|}$$

$$\equiv \sum_{i=1}^{n} H_i + \frac{1}{4\pi\varepsilon_0}\sum_{i<j}^{n}\frac{e^2}{|\mathbf{r}_i - \mathbf{r}_j|} \tag{4.38}$$

(For stationary nuclei, the nuclear repulsion term $Z_A Z_B e^2/4\pi\varepsilon_0|\mathbf{R}_A - \mathbf{R}_B|$ becomes a constant that may be added to the total energy at the end of the calculation.) The diatomic Hamiltonian (4.38) is identical in form to the many-electron atomic Hamiltonian given in Eq. 2.62. One may write a single-determinant many-electron wave function analogous to Eq. 2.61, with each of the spatial wave functions ϕ_i (with $1 \leqslant i \leqslant n/2$) representing a doubly occupied *molecular* orbital. A set of equations analogous to the Hartree-Fock equation (2.67) may then be solved numerically to determine the best obtainable single-determinant wave function in the form of the Slater determinant (2.61). The discrepancy between the resulting SCF electronic energy (given by Eq. 2.70, with H_{ii} denoting the diagonal matrix element of the one-electron operator $\hat{H}_i$ defined in Eq. 4.38) and the true electronic energy is the correlation error. Such a procedure is so cumbersome in diatomics and polyatomics that relatively few such calculations have been performed in species other than atoms [7]. In atoms, the centrosymmetric one-electron Hamiltonian $\hat{H}_i$ (Eq. 2.48) allows factorization of the atomic orbitals into radial and angular parts. The latter can be represented in atoms by spherical harmonics $Y_{lm}(\theta, \phi)$, and the numerical optimization of the Hartree-Fock wave function becomes confined to the radial coordinate. In diatomics, the angular dependence of molecular orbitals under the cylindrically symmetric Hamiltonian is $e^{i\lambda\phi}$ (Section 4.1), so that numerical SCF calculations must be performed over the two remaining coordinates. In nonlinear polyatomics, the molecular orbitals must generally be optimized over a full three-dimensional grid. Compounding this problem is the fact that SCF energies must also be calculated over a grid of molecular geometries (bond lengths and bond angles) to search for an equilibrium geometry.

More commonly, the molecular orbitals in a single-determinant wave function are expressed as linear combinations of atomic orbitals (LCAOs),

$$|\phi_i\rangle = \sum_j |\chi_j\rangle a_{ji} \tag{4.39}$$

where the $|\chi_j\rangle$ are atomic orbitals (AOs) centered on the nuclei and the a_{ji} are expansion coefficients. A *minimal basis* set of AOs includes all AOs that are occupied in the separated constituent atoms. The coefficients a_{ji} may be varied to minimize the energy. Parameters in the AOs themselves may also be varied, but these are frequently fixed at values established by prior experience with the same atoms in similar molecules. STOs (Eq. 2.58) may be used for the AOs in Eq. 4.39. However, numerical calculation of many of the resulting matrix elements $\langle i|\hat{H}_i|i\rangle$, $\langle ij|ij\rangle$, and $\langle ij|ji\rangle$ is slow using STOs, and the use of Gaussian type orbitals (GTOs) of the form

$$|G_{nlm}\rangle = Nr_i^n \exp(-\sigma r_i^2) Y_{lm}(\theta, \phi) \tag{4.40}$$

is far more economical [7,8]. For this reason, LCAO–MO–SCF calculations have sometimes employed AOs obtained by expressing STOs with known parameters as linear combinations of several GTOs. The inconvenience of evaluating the resulting larger number of Hamiltonian matrix elements is more than offset by their efficiency of calculation using GTOs.

Improved accuracy may be obtained by using expanded basis sets in single-determinant wave functions, but such calculations still do not remove the correlation error associated with representing the electron–electron repulsion $1/r_{ij}$ by the time-averaged expectation value $\langle 1/r_{ij}\rangle$. The configuration interaction technique, which is analogous to that described in Section 2.3 for atoms, begins with a many-electron wave function consisting of a superposition of the closed-shell determinant and additional determinants in which electrons are promoted to unoccupied orbitals,

$$\psi(1, 2, \ldots, n) = \sum_N C_N \Delta_N \tag{4.41}$$

The coefficients C_N are obtained in a variational calculation. In a multiconfigurational self-consistent field (MCSCF) calculation, these expansion coefficients are simultaneously varied with the parameters a_{ji} of the basis functions $|\phi_i\rangle$ in Eq. 4.39. For detailed discussions of electronic structure calculations in molecules, the reader is referred to W. H. Flygare, *Molecular Structure and Dynamics* (Prentice-Hall, Englewood Cliffs, NJ, 1978) and references therein.

4.4 ELECTRONIC SPECTRA OF DIATOMICS

It was emphasized in the introduction to this chapter that molecular electronic transitions are generally accompanied by simultaneous changes in vibrational and rotational states. A calculation of the transition energy of a particular spectroscopic line thus requires knowledge of the rovibrational energy for a diatomic with vibrational and rotational quantum numbers v'' and J'' in the lower diatomic state

$$\begin{aligned} E(v'', J'') &= \omega_e''(v'' + \tfrac{1}{2}) - \omega_e'' x_e''(v'' + \tfrac{1}{2})^2 + \cdots \\ &\quad + B_{v''}'' J''(J'' + 1) + D_{v''}'' J''^2(J'' + 1)^2 + \cdots \\ &\equiv G''(v'') + F_v''(J'') \end{aligned} \tag{4.42}$$

along with the total energy

$$E(v', J') = G'(v') + F_{v'}(J') + T_e \tag{4.43}$$

in the upper electronic state. T_e is the energy separation (conventionally in cm^{-1}) between the minima in the two electronic state potential energy curves. (If the upper electronic state is purely repulsive, its separated-atom asymptote is used to calculate T_e.) The transition frequency in cm^{-1} is then

$$\bar{\nu} = T_e + [G'(v') - G''(v'')] + [F_{v'}(J') - F_{v''}(J'')] \tag{4.44}$$

which shows that numerous combinations of (v', J') and (v'', J'') can add rich structure to electronic spectra in molecules. We will see later that the E1 rotational selection rules are reminiscent of the ones we derived for pure rotational and vibration–rotation spectra (although electronic state symmetry must be carefully considered, using the Herzberg diagrams introduced in Section 4.6). Since the upper and lower electronic states have *unrelated* vibrational potentials, it will turn out that *all* $\Delta v = v' - v''$ are $E1$-allowed a priori in electronic transitions.

In the Born-Oppenheimer approximation, the total diatomic wave functions in the upper and lower states are

$$|\Psi'(\mathbf{r}, \mathbf{R})\rangle = |\psi_{\text{el}}'(\mathbf{r}; \mathbf{R})\rangle |\chi_{v'}(R)\chi_{J'}(\hat{R})\rangle \tag{4.45}$$

and

$$|\Psi''(\mathbf{r}, \mathbf{R})\rangle = |\psi_{\text{el}}''(\mathbf{r}, \mathbf{R})\rangle |\chi_{v''}(R)\chi_{J''}(\hat{R})\rangle \tag{4.46}$$

Since changes in rotational and vibrational as well as electronic state are possible, we must consider both the **r**- and **R**- dependence in the total dipole moment operator when calculating E1 transition probabilities,

$$\boldsymbol{\mu} = -\sum_i e\mathbf{r}_i + \sum_N eZ_N\mathbf{R}_N = \boldsymbol{\mu}_{\text{el}} + \boldsymbol{\mu}_{\text{nucl}} \tag{4.47}$$

For E1 transitions we must then have

$$\begin{aligned} \langle \Psi'|\boldsymbol{\mu}|\Psi'' \rangle &= \langle \psi'_{\text{el}}\chi_{v'}|\boldsymbol{\mu}_{\text{el}} + \boldsymbol{\mu}_{\text{nucl}}|\psi''_{\text{el}}\chi_{v''}\rangle \\ &= \langle \psi'_{\text{el}}|\psi''_{\text{el}}\rangle\langle\chi_{v'}|\boldsymbol{\mu}_{\text{nucl}}|\chi_{v''}\rangle + \langle\psi'_{\text{el}}\chi_{v'}|\boldsymbol{\mu}_{\text{el}}|\psi''_{\text{el}}\chi_{v''}\rangle \\ &\neq 0 \end{aligned} \tag{4.48}$$

These expressions ignore rotation, which is considered in Section 4.6. The term proportional to $\langle\psi'_{\text{el}}|\psi''_{\text{el}}\rangle$ vanishes due to orthogonality of the electronic states. Note that $\langle\psi'_{\text{el}}\chi_{v'}|\boldsymbol{\mu}_{\text{el}}|\psi''_{\text{el}}\chi_{v''}\rangle$ does not factor into $\langle\psi'_{\text{el}}|\boldsymbol{\mu}_{\text{el}}|\psi''_{\text{el}}\rangle\langle\chi_{v'}|\chi_{v''}\rangle$, since the electronic states $|\psi'_{\text{el}}\rangle$ and $|\psi''_{\text{el}}\rangle$ depend parametrically on **R** as well as on **r**. Instead, this matrix element is

$$\int dR\chi^*_{v'}(R)\chi_{v''}(R)\int dr\psi'^*_{\text{el}}(\mathbf{r};\mathbf{R})\boldsymbol{\mu}_{\text{el}}\psi''_{\text{el}}(\mathbf{r}\cdot\mathbf{R}) \equiv \int dR\chi^*_{v'}(R)\chi_{v''}(R)\mathbf{M}_e(R) \tag{4.49}$$

where $\mathbf{M}_e(R)$ is the (R-dependent) electronic transition moment *function.* If $\mathbf{M}_e(R)$ varies slowly over the range of R for which $\chi_{v'}(R)$ and $\chi_{v''}(R)$ are substantial, it becomes meaningful to write

$$\begin{aligned} \langle\Psi'(\mathbf{r},\mathbf{R})|\boldsymbol{\mu}|\Psi''(\mathbf{r},\mathbf{R})\rangle &= \overline{\mathbf{M}_e(R)}\int dR\chi^*_{v'}(R)\chi_{v''}(R) \\ &= \overline{\mathbf{M}_e(R)}\langle v'|v''\rangle \end{aligned} \tag{4.50}$$

where $\overline{\mathbf{M}_e(R)}$ is an averaged value of the electronic transition moment function. The probability of the electronic transition is then proportional to

$$|\langle\Psi'(\mathbf{r},\mathbf{R})|\boldsymbol{\mu}|\Psi''(\mathbf{r},\mathbf{R})\rangle|^2 = |\overline{\mathbf{M}_e(R)}|^2|\langle v'|v''\rangle|^2 \tag{4.51}$$

where $|\langle v'|v''\rangle|^2$ is called the *Franck-Condon factor* for the $v'' \to v'$ vibrational band of the electronic transition (we allude to it as a band, since it exhibits rotational structure in the gas phase as shown for Na_2 in Fig. 4.3). Since the vibrational states $|v'\rangle$ and $|v''\rangle$ belong to potential energy functions having different shape, the Franck-Condon factor $|\langle v'|v''\rangle|^2$ can assume *any* value $\leqslant 1$ regardless of v', v''. This is typified by the seemingly random vibrational band intensities in the Na_2 fluorescence spectrum of Fig. 4.1. It is only in the special limit where the upper and lower electronic states have *identical* potential energy curves that one obtains $|\langle v'|v''\rangle|^2 = \delta_{v',v''}$, since it is only then that the electronic states have identical orthonormal sets of vibrational states.

The Franck-Condon factors do obey a sum rule, however. If one sums the Franck-Condon factors for transitions from a particular vibrational level v'' in the lower electronic state to the complete set of levels v' in the upper electronic state, one obtains

$$\sum_{v'} |\langle v'|v''\rangle|^2 = \sum_{v'} \langle v''|v'\rangle\langle v'|v''\rangle = \langle v''|v''\rangle$$
$$= 1 \tag{4.52}$$

by closure. This implies that summing the vibrational band intensities over an electronic band spectrum allows direct measurement of the averaged electronic transition moment function $\mathbf{M}_e(R)$ according to Eq. 4.51.

We now turn to the E1 selection rules embodied in the electronic transition moment

$$\mathbf{M}_e(R) = \langle \psi'_{\text{el}}|\boldsymbol{\mu}_{\text{el}}|\psi''_{\text{el}}\rangle \tag{4.53}$$

To have a nonvanishing matrix element (4.53), it is necessary for the direct product of irreducible representations

$$\Gamma(\psi'_{\text{el}}) \otimes \Gamma(\boldsymbol{\mu}_{\text{el}}) \otimes \Gamma(\psi''_{\text{el}})$$

to contain the totally symmetric irreducible representation (Σ^+ in $C_{\infty v}$, Σ_g^+ in $D_{\infty h}$). Since the components $(\mu_{\text{el}})_{x,y}$ and $(\mu_{\text{el}})_z$ of the dipole moment operator transform as Π_u and Σ_u^+ in $D_{\infty h}$, the direct products

$$\Gamma(\psi'_{\text{el}}) \otimes \begin{pmatrix} \Pi_u \\ \Sigma_u^+ \end{pmatrix} \otimes \Gamma(\psi''_{\text{el}})$$

must contain the Σ_g^+ representation for E1 transitions in homonuclear diatomics. (The g, u subscripts may be dropped to generalize this discussion to heteronuclear diatomics.) In many homonuclear diatomics, the electronic ground state $|\psi''_{\text{el}}\rangle$ is $X^1\Sigma_g^+$. Then electronic transitions from the ground state to state $|\psi''_{\text{el}}\rangle$ are E1-allowed if

$$\Gamma(\psi'_{\text{el}}) \otimes \begin{pmatrix} \Pi_u \\ \Sigma_u^+ \end{pmatrix} \otimes \Sigma_g^+$$

contains Σ_g^+, which happens only when $\Gamma(\psi''_{\text{el}})$ is either Π_u or Σ_u^+. Working out similar direct products for electronic states $|\psi'_{\text{el}}\rangle$, $|\psi''_{\text{el}}\rangle$ of *arbitrary* symmetry would show that the E1 selection rules for electronic transitions in diatomics are

$$\Delta\Lambda = 0, \pm 1$$

$$+ \not\leftrightarrow - \text{ in } \Sigma - \Sigma \text{ transitions}$$

$$\text{u} \longleftrightarrow g,\ g \not\leftrightarrow \text{g},\ \text{u} \not\leftrightarrow \text{u} \qquad \text{(Laporte rule)}$$

Since $\boldsymbol{\mu}_{el}$ does not operate on the spin coordinates, we also have the E1 spin selection rules

$$\Delta S = 0$$

$$\Delta \Sigma = 0$$

for small $\hat{H}_{so}$. When the spin–orbit coupling is large, the selection rule becomes $\Delta\Omega = 0, \pm 1$. Examples of E1-allowed transitions under these selection rules are $^2\Pi_{1/2} - {}^2\Pi_{1/2}$, $^3\Pi_1 - {}^3\Pi_1$, $^2\Delta_{5/2} - {}^2\Delta_{5/2}$, $^3\Pi_2 - {}^3\Delta_3$; a pair of E1-forbidden transitions would be $^3\Pi_1 - {}^3\Pi_2$, $^2\Pi_{3/2} - {}^2\Pi_{1/2}$. When $\Delta\Lambda = 0$, the z component of $\langle\psi'_{el}|\boldsymbol{\mu}_{el}|\psi''_{el}\rangle$ is nonvanishing, and the electronic transition is said to be *parallel* (i.e., polarized along the molecular axis). When $\Delta\Lambda = \pm 1$, the x and/or y components of the transition moment are nonzero, and the transition is termed *perpendicular*. The formalism developed in Chapter 1 implies that to effect an E1 parallel absorptive transition in a diatomic molecule, the electric field vector **E** of the incident electromagnetic wave must have a component along the molecular axis (the E1 transition probability amplitude is proportional to $\mathbf{E}\cdot\langle\psi'_{el}|\boldsymbol{\mu}|\psi''_{el}\rangle$). Perpendicular transitions require the presence of an **E** component normal to the molecular axis.

For M1 electronic transitions in diatomics, we inspect the matrix elements $\langle\psi'_{el}|\mathbf{L}|\psi''_{el}\rangle$, where **L** is the orbital angular momentum operator. Since the components of **L** transform as the rotations (R_x, R_y, R_z), the direct product

$$\Gamma(\psi'_{el}) \otimes \begin{pmatrix} \Gamma(R_x, R_y) \\ \Gamma(R_z) \end{pmatrix} \otimes \Gamma(\psi''_{el}) = \Gamma(\psi'_{el}) \otimes \begin{pmatrix} \Pi_g \\ \Sigma_g^- \end{pmatrix} \otimes \Gamma(\psi''_{el}) \tag{4.54}$$

must contain the totally symmetric irreducible representation for M1-allowed transitions. This is satisfied only if

$$\Delta\Lambda = 0, \pm 1$$

$$\text{u} \longleftrightarrow \text{u},\ \text{g} \longleftrightarrow \text{g},\ \text{u} \not\leftrightarrow \text{g} \qquad \text{(anti-Laporte)}$$

$$+ \longleftrightarrow -,\ + \not\leftrightarrow +,\ - \not\leftrightarrow - \text{ in } \Sigma - \Sigma \text{ transitions}$$

$$\Delta S = 0$$

It may similarly be shown that for E2 electronic transitions in diatomics one has the selection rules

$$\Delta\Lambda = 0, \ \pm 1, \ \pm 2$$

$$u \longleftrightarrow u, \ g \longleftrightarrow g, \ u \not\longleftrightarrow g \qquad \text{(anti-Laporte)}$$

$$\Delta S = 0$$

The $A^1\Sigma_u^+$ and $B^1\Pi_u$ states in Na_2 (Fig. 4.8) are two of the best-characterized diatomic states in the literature because the $A^1\Sigma_u^+ \leftarrow X^1\Sigma_g^+$ and $B^1\Pi_u \leftarrow X^1\Sigma_g^+$ transitions are E1-allowed, because these transitions occur at visible wavelengths easily generated by tunable lasers, and because Na_2 vapor is easily made. The $^3\Sigma_g^+$ and $^1\Pi_g$ Na_2 bound states correlating with $Na(3^2S) + Na(3^2P)$ are less well known, because the E1 transitions to these states from the ground state are symmetry forbidden and because the M1 and E2 transitions (symmetry-allowed for $^1\Pi \leftarrow {}^1\Sigma_g^+$) are far weaker than E1 transitions (Chapter 1). The best-known electronic transitions in I_2 are the $A^3\Pi_{1u} \leftarrow X^1\Sigma_g^+$ (red-IR) and $B^3\Pi_{0u} \leftarrow X^1\Sigma_g^+$ (green) transitions (Fig. 4.7); the latter is responsible for the purple color in I_2. They are E1-forbidden according to the selection rules ($\Delta S \neq 0$), but they gain small E1 intensity because the large spin–orbit coupling in I_2 causes considerable admixtures of singlet character into the A and B states and triplet character into the X state [9]. These transitions do obey the selection rule on $\Delta\Omega$ ($= +1$ and 0, respectively). The $^3\Pi_2 \leftarrow X^1\Sigma_g^+$ transition ($\Delta\Omega = +2$) is extremely weak, on the other hand, because it violates the $\Delta\Omega = 0, \pm 1$ selection rule.

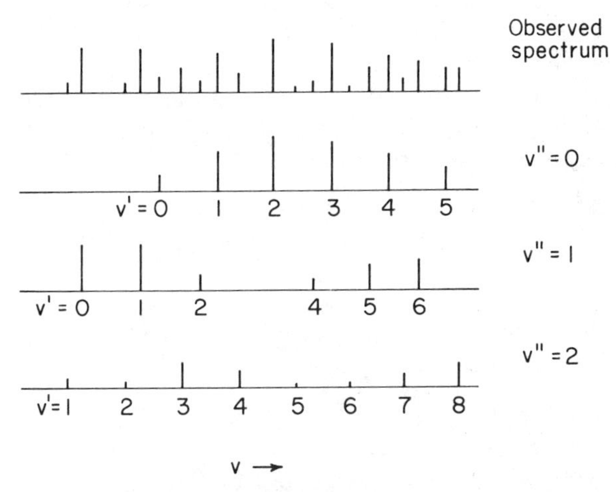

Figure 4.16 The observed $X^1\Sigma_g^+ \rightarrow A^1\Sigma_u^+$ electronic band absorption spectrum of a high-temperature Na_2 vapor is a superposition of spectra originating from $v'' = 0, 1, 2, \ldots$; only the first three series are shown. While the band intensities arising from a particular v'' level show systematic variations (in fact, their intensity envelopes exhibit the same number of nodes as the vibrational state $|v''\rangle$), this regularity is not apparent in the composite absorption spectrum.

The large number of $v' \leftarrow v''$ vibrational bands observed in an electronic transition can complicate the analysis of electronic band spectra, particularly if several of the initial v'' levels have appreciable populations (e.g., when $\omega_e'' \lesssim kT/hc$ at thermal equilibrium). Neglecting rotational fine structure, $\bar{\nu}$ becomes equal to $T_e + G'(v') - G''(v'')$. If we should superimpose the spectra arising from, say, $v'' = 0$, 1, and 2 and recall that the $v' \leftarrow v''$ band intensities are weighted by (seemingly) random Franck-Condon factors, the resulting electronic band spectrum (Fig. 4.16) shows no comprehensible patterns, even though the regularity would be apparent if only a single progression of bands (e.g., from $v'' = 0$) were present. The Ritz combination principle offers a brute-force method of assigning (v', v'') combinations to the bands: one obtains all possible difference frequencies between pairs of band frequencies $\bar{\nu}(v', v'')$ to determine if particular difference frequencies crop up repeatedly. For example, $[\bar{\nu}(3,0) - \bar{\nu}(2,0)]$ and $[\bar{\nu}(3,1) - \bar{\nu}(2,1)]$ and $[\bar{\nu}(3,2) - \bar{\nu}(2,2)]$ must all equal $G'(3) - G'(2)$, and this gives a start in organizing the assignment of the spectrum. This approach is known as a Deslandres analysis (Problem 4.6). A preferable approach is to simplify the spectrum experimentally: the use of supersonic jets produces gases with very low vibrational temperatures, essentially populating only $v'' = 0$ (Fig. 4.3).

4.5 ANGULAR MOMENTUM COUPLING CASES

The expressions we derived for diatomic rotational energy levels in Chapter 3 are applicable only to molecules in $^1\Sigma$ states. In such molecules, the magnitude $|\mathbf{J}|$ of the rotational angular momentum is conserved, and the rotor energy levels ($E_J = BJ(J+1)$ in the absence of centrifugal distortion and vibration–rotation interaction) are those of a freely rotating molecule whose nuclear angular momentum is uncoupled to any other angular momenta. In more general cases where Λ and/or S is nonzero, it is the magnitude of the *total* (electronic plus rotational) rather than just rotational angular momentum that is conserved. Several coupling schemes for spin, orbital, and rotational angular momenta may be identified, depending on the magnitudes of the magnetic field generated by the electrons' orbital motion and the spin–orbit coupling.

Hund's case (a) describes the majority of molecules with $\Lambda \neq 0$ that exhibit small spin–orbit coupling. Since the electronic orbital and spin angular momenta **L** and **S** are not mutually strongly coupled, they precess independently about the quantization axis (the molecular axis) established by the magnetic fields arising from electronic motion. The projections $\Lambda\hbar$ and $\Sigma\hbar$ of **L** and **S** respectively along the molecular axis are then conserved, as is their sum $\Omega\hbar$. In contrast, the parts of **L** and **S** normal to the molecular axis oscillate rapidly; they are denoted $\mathbf{L}_\perp$ and $\mathbf{S}_\perp$ (Fig. 4.17).

For the total angular momentum **J**, the quantity $|\mathbf{J}|^2 = J(J+1)\hbar^2$ is necessarily a constant of motion. (**J** was previously used for rotational angular momentum in $^1\Sigma$ state molecules in Chapter 3; it is now reserved for total

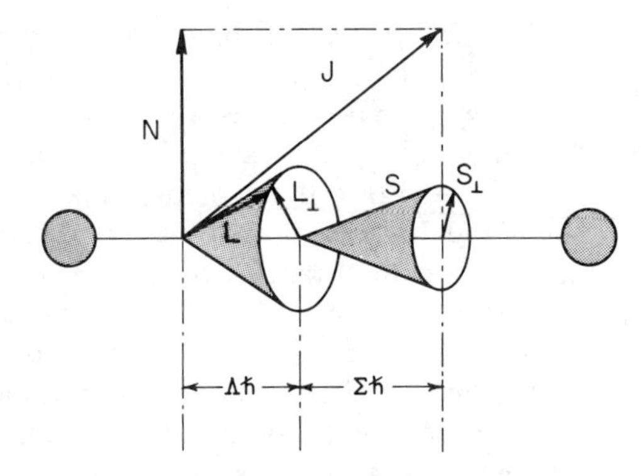

Figure 4.17 Hund's case (a) coupling. The orbital and spin angular momenta **L** and **S** precess rapidly about the molecular axis with fixed projections $\Lambda\hbar$ and $\Sigma\hbar$. The total angular momentum normal to the molecular axis is in effect the rotational angular momentum **N**, since the normal components $\mathbf{L}_\perp$ and $\mathbf{S}_\perp$ of **L** and **S** fluctuate rapidly and their expectation values are zero. The total angular momentum **J** is the vector sum of **N** and **Ω**, the projection of (**L** + **S**) upon the molecular axis. Interactions between $\mathbf{L}_\perp$, $\mathbf{S}_\perp$ and the rotational angular momentum **N** give rise to Λ-doubling (see text).

(electronic plus rotational) angular momentum in diatomics with $\Lambda \neq 0$.) Since the rotational angular momentum vector **N** must be normal to the molecular axis, it has no z component—which implies that the projection of **J** along the molecular axis must be $(\mathbf{L} + \mathbf{S})_z = \Omega\hbar$ (Fig. 4.17). According to the inequality $|\mathbf{J}|^2 \geqslant J_z^2$, we must then have $J(J+1)\hbar^2 \geqslant \Omega^2\hbar^2$, or $J \geqslant \Omega$. Hence for given Ω, the allowed J quantum numbers for the total angular momentum are

$$J = \Omega, \Omega + 1, \Omega + 2, \ldots \tag{4.55}$$

We finally compute the rotational energy $E_{\text{rot}} = |\mathbf{N}|^2/2I$ for a case (a) molecule. Since the portion $\mathbf{J}_\perp$ of J which lies in a plane perpendicular to the molecular axis is

$$\mathbf{J}_\perp = (\mathbf{L} + \mathbf{S})_\perp + \mathbf{N} \tag{4.56}$$

we have

$$\begin{aligned} \mathbf{J}^2 &= \mathbf{J}_\perp^2 + J_z^2 \\ &= (\mathbf{L}_\perp + \mathbf{S}_\perp)^2 + \mathbf{N}^2 + 2\mathbf{N}\cdot(\mathbf{L}_\perp + \mathbf{S}_\perp) + J_z^2 \end{aligned} \tag{4.57}$$

The classical rotational energy is then

$$E_{\text{rot}} = [\mathbf{J}^2 - J_z^2 - (\mathbf{L}_\perp + \mathbf{S}_\perp)^2 - 2\mathbf{N}\cdot(\mathbf{L}_\perp + \mathbf{S}_\perp)]/2I \tag{4.58}$$

The quantum analog of this expression for the rotational energy in cm^{-1} is (in the absence of centrifugal distortion)

$$E_{\text{rot}} = B_v[J(J+1) - \Omega^2 - \langle(\mathbf{L}_\perp + \mathbf{S}_\perp)^2\rangle - 2\langle\mathbf{N}\cdot(\mathbf{L}_\perp + \mathbf{S}_\perp)\rangle] \qquad (4.59)$$

where the brackets $\langle\ \rangle$ denote expectation values. The terms $\Omega^2 + \langle(\mathbf{L}_\perp + \mathbf{S}_\perp)^2\rangle$ are often lumped with the vibronic energy, in which case the rotational energy is

$$E_{\text{rot}} = B_v J(J+1) - 2B_v\langle\mathbf{N}\cdot(\mathbf{L}_\perp + \mathbf{S}_\perp)\rangle \qquad (4.60)$$

The first term resembles the rotational energy of a $^1\Sigma$ molecule, but is subject to the restriction $J \geqslant \Omega$. The second term describes the *electronic–rotational interactions*, which constitute a small perturbation to the rotational energy in molecular states with $\Lambda \neq 0$. In the absence of electronic–rotational interactions, such states are doubly degenerate with components exhibiting + and − behavior under σ_v (Section 4.1). This degeneracy is split by the perturbation $B_v\mathbf{N}\cdot(\mathbf{L}_\perp + \mathbf{S}_\perp)$, producing closely spaced doublets of rotational energy levels with opposite reflection symmetry (Figure 4.18). This phenomenon is known as *Λ-doubling.*

For atoms with Russell-Saunders coupling (those in which $\hat{H}_{\text{so}}$ can be treated as a perturbation) the spin–orbital correction to the energy was seen to behave

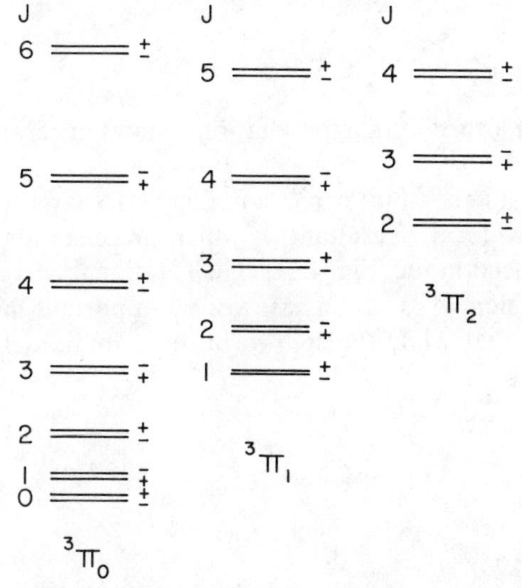

Figure 4.18 Rotational energy levels in Hund's case (a) for a $^3\Pi_0$ state, a $^3\Pi_1$ state, and a $^3\Pi_2$ state. No levels appear with $J < \Omega$. Each J level is split by Λ-doubling into sublevels with opposite reflection symmetry. The Ω-doubling is greatly exaggerated.

as $A[J(J+1) - L(L+1) - S(S+1)]$. The corresponding spin–orbital energy in diatomics has the form $A\Omega^2$, where A is a spin–orbital constant [10]. The total energy in excess of the vibronic energy (cf. Eq. 4.59) then becomes

$$E_{\text{rot}} = B_v[J(J+1) - \langle(\mathbf{L}_\perp + \mathbf{S}_\perp)^2\rangle - 2\langle\mathbf{N}\cdot(\mathbf{L}_\perp + \mathbf{S}_\perp)\rangle] + (A - B_v)\Omega^2 \tag{4.61}$$

The differences arising from the $(A - B_v)\Omega^2$ term in the rotational energies of spin–orbital components with different Ω are reflected in the case (a) energy level diagram in Fig. 4.18.

In *Hund's case* (b) the magnetic fields due to orbiting electrons are so weak that the electron spin angular momentum $\mathbf{S}$ does not precess about the molecular axis. Molecules with $\Lambda = 0$ (Σ states) and $\mathbf{S} \neq 0$ belong to case (a) by default, but light molecules (with few electrons) having $\Lambda \neq 0$ may also exhibit case (b) coupling. For simplicity, we restrict our discussion to case (b) molecules with $\Lambda = 0$. The rotational angular momentum $\mathbf{N}$ is directed normal to the molecule axis. In the absence of orbital angular momentum, the local magnetic field is dominated by molecular rotation, with the result that $\mathbf{S}$ precesses about $\mathbf{J}$ (Fig. 4.19). The rotational energies to zeroth order are

$$E_{\text{rot}} = B_v N(N+1) \tag{4.62}$$

with $N = 0, 1, 2, \ldots$ Each rotational level N may be split by spin–rotation coupling into sublevels with total angular momentum quantum number

$$J = N + S, N + S - 1, \ldots, |N - S| \tag{4.63}$$

A schematic energy level diagram for rotational states in $^2\Sigma$ and $^3\Sigma$ molecules is shown in Fig. 4.20.

In *Hund's case* (c) the spin–orbit coupling is so large that $\mathbf{L}$ and $\mathbf{S}$ are mutually coupled to form a resultant $\mathbf{J}_a$ which precesses about the molecular axis with fixed projection $\hbar\Omega$ (Fig. 4.21). The rotational energy levels are given by the same expression, (4.61), as in case (a). An important distinction between cases (a) and (c) is that while the spin–orbit contribution $A\Omega^2$ to Eq. 4.61 is

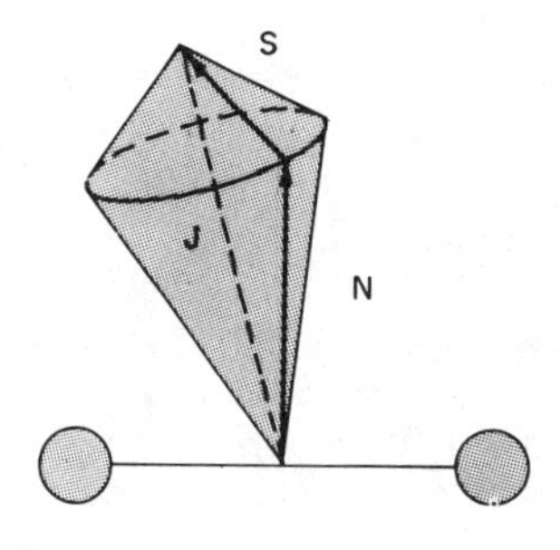

Figure 4.19 Hund's case (b) for $\Lambda = 0$. The spin angular momentum **S** and the rotational angular momentum **N** precess about the total angular momentum **J**.

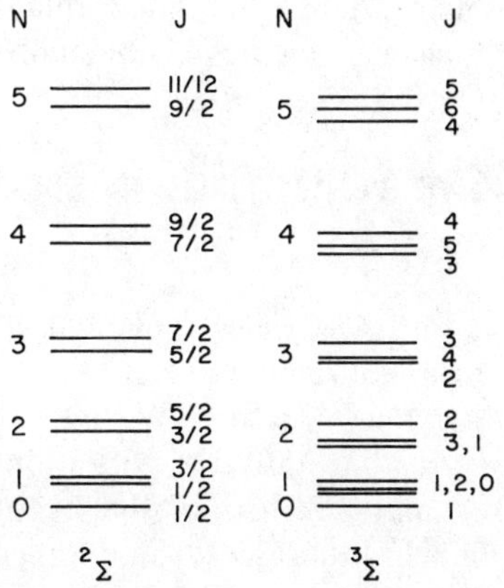

N	J ($^2\Sigma$)	N	J ($^3\Sigma$)
5	11/12, 9/2	5	5, 6, 4
4	9/2, 7/2	4	4, 5, 3
3	7/2, 5/2	3	3, 4, 2
2	5/2, 3/2	2	2, 3,1
1	3/2, 1/2	1	1,2,0
0	1/2	0	1

Figure 4.20 Rotational energy levels in Hund's case (b) for a $^2\Sigma$ state and for a $^3\Sigma$ state. Each N level (Eq. 4.62) is split into $2S + 1$ sublevels through the interaction between the rotational angular momentum **N** and the spin angular momentum **S**.

generally the order of the rotational level spacing or smaller in case (a), it becomes large enough in case (c) to endow the different Ω components of $^{2S+1}\Lambda_\Omega$ states with individual potential energy curves. This is what happens in the $B^3\Pi_{ou}$, $A^3\Pi_{1u}$, and $^3\Pi_{2u}$ states of I_2 (Fig. 4.7), and these are good examples of case (c) coupling in a heavy molecule with large spin–orbit coupling. Since **L** and **S** do not precess independently about the molecular axis in case (c)

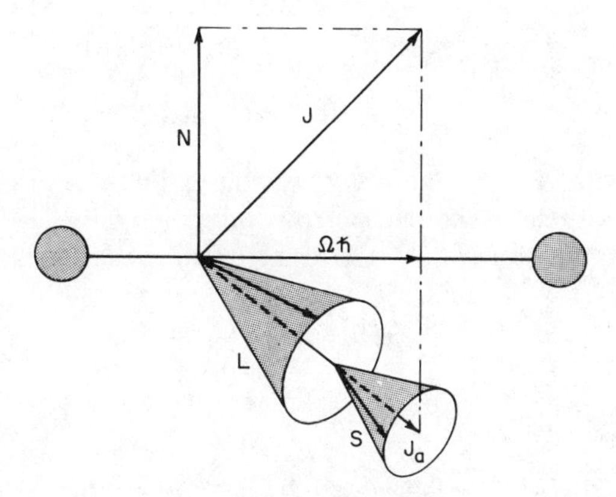

Figure 4.21 Hund's case (c) coupling. The orbital and spin angular momenta **L** and **S** are strongly coupled to form a resultant $\mathbf{J}_a$. The projection $\boldsymbol{\Omega}$ of $\mathbf{J}_a$ on the molecular axis and the rotational angular momentum **N** combine to form the total angular momentum **J**. **L** and **S** clearly do not have well-defined projections on the molecular axis, and so case (c) molecular states cannot be strictly characterized in terms of the quantum numbers Λ and Σ.

molecules, Λ and Σ (unlike Ω) are not good quantum numbers, and the designation of these I_2 states as $^3\Pi$ states is not rigorous.

4.6 ROTATIONAL FINE STRUCTURE IN ELECTRONIC BAND SPECTRA

In Section 4.4 we worked out the E1 electronic and vibrational selection rules for electronic band spectra, and it remains for us to determine the selection rules that govern the rotational fine structure. We have seen that no symmetry selection rule exists for Δv, but that the vibrational band intensities are proportional to Franck-Condon factors in the Born-Oppenheimer approximation. To understand the selection rules for simultaneous changes in electronic and rotational state, we must find how $|\psi_{el}\chi_{rot}\rangle = |\psi_{el}\rangle|JM\rangle$ transforms under the symmetry operations of the molecule. A complication arises here because we have derived our earlier pure rotational and vibration–rotation selection rules using *space-fixed* coordinates (i.e., the rotational state $|JM\rangle$ depends on angles (θ, ϕ) relative to Cartesian coordinates (x, y, z), which are fixed in space and do not rotate with the molecule), whereas our group theoretical E1 electronic selection rules for $\langle\psi'_{el}|\boldsymbol{\mu}|\psi''_{el}\rangle$ were derived using electronic states $|\psi'_{el}\rangle$, $|\psi''_{el}\rangle$, which depend on *molecule-fixed* coordinates (r_{Ai}, r_{Bi}, ϕ) for each electron i; these coordinates do rotate with the molecule. To consider simultaneous rotational and electronic state changes, we must use *one* set of coordinates consistently.

In *heteronuclear* diatomics $|\psi_{el}\chi_{rot}\rangle$ can be classified according to its behavior under molecule-fixed reflection σ_v. We already know that the molecule-fixed states $|\psi_{el}\rangle$ obey

$$\begin{aligned}\sigma_v|\psi_{el}\rangle &= |\psi_{el}\rangle \qquad \text{for } \Sigma^+ \text{ states}\\ &= -|\psi_{el}\rangle \qquad \text{for } \Sigma^- \text{ states}\end{aligned}$$

and that states with $\Lambda \neq 0$ are doubly degenerate (at least prior to Λ-doubling) with components that can be chosen to exhibit $+$ and $-$ behavior under σ_v. While the rotational states $|JM\rangle$ are space-fixed, it can be shown that [11]

$$\sigma_v \text{ (molecule-fixed)} = i \text{ (space-fixed)} \tag{4.64}$$

with the result that [12]

$$\begin{aligned}\sigma_v\text{(molecule-fixed)}|JM\rangle &= i\text{(space-fixed)}Y_{JM}(\theta, \phi) = Y_{JM}(\pi - \theta, \phi + \pi) \\ &= (-1)^J|JM\rangle\end{aligned} \tag{4.65}$$

Consequently,

$$\sigma_v\text{(molecule-fixed)}|\psi_{el}\chi_{rot}\rangle = (\pm)(-^1)^J|\psi_{el}\chi_{rot}\rangle \tag{4.66}$$

where the upper (+) sign applies to Σ^+ states and the lower (−) sign applies to Σ^- states. In the remainder of this section we use (+) and (−) to indicate the behavior of $|\psi_{el}\chi_{rot}\rangle$ under σ_v (molecule-fixed); this should not be confused with the superscripts in the electronic state notations Σ^+, Σ^-. We finally determine how $\boldsymbol{\mu}$ is influenced by σ_v (molecule-fixed). Since this operation is equivalent to i (space-fixed), and the latter operation converts $\boldsymbol{\mu}$ into $-\boldsymbol{\mu}$,

$$\sigma_v \text{ (molecule-fixed)}\boldsymbol{\mu} = -\boldsymbol{\mu} \tag{4.67}$$

It follows that for $\langle\psi'_{el}\chi'_{rot}|\boldsymbol{\mu}|\psi''_{el}\chi''_{rot}\rangle$ to be nonvanishing, the states $|\psi'_{el}\chi'_{rot}\rangle$ and $|\psi''_{el}\chi''_{rot}\rangle$ must exhibit opposite reflection behavior under σ_v (molecule-fixed), and (+) states can combine only with (−) states in the E1 approximation. To keep track of all rotational transitions that are consistent with this rule for a given electronic transition, we set up a *Herzberg diagram* [10] in which the rotational J levels in the relevant electronic states are labeled with their behavior under σ_v (molecule-fixed). For definiteness, suppose we are interested in a $^1\Sigma^+ \rightarrow {}^1\Sigma^+$ electronic transition (which is E1-allowed according to the selection rules obtained in Section 4.4). Using Eq. 4.66, the pertinent Herzberg diagram can be generated as shown in Fig. 4.22, where the arrows show all allowed transitions subject to $\Delta J = 0, \pm 1$ connecting levels that are (+) under σ_v (molecule-fixed) with levels that are (−). This diagram exhibits only R- and P-branch transitions. No Q-branch ($\Delta J = 0$) transitions are possible, since they would connect (+) with (+) or (−) with (−) levels. A more interesting example is the case of a $^1\Sigma^+ \rightarrow {}^1\Pi$ transition. The pertinent Herzberg diagram (Fig. 4.23) reflects the fact that doubly degenerate $^1\Pi$ states can have both (+) and (−) components under σ_v (molecule-fixed) for *any* J, but cannot have $J = 0$. (Recall that we require $J \geqslant \Omega$, and $\Omega = 1$ in a $^1\Pi$ state.) In this case $\Delta J = 0$ transitions are possible as well as $\Delta J = \pm 1$ transitions, so Fig. 4.23 provides an example of the fact that Q branches can appear in electronic transitions involving states with $\Lambda \neq 0$.

In *homonuclear* diatomics an additional symmetry element is needed to classify state symmetries in the $D_{\infty h}$ point group, and we can use i (molecule-fixed) for this purpose. A new complication, peculiar to homonuclear molecules

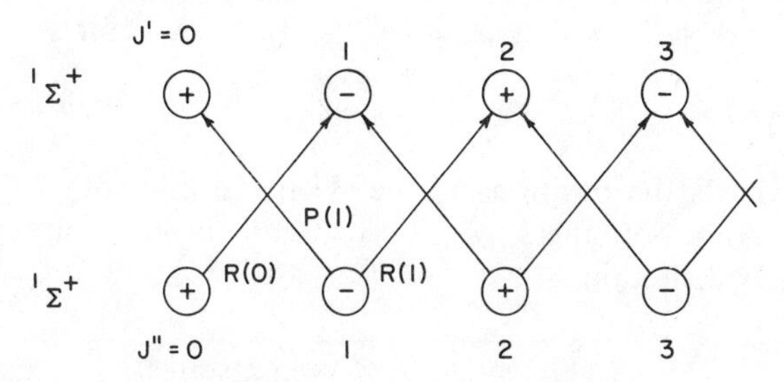

Figure 4.22 Herzberg diagram for allowed rotational transitions accompanying a $^1\Sigma^+ \rightarrow {}^1\Sigma^+$ electronic transition in a heteronuclear diatomic molecule.

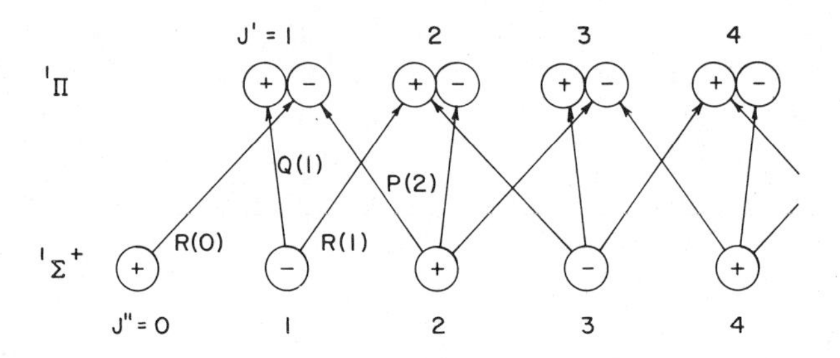

Figure 4.23 Allowed rotational transitions in a $^1\Sigma^+ \rightarrow {}^1\Pi$ electronic transition. Note the absence of the Q(0) and P(1) transitions, since $J' \geqslant 1$ in a Π state.

in which both nuclei are the same isotope, is the effect of *nuclear exchange symmetry* on rotational state populations. Under the space-fixed nuclear exchange operation X_N, the total wave function $|\Psi\rangle = |\psi_{\mathrm{el}}\chi_{\mathrm{vib}}\chi_{\mathrm{rot}}\psi_{\mathrm{nucl.spin}}\rangle$ must behave as [10]

$$\begin{aligned} X_N|\Psi\rangle &= +|\Psi\rangle \qquad \text{for boson nuclei (s)} \\ &= -|\Psi\rangle \qquad \text{for fermion nuclei (a)} \end{aligned} \tag{4.68}$$

Since the vibrational state $|\chi_{\mathrm{vib}}\rangle$ depends only on $|\mathbf{R}| = R$, it is unaffected by exchange of the nuclei. To decide the effect of X_N on $|\psi_{\mathrm{el}}\chi_{\mathrm{rot}}\rangle$, we use the identity [10]

$$X_N = i_1 \cdot i_2 \tag{4.69}$$

where i_1 represents inversion of all particles through the origin and i_2 represents inversion of electrons only through the origin. We have already seen that $i_1|\chi_{\mathrm{rot}}\rangle = i_1|JM\rangle = (-1)^J|\chi_{\mathrm{rot}}\rangle$. For the effect of i_1 on $|\psi_{\mathrm{el}}\rangle$, we note that $i_1 \equiv C_2\sigma_v$ in $D_{\infty h}$, where the latter two operations are applied to all particles in the molecule. C_2 then has no effect on $|\psi_{\mathrm{el}}\rangle$, since $|\psi_{\mathrm{el}}\rangle$ depends only on the electronic coordinates *relative* to the positions of the nuclei. Then

$$\begin{aligned} i_1|\psi_{\mathrm{el}}\rangle = \sigma_v|\psi_{\mathrm{el}}\rangle &= |\psi_{\mathrm{el}}\rangle \qquad \text{for } \Sigma^+ \text{ states} \\ &= -|\psi_{\mathrm{el}}\rangle \qquad \text{for } \Sigma^- \text{ states} \end{aligned} \tag{4.70}$$

because σ_v reflects the positions of the electrons, but not of the nuclei. The inversion i_2 does not affect $|\chi_{\mathrm{rot}}\rangle$ because it inverts only the electronic coordinates. By definition,

$$\begin{aligned} i_2|\psi_{\mathrm{el}}\rangle &= |\psi_{\mathrm{el}}\rangle \qquad \text{in g states} \\ &= -|\psi_{\mathrm{el}}\rangle \qquad \text{in u states} \end{aligned} \tag{4.71}$$

We finally consider the effect of X_N on the nuclear spin function $|\psi_{\text{nucl.spin}}\rangle$. Each nucleus has spin I with $(2I + 1)$ magnetic sublevels, giving a total of $(2I + 1)^2$ diatomic nuclear spin states. Of these, $(2I + 1)(I + 1)$ states are always symmetric (s) with respect to X_N, and $(2I + 1)I$ are antisymmetric (a). For concreteness, consider the H_2 molecule in which both nuclei are ^{1}H with spin $I = \frac{1}{2}$. From the nuclear magnetic sublevels α ($m_I = \frac{1}{2}$) and β ($m_I = -\frac{1}{2}$) on the individual nuclei, we may form the symmetric (s) and antisymmetric (a) nuclear spin states

$$\begin{array}{lc} \alpha(1)\alpha(2) & \text{(s)} \\ \alpha(1)\beta(2) + \alpha(2)\beta(1) & \text{(s)} \\ \beta(1)\beta(2) & \text{(s)} \\ \alpha(1)\beta(2) - \alpha(2)\beta(1) & \text{(a)} \end{array}$$

for the diatomic molecule. For $I = \frac{1}{2}$, there are $(2I + 1)(I + 1) = 3$ and $(2I + 1)I = 1$ nuclear spin states that are symmetric (s) and antisymmetric (a), respectively, under X_N.

The possible combinations of electronic, rotational, and nuclear spin states can now be compiled [10] as shown in Table 4.6. The total wave function $|\Psi\rangle$ must be (s) and (a) under the nuclear exchange X_N for boson and fermion nuclei, respectively. Fermion nuclei exhibit half-integral spin: $I = \frac{1}{2}$ (^{1}H, ^{3}He, ^{13}C), $I = \frac{3}{2}$ (^{23}Na), $I = \frac{5}{2}$ (^{17}O), etc. Bosons include integral-spin nuclei, such as ^{4}He, ^{16}O ($I = 0$) and ^{14}N ($I = 1$).

For $^1H_2$ in its $X^1\Sigma_g^+$ ground state, the nuclei are fermions with $I = \frac{1}{2}$. The exchange symmetry of $|\psi_{\text{el}}\chi_{\text{rot}}\rangle$ under X_N is

$$\begin{aligned} X_N|\psi_{\text{el}}\chi_{\text{rot}}\rangle &= i_1 i_2|\psi_{\text{el}}\chi_{\text{rot}}\rangle \\ &= [(-1)^J(+)][+]|\psi_{\text{el}}\chi_{\text{rot}}\rangle \\ &= (-1)^J|\psi_{\text{el}}\chi_{\text{rot}}\rangle \end{aligned} \tag{4.72}$$

so that $|\psi_{\text{el}}\chi_{\text{rot}}\rangle$ is (s) for even J and (a) for odd J. The total wave function $|\Psi\rangle$ must be (a) for these fermion nuclei. According to Table 4.6, the $X^1\Sigma_g^+$ state

Table 4.6

$\lvert\psi_{\text{el}}\chi_{\text{rot}}\rangle$	$\lvert\psi_{\text{nucl.spin}}\rangle$	$\lvert\Psi\rangle$	Statistical weight
a	a	s	$(2I + 1)I$
s	a	a	$(2I + 1)I$
a	s	a	$(2! + 1)(I + 1)$
s	s	s	$(2I + 1)(I + 1)$

rotational level pupulations in a thermal gas gain extra factors (beyond the Boltzmann factors in Eq. 3.28) of $(2I + 1)I = 1$ for even J and $(2I + 1)(I + 1) = 3$ for odd J. This leads to a nearly 3:1 intensity alternation in rotational state populations of adjacent levels in $X^1\Sigma_g^+\ ^1H_2$. For $^{16}O_2$ in its $X^3\Sigma_g^-$ ground state, the nuclei are bosons with $I = 0$. The exchange symmetry of $|\psi_{el}\chi_{rot}\rangle$ in this case becomes

$$\begin{aligned} i_1 i_2 |\psi_{el}\chi_{rot}\rangle &= [(-1)^J(-)][+]|\psi_{el}\chi_{rot}\rangle \\ &= (-1)^{J+1}|\psi_{el}\chi_{rot}\rangle \end{aligned} \tag{4.73}$$

so that $|\psi_{el}\chi_{rot}\rangle$ is (a) for even J and (s) for odd J. Since $|\Psi\rangle$ must be (s) in this molecule, the rotational state populations gain statistical factors of $(2I + 1)I = 0$ for even J and $(2I + 1)(I + 1) = 1$ for odd J. Consequently, $X^3\Sigma_g^-\ ^{16}O_2$ cannot exist in even $-J$ levels at all. Such levels *can* be populated in other electronic states (e.g., a $^1\Delta_g$) of $^{16}O_2$, however.

Prior to digressing on the subject of nuclear exchange symmetry, we mentioned that a new symmetry element besides σ_v (molecule-fixed) was required to classify electronic–rotational states in homonuclear diatomics. A logical choice is i (molecule-fixed), an operation which belongs to $D_{\infty h}$ but not $C_{\infty v}$. It may be shown that i (molecule-fixed) is equivalent to X_N (space-fixed), and so the procedures worked out in the foregoing discussion may be used to classify $|\psi_{el}\chi_{rot}\rangle$ as either (s) or (a) under X_N in lieu of determining their behavior under molecule-fixed inversion. The dipole moment operator $\boldsymbol{\mu}$ in homonuclear molecules is (s) under X_N [11]. This leads to the conclusion that only states $|\psi_{el}\chi_{rot}\rangle$ with like symmetry under X_N can be connected by E1 transitions in electronic band spectra:

$$\mathrm{s} \leftrightarrow \mathrm{s}$$

$$\mathrm{a} \leftrightarrow \mathrm{a}$$

$$\mathrm{s} \leftarrow\!\times\!\rightarrow \mathrm{a}$$

The Herzberg diagrams for homonuclear diatomics can now be augmented with (s) and (a) labels denoting behavior under X_N. For a $^1\Sigma_g^+ \rightarrow {}^1\Sigma_u^+$ transition, the diagram is shown in Fig. 4.24, where all of the E1-allowed transitions simultaneously obey $(+)\leftrightarrow(-)$, (s)$\leftrightarrow$(s), and (a)$\leftrightarrow$(a). As in $^1\Sigma^+ \rightarrow {}^1\Sigma^+$ transitions for heteronuclear molecules, only R and P branches can appear. It is easy to show that the Herzberg diagrams for $^1\Sigma_g^+ \rightarrow {}^1\Sigma_g^+$ and $^1\Sigma_u^+ \rightarrow {}^1\Sigma_u^+$ transitions forbid *any* transitions from occurring at all (i.e., the Herzberg diagrams have the Laporte rule built into them). For a $^1\Sigma_g^+ \rightarrow {}^1\Pi_u$ transition (Fig. 4.25), all three rotational branches appear. Note that the P(1) and Q(0) lines are absent, since the $J = 0$ level cannot occur in a $^1\Pi_u$ state. A wealth of additional Herzberg diagrams may be found in G. Herzberg's classic *Spectra of Diatomic Molecules* [10].

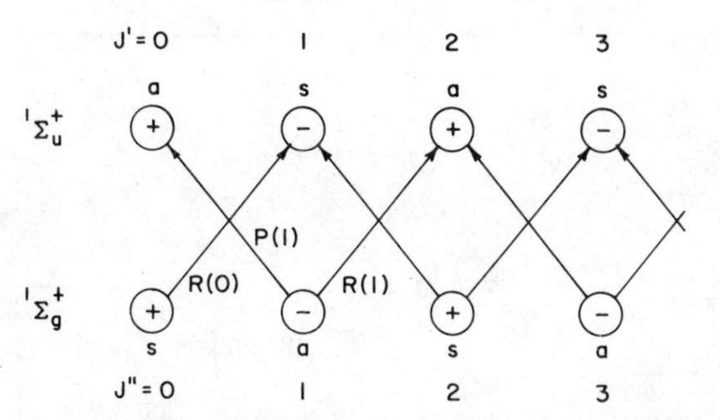

Figure 4.24 Allowed rotational transitions in a $^1\Sigma_g^+ \rightarrow {}^1\Sigma_u^+$ electronic transition in a heteronuclear diatomic molecule. Such Herzberg diagrams automatically incorporate the Laporte g $\leftrightarrow$ u selection rule, since no allowed rotational transitions can be drawn for a $^1\Sigma_g^+ \leftrightarrow {}^1\Sigma_g^+$ or a $^1\Sigma_u^+ \leftrightarrow {}^1\Sigma_u^+$ electronic transition.

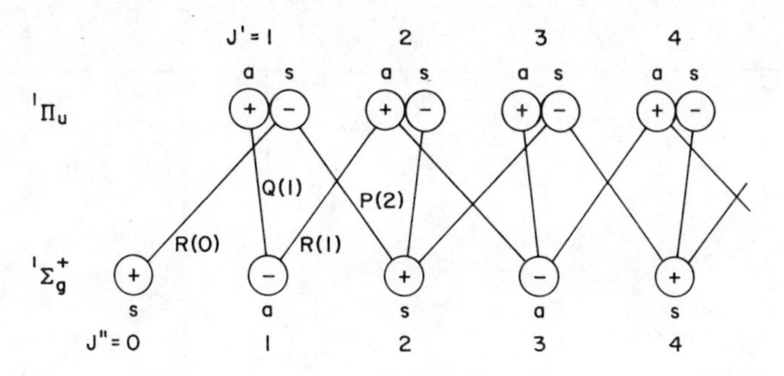

Figure 4.25 Allowed rotational transitions in a $^1\Sigma_g^+ \rightarrow {}^1\Pi_u$ transition.

The Na_2 fluorescence spectra in Fig. 4.26 are excellent examples of the rotational selection rules predicted by the $^1\Sigma_g^+ \rightarrow {}^1\Pi_u$ Herzberg diagram in Fig. 4.25. The first of these spectra was excited using the 4727-Å line from an Ar^+ laser, which matches the energy difference between $^1\Sigma_g^+$ ($v'' = 1$, $J'' = 37$) and $^1\Pi_u$ ($v' = 9$, $J' = 38$) in Na_2. The laser linewidth was considerably broader than the energy separation (due to Λ doubling) between the (a) and (s) $^1\Pi_u$ levels belonging to $J' = 38$ in Na_2. Since the initial $J'' = 37$ level is an (a) level in a Σ_g^+ state, however, the selection rules in Fig. 4.25 show that only the (a) sublevel in $J' = 38$ can be reached from $J'' = 37$ in E1 transition. Subsequent fluorescence transitions from $^1\Pi_u$ ($v' = 9$, $J' = 38$) to the $^1\Sigma$ state in various v' levels can only terminate in $J'' = 37$ or 39 according to Fig. 4.25. Hence, the $^1\Pi_u \rightarrow {}^1\Sigma_g^+$ fluorescence spectrum excited at 4727 A exhibits only P and R branches. The second spectrum was excited by the 4800-Å Ar^+ laser line, which matches the energy separation between $^1\Sigma_g^+$ ($v'' = 3$, $J'' = 43$) and $^1\Pi_u$ ($v' = 6$, $J'' = 43$). Such a transition can populate only the (s) sublevel in $J' = 43$; reasoning similar to that outlined above shows that fluorescence from $v' = 6$, $J' = 43$ can then

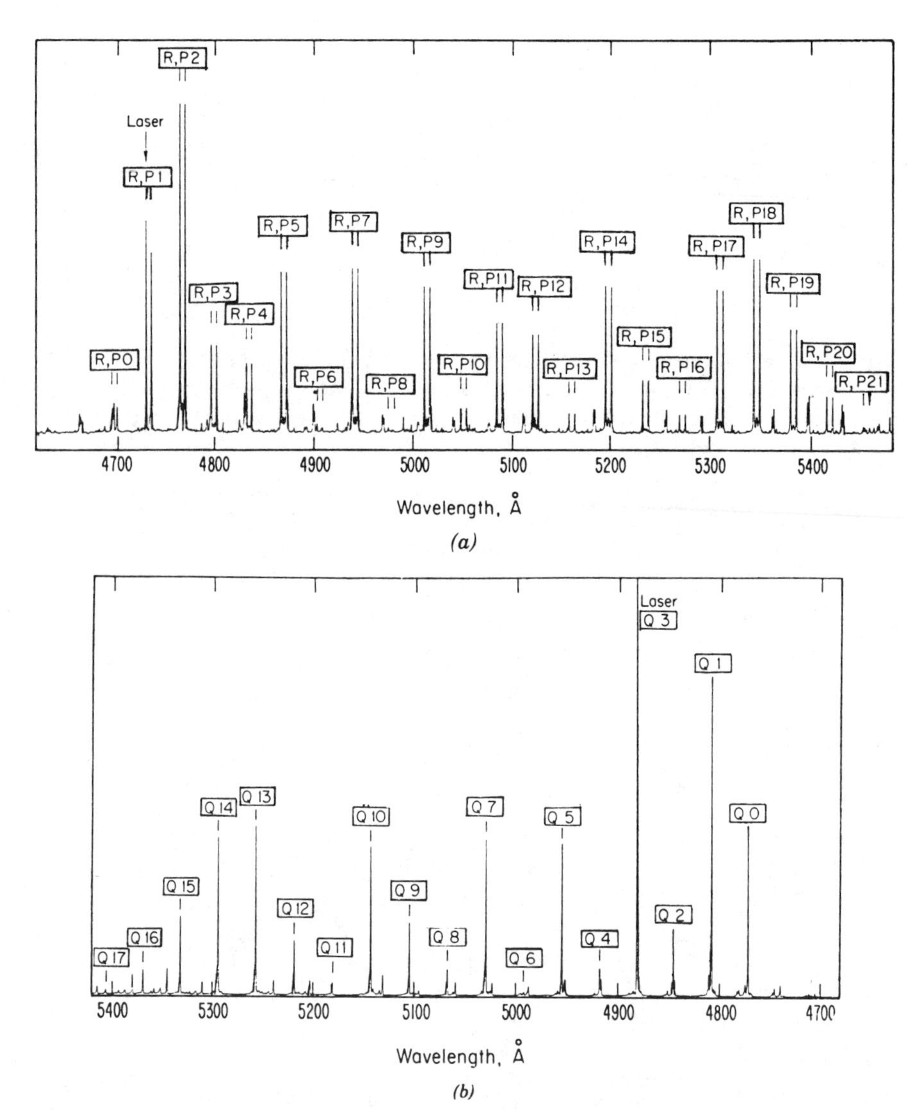

Figure 4.26 $B^1\Pi_u \to X^1\Sigma_g^+$ fluorescence spectra of Na_2 vapor following excitation by an argon ion laser at (a) 4727 Å and (b) 4880 Å. The first spectrum is due to the transitions $v' = 9, J' = 38 \to v'', J'' = 37, 39$; its rotational fine structure exhibits only P and R branches. The second spectrum arises from the transitions $v' = 6$, $J' = 43 \to v', J'' = 43$; only the Q branch appears in its rotational structure. These spectra are excellent examples of the selection rules in Fig. 4.25. Monochromatic laser excitation of an $X^1\Sigma_g^+$ molecule creates a $B^1\Pi_u$ molecule in either an *s* or an *a* level, but not both. According to Fig. 4.25, the level can consequently fluoresce with either (P, R) or Q branches, but not both. The numbers in boxes give v'' for the lower level in fluorescence transitions. Vibrational band intensities are proportional to the Franck-Condon factors $|\langle v'|v''\rangle|^2$. Reproduced with permission from W. Demtröder, M. McClintock, and R. N. Zare, *J. Chem. Phys.* **51**: 5495 (1969).

exhibit only a Q branch. Hence, the (v', v'') fluorescence bands are doublets in the first fluorescence spectrum, but are singlets in the second one. The band intensities are proportional to the pertinent Franck-Condon factors $|\langle v'|v''\rangle|^2$ in both spectra.

Rotational line assignments are easily made by inspection in pure rotational and vibration–rotation spectra (Chapter 3). In the former case, the rotational energies and transition frequencies are

$$E_{\text{rot}}(J)/hc = BJ(J+1) - DJ^2(J+1)^2 \tag{4.74}$$

and

$$\bar{\nu} = 2BJ - 4DJ^3 \tag{4.75}$$

where J denotes the upper level in absorptive transitions ($\Delta J = +1$). Since the centrifugal distortion constant D is generally small compared to B ($D/B \lesssim 10^{-4}$), the rotational lines are very nearly equally spaced in frequency. For vibration–rotation spectra, the P-, Q-, and R-branch line frequencies (ignoring centrifugal distortion) are

$$\begin{aligned}
\bar{\nu}_{\text{P}} &= \bar{\nu}_0 - (B' + B'')J + (B' - B'')J^2 \\
\bar{\nu}_{\text{Q}} &= \bar{\nu}_0 + (B' - B'')J + (B' - B'')J^2 \\
\bar{\nu}_{\text{R}} &= \bar{\nu}_0 + 2B' + (3B' - B'')J + (B' - B'')J^2
\end{aligned} \tag{4.76}$$

where J pertains to the lower level in the transition and $\bar{\nu}_0$ is the vibrational energy difference. Since B' and B'' are nearly the same in vibration–rotation spectra (because B varies weakly with v within a given electronic state), Eqs. 4.76 imply that rotational lines in such spectra will also be roughly equally spaced in frequency for low J (Fig. 3.3).

For rotational fine structure in electronic band spectra, the P- and R-branch line positions are still given by Eqs. 4.76, except that $\bar{\nu}_0$ now becomes $T_e + G'(v') - G''(v'')$. The important physical difference here is that B' and B'' are often *grossly* different in transitions between different electronic states. For example, B'_e and B''_e are 0.029 and 0.037 cm^{-1}, respectively, for the $B^3\Pi_{ou} \leftarrow X^1\Sigma_g^+$ transition in I_2 (Fig. 4.7). The rotational line positions in the P and R branches are no longer even approximately equally spaced. For that matter, $\bar{\nu}_{\text{P}}$ and $\bar{\nu}_{\text{R}}$ both vary as $(B' - B'')J^2$ for large J—since they are then dominated by terms quadratic in J—and hence they run in the same direction as functions of J. In contrast, $\bar{\nu}_{\text{P}}$ and $\bar{\nu}_{\text{R}}$ run in opposite directions as functions of J for small J, where the linear terms dominate. Hence, either the P or the R branch *turns around* as a function of J (depending on whether $B' < B''$ or $B' > B''$) at the *bandhead* as shown in Fig. 4.27. The approximate value of J at which one of these branches turns around can be found by differentiating the

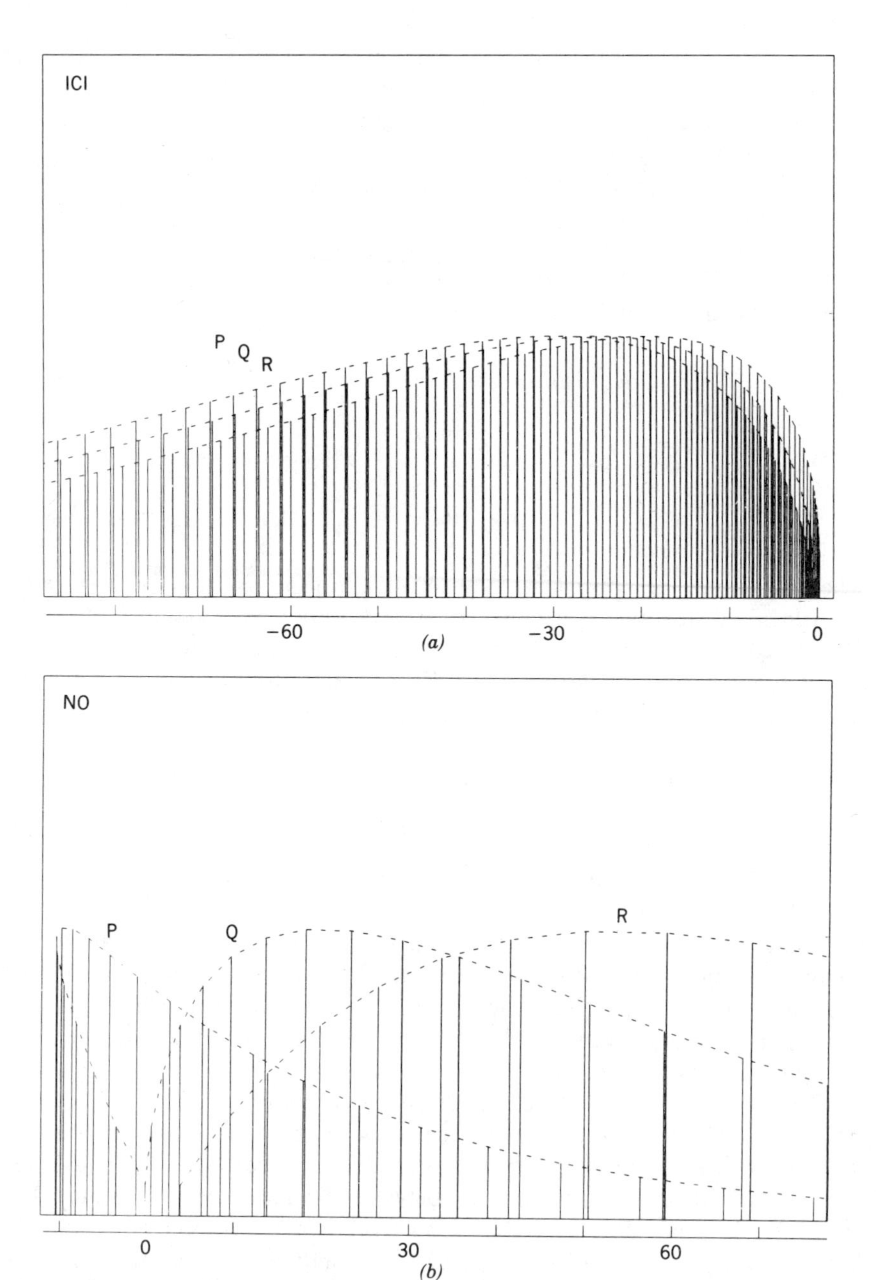

Figure 4.27 Rotational fine structures at 300 K for the $v'' = 0 \rightarrow v' = 0$ absorptive transitions in (a) the $X^1\Sigma^+ \rightarrow B^3\Pi_0$ transition in ICl, $B''_e = 0.1142$ cm^{-1}, $B'_e = 0.0872$ cm^{-1}; and (b) the $X^2\Pi_r \rightarrow A^2\Sigma^+$ transition in NO, $B''_e = 1.67195$ cm^{-1}, $B'_e = 1.9965$ cm^{-1}. The rotational line spacings are smaller for ICl, owing to its larger reduced mass. The ICl spectrum is shaded to the red, and typifies the usual situation in which $B''_e > B'_e$. The reverse is true in the $X^2\Pi_r \rightarrow A^2\Sigma^+$ transition of NO, in which the rotational structure is shaded to the blue. Horizontal energy scale is in cm^{-1} for both spectra.

equations for $\bar{\nu}_P$ and $\bar{\nu}_R$ with respect to J,

$$\frac{d\bar{\nu}_P}{dJ} = -(B' + B'') + 2(B' - B'')J = 0$$

$$\frac{d\bar{\nu}_R}{dJ} = 3B' - B'' + 2(B' - B'')J = 0 \tag{4.77}$$

Then the J values at which these branches turn around are

$$J_P^* = (B' + B'')/2(B' - B'')$$

$$J_R^* = (B'' - 3B')/2(B' - B'') \tag{4.78}$$

Only one of these values is physical (positive). Since the upper electronic state is commonly more weakly bound than the lower state ($R_e' > R_e''$), one usually observes $B_e' < B_e''$—in which case the R branch is the one that turns around ($J_R^* > 0$). Both the P and R branches then run to lower frequencies for large J (since $(B' - B'') < 0$), and the band is said to be *shaded to the red.* When $R_e' < R_e''$, the band is *shaded to the blue.* The rotational fine structure is barely resolved in the Na_2 fluorescence excitation spectrum of Fig. 4.3, but the asymmetric shading of the vibrational bands to the red is clearly asserted. It occurs because the equilibrium separation of Na_2 is considerably larger in the $A^1\Sigma_u^+$ than in the $X^1\Sigma_g^+$ state, as is apparent in Fig. 4.8.

4.7 POTENTIAL ENERGY CURVES FROM ELECTRONIC BAND SPECTRA

The spectroscopic techniques described in this and the preceding chapters yield an impressive array of molecular constants, in terms of which the rovibrational energy levels may be expanded to desired accuracy via

$$E/hc = \omega_e(v + \tfrac{1}{2}) - \omega_e x_e(v + \tfrac{1}{2})^2 + \omega_e y_e(v + \tfrac{1}{2})^3 + \cdots$$
$$+ B_e J(J + 1) - D_e J^2(J + 1)^2 - \alpha_e J(J + 1)(v + \tfrac{1}{2}) + \cdots \tag{4.79}$$

For example, spectral lines in vibration–rotation spectra of heteronuclear diatomics are readily assigned to particular $(v'J') \leftarrow (v''J'')$ transitions by inspection. Their positions may be analyzed using Eqs. 3.64, 3.65, 3.68, and 3.81 to extract the ground-state constants ω_e'', $\omega_e''x_e''$, $\omega_e''y_e''$, B_e'', D_e'', α_e'', ... In homonuclear diatomics (which are infrared-inactive), analysis of rotational fine structure in fluorescence spectra yields the rotational constants B_e'', D_e'', α_e'', ... and B_e', D_e', α_e', ... for the upper and lower states. The vibrational band positions in such spectra (e.g., the Na_2 fluorescence spectrum in Fig. 4.1) may be analyzed to obtain the vibrational constants ω_e'', $\omega_e''x_e''$, $\omega_e''y_e''$, ... in the lower state. Similar treatment of the absorption or fluorescence excitation spectrum (cf. Fig. 4.3) yields the corresponding upper state vibrational constants ω_e', $\omega_e'x_e'$, $\omega_e'y_e'$, ... Conclusive vibrational assignments in electronic band spectra can be difficult to establish when the electronic energy separation T_e between upper and lower states is not independently known. Failure to observe the $v'' = 0 \to v' = 0$ band due to a small Franck-Condon factor $|\langle v'|v''\rangle|^2$, for example, can result in misnumbering of the true levels $v'' = 1, 2, \ldots$, as $v'' = 0, 1, \ldots$ [13]. A strategem for confirming vibrational assignments of electronic band spectra is described later in this section. Huber and Herzberg [14] have compiled molecular constants for over 900 diatomic molecules and molecule-ions, based on critical examination of the literature up to 1978.

In many applications, it is desirable to know the detailed potential energy curves $U_{kk}(R)$ for the upper and lower states. Such information would be required to predict spectral line intensities of heretofore unobserved vibronic transitions (e.g., for investigation as possible laser transitions). We have already shown that the radial Schrödinger equation (3.30) can be solved under a given potential $U_{kk}(R)$ to obtain the vibrational eigenvalues for $J = 0$. We are now concerned with the reverse procedure: Given a set of spectroscopically determined vibrational levels, can the detailed potential energy curve $U_{kk}(R)$ be reconstructed? For nonpathological potentials, the intuitive answer is yes. When the vibrational levels are equally spaced in energy, $U_{kk}(R)$ is a parabola with curvature determined by the level spacing. Nonuniformities in level spacing (manifested by nonvanishing anharmonic constants $\omega_e x_e$, $\omega_e y_e$, ...) should deform the parabola in predictable directions.

The current technique for derivation of experimental potential energy curves, developed in the 1930s but popularized only with the advent of high-speed computers, is based on the Sommerfeld condition [15] for quantization of action in systems undergoing periodic motion,

$$\oint p\,dq = (v + \tfrac{1}{2})h \qquad v = 0, 1, 2 \tag{4.80}$$

The generalized coordinate q and its conjugate momentum p vary periodically, and the line integral is evaluated over one cycle. (A similar quantization condition on angular momentum led to the famous Bohr postulate $L = n\hbar$ for the hydrogen atom in the old quantum theory [16].) For vibrational motion subject to a potential $U(R)$ in a diatomic, the classical energy is $E = p^2/2\mu + U(R)$. The integral in (4.80) then translates into

$$2\int_{R_-}^{R_+} \sqrt{E - U(R)}\,dR = (v + \tfrac{1}{2})h \tag{4.81}$$

for periodic motion between the *classical turning points* R_- and R_+ (Fig. 4.28). Rydberg, Klein, and Rees demonstrated that this semiclassical procedure for deriving the allowed vibrational energies E from a given potential $U(R)$ may be inverted. If the rovibrational energy (4.79) is rewritten in terms of the vibrational

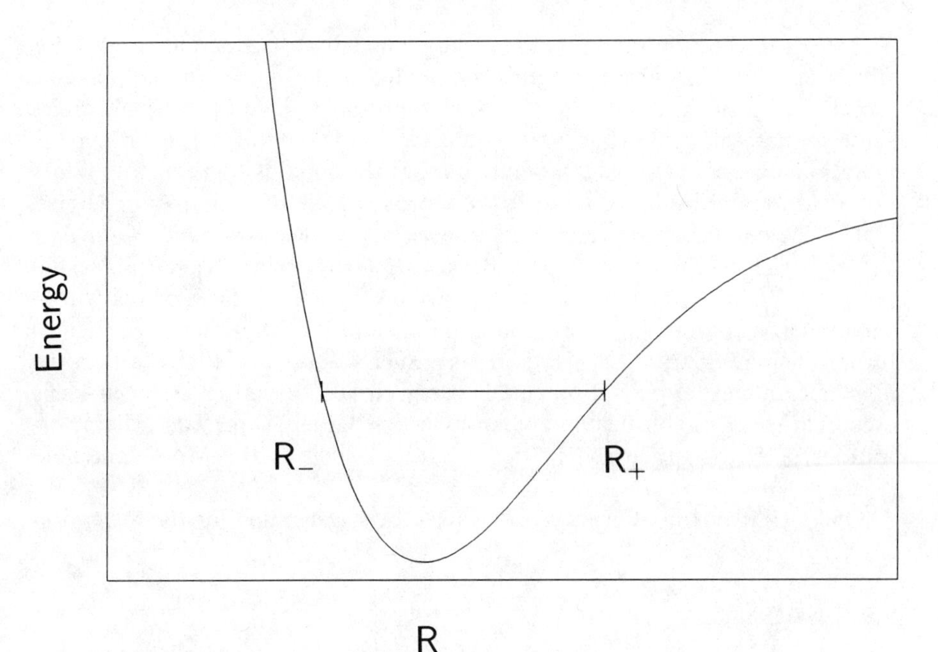

Figure 4.28 The classical turning points R_- and R_+ for a given vibrational level. For this vibrational energy, nuclear motion is classically forbidden for $R < R_-$ and for $R > R_+$.

and rotational action variables $I = h(v + \frac{1}{2})$ and $\kappa = J(J+1)\hbar^2/2\mu$ as

$$E(I, \kappa)/hc = I\omega_e/h - I^2\omega_e x_e/h^2 + I^3\omega_e y_e/h^3 + \kappa/hcR_e^2 - \kappa^2 D_e/h^2c^2B_e^2R_e^4 - I\kappa\alpha_e/h^2cB_eR_e^2 \tag{4.82}$$

it may be shown [17] that the inner and outer turning points $R_\pm$ of a vibrational state with known energy E_v are given by

$$R_\pm = (f/g + f^2)^{1/2} \pm f \tag{4.83}$$

with

$$f = \frac{h}{2\pi(2\mu)^{1/2}} \int_0^{I'} [E_v - E(I, \kappa)]^{-1/2} dI \tag{4.84}$$

and

$$g = \frac{h}{2\pi(2\mu)^{1/2}} \int_0^{I'} \frac{\partial E}{\partial \kappa} [E_v - E(I, \kappa)]^{-1/2} dI \tag{4.85}$$

The upper integration limit I' is the value of the vibrational action variable for which $E(I', \kappa) = E_v$. These integrals are evaluated for $\kappa = 0$ to compute the vibrational turning points $R_\pm$ for the nonrotating ($J = 0$) molecule. Rees showed that these integrals are expressible in closed form [18] when the vibrational energy $E(I, 0)$ is quadratic in I, but this level of approximation does not yield accurate turning points over a broad range of vibrational energies. Hence, these integrals are calculated numerically to yield two points $R_\pm(E_v)$ on the potential energy curve $U_{kk}(R)$ at each spectroscopic energy E_v. The full potential $U_{kk}(R)$ is then constructed by connecting the turning points with a smooth curve. Figure 4.29 shows the experimental Rydberg-Klein-Rees (RKR) turning points for the $X^1\Sigma_g^+$ state in Na_2. Such RKR calculations furnish the most accurate experimental potential energy curves for diatomics. Their facile execution using established computer codes has largely superseded characterizations of $U_{kk}(R)$ by analytic fitted potentials such as the Morse potential (Section 3.6).

Once RKR potential energy curves have been generated for the lower and

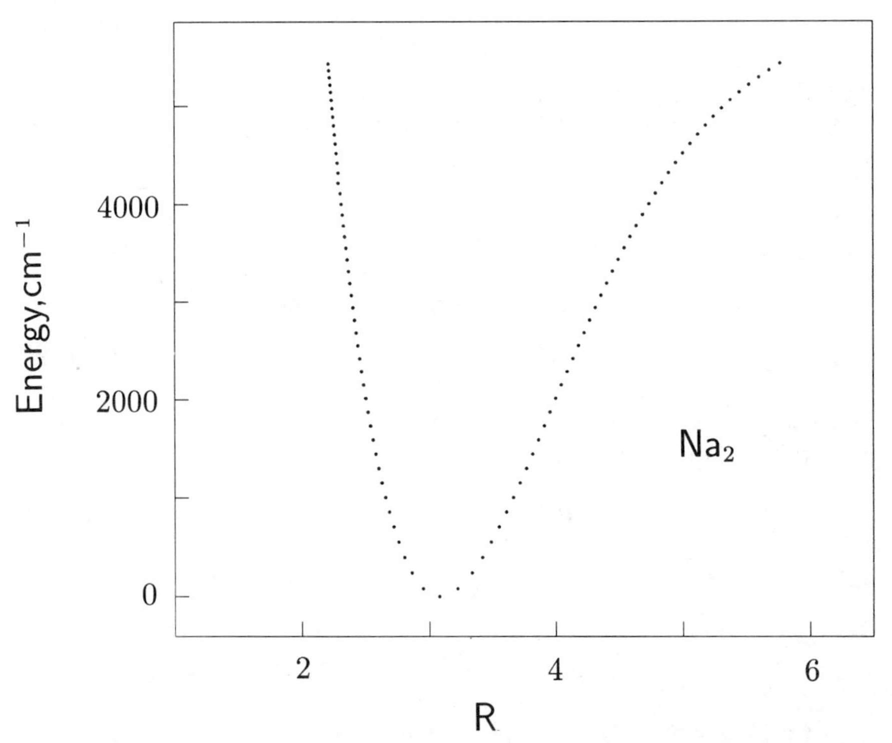

Figure 4.29 RKR points for the $X^1\Sigma_g^+$ state of Na_2. These are the classical turning points R_-, R_+ calculated for $v'' = 0$ through 45 using spectroscopically determined rovibrational energy levels from Na_2 fluorescence spectra. Separations are given in Å. Data are taken from P. Kusch and M. M. Hessel, *J. Chem. Phys.* **68**: 2591 (1978).

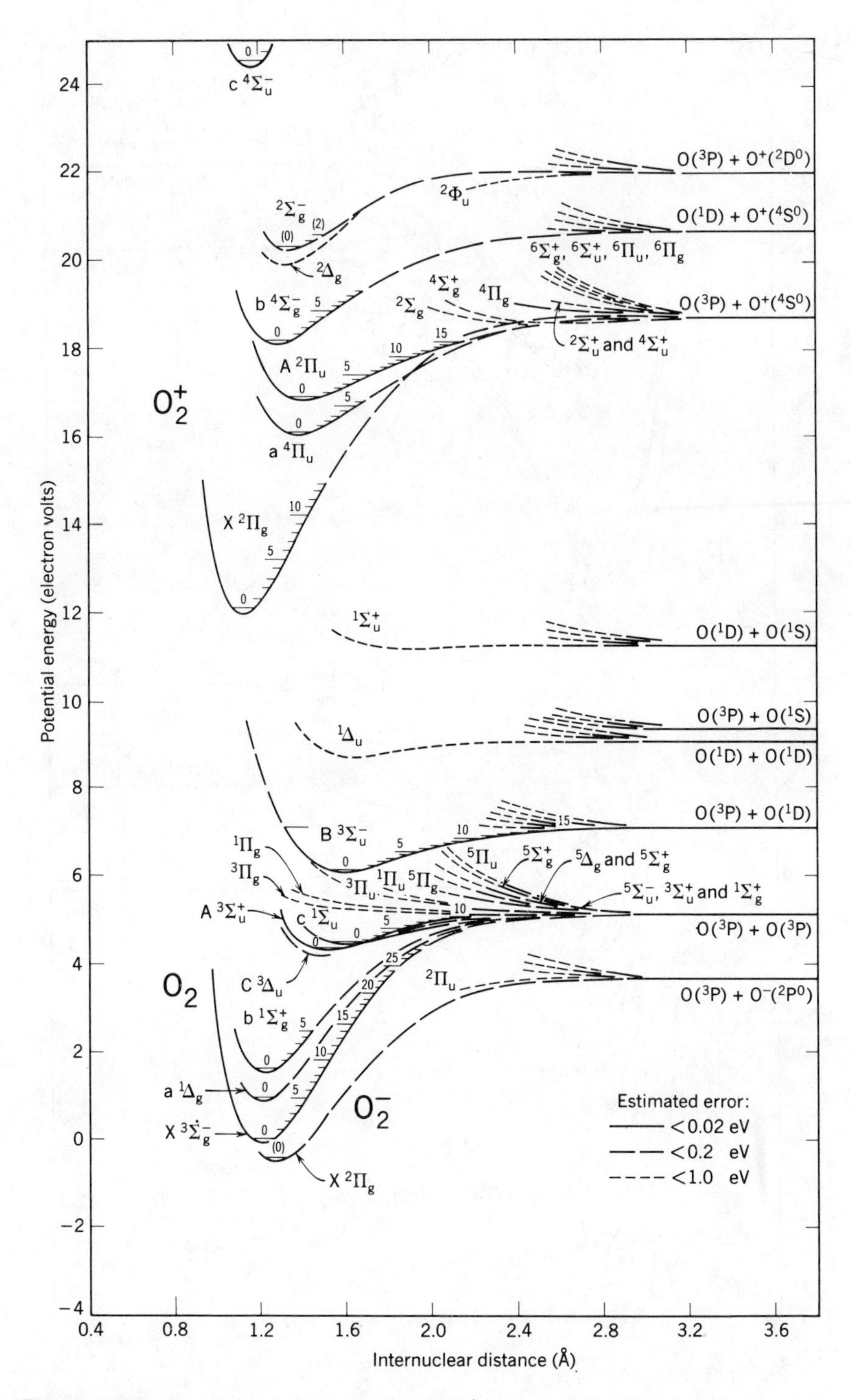

Figure 4.30 Potential energy curves for O_2^+, O_2, and O_2^-. Used with permission from F. R. Gilmore, RAND Corporation Memorandum R-4034-PR (June 1964).

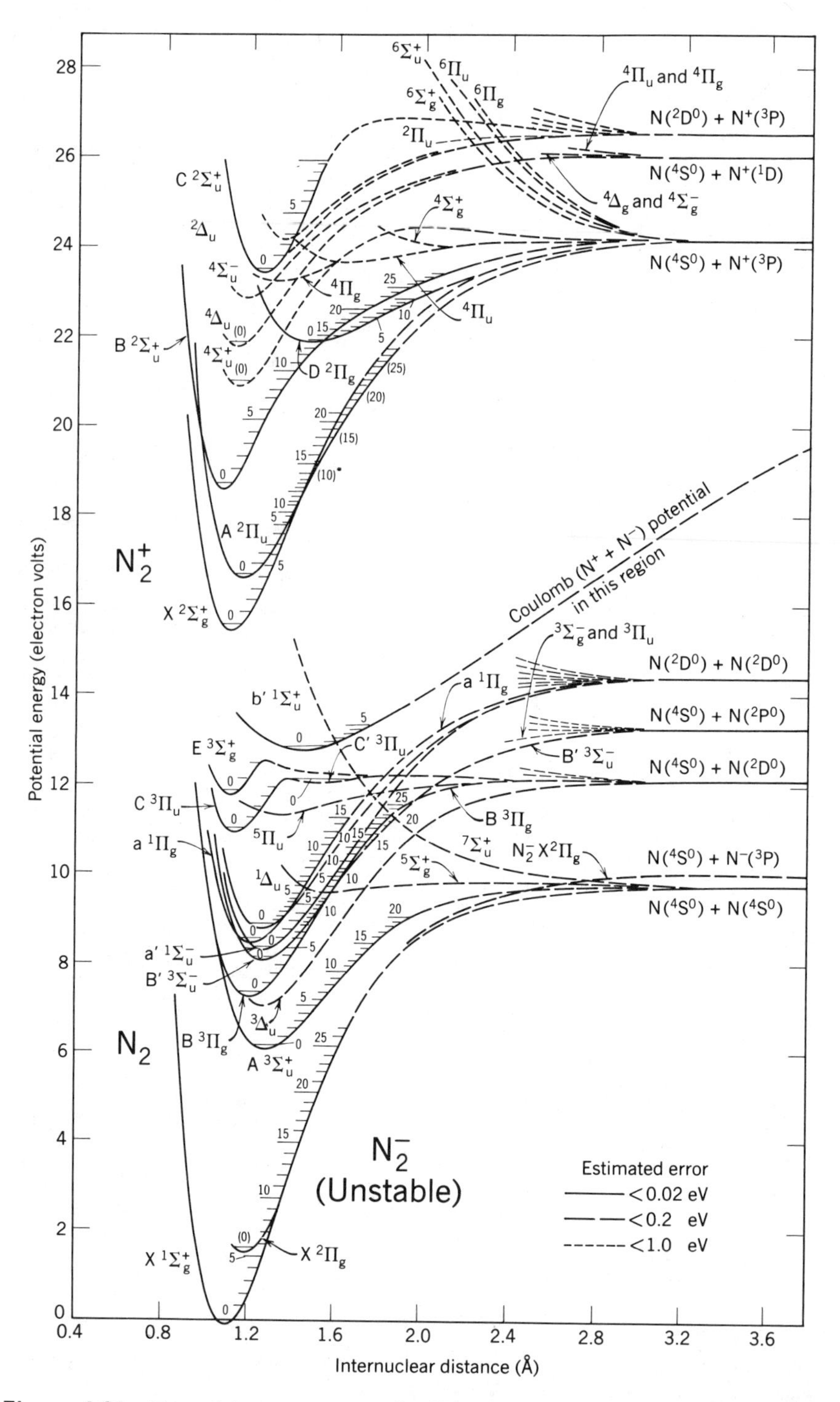

Figure 4.31 Potential energy curves for N_2^+, N_2, and N_2^-. Used with permission from F. R. Gilmore, RAND Corporation Memorandum R-4034-PR (June 1964).

upper states (e.g., the $X^1\Sigma_g^+$ and $B^3\Pi_{ou}$ states in I_2, Fig. 4.7), the radial Schrödinger equation (3.30) can be solved numerically to obtain the corresponding vibrational states $|v''\rangle$ and $|v'\rangle$ in the respective electronic states. It is then straightforward to calculate the predicted Franck-Condon factors $|\langle v'|v''\rangle|^2$ for all vibrational bands in the electronic transition. The band intensities observed in an absorption or fluorescence spectrum should be proportional to these calculated Franck-Condon factors (Section 4.4). If they are not, the vibrational bands in the electronic spectrum may have been incorrectly assigned. In this manner, Zare found it necessary to reassign the v' quantum numbers previously attributed to vibrational bands in the $X^1\Sigma_g^+ \rightarrow B^3\Pi_{ou}$ spectrum [13]. RKR calculations are, of course, possible only for spectroscopically accessible electronic states for which a reasonable number of the molecular constants in Eq. 4.82 are known. For this reason, only the $X^1\Sigma_g^+$, $A^3\Pi_{1u}$, and $B^3\Pi_{o+u}$ potential energy curves for I_2 in Fig. 4.7 are well known. The lowest-lying $X^1\Sigma_g^+ \rightarrow {}^3\Pi_{2u}$ transition ($\Delta\Omega = 2$) is so weak that no RKR curve has been generated for that excited state to the author's knowledge. RKR calculations are not applicable to purely repulsive states, and the shapes (but not the asymptotes) of the repulsive potential energy curves depicted in this chapter are largely conjectural. Methods for predicting intensities of transitions to repulsive states ("bound-continuum transitions") have recently been developed [19].

A number of potential energy curves is shown for O_2, O_2^-, and O_2^+ in Fig. 4.30, and for N_2, N_2^-, and N_2^+ in Fig. 4.31. The lowest three bound states in neutral O_2 ($X^3\Sigma_g^-$, $a^1\Delta_g$, and $b^1\Sigma_g^+$) arise from the equivalent π^2 configuration (Section 4.3). E1 transitions among these three states are forbidden, in accordance with the general rule that no E1 transitions exist among states generated from the same electron configuration. The better-characterized excited states (i.e., the ones with more labeled vibrational levels) tend to be ones which are E1-connected to the ground-state molecule or molecule-ion. As examples, the $B^3\Sigma_u^-$ excited-state potential in O_2 is known up to nearly its dissociation limit, but the $C^3\Delta_u$ state is not (Fig. 4.30). In N_2, the $X^1\Sigma_g^+$ ground state is a closed-shell configuration; its excited states are among the most thoroughly studied in any diatomic. Since it is not connected to any lower electronic state by E1 transition, the $A^3\Sigma_u^+$ state is metastable, and E1 absorptive transitions from this state to higher states (e.g., $C^3\Pi_u$) can easily be observed. The well-known N_2 laser, which operates at 337 nm, stems from the $v' = 0 \rightarrow v'' = 0$ transition between the $C^3\Pi_u$ and $B^3\Pi_g$ states.

REFERENCES

1. D. S. Schonland, *Molecular Symmetry*, Van Nostrand, London, 1965.
2. L. D. Landau and E. M. Lifshitz, *Quantum Mechanics: Non-Relativistic Theory*, Addison-Wesley, Reading, MA, 1958.
3. B. H. Mahan, *College Chemistry*, Addison-Wesley, Reading, MA, 1966.
4. R. S. Berry, S. A. Rice, and J. Ross, *Physical Chemistry*, Wiley, New York, 1980.

5. K. Ruedenberg, *Rev. Mod. Phys.* **34**: 326 (1962); M. J. Feinberg, K. Ruedenberg, and E. L. Mehler, *Adv. Quantum Chem.* **5**: 27 (1970).
6. J. G. Verkade, *A Pictorial Approach to Molecular Bonding*, Springer-Verlag, New York, 1986.
7. W. H. Flygare, *Molecular Structure and Dynamics*, Prentice-Hall, Englewood Cliffs, NJ, 1978.
8. H. F. Schaefer III, *The Electronic Structure of Atoms and Molecules: A survey of Rigorous Quantum Mechanical Results*, Addison-Wesley, Reading, MA, 1972.
9. R. S. Mulliken, *J. Chem. Phys.* **55**: 289 (1971).
10. G. Herzberg, *Molecular Spectra and Molecular Structure; I. Spectra of Diatomic Molecules*, Van Nostrand, Princeton, NJ, 1950.
11. J. Hougen, *The Calculation of Rotational Energy Levels and Rotational Line Intensities in Diatomic Molecules*, National Bureau of Standards Monograph 115, U.S. Government Printing Office, Washington, DC, 1970.
12. K. Gottfried, *Quantum Mechamics*, Vol. 1, W. A. Benjamin, New York, 1966.
13. R. N. Zare, *J. Chem. Phys.* **40**: 1934 (1964).
14. K. P. Huber and G. Herzberg, *Molecular Spectra and Molecular Structure: Constants of Diatomic Molecules*, Van Nostrand Reinhold, New York, 1979.
15. A. Sommerfeld, *Ann. Phys.* **57**: 1 (1916).
16. M. Jammer, *The Conceptual Development of Quantum Mechanics*, McGraw-Hill, New York, 1966.
17. R. Rydberg, *Ann. Phys.* **73**: 376 (1931); O. Klein, *Z. Phys.* **76**: 226 (1932).
18. A. L. G. Rees, *Proc. Phys. Soc.* (*London*) *A* **59**: 998 (1947).
19. W. T. Zemke, K. K. Verma, and W. C. Stwalley, Proc. Intern. Conf. on Lasers '81, 216 (1981).

PROBLEMS

1. For a heteronuclear diatomic molecule, determine the electronic term symbols that correlate with ^{2}D + ^{2}D atoms.

2. Write down the electron configurations for the ground states of B_2, C_2, N_2, and O_2. What term symbol corresponds to the ground state in each of these molecules?

3. For a homonuclear diatomic molecule, what term symbols $^{2S+1}\Lambda_\Omega$ can arise (a) from an equivalent δ^2 configuration and (b) from a nonequivalent π^2 (e.g., $\pi_u^1\pi_g^1$) configuration?

4. Determine the relative statistical weights of the even- and odd-J levels in $^{85}Rb_2$ and $^{40}K_2$, which have nuclear spin $I = \frac{5}{2}$ and 4, respectively.

5. Using the potential energy curves in Fig. 4.31 for neutral N_2, decide which of the shown states are E1-metastable. Are *all* of the diatomic states which correlate with N(^{4}S) + N(^{4}S) shown? Which of the shown excited states are theoretically accessible by M1 transitions from ground-state N_2?

6. The $C^1\Pi_u \leftarrow X^1\Sigma_g^+$ absorption spectrum of Na_2 in the near-ultraviolet region of the spectrum exhibits bands at the following frequencies in cm^{-1}:

28871	29182	29338	29479	29682
28986	29213	29368	29524	29745
29026	29255	29411	29569	29793
29100	29297	29434	29589	29903
29141	29324	29455	29635	

Starting from the assumption that one of these frequencies corresponds to the $v'' = 0 \rightarrow v' = 0$ transition, make the vibrational assignments by ordering the observed band frequencies $\bar{\nu}(v'', v')$ into a *Deslandres table*:

$\bar{\nu}(0, 0)$	Δ'_{01}	$\bar{\nu}(0, 1)$	Δ'_{12}	$\bar{\nu}(0, 2)$	$\cdots$
Δ''_{01}		Δ''_{01}		Δ''_{01}	
$\bar{\nu}(1, 0)$	Δ'_{01}	$\bar{\nu}(1, 1)$	Δ'_{12}	$\bar{\nu}(1, 2)$	$\cdots$
Δ''_{12}		Δ''_{12}		Δ''_{12}	
$\bar{\nu}(2, 0)$	Δ'_{01}	$\bar{\nu}(2, 1)$	Δ'_{12}	$\bar{\nu}(2, 2)$	$\cdots$
$\vdots$		$\vdots$		$\vdots$	

The numbers Δ'_{mn} and Δ''_{mn} represent the numerical differences between adjacent frequencies in the table. (In a correctly ordered table, all differences denoted by the same label (e.g., Δ''_{12}) must exhibit the same value to within experimental error, in accordance with the Ritz combination principle. The value found for $\Delta'_{mn}(\Delta''_{mn})$ is the observed energy separation between vibrational levels m and n in the upper (lower) electronic state.) Determine the vibrational constants ω''_e, ω'_e, $\omega''_e x''_e$, $\omega'_e x'_e$ and (if possible) $\omega''_e y''_e$, $\omega'_e y'_e$ from these data.

7. The 0–3 absorption band of an allowed E1 electronic transition originates from the $X^3\Sigma_g^-$ state in O_2. Rotational fine structure lines occur at the following wave numbers:

51354.37	$R(1)$
51353.99	
51353.50	$R(2)$
51351.37	
51349.46	
51348.01	$R(4)$
51345.31	$P(2)$
51343.38	$R(5)$
51339.92	

(a) Complete the assignments of the four unlabeled lines. Is this band shaded to the red or violet? Is B' greater than or less than B''?
(b) Determine the numerical values of the rotational constants B' and B''.
(c) Recognizing that this is an allowed transition in the relatively light molecule O_2, specify the term symbol of the upper state.

8. The analytic expressions 4.24 and 4.34 for the overlap, Coulomb, and exchange integrals in H_2^+ may be obtained using the ellipsoidal coordinates λ, μ, and ϕ, where

$$\lambda = (r_A + r_B)/R$$

$$\mu = (r_A - r_B)/R$$

and the angle ϕ has its usual meaning in a diatomic molecule. These coordinates exhibit the range

$$1 \leqslant \lambda < \infty$$

$$-1 \leqslant \mu \leqslant 1$$

$$0 \leqslant \phi < 2\pi$$

and the volume element is $R^3(\lambda^2 - \mu^2)d\lambda\, d\mu\, d\phi/8$. Use this information to confirm that S_{AB}, J, and K are correctly given by Eqs. 4.24 and 4.34.

5

POLYATOMIC ROTATIONS

In this chapter, we extend our treatment of rotation in diatomic molecules to nonlinear polyatomic molecules. A traditional motivation for treating polyatomic rotations quantum mechanically is that they form a basis for experimental determination for bond lengths and bond angles in gas-phase molecules. Microwave spectroscopy, a prolific area in chemical physics since 1946, has provided the most accurate available equilibrium geometries for many polar molecules. A background in polyatomic rotations is also a prerequisite for understanding rotational fine structure in polyatomic vibrational spectra (Chapter 6). The shapes of rotational contours (i.e., unresolved rotational fine structure) in polyatomic electronic band spectra are sensitive to the relative orientations of the principal rotational axes and the electronic transition moment (Chapter 7). Rotational contour analysis has thus provided an invaluable means of assigning symmetries to the electronic states involved in such spectra.

We begin this chapter with a derivation of the classical Hamiltonian for a rigid, freely rotating polyatomic molecule. Such a polyatomic rotor may be classified according to its point group symmetry (Section 5.1). We formulate the quantum mechanical Hamiltonian and rotational energy eigenvalues for molecules having at least one C_n axis with $n \geqslant 3$. Molecules with lower symmetry (e.g., H_2O) are far more difficult to treat, and readers are referred to specialized sources on rotational spectroscopy for these cases. E1 selection rules are obtained for pure rotational transitions, and examples are given of structural information yielded by microwave spectra.

5.1 CLASSICAL HAMILTONIAN AND SYMMETRY CLASSIFICATION OF RIGID ROTORS

Consider a rigid body consisting of a collection of point masses m_N located at positions $\mathbf{R}_N \equiv (X_N, Y_N, Z_N)$ in a Cartesian coordinate system which rotates with the molecule. The origin of this coordinate system coincides with the body's center of mass (Fig. 5.1). If the body is caused to rotate with angular velocities $\omega_X, \omega_Y, \omega_Z$ about the molecule-fixed X, Y, and Z axes, respectively, its rotational kinetic energy becomes [1]

$$T = \frac{1}{2} \sum_{ij} \omega_i I_{ij} \omega_j \equiv \tfrac{1}{2}\boldsymbol{\omega} \cdot \mathbf{I} \cdot \boldsymbol{\omega} \tag{5.1}$$

where the indices i, j run over X, Y, and Z. The matrix $\mathbf{I}$ is the *rotational inertia tensor* [1]

$$\mathbf{I} = \begin{bmatrix} \sum_N m_N(Y_N^2 + Z_N^2) & -\sum_N m_N X_N Y_N & -\sum_N m_N X_N Z_N \\ -\sum_N m_N Y_N X_N & \sum_N m_N(X_N^2 + Z_N^2) & -\sum_N m_N Y_N Z_N \\ -\sum_N m_N Z_N X_N & -\sum_N m_N Z_N Y_N & \sum_N m_N(X_N^2 + Y_N^2) \end{bmatrix}$$

$$= \begin{bmatrix} I_{XX} & I_{XY} & I_{XZ} \\ I_{YX} & I_{YY} & I_{YZ} \\ I_{ZX} & I_{ZY} & I_{ZZ} \end{bmatrix} \tag{5.2}$$

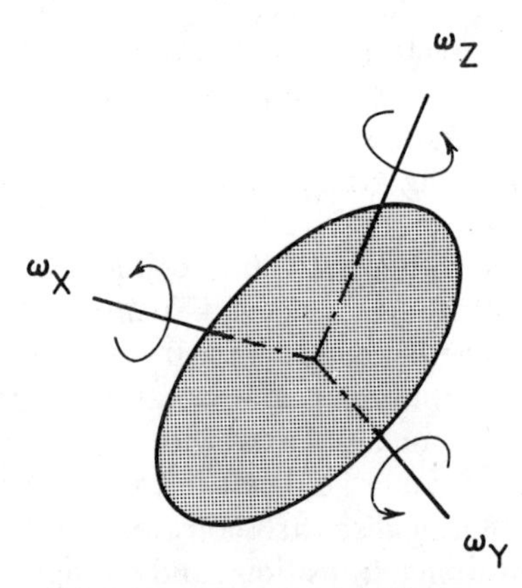

Figure 5.1 The rotational kinetic energy of a rigid body may be specified in terms of angular velocities ω_X, ω_Y, ω_Z about arbitrary orthonormal axes X, Y, Z intersecting at the center of mass.

In general, the classical rotational kinetic energy (5.1) contains three diagonal terms (proportional to ω_X^2, ω_Y^2, and ω_Z^2). It also contains six off-diagonal terms (proportional to $\omega_X\omega_Y$, etc.). Since the inertia tensor is necessarily a symmetric matrix according to Eq. 5.2, only three of these off-diagonal terms are independent.

It always proves possible to rotate the XYZ axes into a new body-fixed coordinate system with orthonormal axes a, b, and c (called the *principal axes*) which diagonalize the inertia tensor,

$$I = \begin{bmatrix} \sum_N M_N(b_N^2 + c_N^2) & 0 & 0 \\ 0 & \sum_N m_N(a_N^2 + c_N^2) & 0 \\ 0 & 0 & \sum_N m_N(a_N^2 + b_N^2) \end{bmatrix}$$

$$\equiv \begin{bmatrix} I_a & 0 & 0 \\ 0 & I_b & 0 \\ 0 & 0 & I_c \end{bmatrix} \tag{5.3}$$

By virtue of Eq. 5.1, these new coordinates also diagonalize the rotational kinetic energy,

$$T = \tfrac{1}{2}(I_a\omega_a^2 + I_b\omega_b^2 + I_c\omega_c^2) \tag{5.4}$$

where ω_a, ω_b, ω_c are the rigid body's angular velocities about the principal axes (Fig. 5.2). Since the body-fixed components of rotational angular momentum about the principal axes are

$$\begin{aligned} J_a &= I_a\omega_a \\ J_b &= I_b\omega_b \\ J_c &= I_c\omega_c \end{aligned} \tag{5.5}$$

the classical Hamiltonian for a rigid free rotor becomes

$$\begin{aligned} H &= \frac{J_a^2}{2I_a} + \frac{J_b^2}{2I_b} + \frac{J_c^2}{2I_c} \\ &\equiv \frac{4\pi^2 c}{h}(AJ_a^2 + BJ_b^2 + CJ_c^2) \end{aligned} \tag{5.6}$$

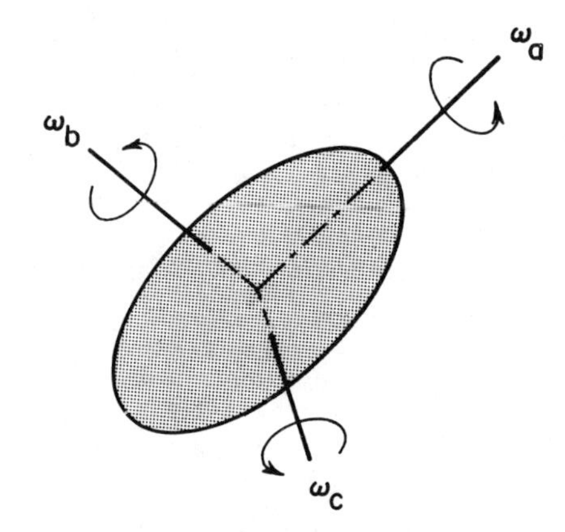

Figure 5.2 For a special choice of body-fixed axes $(X, Y, Z) = (a, b, c)$, the rotational kinetic energy assumes the diagonal form in Eq. 5.4. In the ellipsoid illustrated, one of the principal axes a lies along the C_∞ axis, while the other two principal axes b and c are any pair of mutually orthogonal axes which are perpendicular to a.

Equation 5.6 defines three rotational constants having units of cm^{-1},

$$
\begin{aligned}
A &= h/8\pi^2 cI_a \\
B &= h/8\pi^2 cI_b \qquad (5.7) \\
C &= h/8\pi^2 cI_c
\end{aligned}
$$

The principal axes a, b, c are conventionally labeled so that the rotational constants are ordered in decreasing magnitude, $A \geqslant B \geqslant C$. Four classes of nonlinear rigid rotors may then be distinguished:

$A = B = C$	spherical top
$A > B = C$	prolate symmetric top
$A = B > C$	oblate symmetric top
$A > B > C$	asymmetric top

The orientations of the principal axes are easily obtained by inspection in molecules with sufficient symmetry. Any C_n symmetry axis $(n \geqslant 2)$ coincides with a principal axis. In the D_{2h} ethylene molecule, for example, the $C_2(x)$, $C_2(y)$, and $C_2(z)$ twofold axes are principal axes. All of the principal moments of inertia in ethylene are different in magnitude, so this molecule is an asymmetric top. All molecules with one C_3 or higher-order axis are symmetric tops, because the principal moments about two axes normal to a C_n axis with $n \geqslant 3$ are necessarily equal. Such molecules include BF_3 (which belongs to the D_{3h} point group), NH_3 and CH_3I (C_{3v}), and benzene (D_{6h}). Molecules with two or more C_3 or higher order rotation axes are spherical tops; examples of these are CCl_4 (T_d) and SF_6 (O_h). It is possible in principle for a molecule lacking a C_n axis ($n \geqslant 3$) to exhibit two equal principal moments; such a molecule is termed an *accidental* symmetric top. The frequency resolution afforded by microwave technology can detect such small differences between two moments of inertia that very few molecules pass for accidental symmetric tops under microwave spectroscopy.

In symmetric tops, two of the rotational moments of inertia are equal. The third moment is associated with rotation about the axis of highest symmetry (called the *figure axis*). In prolate symmetric tops, the figure axis exhibits the largest rotational constant (and the smallest principal moment of inertia). Such molecules concentrate most of their nuclear mass along the figure axis (the a axis), and tend to be cigar-shaped. Eclipsed and staggered ethane molecules are both prolate symmetric tops, as is 2-butyne (Fig. 5.3). In oblate symmetric tops, the nuclear mass tends to concentrate at an appreciable distance from the figurc axis (the c axis), endowing the figure axis with the smallest rotational constant and the largest principal moment of inertia. Oblate tops (e.g., benzene) thus resemble disks rather than cigars. Linear polyatomic molecules are a special case of prolate symmetric tops, in which the rotational moment about the figure axis (the C_∞ axis) is negligible. For such species (e.g., OCS, HCN), the rotational Hamiltonian is formally identical to that for a diatomic molecule.

The classical Hamiltonian (5.6) for a rigid rotor becomes simplified when two or more of the principal moments are equal. In the case of the *spherical top* ($A = B = C$), it is

$$H = \frac{4\pi^2 c}{h} B(J_a^2 + J_b^2 + J_c^2) = \frac{4\pi^2 c}{h} BJ^2 \tag{5.8}$$

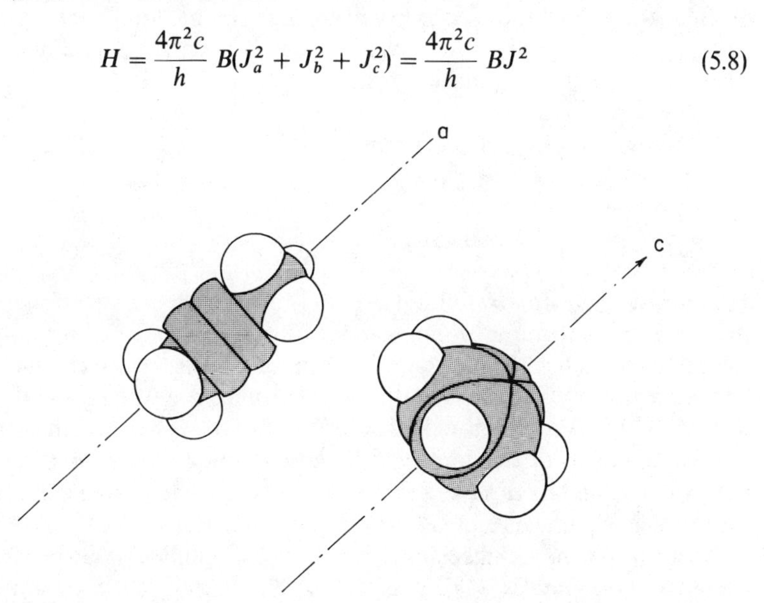

Figure 5.3 Examples of a prolate symmetric top, 2-butyne (left) and an oblate symmetric top, benzene (right). By convention, the figure axis is labeled the *a* axis and the *c* axis in prolate and oblate tops, respectively.

For the *prolate top* ($A > B = C$), the Hamiltonian becomes

$$H = \frac{4\pi^2 c}{h}\,[AJ_a^2 + B(J_b^2 + J_c^2)]$$

$$= \frac{4\pi^2 c}{h}\,[(A - B)J_a^2 + BJ^2] \tag{5.9}$$

while for the *oblate top* ($A = B > C$) it is

$$H = \frac{4\pi^2 c}{h}\,[BJ^2 + (C - B)J_c^2] \tag{5.10}$$

No such simplification is possible for the asymmetric top, where $A > B > C$. We will see that this fact considerably complicates the derivation of the quantum mechanical energy levels for an asymmetric top.

5.2 RIGID ROTOR ANGULAR MOMENTA

Simple as they may appear, the classical Hamiltonians developed for rigid rotors in the preceding section are conceptually new. In our discussion of diatomic rotations in Chapter 3, the rotational states $|JM\rangle$ were obtained as eigenfunctions of the *space-fixed* angular momentum operators $\hat{J}_z$ and $\hat{J}^2 = \hat{J}_x^2 + \hat{J}_y^2 + \hat{J}_z^2$. The space-fixed angular momentum components $\hat{J}_x, \hat{J}_y, \hat{J}_z$ obey the familiar commutation rules

$$[\hat{J}_x, \hat{J}_y] = i\hbar\hat{J}_z$$

$$[\hat{J}_z, \hat{J}^2] = 0 \tag{5.11}$$

and cyclic permutations of x, y, z

In contrast, the rotor Hamiltonian (5.6) is expressed as a function of the *body-fixed* angular momentum components $\mathbf{J}_a, \mathbf{J}_b, \mathbf{J}_c$ associated with rotations about the principal axes a, b, and c, which themselves rotate with the molecule. It is unclear a priori what commutation relationships are obeyed by $\mathbf{J}_a, \mathbf{J}_b, \mathbf{J}_c$ and $\mathbf{J}^2 = \mathbf{J}_a^2 + \mathbf{J}_b^2 + \mathbf{J}_c^2$. It would also be useful to know whether any space-fixed components of $\mathbf{J}$ commute with particular body-fixed components of $\mathbf{J}$, so that commuting sets of observables may be constructed as an aid in visualizing the physical significance of rigid rotor wave functions.

An arbitrary three-dimensional rotation of a rigid body may be described [1] using the *Euler angles* ϕ, θ, χ (Fig. 5.4). The body-fixed a, b, and c axes are initially aligned with the space-fixed x, y, and z axes, respectively. The body is first rotated counterclockwise by the angle ϕ about the z axis; this rotation does not affect the orientation of the c axis, but rotates the a and b axes in the xy

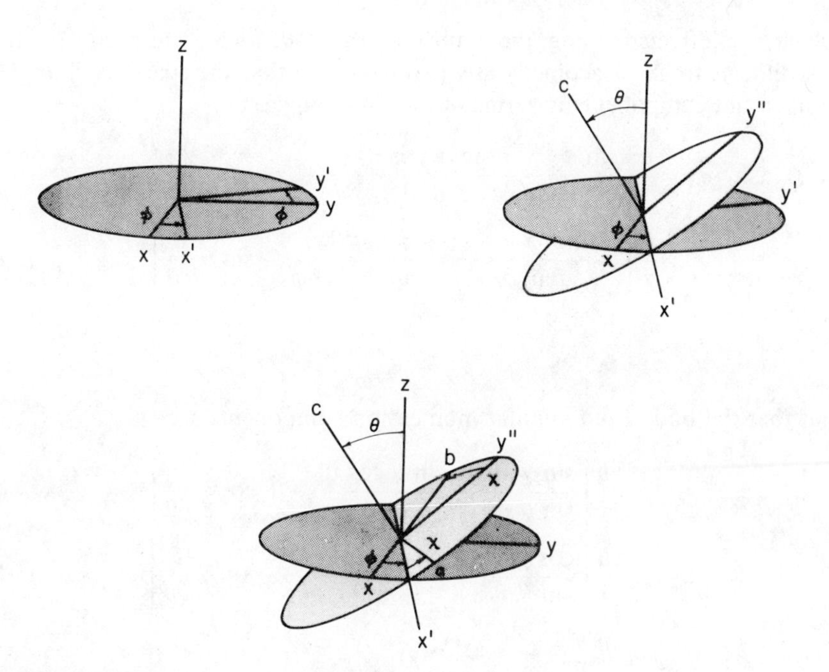

Figure 5.4 Euler angles (ϕ, θ, χ) describing the orientation of body-fixed axes (a, b, c) with respect to space-fixed axes (x, y, z). The molecule is first rotated by an angle ϕ about the space-fixed z axis, causing molecule-fixed axes initially pointing along x and y to be rotated into x' and y', respectively. The molecule is then rotated by the angle θ about the new x' axis, displacing the body-fixed axis initially pointing along z into the direction c and the y' axis into direction y''. The molecule is finally rotated by the angle χ about the c axis, rotating the x' and y'' axes into the a and b axes, respectively.

plane. The second rotation is a counterclockwise rotation by the angle θ about the new orientation of the a axis; it rotates the body-fixed c axis by the angle θ from the space-fixed z axis. The body is finally rotated counterclockwise by the angle χ about the new c axis. The components of the rigid rotor's angular momentum may be specified as projections of $\mathbf{J}$ along either the space-fixed axes (J_x, J_y, J_z) or body-fixed axes (J_a, J_b, J_c). They may also be expressed in terms of the three independent angular momentum components J_ϕ, J_θ, J_χ associated with the Euler rotation axes,

$$
\begin{aligned}
\hat{J}_\phi &= \frac{\hbar}{i}\frac{\partial}{\partial\phi} \\
\hat{J}_\theta &= \frac{\hbar}{i}\frac{\partial}{\partial\theta} \qquad (5.12) \\
\hat{J}_\chi &= \frac{\hbar}{i}\frac{\partial}{\partial\chi}
\end{aligned}
$$

which are directed along the unit vectors $\hat{\phi}$, $\hat{\theta}$, and $\hat{\chi}$, respectively. It is not difficult to show geometrically (Problem 5.1) that the space-fixed angular momentum components in terms of the Euler angles are

$$\begin{aligned}
\hat{J}_x &= \frac{\hbar}{i}\left[\frac{\sin\phi}{\sin\theta}\frac{\partial}{\partial\chi} - \frac{\sin\phi\cos\theta}{\sin\theta}\frac{\partial}{\partial\phi} + \cos\phi\frac{\partial}{\partial\theta}\right] \\
\hat{J}_y &= \frac{\hbar}{i}\left[-\frac{\cos\phi}{\sin\theta}\frac{\partial}{\partial\chi} + \frac{\cos\phi\cos\theta}{\sin\theta}\frac{\partial}{\partial\phi} - \sin\theta\frac{\partial}{\partial\theta}\right] \\
\hat{J}_z &= \frac{\hbar}{i}\frac{\partial}{\partial\phi}
\end{aligned} \tag{5.13}$$

and that the body-fixed angular momentum components are

$$\begin{aligned}
\hat{J}_a &= \frac{\hbar}{i}\left[\frac{\sin\chi}{\sin\theta}\frac{\partial}{\partial\phi} - \frac{\sin\chi\cos\theta}{\sin\theta}\frac{\partial}{\partial\chi} + \cos\chi\frac{\partial}{\partial\theta}\right] \\
\hat{J}_b &= \frac{\hbar}{i}\left[\frac{\cos\chi}{\sin\theta}\frac{\partial}{\partial\phi} - \frac{\cos\chi\cos\theta}{\sin\theta}\frac{\partial}{\partial\chi} - \sin\chi\frac{\partial}{\partial\theta}\right] \\
\hat{J}_c &= \frac{\hbar}{i}\frac{\partial}{\partial\chi}
\end{aligned} \tag{5.14}$$

The well-known space-fixed commutation relationships (5.11) may readily be confirmed using Eqs. 5.13 for $\hat{J}_x$, $\hat{J}_y$, $\hat{J}_z$ in terms of the Euler angles. Use of Eqs. 5.14 for the body-fixed components $\hat{J}_a$, $\hat{J}_b$, $\hat{J}_c$ quickly leads to the somewhat surprising commutation rules:

$$[\hat{J}_a, \hat{J}_b] = -i\hbar\hat{J}_c \quad \text{and cyclic permutations of } a, b, c \tag{5.15}$$

In analogy to space-fixed angular momenta, one consequently obtains

$$[\hat{J}_c, \hat{J}^2] \equiv [\hat{J}_c, \hat{J}_a^2 + \hat{J}_b^2 + \hat{J}_c^2] = 0 \quad \text{and cyclic permutations of } a, b, c \tag{5.16}$$

It is manifest from Eqs. 5.13 and 5.14 that

$$[\hat{J}_z, \hat{J}_c] = 0 \tag{5.17}$$

because the Euler angles ϕ and χ associated with $\hat{J}_z$ and $\hat{J}_c$ respectively are independent variables. It may also be shown [2] that each space-fixed component of **J** commutes with any body-fixed component of **J**,

$$[\hat{J}_i, \hat{J}_\alpha] = 0 \qquad \begin{aligned} i &= x, y, \text{ or } z \\ \alpha &= a, b, \text{ or } c \end{aligned} \tag{5.18}$$

5.3 RIGID ROTOR STATES AND ENERGY LEVELS

We are now in a position to determine the rigid rotor energy levels for tops in order of decreasing symmetry. The Hamiltonian (5.8) for a *spherical top* is functionally indistinguishable from that of a diatomic rotor,

$$\hat{H} = \frac{4\pi^2 c}{h} B\hat{J}^2 = \frac{4\pi^2 c}{h} B(\hat{J}_x^2 + \hat{J}_y^2 + \hat{J}_z^2) \tag{5.19}$$

with the consequence that spherical top energy levels in cm^{-1} are given by (cf. Eq. 3.24)

$$E_J = BJ(J+1) \qquad J = 0, 1, 2, \ldots \tag{5.20}$$

Owing to the spherical symmetry of its inertia tensor ($I_a = I_b = I_c$), the spherical top's rotational levels depend only on the single quantum number J.

In the *oblate symmetric top*, the rotational Hamiltonian is given by Eq. 5.10. The c axis is denoted the figure axis. According to the commutation rules obtained in Section 5.3, one possible commuting set of observables is J_z, J_c, and J^2. It is then possible to formulate rotational states $|JKM\rangle$ which simultaneously obey the eigenvalue equations

$$\hat{J}^2|JKM\rangle = J(J+1)\hbar^2|JKM\rangle \qquad J = 0, 1, 2, \ldots \tag{5.21}$$

$$\hat{J}_z|JKM\rangle = M\hbar|JKM\rangle \qquad -J \leqslant M \leqslant J \tag{5.22}$$

$$\hat{J}_c|JKM\rangle = K\hbar|JKM\rangle \qquad -J \leqslant K \leqslant J \tag{5.23}$$

This implies that the eigenstates $|JKM\rangle$ must behave as

$$|JKM\rangle = e^{iM\phi} f_{KM}^{(J)}(\theta) e^{iK\chi} \tag{5.24}$$

where M and K are both integers ranging from $-J$ to J. The θ-dependent functions $f_{KM}^{(J)}(\theta)$ prove to be hypergeometric functions of $\sin^2(\theta/2)$ [3,4]. These are rarely given explicitly in modern spectroscopy texts, because no knowledge of these functions is necessary to obtain either the energy levels or the selection rules for spectroscopic transitions in symmetric tops. By combining Eqs. 5.10, 5.21, and 5.23, we obtain the energy levels for the oblate symmetric top in cm^{-1},

$$E_{JK} = BJ(J+1) + (C - B)K^2 \tag{5.25}$$

A schematic energy level diagram is given for the oblate top in Fig. 5.5.

The quantum number M measures the projection of the rotational angular momentum $\mathbf{J}$ upon the space-fixed z axis (Eq. 5.22). Since the rotational energy is independent of the orientation of $\mathbf{J}$ in isotropic space, E_{JK} is independent of M (Eq. 5.25). The projection $K\hbar$ of $\mathbf{J}$ on the body-fixed c axis does influence the

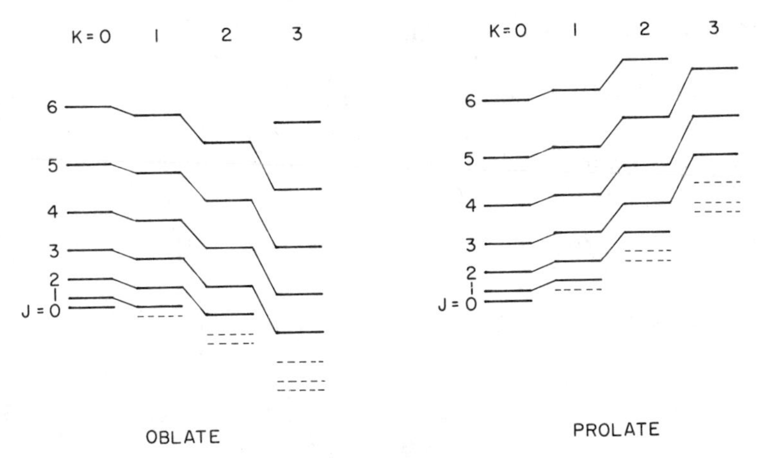

Figure 5.5 Rotational energy levels for oblate and prolate symmetric tops. Note that levels do not exist for $|K| > J$ (dashed lines). For given J, the rotational energy is a decreasing (increasing) function of $|K|$ for oblate (prolate) tops.

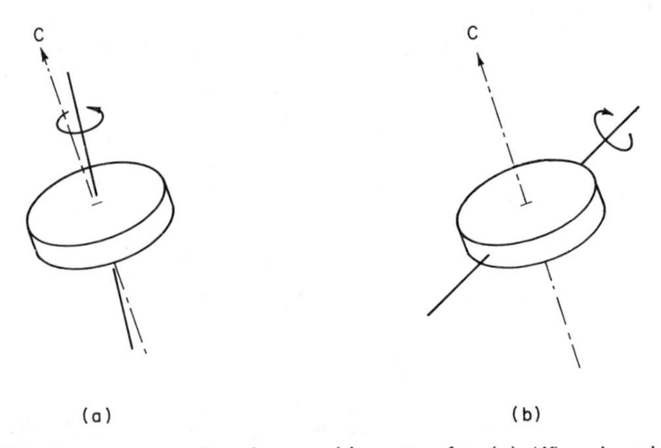

Figure 5.6 Rotational motion in an oblate top for (a) $|K| = J$ and (b) $K = 0$. According to Eq. 5.25, the rotational energy for given J is smaller in case (a) than in case (b). This is in agreement with the classical result, where the rotational energy $J^2/2I$ depends on the total angular momentum **J** and the moment of inertia I about the axis of rotation: I is clearly larger in case (a) than in case (b).

rotational energy, as is illustrated in Fig. 5.6 for an oblate top. When $|K|$ attains its maximum value ($=J$) for given J, the molecule is rotating about an axis that is nearly parallel to the figure axis. (It cannot rotate purely about the c axis: In such a state, one would have $J_a = J_b = 0$, which would violate the uncertainty principle embodied in Eq. 5.15.) When K is small, the molecule undergoes a tumbling motion dominated by rotation about the a and/or b axis. Since numerically different rotational constants are associated with rotations about the figure axis and the normal axes, the rotational energy must depend on $|K|$. It is insensitive to the sign of K, which only controls the direction of rotation about the figure axis.

In a *prolate symmetric top*, the a axis rather than the c axis becomes the figure axis. The body-fixed a, b, c axes depicted in Fig. 5.4 are replaced by the b, c, and a axes, respectively. The mutually commuting set of observables becomes J_z, J_a, and J^2. With these modifications, the oblate top eigenvalue equations and wave functions (5.21)–(5.24) become applicable to the prolate top as well. The pertinent rotational energies in cm^{-1} are

$$E_{JK} = BJ(J + 1) + (A - B)K^2 \tag{5.26}$$

The energy levels for the prolate top are contrasted with those for the oblate top in Fig. 5.5. Since $(C - B) < 0$ in the oblate top and $(A - B) > 0$ in the prolate

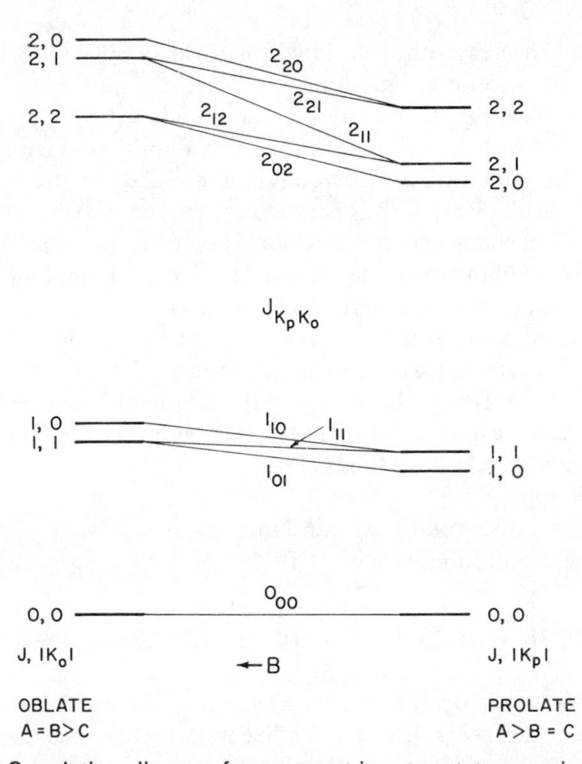

Figure 5.7 Correlation diagram for asymmetric rotor state energies between the oblate ($A = B > C$) and prolate ($A > B = C$) limits. Here the rotational constants A and C are fixed, while B (the horizontal coordinate) is varied continously between A and C. The oblate and prolate top energies are given by Eqs. 5.25 and 5.26, respectively; the asymmetric top energies for intermediate rotational constants B must be found by explicit diagonalization of the asymmetric top Hamiltonian [2, 6].

Table 5.1 Rotational constants in cm^{-1} for some asymmetric, symmetric, and spherical top molecules

Species	Point group	A	B	C
H_2O	C_{2v}	27.877	14.512	9.285
H_2CO	C_{2v}	9.4053	1.2953	1.1342
HCO_2H	C_s	2.58548	0.402112	0.347447
$CH_3{}^{35}Cl$	C_{3v}	5.09	0.443401	—
$CH_3{}^{127}I$	C_{3v}	5.11	0.250217	—
C_2H_6	D_{3d}	2.681	0.6621	—
NH_3	C_{3v}	—	9.4443	6.196
C_6H_6	D_{6h}	—	0.1896	0.0948
CH_4	T_d	—	5.2412	—

Data taken from G. Herzberg, *Molecular Structure and Molecular Spectra, III. Electronic Spectra and Electronic Structure of Polyatomic Molecules*, Van Nostrand Reinhold, New York, 1966.

top, increasing K decreases the rotational energy in an oblate top, but increases the rotational energy in a prolate top.

In the *asymmetric top*, the rotational Hamiltonian (5.6) is proportional to $A\hat{J}_a^2 + B\hat{J}_b^2 + C\hat{J}_c^2$ with $A > B > C$. Since no two body-fixed components of **J** commute (Eq. 5.15), no wave function can simultaneously be an eigenfunction of any two (let alone all three) of the operators $\hat{J}_a^2$, $\hat{J}_b^2$, and $\hat{J}_c^2$. It is possible to find the rotational eigenstates and energies of an asymmetric top by diagonalizing the Hamiltonian (5.6) in the $|JKM\rangle$ basis. This has been done in several texts [2], and it will not be repeated here. An asymmetric top (in which $A > B > C$) may be considered an intermediate case in which the value of the rotational constant B lies between the extremes exhibited in a prolate top ($A > B = C$) and in an oblate top ($A = B > C$). In this spirit, the energies of the asymmetric rotor may be visualized on a qualitative energy level diagram showing the correlations between the prolate and oblate limits (Fig. 5.7). The vast majority of molecules are asymmetric tops.

Some representative rotational constants are listed for asymmetric, symmetric, and spherical top molecules in Table 5.1.

5.4 SELECTION RULES FOR PURE ROTATIONAL TRANSITIONS

We begin this section by deriving the E1 selection rules for rotational transitions in symmetric tops, since spherical and linear molecules may be considered special cases of symmetric tops. Fo: E1-allowed transitions from state $|JKM\rangle$ to state $|J'K'M'\rangle$, we require a nonzero electric dipole transition moment

$$\langle JKM|\boldsymbol{\mu}|J'K'M'\rangle = \langle JKM| \begin{bmatrix} \mu_x \\ \mu_y \\ \mu_z \end{bmatrix} |J'K'M'\rangle \neq 0 \tag{5.27}$$

Here $\boldsymbol{\mu}$ is the permanent molecular electric dipole moment. An intuitive argument will suffice for the selection rule on ΔK. Since $\boldsymbol{\mu}$ must be parallel to the figure axis by symmetry in any prolate or oblate top, and since K controls the velocity of rotation about the figure axis, changing K has no effect on the motion of the molecule's permanent dipole moment. Accordingly, the presence of an oscillating external electric field cannot influence K, and we have the selection rule $\Delta K = 0$.

To obtain the selection rules on ΔJ and ΔM, we exploit the properties of *vector operators.* All quantities that transform like vectors under three-dimensional rotations have operators exhibiting commutation rules that are identical to those shown by the space-fixed angular momentum operators $\hat{J}_x$, $\hat{J}_y$, $\hat{J}_z$. Such operators, which we will denote $\hat{\mathbf{V}} \equiv (\hat{V}_x, \hat{V}_y, \hat{V}_z)$, exhibit the commutation rules

$$
\begin{aligned}
&[\hat{J}_x, \hat{V}_x] = 0 \\
&[\hat{J}_x, \hat{V}_y] = i\hbar\hat{V}_z = -[\hat{J}_y, \hat{V}_x] \\
&\text{and cyclic permutations of } x, y, z
\end{aligned}
\tag{5.28}
$$

These operators are termed vector operators with respect to $\hat{\mathbf{J}}$. The space-fixed angular momentum components $\hat{J}_x$, $\hat{J}_y$, $\hat{J}_z$ are obviously vector operators with respect to themselves (cf. Eq. 5.11). The position and linear momentum $\mathbf{r} = (x, y, z)$ and $\mathbf{p} = (\hat{p}_x, \hat{p}_y, \hat{p}_z)$ are also vector operators, as is the electric dipole moment operator $\boldsymbol{\mu}$. Since we have from Eq. 5.28 that for any vector operator $\hat{\mathbf{V}}$

$$
\begin{aligned}
&[\hat{J}_z, \hat{V}_x] = i\hbar\hat{V}_y \\
&[\hat{J}_z, \hat{V}_y] = i\hbar\hat{V}_x
\end{aligned}
\tag{5.29}
$$

it follows that

$$
[\hat{J}_z, \hat{V}_x + i\hat{V}_y] = \hbar(\hat{V}_x + i\hat{V}_y]
\tag{5.30}
$$

Using the notation $\hat{V}_+ \equiv \hat{V}_x + i\hat{V}_y$, this implies that $[\hat{J}_z, \hat{V}_+] = \hat{J}_z\hat{V}_+ - \hat{V}_+\hat{J}_z = \hbar\hat{V}_+$. We consequently find that for any vector operator

$$
\begin{aligned}
&\langle JKM|\hat{J}_z\hat{V}_+ - \hat{V}_+\hat{J}_z - \hbar\hat{V}_+|J'K'M'\rangle \\
&\quad = (M - M' - 1)\hbar\langle JKM|\hat{V}_+|J'K'M'\rangle = 0
\end{aligned}
\tag{5.31}
$$

Proceeding similarly with the operator $\hat{V}_- \equiv \hat{V}_x - i\hat{V}_y$, we find that $[\hat{J}_z, \hat{V}_-] = -\hbar\hat{V}_-$, with the result that

$$
\begin{aligned}
&\langle JKM|\hat{J}_z\hat{V}_- - \hat{V}_-\hat{J}_z + \hbar\hat{V}_-|J'K'M'\rangle \\
&\quad = (M - M' + 1)\hbar\langle JKM|\hat{V}_-|J'K'M'\rangle = 0
\end{aligned}
\tag{5.32}
$$

Since $[\hat{J}_z, \hat{V}_z] = 0$, we also have

$$\begin{aligned}\langle JKM|\hat{J}_z\hat{V}_z - \hat{V}_z\hat{J}_z|J'K'M'\rangle \\ = (M - M')\hbar\langle JKM|\hat{V}_z|J'K'M'\rangle = 0\end{aligned} \quad (5.33)$$

Recognizing that the dipole moment operator $\boldsymbol{\mu}$ is a vector operator, we see that Eqs. 5.31 and 5.32 give the selection rule $\Delta M = \pm 1$ for the x and y components of the E1 transition moment. From Eq. 5.33, it is clear that $\langle JKM|\mu_z|J'K'M'\rangle$ is zero unless $\Delta M = 0$.

For the selection rule on ΔJ, we may use the more complicated vector operator identity [5]

$$[\hat{J}^2, [\hat{J}^2, \hat{\mathbf{V}}]] = 2\hbar^2[\hat{J}^2\hat{\mathbf{V}} + \hat{\mathbf{V}}\hat{J}^2 - (\hat{\mathbf{J}}\cdot\hat{\mathbf{V}})\hat{\mathbf{J}}] \quad (5.34)$$

This leads to the condition

$$\begin{aligned}&\langle JKM|[\hat{J}^2, [\hat{J}^2, \hat{\mathbf{V}}] - 2\hbar^2[\hat{J}^2\hat{\mathbf{V}} + \hat{\mathbf{V}}\hat{J}^2 - 2(\hat{\mathbf{J}}\cdot\hat{\mathbf{V}})\hat{\mathbf{J}}]|J'K'M'\rangle \\ &= \hbar^4[(J - J')^2 - 1][(J + J' + 1)^2 - 1]\langle JKM|\hat{\mathbf{V}}|J'K'M'\rangle \\ &= 0\end{aligned} \quad (5.35)$$

Thus, either $\langle JKM|\hat{\mathbf{V}}|J'K'M'\rangle$ vanishes, or $J' = J \pm 1$ (the quantity in the second set of square brackets cannot vanish when $J \neq J'$ and $J, J' \geqslant 0$). Hence we find the selection rule $\Delta J = \pm 1$ for pure rotational transitions in a symmetric top. The symmetric top E1 selection rules can therefore be summarized as $\Delta J = \pm 1$, $\Delta K = 0$. In the presence of strong external electric fields, the rotational energy levels depend on M as well as on J and K (via the Stark affect); the selection rules $\Delta M = 0$ (μ_z) and $\Delta M = \pm 1$ (μ_x, μ_y) then become relevant. For freely rotating spherical tops and linear polyatomics, the K dependence drops out of the expressions for the rotational Hamiltonian and energy levels; their selection rules reduce to $\Delta J = \pm 1$, $\Delta M = 0, \pm 1$ as in diatomics (Chapter 3). It is worth remarking that all of the selection rules derived in this section emerged as consequences of vector operator properties; consideration of the specific rotational wave functions (5.24) was unnecessary.

5.5 MICROWAVE SPECTROSCOPY OF POLYATOMIC MOLECULES

Rotational transitions in most polyatomic molecules occur in the microwave region of the electromagnetic spectrum ($\lambda = 1$ mm to 1 cm). In a prototype microwave absorption experiment, microwaves are generated in a specialized oscillator (called a *klystron*), propagated through a hollow rectangular metal pipe (a waveguide) containing the sample gas at low pressure, and finally detected at a silicon crystal. At gas pressures low enough to render collisional broadening negligible (Chapter 8), the frequency widths of rotational absorption lines are narrowed to the point where their frequency positions may be determined with high accuracy. This fact, coupled with the available frequency stability of klystron sources (better than one part in 10^6), endows microwave spectroscopy with the capability of determining the rotational constants extremely to high precision. A drawback of obtaining microwave spectra at such low gas pressures (10^{-3} to 10^{-4} torr) is loss of sensitivity, since the absorbance (Appendix D) is proportional to the molecule number density. This problem is addressed by the technique of Stark-modulated microwave spectroscopy [6]. The waveguide used for microwave propagation through the sample gas is bisected by an insulated, conducting septum running down its length. Application of a square-wave voltage (typical frequency ~ 100 kHz) to the septum generates a modulated electric field in the gas, modifying its rotational energies by the Stark effect. The rotational line frequencies in the presence of the field are shifted from their zero-field positions, because the level Stark shifts depend on J, K, and M. Hence the absorption lines may be switched in and out of resonance with the incident microwave frequency from the klystron, allowing selective detection of the molecule-specific absorption signal with a lock-in amplifier. This widely used method greatly enhances the sensitivity and resolution of microwave spectroscopy.

As examples of the use of microwave spectroscopy to determine equilibrium geometry, consider the molecules NF_3 and CH_3Cl, which both have C_{3v} symmetry. If the N–F bond length is denoted l and the bond angle is θ, the use of Eq. 5.3 leads to the expression

$$I_c = 2m_F l^2(1 - \cos\theta) \tag{5.36}$$

for the moment of inertia about the figure (C_3) axis. Similarly, the two moments of inertia about axes normal to the figure axis are both

$$I_b = m_F l^2(1 - \cos\theta) + \frac{m_N m_F l^2}{3m_F + m_N}(1 + 2\cos\theta) \tag{5.37}$$

Since the symmetric top selection rules are $\Delta J = \pm 1$, $\Delta K = 0$, absorptive transitions will be observed from state $|J - 1, KM\rangle$ to states $|JKM\rangle$ and $|JKM \pm 1\rangle$, occurring at frequencies $2BJ$ according to Eq. 5.25. Hence, the microwave spectrum of NF_3 yields no measurement of the rotational constant C. Since two structural parameters (l and θ) enter in the rotational moment I_b, a single microwave spectrum cannot determine the geometry of NF_3. However, spectra of the isotopic species $^{14}NF_3$ and $^{15}NF_3$ may be combined to give two equations similar to Eq. 5.37, assuming the isotopes have identical geometry. In this manner, l and θ were determined to be 1.71 Å and 102°9′, respectively [6]. In a similar vein, the perpendicular moments of inertia in C_{3v} CH_3Cl are given by

$$I_b = m_H l_1^2(1 - \cos\theta) + \frac{1}{3m_H + m_C + m_{Cl}} \times \left\{ m_H(m_C + m_{Cl})l_1^2(1 + 2\cos\theta) + m_{Cl}l_2\left[(m_C + 3m_H)l_2 + 6m_H l_1\left(\frac{1 + 2\cos\theta}{3}\right)^{1/2}\right]\right\} \tag{5.38}$$

where l_1 is the C–H bond length, l_2 is the C–Cl bond length, and θ is the H–C–H bond angle. (The expression for the moment about the figure axis is analogous to that for NF_3 in Eq. 5.36; the $\Delta K = 0$ selection rate prevents its measurement by microwave spectroscopy.) In this case, microwave spectra of three isotopic species must be made in principle to determine l_1, l_2, and θ. The assumption that molecular geometries are insensitive to isotopic substitution is not always justified; the C–H bond distance in CH_3Cl is in fact some 0.009 Å larger than the C–D bond distance in the deuterated compound [6]. Hydrogen atom positions appear to be especially prone to isotopic geometry variation.

REFERENCES

1. J. B. Marion, *Classical Dynamics of Particles and Systems*, Academic, New York, 1965.
2. W. H. Flygare, *Molecular Structure and Dynamics*, Prentice-Hall, Englewood Cliffs, NJ, 1978.
3. G. Herzberg, *Molecular Spectra and Molecular Structure, II. Infrared and Raman Spectra of Polyatomic Molecules*, Van Nostrand, Princeton, NJ, 1945.
4. L. Pauling and E. B. Wilson, *Introduction to Quantum Mechanics*, McGraw-Hill, New York, 1935.
5. E. U. Condon and G. H. Shortley, *The Theory of Atomic Spectra*, Cambridge Univ. Press, London, 1935.
6. C. H. Townes and A. L. Schawlow, *Microwave Spectroscopy*, Dover, New York, 1975.

PROBLEMS

1. In this problem, we derive the expressions given in Eqs. 5.13 and 5.14 for the space-fixed and molecule-fixed angular momentum components in terms of the Euler angles ϕ, θ, and χ. It is already clear from Fig. 5.4 that $\hat{J}_z = \hat{J}_\phi = (\hbar/i)\partial/\partial\phi$ and $\hat{J}_c = \hat{J}_\chi = (\hbar/i)\partial/\partial\chi$. Hence, the problem reduces to finding $\hat{J}_x$, $\hat{J}_y$, $\hat{J}_a$, and $\hat{J}_b$ in terms of the Euler angles.

(a) Show geometrically that the projection of the unit vector $\hat{\chi}$ in the xy plane is $\hat{\chi} - (\hat{\chi}\cdot\hat{\phi})\hat{\phi}$. Then show that this projection, normalized to unit length, becomes

$$\hat{\chi}_\perp = \frac{\hat{\chi} - (\hat{\chi}\cdot\hat{\phi})\hat{\phi}}{[1 - (\hat{\chi}\cdot\hat{\phi})^2]^{1/2}} = \frac{\hat{\chi} - \cos\theta\hat{\phi}}{\sin\theta}$$

(b) Noting that $\hat{\chi}_\perp$ and $\hat{\theta}$ are orthogonal unit vectors in the xy plane, show that

$$\hat{x} = (\hat{\chi}_\perp\cdot\hat{x})\hat{\chi}_\perp + (\hat{\theta}\cdot\hat{x})\hat{\theta} = \sin\phi\cdot\hat{\chi}_\perp + \cos\phi\cdot\hat{\theta}$$

and similarly that

$$\hat{y} = -\cos\phi\cdot\hat{\chi}_\perp + \sin\phi\cdot\hat{\theta}$$

(c) Using the fact that

$$\hat{J}_x = \hat{x}\cdot\hat{\mathbf{J}} = \frac{\hbar}{i}\,\hat{x}\cdot\left(\hat{\phi}\,\frac{\partial}{\partial\phi} + \hat{\theta}\,\frac{\partial}{\partial\theta} + \hat{\chi}\,\frac{\partial}{\partial\chi}\right)$$

$$\hat{J}_y = \hat{y}\cdot\hat{\mathbf{J}} = \frac{\hbar}{i}\,\hat{y}\cdot\left(\hat{\phi}\,\frac{\partial}{\partial\phi} + \hat{\theta}\,\frac{\partial}{\partial\theta} + \hat{\chi}\,\frac{\partial}{\partial\chi}\right)$$

derive Eqs. 5.13 for J_x and J_y.

(d) Using an analogous procedure, derive Eqs. 5.14 for $\hat{J}_a$ and $\hat{J}_b$. (*Hint*: The unit vectors $\hat{a}$ and $\hat{b}$ lie in a tilted plane perpendicular to $\hat{\chi}$. Express $\hat{a}$ and $\hat{b}$ in terms of $\hat{\theta}$ and the projection $\hat{\phi}_\perp$ of $\hat{\phi}$ upon this plane. The desired quantities are given by $J_a = \hat{a}\cdot\mathbf{J}$ and $J_b = b\cdot\mathbf{J}$.)

2. Obtain the angular momentum commutation relations (5.15) and (5.17) directly from Eqs. 5.13–5.14 for the space-fixed and body-fixed angular momentum components.

3. Classify each of the following molecules by point group and by rotor type (spherical, symmetric, or asymmetric). Which of them will exhibit a microwave spectrum?

(a) SF_6
(b) Allene, C_3H_4
(c) Fluorobenzene, C_6H_5F
(d) NH_3
(e) CH_2Cl_2

4. The microwave spectroscopic bond lengths l and bond angles θ for the C_{3v} molecules PF_3 and $P^{35}Cl_3$ are given below. Determine whether each molecule is

a prolate or an oblate symmetric top.

	l(Å)	θ(°)
PF_3	1.55	102°
$P^{35}Cl_3$	2.043	100°6′

5. The microwave spectrum of NH_3 yields the rotational constant $B = 9.933\ \text{cm}^{-1}$. If the N–H bond length is independently known to be 1.014 Å in NH_3, what is the H–N–H bond angle?

6. Consider a hypothetical molecule AB_5C which is known to exhibit C_{4v} geometry (it would be an O_h molecule if atoms B and C were identical and all bond lengths were equal).

(a) Obtain expressions for the principal moments of inertia.
(b) In view of the symmetric top selection rules, for how many isotopic species must microwave spectra be obtained to specify its geometry?

7. For what H–N–H bond angle would C_{3v} NH_3 become an accidental spherical top?

8.

(a) Show that the moment of inertia in the linear OCS molecule is given by

$$I_b = [m_O m_C l_{CO}^2 + m_C m_S l_{CS}^2 + m_O m_S (l_{CO} + l_{CS})^2]/m$$

where m is the total mass $m_O + m_C + m_S$.
(b) The pure rotational spectrum of $^{16}O^{12}C^{32}S$ exhibits adjacent lines at the frequencies 48651.7 and 60814.1 MHz. Assign the J values for these transitions and calculate the rotational constant B for this isotope.
(c) The frequencies of one of the lines in the spectrum of $^{16}O^{12}C^{34}S$ is 23731.3 MHz. Assign this line, and compute the experimental values of l_{CO} and l_{CS}.

9. The FNO molecule is a bent triatomic with $l_{NF} = 1.52$ Å, $l_{NO} = 1.13$Å; the F–N–O bond angle is 110°.

(a) Locate the center of mass, and evaluate the moment of inertia tensor in a right-handed Cartesian coordinate system in which $\hat{x}$ points along the N–F bond, $\hat{y}$ lies in the molecular plane, and $\hat{z}$ is perpendicular to the plane.
(b) Diagonalize the inertia tensor to obtain the principal moments of inertia I_a, I_b, I_c. Determine the directions $\hat{a}$, $\hat{b}$, $\hat{c}$ of the principal axes.

6

POLYATOMIC VIBRATIONS

In a polyatomic molecule with N nuclei, $3N$ independent coordinates are required to specify all of the nuclear positions in space. We have already seen in the preceding chapter that rotations of nonlinear polyatomics about their center of mass may be described in terms of the three Euler angles ϕ, θ, and χ. Three additional coordinates are required to describe spatial translation of a molecule's center of mass. Hence, there will be $3N - 6$ independent vibrational coordinates in a nonlinear polyatomic molecule. In a linear polyatomic molecule, the orientation may be given in terms of two independent angles θ and ϕ. Linear polyatomics therefore exhibit $3N - 5$ rather than $3N - 6$ independent vibrational coordinates.

By their nature, such vibrational coordinates involve collective, oscillatory nuclear motions that leave the molecule's center of mass undisplaced. It is of interest to know the relative nuclear displacements and vibrational frequencies associated with these coordinates, because they are instrumental in predicting band positions and intensities in vibrational spectroscopy. An important question arises as to whether vibrational motion occurs in modes that are *dynamically uncoupled.* If it does, an isolated molecule initially having several quanta of vibrational energy in a particular mode will not spontaneously redistribute this energy into some of its other modes, even though such a process may conserve energy. Hence, the form of the modes has implications not only for vibrational structure (as manifested by energy levels and selection rules in infrared spectra), but also for understanding dynamical processes like vibrational energy transfer in collisions and intramolecular vibrational relaxation (IVR). These phenomena are currently well-pursued research areas, and can only be understood with a seasoned physical appreciation of vibrational modes.

This chapter begins with a classical treatment of vibrational motion, because most of the important concepts that are specific to vibrations in polyatomics carry over naturally from the classical to the quantum mechanical description. In molecules with *harmonic* potential energy functions, vibrational motion occurs in *normal modes* that are mutually uncoupled. Coupling between vibrational modes inevitably occurs in the presence of anharmonic potentials (potentials exhibiting cubic and/or higher order terms in the nuclear coordinates). In molecules with sufficient symmetry, the use of group theory simplifies the procedure of obtaining the normal mode frequencies and coordinates. We obtain E1 selection rules for vibrational transitions in polyatomics, and consider the rotational fine structure of vibrational bands. We finally treat breakdown of the normal mode approximation in real molecules, and discuss the local mode formulation of vibrational motion in polyatomics.

6.1 CLASSICAL TREATMENT OF VIBRATIONS IN POLYATOMICS

The positions of all N nuclei in a polyatomic molecule may be specified using the $3N$ Cartesian coordinates $\xi_1, \xi_2, \ldots, \xi_{3N}$. In terms of these, the nuclear kinetic energy is given by

$$T = \frac{1}{2} \sum_{i=1}^{3N} m_i \dot{\xi}_i^2 \equiv T(\dot{\xi}_1, \ldots, \dot{\xi}_{3N}) \tag{6.1}$$

where m_i is the mass of the nucleus with which coordinate ξ_i is associated. In the Born-Oppenheimer approximation, the eigenvalues of the electronic Hamiltonian will act as a conservative potential energy function $V(\xi_1, \ldots, \xi_{3N})$ for vibrational motion (in reality, V will only depend on $3N - 6$ independent coordinates in nonlinear polyatomics). If one forms the Lagrangian function $L = T - V$, the nuclear Cartesian coordinates will obey the equations of motion

$$\frac{d}{dt}\frac{\partial L}{\partial \dot{\xi}} - \frac{\partial L}{\partial \xi_i} = 0 \qquad i = 1, \ldots, 3N \tag{6.2}$$

which lead to

$$\frac{d}{dt}(m_i \dot{\xi}_i) + \frac{\partial V}{\partial \xi_i} = 0 \qquad i = 1, \ldots, 3N \tag{6.3}$$

These equations may be solved for a specified potential energy function V. In analogy to what was done for diatomics (Eq. 3.32), we may expand the vibrational potential V as a Taylor series about the equilibrium geometry,

$$V = V_0 + \sum_{i}^{3N} \left(\frac{\partial V}{\partial \xi_i}\right)_0 \xi_i + \frac{1}{2} \sum_{ij}^{3N} \left(\frac{\partial^2 V}{\partial \xi_i \partial \xi_j}\right)_0 \xi_i \xi_j + \cdots \tag{6.4}$$

Arbitrarily setting $V_0 = 0$ and recognizing that $(\partial V/\partial \xi_i) = 0$ for all coordinates ξ_i at the equilibrium geometry, we obtain the general potential energy function

$$V = \frac{1}{2} \sum_{ij}^{3N} c_{ij} \xi_i \xi_j + \cdots \tag{6.5}$$

where we have introduced the notation $c_{ij} = (\partial^2 V/\partial \xi_i \partial \xi_j)_0$. If we now make the *harmonic approximation* by ignoring third- and higher order terms in Eq. 6.5, the equations of motion become

$$\frac{d}{dt}(m_i \dot{\xi}_i) + \sum_j c_{ij} \xi_j = 0 \tag{6.6}$$

In terms of the *mass-weighted coordinates*

$$\eta_i = \xi_i \sqrt{m_i} \tag{6.7}$$

the kinetic energy assumes the simpler form

$$T = \frac{1}{2} \sum_{i}^{3N} \dot{\eta}_i^2 \tag{6.8}$$

The equations of motion then become transformed into

$$\frac{d}{dt}(\dot{\eta}_i) + \sum_j b_{ij} \eta_j = 0 \tag{6.9}$$

with $b_{ij} = c_{ij}/(m_i m_j)^{1/2}$. Note that motion in any mass-weighted coordinate η_i is coupled to motion in all other coordinates η_j when $b_{ij} \neq 0$, so that polyatomic vibrations generally involve all of the nuclei moving *simultaneously* in collective motions. Since the solution to the classical equation of motion for an undamped one-dimensional harmonic oscillator is a sinusoidal function of time [1], it is physically reasonable to try solutions of the form

$$\eta_i = \eta_i^0 \sin(t\sqrt{\lambda} + \delta) \qquad i = 1, \ldots, 3N \tag{6.10}$$

for the coupled homogeneous second-order differential equations (6.9). Differentiating this twice with respect to time gives

$$\ddot{\eta}_i = -\lambda \eta_i^0 \sin(t\sqrt{\lambda} + \delta) \tag{6.11}$$

and then substituting Eqs. 6.10 and 6.11 into the coupled equations (6.9) yields

$$-\lambda \eta_i^0 + \sum_j b_{ij} \eta_j^0 = 0 \qquad i, j = 1, \ldots, 3N \tag{6.12}$$

because the function $\sin(t\sqrt{\lambda} + \delta)$ cancels throughout. Writing the trial solutions η_i in the form of Eq. 6.10 is tantamount, by the way, to assuming that *all* of the nuclear motions in a given vibrational mode oscillate with the *same frequency* $\nu = \sqrt{\lambda}/2\pi$ and at the *same phase* δ. The simultaneous equations (6.12) for the $3N$ oscillation amplitudes η_i^0 can have nontrivial solutions only when [1]

$$\begin{vmatrix} b_{11} - \lambda & b_{12} & b_{1,3N} \\ b_{21} & b_{22} - \lambda & \vdots \\ \vdots & & \\ b_{3N,1} & \cdots & b_{3N,3N} - \lambda \end{vmatrix} = 0 \qquad (6.13)$$

The $3N$ roots $\lambda_1, \lambda_2, \ldots, \lambda_{3N}$ of this secular determinant are related to the allowed vibrational frequencies ν_i by $\lambda_i = 4\pi^2\nu_i^2$. They depend on the coefficients b_{ij}, which are in turn governed by the potential energy function (via $c_{ij} = (\partial^2 V/\partial\xi_i\partial\xi_i)$) and the nuclear masses (via $b_{ij} = c_{ij}/(m_i m_j)^{1/2}$). It turns out that six of the eigenvalues of (6.13) will be zero in general (five in linear molecules), corresponding to the nonoscillating center-of-mass translational and rotational motions. The only approximation we have used in this treatment was to break off the Taylor series (6.4) at the second-order (harmonic) approximation; otherwise it is classically exact. The principal difficulty with this approach is that except in molecules with special symmetry, the coefficients $c_{ij} = (\partial^2 V/\partial\xi_i\partial\xi_j)$ are not easily evaluated in terms of the Cartesian coordinates ξ_i.

We will now apply this formalism to the example of a linear ABC molecule in which the nuclear motion is artificially restricted to motion along the internuclear axis for simplicity. In a full three-dimensional treatment, such a molecule would have three translations, two rotations, and four vibrational modes. Constraining the nuclei to one-dimensional motion along the axis will eliminate two of the translations, both rotations, and two bending vibrational modes (which involve nuclear displacements perpendicular to the axis). The nuclear masses and Cartesian coordinates are shown in Fig. 6.1. For this system, the nuclear kinetic energy is rigorously given by

$$\begin{aligned} 2T &= m_A\dot{\xi}_1^2 + m_B\dot{\xi}_2^2 + m_C\dot{\xi}_3^2 \\ &= \dot{\eta}_1^2 + \dot{\eta}_2^2 + \dot{\eta}_3^2 \end{aligned} \qquad (6.14)$$

We may use a harmonic approximation for the potential energy function

$$2V = k_1(\xi_2 - \xi_1)^2 + k_2(\xi_3 - \xi_2)^2 \qquad (6.15)$$

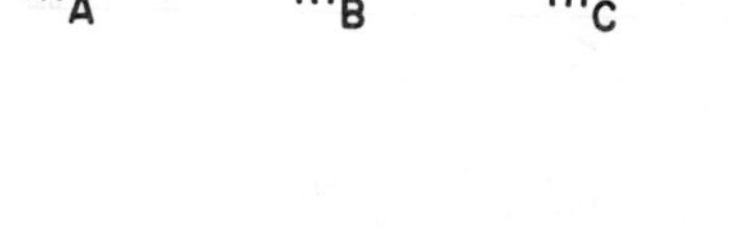

Figure 6.1 Nuclear masses and Cartesian displacement coordinates for linear ABC.

which attributes Hooke's law forces to the A–B and B–C bonds with force constants k_1 and k_2, respectively. (A more accurate potential could also include some interaction between the end atoms, and so depend on $(\xi_3 - \xi_1)$ as well). In mass-weighted coordinates, the potential becomes

$$\begin{aligned} 2V = {} & k_1\eta_1^2/m_A + \eta_2^2(k_1 + k_2)/m_B + k_2\eta_3^2/m_C \\ & - 2k_1\eta_1\eta_2/\sqrt{m_A m_B} - 2k_2\eta_2\eta_3/\sqrt{m_B m_C} \end{aligned} \tag{6.16}$$

The secular determinant (6.13) then becomes

$$\begin{vmatrix} k_1/m_A - \lambda & -k_1/\sqrt{m_A m_B} & 0 \\ -k_1/\sqrt{m_A m_B} & (k_1 + k_2)/m_B - \lambda & -k_2/\sqrt{m_B m_C} \\ 0 & -k_2/\sqrt{m_B m_C} & k_2/m_C - \lambda \end{vmatrix}$$

$$\begin{aligned} &= \lambda \left\{ \lambda^2 - \lambda \left[k_1 \left(\frac{1}{m_A} + \frac{1}{m_B} \right) + k_2 \left(\frac{1}{m_B} + \frac{1}{m_C} \right) \right] \right. \\ &\quad \left. + k_1 k_2 (m_A + m_B + m_C)/m_A m_B m_C \right\} = 0 \end{aligned} \tag{6.17}$$

Hence the three roots of the secular determinant are given by

$$\begin{aligned} \lambda_1 + \lambda_2 &= k_1 \left(\frac{1}{m_A} + \frac{1}{m_B} \right) + k_2 \left(\frac{1}{m_B} + \frac{1}{m_C} \right) \\ \lambda_1 \lambda_2 &= (m_A + m_B + m_C) k_1 k_2 / m_A m_B m_C \\ \lambda_3 &= 0 \end{aligned} \tag{6.18}$$

These roots λ_i are related to the allowed vibrational frequencies ν_i by $\lambda_i = 4\pi^2\nu_i^2$. The third root $\lambda_3 = 0$ is associated with the zero-frequency translation of molecule's center of mass along the molecular axis; the nonzero roots λ_1 and λ_2 correspond to vibrational modes involving stretching of the A–B and B–C bonds.

We can gain considerably more insight by recasting this treatment in the form of a matrix eigenvalue problem, because then we can exploit several well-established theorems from matrix algebra [2]. The coefficients b_{ij} in the potential energy function (6.9) may be organized into the matrix

$$\mathbf{B} = \begin{bmatrix} b_{11} & b_{12} & \cdots & b_{1,3N} \\ b_{21} & & & \vdots \\ \vdots & & & \\ b_{3N,1} & \cdots & & b_{3N,3N} \end{bmatrix} \tag{6.19}$$

and the set of mass-weighted coordinates η_i can be written as the column vector

$$\boldsymbol{\eta} = \begin{bmatrix} \eta_1 \\ \eta_2 \\ \vdots \\ \eta_{3N} \end{bmatrix} \tag{6.20}$$

with transpose

$$\boldsymbol{\eta}^t = [\eta_1 \quad \eta_2 \quad \cdots \quad \eta_{3N}] \tag{6.21}$$

The secular determinant equation (6.13) then becomes

$$|\mathbf{B} - \lambda \mathbf{E}| = 0 \tag{6.22}$$

where $\mathbf{E}$ is the unit matrix. Each of the roots λ_k of this secular determinant can be substituted back into Eqs. 6.12 to give a new set of equations. These give the relative amplitudes η_{ik}^0 of oscillation in the coordinates η_i corresponding to the kth allowed vibrational frequency $\nu_k = \sqrt{\lambda_k}/2\pi$:

$$\begin{aligned} b_{11}\eta_{1k}^0 + b_{12}\eta_{2k}^0 + \cdots - \lambda_k \eta_{1k}^0 &= 0 \\ &\vdots \\ b_{3N,1}\eta_{1k}^0 + b_{3N,2}\eta_{2k}^0 + \cdots - \lambda_k \eta_{3N,k}^0 &= 0 \end{aligned} \tag{6.23}$$

For a given vibrational frequency ν_k, these relative displacements may be expressed in a column vector

$$\boldsymbol{\eta}_k^0 = \begin{bmatrix} \eta_{1k}^0 \\ \eta_{2k}^0 \\ \vdots \\ \eta_{3N,k}^0 \end{bmatrix} \tag{6.24}$$

These column vectors may be normalized,

$$l_{ik} = C_k \eta_{ik}^0 \tag{6.25}$$

so that

$$\sum_i l_{ik}^2 = 1 \tag{6.26}$$

In matrix notation, the normalized column vectors

$$\mathbf{l}_k = \begin{bmatrix} l_{1k} \\ l_{2k} \\ \vdots \\ l_{3N,k} \end{bmatrix} \tag{6.27}$$

obey the condition $\mathbf{l}_k^t \cdot \mathbf{l}_k = 1$. All of the information about the form of the vibrational motions is contained in the normalized eigenvectors $\mathbf{l}_k$, whose elements reflect the *relative* displacements in each of the mass-weighted coordinates. The number C_k in Eq. 6.25 is just a constant fixing the vibrational energy in the mode with frequency ν_k; C_k^{-2} is proportional to the total energy $T + V$ in mode k, and forms an allowed vibrational energy continuum in the classical picture.

Finding the roots $\lambda_1, \ldots, \lambda_{3N}$ of the secular determinant (6.13) is mathematically equivalent to finding a similarity transformation which diagonalizes the **B** matrix,

$$\mathbf{L}^{-1} \cdot \mathbf{B} \cdot \mathbf{L} = \begin{bmatrix} \lambda_1 & 0 & 0 & 0 & \cdots & 0 \\ 0 & \lambda_2 & 0 & 0 & \cdots & 0 \\ 0 & 0 & \lambda_3 & 0 & \cdots & 0 \\ \vdots & \vdots & \vdots & \ddots & & \vdots \\ 0 & 0 & 0 & \cdots & & \lambda_{3N} \end{bmatrix} \equiv \boldsymbol{\Lambda} \tag{6.28}$$

where $\boldsymbol{\Lambda}$ is a diagonalized matrix of eigenvalues whose elements are the roots. Each of the column vectors $\mathbf{l}_k$ (Eq. 6.27) is an *eigenvector* of **B** with eigenvalue λ_k,

$$\mathbf{B} \cdot \mathbf{l}_k = \lambda_k \mathbf{l}_k$$

According to the theory of the matrix eigenvalue problem [2], the matrix **L** required for the similarity transformation (6.28) is equivalent to the *matrix of eigenvectors*

$$\mathbf{L} \equiv \begin{bmatrix} l_{11} & l_{12} & \cdots & l_{1,3N} \\ l_{21} & l_{22} & \cdots & l_{2,3N} \\ \vdots & & & \vdots \\ l_{3N,1} & & \cdots & l_{3N,3N} \end{bmatrix} \tag{6.29}$$

obtained by stacking up the eigenvectors $\mathbf{l}_1, \mathbf{l}_2, \ldots, \mathbf{l}_{3N}$ side by side. When **B** is a symmetric matrix ($b_{ij} = b_{ji}$)—which it must be in our problem because $b_{ij} = (m_i m_j)^{-1/2} \partial^2 V / \partial\xi_i \partial\xi_j$—the eigenvectors $\mathbf{l}_k$ form an orthonormal set

$$\mathbf{l}_k^t \cdot \mathbf{l}_{k'} = \delta_{kk'} \tag{6.30}$$

and the inverse $\mathbf{L}^{-1}$ of the matrix of eigenvectors is given by

$$\mathbf{L}^{-1} = \mathbf{L}^{t} \tag{6.31}$$

where the elements $(\mathbf{C}^{t})_{ij}$ of the transpose $\mathbf{C}^{t}$ of any matrix $\mathbf{C}$ are defined as $(\mathbf{C}^{t})_{ij} = c_{ji}$.

To illustrate how we obtain the normalized eigenvectors $\mathbf{l}_k$ for molecular vibrations, we resume our discussion of the linear ABC molecule. When specialized to the more symmetrical case of linear ABA in which the end masses and force constants are equal ($m_A = m_C$ and $k_1 = k_2 = k$), the secular determinant roots λ_k in Eq. 6.18 become

$$\begin{aligned} \lambda_1 &= k/m_A \\ \lambda_2 &= k/m_A + 2k/m_B \\ \lambda_3 &= 0 \end{aligned} \tag{6.32}$$

Substitution of the root $\lambda_3 = 0$ into one of Eqs. 6.23 then yields

$$(k/m_A)\eta_{13}^0 + (-k/\sqrt{m_A m_B})\eta_{23}^0 = 0 \tag{6.33}$$

since $b_{13} = 0$ and $\lambda_3 = 0$. Similarly,

$$b_{31}\eta_{13}^0 + b_{32}\eta_{23}^0 + b_{33}\eta_{33}^0 - \lambda_3\eta_{33}^0 = 0 \tag{6.34}$$

leads to

$$(-k/\sqrt{m_A m_B})\eta_{23}^0 + (k/m_A)\eta_{33}^0 = 0 \tag{6.35}$$

Equations 6.34 and 6.35 imply that

$$\begin{aligned} \eta_{13}^0/\eta_{23}^0 &= (m_A/m_B)^{1/2} \quad \text{and} \quad \xi_{13}^0/\xi_{23}^0 = 1 \\ \eta_{33}^0/\eta_{23}^0 &= (m_A/m_B)^{1/2} \quad \text{and} \quad \xi_{33}^0/\xi_{23}^0 = 1 \end{aligned} \tag{6.36}$$

Hence it is apparent that $\xi_{13}^0 = \xi_{23}^0 = \xi_{33}^0$, so that the zero-frequency mode ($\lambda_3 = 0$) is associated with overall molecular translation parallel to the molecular axis (Fig. 6.2). The normalized eigenvector for mode 3 can now easily be shown to be

$$\mathbf{l}_3 = \begin{bmatrix} (m_A/m)^{1/2} \\ (m_B/m)^{1/2} \\ (m_A/m)^{1/2} \end{bmatrix} \tag{6.37}$$

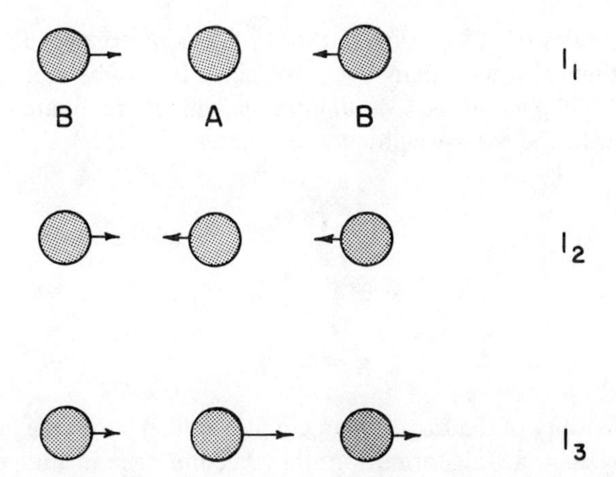

Figure 6.2 Eigenvectors $\mathbf{l}_1$, $\mathbf{l}_2$, $\mathbf{l}_3$ of the matrix **B** for linear AB_2. Lengths of arrows give relative displacements in mass-weighted (not Cartesian) coordinates.

where $m = 2m_A + m_B$ is the total nuclear mass. In a similar way, the normalized eigenvectors for the other two modes can be derived as

$$\mathbf{l}_1 = \begin{bmatrix} 1/\sqrt{2} \\ 0 \\ -1/\sqrt{2} \end{bmatrix} \tag{6.38}$$

$$\mathbf{l}_2 = \begin{bmatrix} (m_B/2m)^{1/2} \\ -2(m_A/2m)^{1/2} \\ (m_B/2m)^{1/2} \end{bmatrix} \tag{6.39}$$

It may be verified that these normalized eigenvectors obey the normalization and orthogonality conditions (6.26) and (6.30). It is apparent from Eq. 6.38 that mode 1 is a symmetric stretching mode (Fig. 6.2), because the coordinate η_2 of atom B has zero coefficient in eigenvector $\mathbf{l}_1$, and the eigenvalue λ_1 has no dependence on m_B—as would be expected if nucleus B were motionless. It is clear from the form of eigenvector $\mathbf{l}_2$ that mode 2 is an asymmetric stretching mode (Fig. 6.2) in which the A nuclei move together in a direction opposite to that of nucleus B. Since all of the nuclei move in mode 2, η_2 depends on m_B as well as on m_A.

6.2 NORMAL COORDINATES

To a good approximation, the restoring forces responsible for molecular vibrations are directed along chemical bonds. The potential energy function (6.4) is therefore not easily expressed in Cartesian coordinates except for molecules having bond angles of 90° and/or 180° (e.g., linear molecules, octahedral SF_6, and the hypothetical square-planar A_4 molecule). It may be more naturally described using $3N$ *generalized coordinates* q_i which are related by a linear transformation to the mass-weighted coordinates,

$$\eta_i = \sum_j^{3N} M_{ij} q_j$$

or

$$\boldsymbol{\eta} = \mathbf{M} \cdot \mathbf{q} \tag{6.40}$$

The enormous utility of the Lagrangian equations (6.2) lies in the fact that if $\mathbf{q}$ is related to $\boldsymbol{\eta}$ by such a transformation, they become true in the generalized as well as in Cartesian coordinates:

$$\frac{d}{dt}\frac{\partial L}{\partial \dot{q}_j} - \frac{\partial L}{\partial q_j} = 0 \tag{6.41}$$

The potential function in the harmonic approximation now becomes

$$\begin{aligned} 2V &\equiv \boldsymbol{\eta}^t \cdot \mathbf{B} \cdot \boldsymbol{\eta} = (\mathbf{M} \cdot \mathbf{q})^t \cdot \mathbf{B} \cdot (\mathbf{M} \cdot \mathbf{q}) \\ &= \mathbf{q}^t \cdot \mathbf{B}' \cdot \mathbf{q} \end{aligned} \tag{6.42}$$

where the matrix $\mathbf{B}$ in the Cartesian basis has become transformed into

$$\mathbf{B}' = \mathbf{M}^t \cdot \mathbf{B} \cdot \mathbf{M} \tag{6.43}$$

in the basis of generalized coordinates. Under this linear transformation, $2V$ remains harmonic, because Eq. 6.42 contains no higher than second-order terms in $\mathbf{q}_i$. The kinetic energy now becomes

$$\begin{aligned} 2T &\equiv \dot{\boldsymbol{\eta}}^t \cdot \dot{\boldsymbol{\eta}} = (\mathbf{M} \cdot \dot{\mathbf{q}})^t \cdot (\mathbf{M} \cdot \dot{\mathbf{q}}) \\ &= \dot{\mathbf{q}}^t \cdot (\mathbf{M}^t \cdot \mathbf{M}) \cdot \dot{\mathbf{q}} \end{aligned} \tag{6.44}$$

With this change of basis, the secular determinant equation $|\mathbf{B} - \lambda \mathbf{E}| = 0$ becomes replaced by

$$|\mathbf{B}' - \lambda \mathbf{M}^t \cdot \mathbf{M}| = 0 \tag{6.45}$$

with $\mathbf{B}' = \mathbf{M}^t \cdot \mathbf{B} \cdot \mathbf{M}$. However, the linear transformation (6.40) leaves the eigenvalues λ_k and the physical eigenvectors unchanged [2], so that the secular equations (6.22) and (6.45) yield precisely the same results; the transformation simply creates leeway for choosing convenient coordinates with which to express the potential energy function.

We now assert that for *harmonic* potentials (potentials containing no terms beyond the quadratic terms proportional to $q_i q_j$ in Eq. 6.42) it is always possible to find *normal coordinates* $\mathbf{Q}$, related to $\boldsymbol{\eta}$ by a linear transformation

$$\boldsymbol{\eta} = \mathbf{M}_0 \cdot \mathbf{Q} \tag{6.46}$$

which diagonalize both the kinetic and potential energy. In particular, we claim that these energies become

$$2T = \dot{\mathbf{Q}}^{\mathrm{t}} \cdot \dot{\mathbf{Q}} \equiv \sum_i \dot{Q}_i^2$$

$$2V = \mathbf{Q}^{\mathrm{t}} \cdot \boldsymbol{\Lambda} \cdot \mathbf{Q} \equiv \sum_i \lambda_i Q_i^2, \tag{6.47}$$

In contrast to expressions (6.44) and (6.42) for $2T$ and $2V$ in arbitrary generalized coordinates, these equations contain *no* cross terms in $\dot{Q}_i \dot{Q}_j$ or $Q_i Q_j$. We may appreciate the physical significance of such normal coordinates by substituting Eqs. 6.47 into the Lagrangian equations

$$\frac{d}{dt}\frac{\partial L}{\partial \dot{Q}_i} - \frac{\partial L}{\partial Q_i} = 0 \qquad i = 1, \ldots, 3N \tag{6.48}$$

to yield

$$\ddot{Q}_i + \lambda_i Q_i = 0 \qquad i = 1, \ldots, 3N \tag{6.49}$$

In contrast to the corresponding coupled equations $\ddot{\eta}_i + \sum b_{ij}\eta_j = 0$ in mass-weighted coordinates, Eq. 6.49 shows that *each normal coordinate* Q_i *oscillates independently* with motion which is uncoupled to that in other normal coordinates Q_j. This separation of motion into noninteracting normal coordinates is possible only if V contains no cubic or higher-order terms in Eq. 6.4. Anharmonicity will inevitably couple motion between different vibrational modes, and then the concept of normal modes will break down. In the normal mode approximation, no vibrational energy redistribution can take place in an isolated molecule.

The formal solutions to the second-order differential equations (6.49) are

$$Q_i = Q_i^0 \sin(t\sqrt{\lambda_i} + \delta) \tag{6.50}$$

so that each normal coordinate oscillates with frequency $\nu_i = \sqrt{\lambda_i}/2\pi$. These frequencies are identical to those found by diagonalizing the secular determinant (6.13) in the mass-weighted coordinate basis. To find the actual form of the normal coordinates $\mathbf{Q}$, we note that from Eq. 6.46

$$2V = \boldsymbol{\eta}^{\mathrm{t}} \cdot \mathbf{B} \cdot \boldsymbol{\eta} = \mathbf{Q}^{\mathrm{t}} \cdot (\mathbf{M}_0^{\mathrm{t}} \cdot \mathbf{B} \cdot \mathbf{M}_0) \cdot \mathbf{Q} \tag{6.51}$$

Comparing this with Eq. 6.47 for $2V$ then implies that

$$\mathbf{M}_0^t \cdot \mathbf{B} \cdot \mathbf{M}_0 = \Lambda \tag{6.52}$$

In the light of Eqs. 6.28 and 6.31, it then follows that

$$\mathbf{M}_0 = \mathbf{L} \tag{6.53}$$

and that

$$\boldsymbol{\eta} = \mathbf{L} \cdot \mathbf{Q} \tag{6.54}$$

where $\mathbf{L}$ is the matrix of eigenvectors of $\mathbf{B}$. The normal coordinates $\mathbf{Q}$ may then be obtained via

$$\mathbf{Q} = \mathbf{L}^{-1} \cdot \boldsymbol{\eta} = \mathbf{L}^t \cdot \boldsymbol{\eta} \tag{6.55}$$

6.3 INTERNAL COORDINATES AND THE FG-MATRIX METHOD

We pointed out in the preceding section that a realistic potential energy function may not be easily expressible in Cartesian coordinates, but may be written more naturally in terms of $3N$ generalized coordinates related to the mass-weighted coordinates by a linear transformation (6.40). In fact, only $3N - 6$ such coordinates are required to fully specify the potential ($3N - 5$ in linear molecules): $2V$ is not sensitive to the center-of-mass position or molecular orientation in space, and a polyatomic molecule exhibits only $3N - 6$ ($3N - 5$) independent bond lengths and bond angles. Such a truncated set of $3N - 6$ ($3N - 5$) generalized coordinates is called an *internal coordinate* basis, and is commonly denoted $\mathbf{S}$. To illustrate how an internal coordinate basis may be used to evaluate normal modes, we consider the bent H_2O molecule in Fig. 6.3. The three internal coordinates are conveniently chosen to be

$$\begin{aligned} S_1 &= r_B - r_0 \\ S_2 &= r_C - r_0 \\ S_3 &= \phi - \phi_0 \end{aligned} \tag{6.56}$$

(where the subscripts 0 denote equilibrium values), because to a good first approximation the potential energy function in water has the form

$$2V = k_1(S_1^2 + S_2^2) + k_2 S_3^2 \tag{6.57}$$

This potential assumes that independent Hooke's law restoring forces are

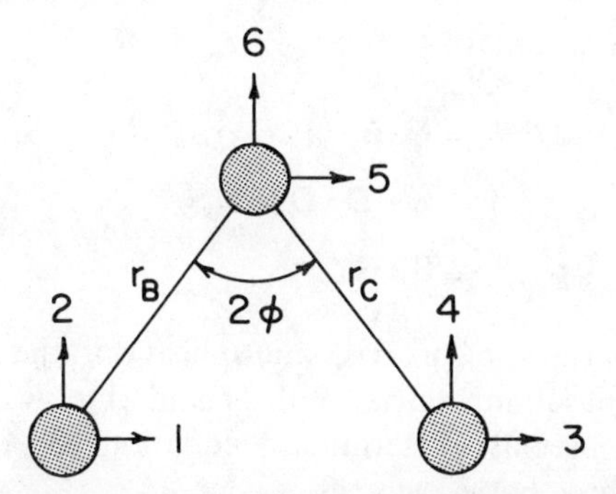

Figure 6.3 Internal coordinates r_B, r_C, and ϕ for the H_2O molecule. Arrows show orientations of the Cartesian basis vectors ξ_1 through ξ_6 used in Eqs. 6.64 and 6.65.

operative in each of the O–H bonds as well as in the bond angle ϕ. The $3N - 6 = 3$ internal coordinates are displacements in two bond lengths and a bond angle, rather than coordinates of individual nuclei as in the mass-weighted basis. The matrix formulation of the vibrational problem in Section 6.2 is nonetheless still applicable, because the S_i can be related to the η_i by a linear transformation. Equation 6.57 implicitly defines a force constant matrix **F**, since

$$2V = \mathbf{S}^t \cdot \mathbf{F} \cdot \mathbf{S} \tag{6.58}$$

with

$$\mathbf{F} = \begin{bmatrix} k_1 & 0 & 0 \\ 0 & k_1 & 0 \\ 0 & 0 & k_2 \end{bmatrix} \tag{6.59}$$

The force constant matrix need not be diagonal (i.e., more sophisticated potentials may be used). For a bent molecule with three nuclei, an arbitrarily accurate force constant matrix will still be a $(3N - 6) \times (3N - 6) = 3 \times 3$ matrix.

To obtain a secular determinant equation analogous to (6.22) in the internal coordinate basis, both $2T$ and $2V$ must be expressed in terms of **S**. Since $2T$ is readily given in terms of mass-weighted coordinates $\boldsymbol{\eta}$, we need a transformation of the form

$$\mathbf{S} = \mathbf{D} \cdot \boldsymbol{\eta} \tag{6.60}$$

Note that because **S** and $\boldsymbol{\eta}$ have $3N - 6$ and $3N$ elements respectively, **D** is a rectangular (rather than square) matrix with $3N - 6$ rows and $3N$ columns. The kinetic energy then becomes

$$
\begin{aligned}
2T = &= \dot{\boldsymbol{\eta}}^{t} \cdot \dot{\boldsymbol{\eta}} = (\mathbf{D}^{-1} \cdot \dot{\mathbf{S}})^{t} \cdot (\mathbf{D}^{-1} \cdot \dot{\mathbf{S}}) \\
&= \dot{\mathbf{S}}^{t} \cdot (\mathbf{D} \cdot \mathbf{D}^{t})^{-1} \cdot \dot{\mathbf{S}} \\
&\equiv \dot{\mathbf{S}}^{t} \cdot \mathbf{G}^{-1} \cdot \dot{\mathbf{S}}
\end{aligned} \tag{6.61}
$$

According to the rules of matrix multiplication, the product $\mathbf{D} \cdot \mathbf{D}$ is a $(3N - 6) \times (3N - 6)$ square matrix. With $2T$ and $2V$ now consistently expressed in the **S** basis, the secular determinant equation $|\mathbf{B}' - \lambda \mathbf{M}^{t} \cdot \mathbf{M}| = 0$ in the generalized coordinate basis becomes

$$
|\mathbf{F} - \lambda \mathbf{G}^{-1}| = 0 \tag{6.62}
$$

in the internal coordinate basis. This is equivalent to

$$
|\mathbf{F} \cdot \mathbf{G} - \lambda \mathbf{E}| = 0 \tag{6.63}
$$

so that finding the normal mode frequencies and coordinates reduces to evaluating the matrix $\mathbf{F} \cdot \mathbf{G}$, and then diagonalizing it [3]. This formalism has become known as the "*FG* matrix method." Since a physically reasonable force constant matrix **F** is often expressible in internal coordinates, most of the labor is incurred in evaluating the kinetic energy (**G**) matrix. Wilson, Decius, and Cross [3] have provided extensive tabulations of **G** for a number of molecular geometries.

The internal coordinates $S_1 = r_B - r_0$ and $S_2 = r_C - r_0$ for H_2O are related to the Cartesian displacement coordinates ξ_1 through ξ_6 in Figure 6.3 by

$$
\begin{aligned}
S_1 &= -(\xi_1 \sin \phi + \xi_2 \cos \phi) + \xi_5 \sin \phi + \xi_6 \cos \phi \\
S_2 &= \xi_3 \sin \phi - \xi_4 \cos \phi - \xi_5 \sin \phi + \xi_6 \cos \phi
\end{aligned} \tag{6.64}
$$

To obtain the bending coordinate S_3, we recognize that small displacement Δr of any nucleus in a direction perpendicular to a bond of length r_0 produces a change $\Delta\phi = \Delta r / r_0$ in the bond angle. Consideration of the effects of each of the Cartesian displacements on the bond angle then leads to

$$
S_3 = \frac{1}{r_0}(-\xi_1 \cos \phi + \xi_2 \sin \phi + \xi_3 \cos \phi + \xi_4 \sin \phi) \tag{6.65}
$$

Equations 6.64 and 6.65 now give us **D** directly. Recalling that $\eta_i = \xi_i \sqrt{m_i}$ and that

$$
\begin{bmatrix} S_1 \\ S_2 \\ S_3 \end{bmatrix} = \begin{bmatrix} D_{11} & \cdots & D_{16} \\ \vdots & & \vdots \\ D_{31} & \cdots & D_{36} \end{bmatrix} \begin{bmatrix} \eta_1 \\ \vdots \\ \eta_6 \end{bmatrix} \tag{6.66}
$$

the **D** matrix is by inspection

$$\mathbf{D} = \begin{bmatrix} \dfrac{-\sin\phi}{\sqrt{m_H}} & \dfrac{-\cos\phi}{\sqrt{m_H}} & 0 & 0 & \dfrac{\sin\phi}{\sqrt{m_O}} & \dfrac{\cos\phi}{\sqrt{m_O}} \\ 0 & 0 & \dfrac{\sin\phi}{\sqrt{m_H}} & \dfrac{-\cos\phi}{\sqrt{m_H}} & \dfrac{-\sin\phi}{\sqrt{m_O}} & \dfrac{\cos\phi}{\sqrt{m_O}} \\ \dfrac{-\cos\phi}{r_O\sqrt{m_H}} & \dfrac{\sin\phi}{r_O\sqrt{m_H}} & \dfrac{\cos\phi}{r_O\sqrt{m_H}} & \dfrac{\sin\phi}{r_O\sqrt{m_H}} & 0 & 0 \end{bmatrix} \tag{6.67}$$

The **G** matrix then becomes

$$\mathbf{G} = \mathbf{D}\cdot\mathbf{D}^t = \begin{bmatrix} \dfrac{1}{m_H} + \dfrac{1}{m_O} & \dfrac{\cos 2\phi}{m_O} & 0 \\ \dfrac{\cos 2\phi}{m_O} & \dfrac{1}{m_H} + \dfrac{1}{m_O} & 0 \\ 0 & 0 & \dfrac{2}{r_O^2 m_H} \end{bmatrix} \tag{6.68}$$

This can be multiplied by the **F** matrix (Eq. 6.59) and $\mathbf{F}\cdot\mathbf{G}$ can be diagonalized to find the eigenvalues λ_1 through λ_3. Such an algebraic procedure can readily be computerized to vary the force constant matrix in order to optimize the closeness of fitted vibrational frequencies $\nu_k^2 = \lambda_k/4\pi^2$ to spectroscopic frequencies.

6.4 SYMMETRY CLASSIFICATION OF NORMAL MODES

When expressed in terms of normal coordinates, the classical kinetic and potential energies associated with vibration in a polyatomic molecule are both diagonalized (cf. Eqs. 6.47). The classical vibrational Hamiltonian becomes

$$H = T + V = \sum_i^{3N-6} (\dot{Q}_i^2 + \lambda_i Q_i^2) \tag{6.69}$$

Under the correspondence principle, the quantum mechanical Hamiltonian for $3N - 6$ independently oscillating normal modes in then

$$\hat{H}_{\text{vib}} = \sum_i^{3N-6} \left(-\frac{\hbar^2}{2} \frac{\partial^2}{\partial Q_i^2} + \lambda_i Q_i^2 \right) \tag{6.70}$$

The eigenfunctions of $\hat{H}_{\text{vib}}$ are

$$|\psi_{\text{vib}}(Q_1, \ldots, Q_{3N-6})\rangle = \prod_i^{3N-6} N_{v_i} H_{v_i}(\zeta_i) \exp(-\zeta_i^2/2) \tag{6.71}$$

with

$$\zeta_i = \lambda_i^{1/4} Q_i / \sqrt{\hbar} \tag{6.72}$$

and the eigenvalues are

$$E_{\text{vib}} = \sum_i^{3N-6} h\nu_i(v_i + \tfrac{1}{2}) \tag{6.73}$$

In Eq. 6.71, the functions $H_{v_i}(\zeta_i)$ are Hermite polynomials of order v_i in ζ_i, and the N_{v_i} are normalization constants. The factorization of $|\psi_{\text{vib}}\rangle$ into independent functions of the normal coordinates Q_i is possible only when the potential energy function is harmonic.

To derive the selection rules for vibrational transitions, it is necessary to take the symmetry of the normal modes into account. We assume that the polyatomic molecule belongs to some point group G of symmetry operations $\hat{R}$. By definition, all $\hat{R}$ in G leave the vibrational Hamiltonian invariant, so that $[\hat{R}, \hat{H}_{\text{vib}}] = 0$. If each normal mode transforms as an irreducible representation (IR) of G, one can set up matrices $\mathbf{R}$ with elements R_{ij} and $\mathbf{H}$ with elements H_{ij} in a basis of vibrational states which are eigenvectors of $\hat{R}$ (i.e., vibrational states that transform as IRs of G). Since $[\hat{R}, \hat{H}_{\text{vib}}] = 0$, it follows that [4]

$$[\mathbf{R} \cdot \mathbf{H}]_{ik} = \sum_j R_{ij} H_{jk} = [\mathbf{H} \cdot \mathbf{R}]_{ik} = \sum_j H_{ij} R_{jk} \tag{6.74}$$

However, if the matrix $\mathbf{R}$ is evaluated in a basis of eigenvectors of $\hat{R}$, it follows by definition that $R_{ij} = R_{ii}\delta_{ij}$. Then Eq. 6.74 implies that

$$R_{ii} H_{ik} = H_{ik} R_{kk}$$

and

$$(R_{ii} - R_{kk}) H_{ik} = 0 \tag{6.75}$$

This means that if $i \neq k$ (i.e., vibrational states i and k belong to different IRs of G), $H_{ik} = 0$. So $\hat{H}_{\text{vib}}$ has no nonvanishing matrix elements connecting states transforming as two different IRs i and k.

Since the molecule is physically unchanged by any symmetry operation $\hat{R}$ in G, $\hat{R}Q_i$ must be a normal mode having the same frequency as Q_i itself. Hence, if Q_i is a nondegenerate mode, $\hat{R}Q_i$ must equal $\pm Q_i$ for all operations $\hat{R}$ in G—so that Q_i must form the basis of a one-dimensional IR in G. If Q_i is degenerate, then

$$\hat{R}Q_i = \sum_k D_{ik}(\hat{R}) Q_k \tag{6.76}$$

where k runs over all modes, including Q_i, that are degenerate with Q_i. This is the case because all of the Q_k in this set of modes will oscillate at the same frequency as Q_i.

Since the normal coordinates $\mathbf{Q}$ are related by a linear transformation to the Cartesian coordinates $\boldsymbol{\xi}$, the matrices for the transformation properties will have the same characters (traces) in either coordinate basis under all group operations $\hat{R}$ [5]. We can see this by assuming that $\boldsymbol{\xi} = \mathbf{N} \cdot \mathbf{Q}$ by hypothesis, and that the effect of a group operation $\hat{R}$ on the Cartesian basis is

$$\hat{R}\boldsymbol{\xi} = \mathbf{P}(\hat{R}) \cdot \boldsymbol{\xi} \tag{6.77}$$

Then the effect of the operation $\hat{R}$ on the normal coordinate set $\mathbf{Q}$ is

$$\begin{aligned} \hat{R}\mathbf{Q} &= \hat{R}(\mathbf{N}^{-1} \cdot \boldsymbol{\xi}) = \mathbf{N}^{-1} \cdot \hat{R}\boldsymbol{\xi} \\ &= \mathbf{N}^{-1} \cdot \mathbf{P}(R) \cdot \boldsymbol{\xi} = [\mathbf{N}^{-1} \cdot \mathbf{P}(R) \cdot \mathbf{N}] \cdot \mathbf{Q} \\ &\equiv \mathbf{P}'(R) \cdot \mathbf{Q} \end{aligned} \tag{6.78}$$

Since the character of a matrix is unaffected by a similarity transformation [2] and $\mathbf{P}'(R) = \mathbf{N}^{-1} \cdot \mathbf{P}(R) \cdot \mathbf{N}$, the transformation matrices $\mathbf{P}'(R)$ and $\mathbf{P}(R)$ in the normal coordinate and Cartesian bases respectively have the same character for all group operations $\hat{R}$ in G. This fact considerably simplifies the determination of the IRs to which the normal coordinates belong, since the behavior of the nuclear Cartesian displacements under the group operations reveals this information even if the form of the normal coordinates is unknown.

For concreteness, consider the planar, nearly T-shaped ClF_3 molecule, which belongs to the C_{2v} point group (Fig. 6.4). The group elements σ_v and σ_v' denote the out-of-plane and in-plane reflection operations. Any symmetry operation on the Cartesian basis vectors ξ_1 through ξ_{12} is expressible using a 12×12 matrix. Under the σ_v' operation, for example, the in-plane vectors ξ_1 through ξ_8 will be unaffected, whereas ξ_9 through ξ_{12} will reverse sign. Hence in the Cartesian basis

$$\boldsymbol{\sigma}_v' \begin{bmatrix} \xi_1 \\ \\ \\ \\ \\ \vdots \\ \\ \\ \\ \\ \\ \xi_{12} \end{bmatrix} = \begin{bmatrix} 1 & 0 & 0 & 0 & 0 & 0 & 0 & 0 & 0 & 0 & 0 & 0 \\ 0 & 1 & 0 & 0 & 0 & 0 & 0 & 0 & 0 & 0 & 0 & 0 \\ 0 & 0 & 1 & 0 & 0 & 0 & 0 & 0 & 0 & 0 & 0 & 0 \\ 0 & 0 & 0 & 1 & 0 & 0 & 0 & 0 & 0 & 0 & 0 & 0 \\ 0 & 0 & 0 & 0 & 1 & 0 & 0 & 0 & 0 & 0 & 0 & 0 \\ 0 & 0 & 0 & 0 & 0 & 1 & 0 & 0 & 0 & 0 & 0 & 0 \\ 0 & 0 & 0 & 0 & 0 & 0 & 1 & 0 & 0 & 0 & 0 & 0 \\ 0 & 0 & 0 & 0 & 0 & 0 & 0 & 1 & 0 & 0 & 0 & 0 \\ 0 & 0 & 0 & 0 & 0 & 0 & 0 & 0 & -1 & 0 & 0 & 0 \\ 0 & 0 & 0 & 0 & 0 & 0 & 0 & 0 & 0 & -1 & 0 & 0 \\ 0 & 0 & 0 & 0 & 0 & 0 & 0 & 0 & 0 & 0 & -1 & 0 \\ 0 & 0 & 0 & 0 & 0 & 0 & 0 & 0 & 0 & 0 & 0 & -1 \end{bmatrix} \begin{bmatrix} \xi_1 \\ \\ \\ \\ \\ \vdots \\ \\ \\ \\ \\ \\ \xi_{12} \end{bmatrix} \tag{6.79}$$

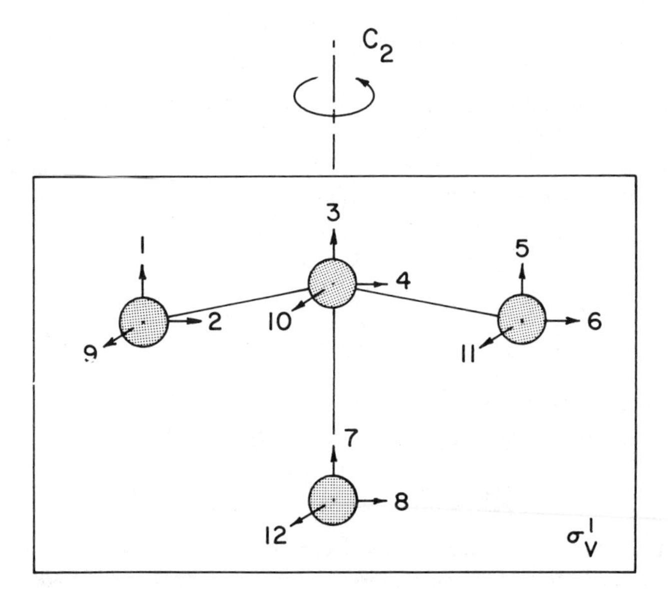

Figure 6.4 Cartesian coordinates for planar T-shaped molecule of C_{2v} symmetry. ξ_1 through ξ_8 lie in the σ_v' reflection plane, while ξ_9 through ξ_{12} point normal to this plane. The σ_v plane (not shown) is normal to the σ_v' plane; the twofold rotation axis C_2 lies at the intersection of the σ_v and σ_v' planes.

and the character $\chi(\boldsymbol{\sigma}_v')$ of the matrix $\boldsymbol{\sigma}_v'$ is $8 - 4 = 4$. Under the identity operation, all of ξ_1 through ξ_{12} are unchanged, so that $\chi(\mathbf{E}) = 12$. The C_2 operation is more interesting in that two of the nuclei are displaced:

$$\mathbf{C}_2 \begin{bmatrix} \xi_1 \\ \\ \\ \\ \\ \vdots \\ \\ \\ \\ \\ \\ \xi_{12} \end{bmatrix} = \begin{bmatrix} 0 & 0 & 0 & 0 & 1 & 0 & 0 & 0 & 0 & 0 & 0 & 0 \\ 0 & 0 & 0 & 0 & 0 & -1 & 0 & 0 & 0 & 0 & 0 & 0 \\ 0 & 0 & 1 & 0 & 0 & 0 & 0 & 0 & 0 & 0 & 0 & 0 \\ 0 & 0 & 0 & -1 & 0 & 0 & 0 & 0 & 0 & 0 & 0 & 0 \\ 1 & 0 & 0 & 0 & 0 & 0 & 0 & 0 & 0 & 0 & 0 & 0 \\ 0 & -1 & 0 & 0 & 0 & 0 & 0 & 0 & 0 & 0 & 0 & 0 \\ 0 & 0 & 0 & 0 & 0 & 0 & 1 & 0 & 0 & 0 & 0 & 0 \\ 0 & 0 & 0 & 0 & 0 & 0 & 0 & -1 & 0 & 0 & 0 & 0 \\ 0 & 0 & 0 & 0 & 0 & 0 & 0 & 0 & 0 & 0 & -1 & 0 \\ 0 & 0 & 0 & 0 & 0 & 0 & 0 & 0 & 0 & -1 & 0 & 0 \\ 0 & 0 & 0 & 0 & 0 & 0 & 0 & 0 & -1 & 0 & 0 & 0 \\ 0 & 0 & 0 & 0 & 0 & 0 & 0 & 0 & 0 & 0 & 0 & -1 \end{bmatrix} \begin{bmatrix} \xi_1 \\ \\ \\ \\ \\ \vdots \\ \\ \\ \\ \\ \\ \xi_{12} \end{bmatrix} \tag{6.80}$$

For this matrix, the sum of diagonal elements is $\chi(\mathbf{C}_2) = -2$. It can similarly be shown that $\chi(\boldsymbol{\sigma}_v) = 2$. These characters for the transformation of $\boldsymbol{\xi}$ under the C_{2v}

group operations may then be summarized as

$$\begin{array}{lcccc} & E & C_2 & \sigma_v & \sigma_v' \\ \chi = & 12 & -2 & 2 & 4 \end{array} = 4a_1 \oplus a_2 \oplus 3b_1 \oplus 4b_2$$

According to the C_{2v} character table, ClF_3 will have translations transforming as a_1, b_1, b_2 and rotations transforming as a_2, b_1, b_2. Subtracting these IRs from the above direct sum yields $3a_1 \oplus b_1 \oplus 2b_2$. Hence ClF_3 will have three normal modes of a_1 symmetry, one of b_1 symmetry, and two of b_2 symmetry. We have determined this without determining what the normal mode coordinates Q_1 through Q_6 actually are; this is possible because the transformation matrices $\mathbf{P}(\hat{R})$ and $\mathbf{P}'(\hat{R})$ have the same character in the ξ and $\mathbf{Q}$ bases.

Setting up transformation matrices like those in Eqs. 6.79 and 6.80 is laborious, and a worthwhile simplification results if one sees [6] that only the Cartesian coordinates attached to nuclei that are *undisplaced* by a symmetry operation $\hat{R}$ contribute to the character $\chi(\hat{R})$. In particular, the contributions to the character are $\chi(\boldsymbol{\sigma}) = +1$, $\chi(\mathbf{C}_n^m) = 1 + 2\cos(2m\pi/n)$, $\chi(\mathbf{S}_n^m) = -1 + 2\cos(2m\pi/n)$, and $\chi(\mathbf{i}) = -3$ for each nucleus that is undisplaced by the symmetry operations σ, C_n^m, S_n^m, and i, respectively. The character $\chi(\mathbf{E})$ for the identity operation is always $3N$. These rules are independent of the particular choice of orientation of the Cartesian axes.

The identification of the IRs according to which the normal coordinates transform can greatly reduce the computational labor associated with implementing the FG matrix method. It is frequently easy to set up *symmetry coordinates*, as linear combinations of internal coordinates, which transform according to IRs of the point group G. For ClF_3, one choice of symmetry coordinates would be

$$\begin{aligned} S_1(a_1) &= (r_1 - r_1^0) + (r_2 - r_2^0) \\ S_2(a_1) &= (r_3 - r_3^0) \\ S_3(a_1) &= (\phi_1 - \phi_1^0) + (\phi_2 - \phi_2^0) \\ S_4(b_1) &= \delta \\ S_5(b_2) &= (r_1 - r_1^0) - (r_2 - r_2^0) \\ S_6(b_2) &= (\phi_1 - \phi_1^0) - (\phi_2 - \phi_2^0) \end{aligned} \tag{6.81}$$

where the six independent internal coordinates (bond lengths and bond angles) are defined in Fig. 6.5. In such a basis, the $\mathbf{F \cdot G}$ matrix reduces to *block diagonal* form, with subblocks allocated to symmetry coordinates transforming as particular IRs of G (because $\hat{H}_{vib}$ has no matrix elements connecting normal coordinates belonging to different IRs). The block-diagonal form of $\mathbf{F \cdot G}$ for ClF_3 is shown in Fig. 6.6.

We will now consolidate some of the ideas introduced in this chapter by deriving the vibrational frequencies of the linear acetylene molecule C_2H_2, in

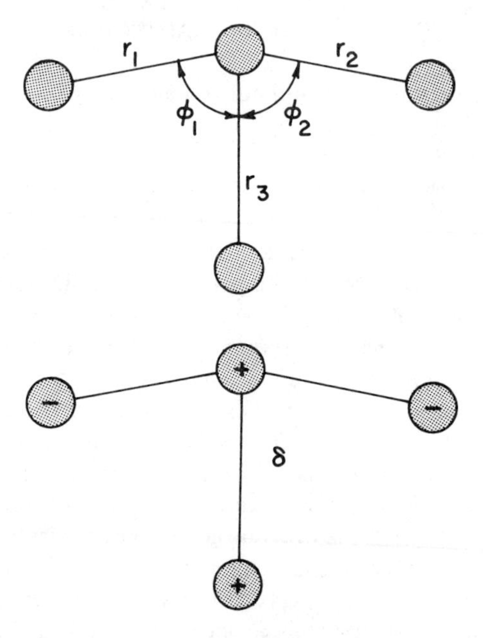

Figure 6.5 The six internal coordinates for a planar T-shaped molecule. The coordinate δ is an out-of-plane bending coordinate; + and − indicate nuclear motions above and below the plane of the paper.

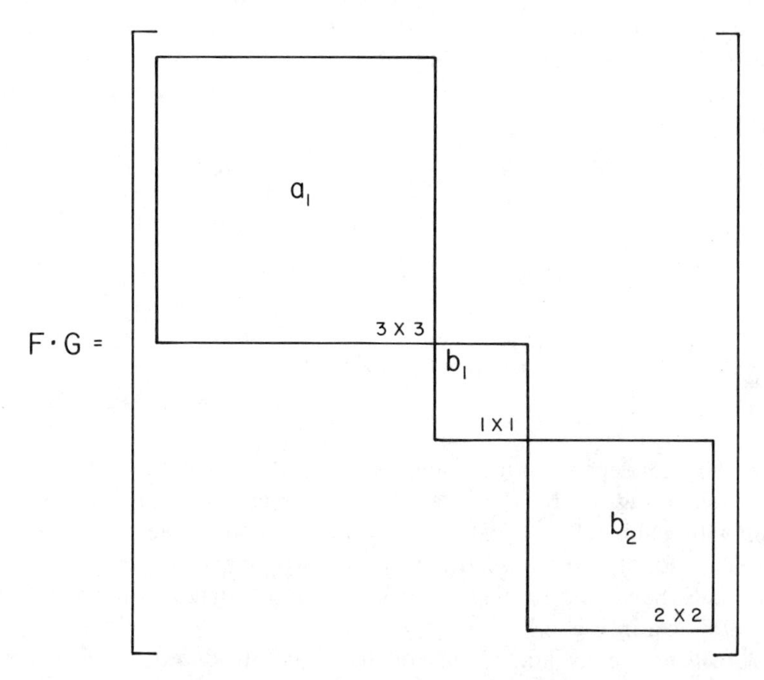

Figure 6.6 The block diagonal **F · G** matrix for ClF_3.

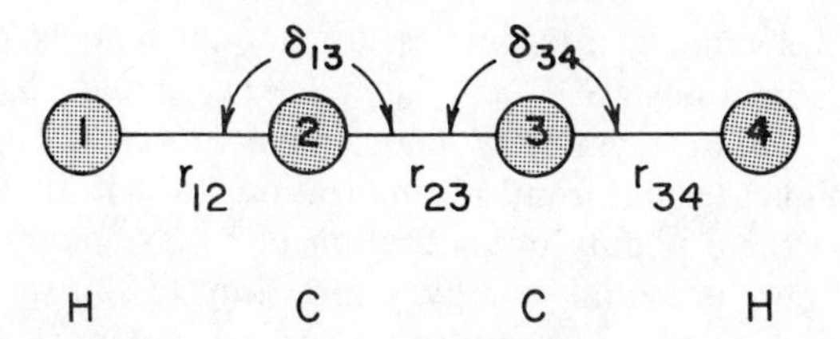

Figure 6.7 Five of the seven independent internal coordinates for $D_{\infty h}$ acetylene. The remaining two internal coordinates are bending motions normal to the plane of the paper, and are not considered in our treatment.

which the nuclei are numbered 1 through 4 from left to right (Fig. 6.7). In terms of the bond radii r_{ij} and bending angles δ_{ij}, we will assume that the potential energy function is given by

$$2V = k_1 r_{23}^2 + k_2(r_{12}^2 + r_{34}^2) + k_\delta(\delta_{13}^2 + \delta_{24}^2) \tag{6.82}$$

(A more detailed potential function, incorporating interaction between non-neighboring atoms, for example, could be used to obtain better fits of derived vibrational frequencies to experimental frequencies.) Here k_1 and k_2 are harmonic force constants for C–C and C–H bond stretching, respectively, while k_δ is a bending force constant. As a first step, we determine the IRs to which the normal modes will belong in linear C_2H_2, using the rules for contributions to the character by nuclei undisplaced by the symmetry operations in $D_{\infty h}$:

	E	$2C\phi$	$\infty\sigma_v$	i	$2S_\phi$	∞C_2
$\chi =$	12	$4 + 8\cos\phi$	4	0	0	0
$\chi_{\text{trans}} =$	3	$1 + 2\cos\phi$	1	-3	$-1 + 2\cos\phi$	$-1\ (=\sigma_u^+ \oplus \pi_u)$
$\chi_{\text{rot}} =$	2	$2\cos\phi$	0	2	$-2\cos\phi$	$0\ (=\pi_g)$
$\chi_{\text{vib}} =$	7	$3 + 4\cos\phi$	3	1	1	$1\ =\sigma_u^+ \oplus 2\sigma_g^+ \oplus \pi_u \oplus \pi_g$

Acetylene therefore has three nondegenerate vibrations (one has σ_u^+ symmetry and two have σ_g^+ symmetry), and two doubly degenerate vibrations each of π_u and π_g symmetry. It is now possible to form symmetry coordinates from linear combinations of the bond lengths r_{ij} and the bond angles δ_{ij}:

$$\begin{aligned} S_1(\sigma_g^+) &= r_{12} + r_{34} \\ S_2(\sigma_g^+) &= r_{23} \\ S_3(\sigma_u^+) &= r_{12} - r_{34} \\ S_4(\pi_u) &= \delta_{13} + \delta_{24} \\ S_5(\pi_g) &= \delta_{13} - \delta_{24} \end{aligned} \tag{6.83}$$

These choices of symmetry coordinates are not all unique; one could take orthogonal combinations $S_1' = a(r_{12} + r_{34}) + br_{23}$ and $S_2' = a'(r_{12} + r_{34}) + b'r_{23}$ as the two σ_g^+ symmetry coordinates instead. The choice $S_3 = r_{12} - r_{34}$ is unique (aside from normalization, which does not concern us here), because there is only one vibration of σ_u^+ symmetry and it is nondegenerate. In our potential energy expression (6.82) we have ignored bending motions which are analogous to δ_{13} and δ_{24}, but move in and out of the paper. These bending motions are simply the degenerate, perpendicular counterparts of the π_u and π_g coordinates listed in Eqs. 6.83. Treating them would only give us redundant information about the π_u and π_g modes, and so they are omitted. With these five symmetry coordinates, the potential energy expression becomes

$$2V = k_1 S_2^2 + \frac{k_2}{2}(S_1^2 + S_3^2) + \frac{k_\delta}{2}(S_4^2 + S_5^2) \tag{6.84}$$

and the **F** matrix is

$$\mathbf{F} = \begin{bmatrix} k_2/2 & 0 & 0 & 0 & 0 \\ 0 & k_1 & 0 & 0 & 0 \\ 0 & 0 & k_2/2 & 0 & 0 \\ 0 & 0 & 0 & k_\delta/2 & 0 \\ 0 & 0 & 0 & 0 & k_\delta/2 \end{bmatrix} \tag{6.85}$$

(Normally the **F** matrix should be $(3N - 5) \times (3N - 5) = 7 \times 7$ in the **S** basis, but we have left out one of the π_u and one of the π_g coordinates.) To derive the **G** matrix, we begin with the Cartesian basis shown in Fig. 6.8. (We could include four additional Cartesian vectors ξ_9 through ξ_{12} pointing out of the paper, but these vectors have projections only along the omitted π_u and π_g modes.) In terms of these, the symmetry coordinates are

$$\begin{aligned} S_1 &= \xi_2 - \xi_1 + \xi_4 - \xi_3 \qquad (r_{12} \text{ and } r_{34} \text{ expand simultaneously}) \\ S_2 &= \xi_3 - \xi_2 \\ S_3 &= \xi_2 - \xi_1 - \xi_4 + \xi_3 \qquad (r_{12} \text{ and } r_{34} \text{ expand out-of-phase}) \\ S_4 &= \frac{1}{r_H}(\xi_6 + \xi_7 - \xi_5 - \xi_8) \\ S_5 &= \frac{1}{r_H}(\xi_6 + \xi_8 - \xi_5 - \xi_7) + \frac{2}{r_C}(\xi_6 - \xi_7) \end{aligned} \tag{6.86}$$

where r_H and r_C are the equilibrium C–H and C–C bond distances, respectively. (The expressions for S_4 and S_5 arise from repeated application of the relationship $\Delta\delta = \Delta r/r$, bearing in mind the effects of increments in the Cartesian coordinates ξ_5 through ξ_8 on the sign of changes in the angles δ_{13} and δ_{24}.) We

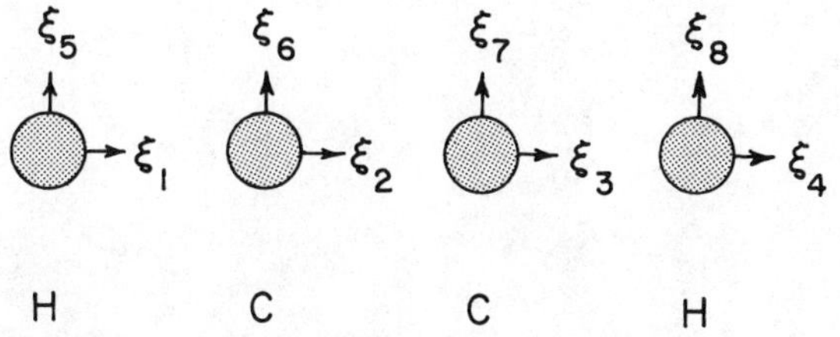

Figure 6.8 Cartesian displacement coordinates for acetylene. Motions perpendicular to the plane of the paper are not considered.

may now construct the matrix **D**, since

$$\begin{bmatrix} S_1 \\ \vdots \\ S_5 \end{bmatrix} = \begin{bmatrix} D_{11} & \cdots & D_{18} \\ \vdots & & \vdots \\ D_{15} & \cdots & D_{58} \end{bmatrix} \cdot \begin{bmatrix} \eta_1 \\ \vdots \\ \eta_8 \end{bmatrix}. \tag{6.87}$$

Recalling that $\eta_i = \sqrt{m_i}\xi_i$ and using the abbreviations $C = m_{\mathrm{C}}$ and $H = m_{\mathrm{H}}$,

$$\mathbf{D} =$$

$$\begin{bmatrix} -1/\sqrt{H} & 1/\sqrt{C} & -1/\sqrt{C} & 1/\sqrt{H} & 0 & 0 & 0 \\ 0 & -1/\sqrt{C} & 1/\sqrt{C} & 0 & 0 & 0 & 0 \\ -1/\sqrt{H} & 1/\sqrt{C} & 1/\sqrt{C} & -1/\sqrt{H} & 0 & 0 & 0 \\ 0 & 0 & 0 & 0 & \dfrac{-1}{r_{\mathrm{H}}\sqrt{H}} & \dfrac{1}{r_{\mathrm{H}}\sqrt{C}} & \dfrac{1}{r_{\mathrm{H}}\sqrt{C}} & \dfrac{-1}{r_{\mathrm{H}}\sqrt{H}} \\ 0 & 0 & 0 & 0 & \dfrac{-1}{r_{\mathrm{H}}\sqrt{H}} & \dfrac{1}{\sqrt{C}}\left(\dfrac{2}{r_{\mathrm{C}}} + \dfrac{1}{r_{\mathrm{H}}}\right) & -\dfrac{1}{\sqrt{C}}\left(\dfrac{1}{r_{\mathrm{H}}} + \dfrac{2}{r_{\mathrm{C}}}\right) & \dfrac{1}{r_{\mathrm{H}}\sqrt{H}} \end{bmatrix} \tag{6.88}$$

and so

$$\mathbf{G} = \mathbf{D} \cdot \mathbf{D}^{\mathrm{t}} =$$

$$\begin{bmatrix} \dfrac{2}{H} + \dfrac{2}{C} & -\dfrac{2}{C} & 0 & 0 & 0 \\ -\dfrac{2}{C} & \dfrac{2}{C} & 0 & 0 & 0 \\ 0 & 0 & \left(\dfrac{2}{H} + \dfrac{2}{C}\right) & 0 & 0 \\ 0 & 0 & 0 & \dfrac{2}{r_{\mathrm{H}}^2}\left(\dfrac{1}{H} + \dfrac{1}{C}\right) & 0 \\ 0 & 0 & 0 & 0 & \left[\dfrac{1}{r_{\mathrm{H}}^2 H} + \dfrac{2}{C}\left(\dfrac{1}{r_{\mathrm{H}}} + \dfrac{2}{r_{\mathrm{C}}}\right)^2\right] \end{bmatrix} \tag{6.89}$$

The **F·G** matrix is then

$$\mathbf{F}\cdot\mathbf{G} =$$

$$\begin{bmatrix} k_2\left(\frac{1}{H}+\frac{1}{C}\right) & -k_2/C & 0 & 0 & 0 \\ -k_1/C & 2k_1/C & 0 & 0 & 0 \\ 0 & 0 & k_2\left(\frac{1}{H}+\frac{1}{C}\right) & 0 & 0 \\ 0 & 0 & 0 & \frac{k_\delta}{r_{\mathrm{H}}^2}\left(\frac{1}{H}+\frac{1}{C}\right) & 0 \\ 0 & 0 & 0 & 0 & k_\delta\left[\frac{1}{r_{\mathrm{H}}^2 H}+\frac{1}{C}\left(\frac{1}{r_{\mathrm{H}}}+\frac{2}{r_{\mathrm{C}}}\right)^2\right] \end{bmatrix} \tag{6.90}$$

Note that this matrix is block-diagonal, with the subblocks $\sigma_g^+(2 \times 2)$, $\sigma_u^+(1 \times 1)$, $\pi_u(1 \times 1)$, and $\pi_g(1 \times 1)$. Hence, we could have evaluated the **F**, **G**, and **F·G** matrices in separate bases of symmetry coordinates transforming as one IR at a time. The eigenvalues in the last three modes can now be read directly from the 1×1 subblocks in the **F·G** matrix,

$$\begin{aligned} \lambda_3 &= k_2\left(\frac{1}{H}+\frac{1}{C}\right) = 4\pi^2\nu_3^2 & \sigma_u^+ \\ \lambda_4 &= \frac{k_\delta}{r_{\mathrm{H}}^2}\left(\frac{1}{H}+\frac{1}{C}\right) = 4\pi^2\nu_4^2 & \pi_u \\ \lambda_5 &= k_\delta\left[\frac{1}{r_{\mathrm{H}}^2 H}+\frac{1}{C}\left(\frac{1}{r_{\mathrm{H}}}+\frac{2}{r_{\mathrm{C}}}\right)^2\right] & \pi_g \end{aligned} \tag{6.91}$$

The σ_g^+ eigenvalues λ_1 and λ_2 are the roots of the 2×2 secular determinant

$$\begin{bmatrix} k_2\left(\frac{1}{H}+\frac{1}{C}\right)-\lambda & -k_2/C \\ -k_1/C & 2k_1/C-\lambda \end{bmatrix} = 0 \tag{6.92}$$

yielding

$$\lambda^2 - \lambda\left[k_2\left(\frac{1}{H}+\frac{1}{C}\right)+\frac{2k_1}{C}\right]+\frac{2k_1k_2}{C}\left(\frac{1}{H}+\frac{1}{C}\right)-k_1k_2/C^2 = 0 \tag{6.93}$$

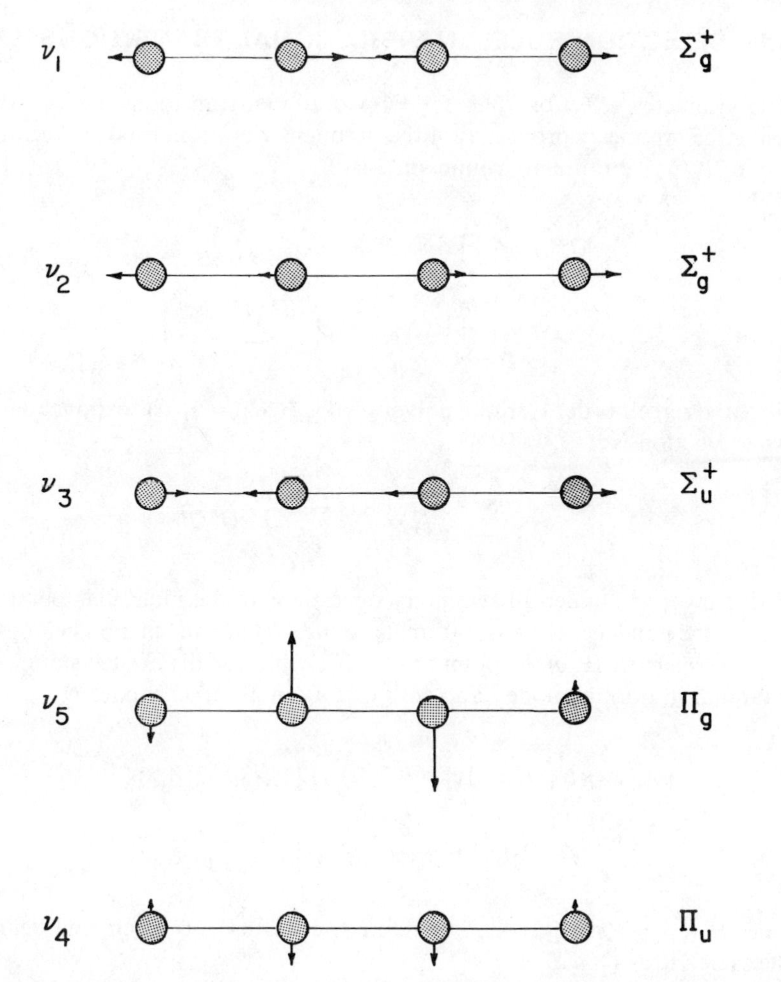

Figure 6.9 Cartesian displacements in the normal coordinates of acetylene. Each of the modes labeled π_g and π_u has a degenerate partner mode involving nuclear motion in and out of the paper.

The eigenvalues λ_1 and λ_2 then obey

$$\lambda_1 + \lambda_2 = k_2\left(\frac{1}{H} + \frac{1}{C}\right) + \frac{2k_1}{C} = 4\pi^2(\nu_1^2 + \nu_2^2) \qquad \sigma_g^+$$

$$\lambda_1\lambda_2 = \frac{2k_1k_2}{C}\left(\frac{1}{H} + \frac{1}{C}\right) - k_1k_2/C^2 = 16\pi^4\nu_1^2\nu_2^2 \quad \sigma_g^+ \qquad (6.94)$$

It may be seen from the expressions for the symmetry coordinates (Eqs. 6.86) that the qualitative normal coordinates in acetylene are those shown in Fig. 6.9.

6.5 SELECTION RULES IN VIBRATIONAL TRANSITIONS

The symmetry selection rules for E1 vibrational transitions can be obtained using the symmetry properties of the pertinent vibrational states. According to Eq. 6.71, the vibrational ground state is

$$
\begin{aligned}
|\psi_{\mathrm{vib}}^{0}\rangle &= \prod_{i}^{3N-6} N_{0i} H_0(\zeta_i) \exp(-\zeta_i^2/2) \\
&= \left(\prod_{i}^{3N-6} N_{0i}\right) \exp\left(-\sum_{i}^{3N-6} \zeta_i^2/2\right)
\end{aligned}
\tag{6.95}
$$

since the zeroth-order Hermite polynomial is $H_0(\zeta_i) = 1$. The exponential in this wave function is

$$
-\sum_{i}^{3N-6} \zeta_i^2/2 = -\sum_{i}^{3N-6} 2\pi v_i Q_i^2/2\hbar \tag{6.96}
$$

and is invariant under all symmetry operations of the molecular point group. Hence, the nondegenerate vibrational ground-state wave function belongs to the *totally symmetric* IR of the point group. The vibrationally excited state with one quantum in normal mode j and zero quanta in all other modes is

$$
\begin{aligned}
|\psi_{\mathrm{vib}}^{j}\rangle &= N_{1j} H_1(\zeta_j) \exp(-\zeta_j^2/2) \prod_{i \neq j}^{3N-6} N_{0i} H_0(\zeta_i) \exp(-\zeta_i^2/2) \\
&= H_1(\zeta_j) \left[\prod_{i}^{3N-6} \exp(-\zeta_i^2/2)\right] N_{ij} \prod_{i \neq j}^{3N-6} N_{0i}
\end{aligned}
\tag{6.97}
$$

Since $H_1(\zeta_j) = 2\zeta_j \propto Q_j$, $|\psi_{\mathrm{vib}}^{j}\rangle$ therefore transforms as Q_j itself, and belongs to the same IR as Q_j.

The symmetries of overtone and combination levels in which no degenerate modes are multiply excited are straightforwardly obtained from direct products of IRs $\Gamma(Q_i)$ according to which the normal coordinates Q_i transform. In particular, the symmetry of the vibrational state with v_1 quanta in mode 1, v_2 quanta in mode 2, etc., is given by

$$
\underbrace{[\Gamma(Q_1) \otimes \Gamma(Q_1) \otimes \cdots]}_{v_1 \text{ factors}} \otimes \underbrace{[\Gamma(Q_2) \otimes \Gamma(Q_2) \otimes \cdots]}_{v_2 \text{ factors}} \otimes \cdots
$$

$$
\otimes \underbrace{(\Gamma(Q_{3N-6}) \otimes \Gamma(Q_{3N-6}) \otimes \cdots]}_{v_{3N-6} \text{ factors}}
$$

Levels involving overtones in *degenerate* modes must be handled with more caution. Consider the two degenerate π_u modes Q_4 and Q_4' corresponding to the bending frequency v_4 in acetylene (Section 6.4). These form a basis for the π_u IR in $D_{\infty h}$. We wish to obtain the symmetries of the vibrational states in which v_4 and v_4' quanta are placed in the respective modes, under the condition that $v_4 + v_4' = 2$. Since both modes have π_u symmetry, using the above procedure would imply (incorrectly) that the resulting vibrational states should have the symmetries $\pi_u \otimes \pi_u = \sigma_g^+ \oplus \sigma_g^- \oplus \delta_g$. This would mean that four vibrational states supposedly arise from distributing two quanta between the degenerate π_u modes (one doubly degenerate δ_g pair of states, and one state each of σ_g^+ and σ_g^- symmetry). However, there are only three distinct (v_4, v_4') combinations possible: (2, 0), (1, 1), and (0, 2). Evaluating the ordinary direct product therefore overcounts the resulting vibrational levels when two quanta are placed in the degenerate π_u vibrations. The symmetries of states arising from multiple excitation of *degenerate* modes Q_i are instead found by evaluating the *symmetric product* $[\Gamma(Q_i) \otimes \Gamma(Q_i)]^+$ [7]. For a doubly excited π_u mode, the pertinent symmetric product is $(\pi_u \otimes \pi_u)^+ = \sigma_g^+ \otimes \delta_g$ rather than $\sigma_g^+ \oplus \sigma_g^- \oplus \delta_g$ [7]. If two vibrational quanta are distributed between *nonequivalent* degenerate modes (e.g., a π_u and a π_g mode, or two π_u modes oscillating at different frequencies), conventional direct products yield the correct vibrational state symmetries.

The E1 selection rules for vibrational transitions $|\psi_{vib}\rangle \rightarrow |\psi'_{vib}\rangle$ can be obtained by expanding the dipole moment $\boldsymbol{\mu}$ in a Taylor series in the normal coordinates,

$$\langle \psi'_{vib}|\boldsymbol{\mu}|\psi_{vib}\rangle$$

$$= \langle \psi'_{vib}|\boldsymbol{\mu}_0 + \sum_i^{3N-6} \left(\frac{\partial \boldsymbol{\mu}}{\partial Q_i}\right)_0 Q_i + \frac{1}{2} \sum_{ij}^{3N-6} \left(\frac{\partial^2 \boldsymbol{\mu}}{\partial Q_i \partial Q_j}\right)_0 Q_i Q_j + \cdots |\psi_{vib}\rangle \qquad (6.98)$$

The leading term is $\boldsymbol{\mu}_0 \langle \psi'_{vib}|\psi_{vib}\rangle$, which vanishes by orthogonality. Group theoretically, the terms $(\partial \boldsymbol{\mu}/\partial Q_i)_0 Q_i$, $(\partial^2 \boldsymbol{\mu}/\partial Q_i \partial Q_j)_0$ $Q_i Q_j$, etc., *all* transform as vector components; in effect, the transformation properties of $Q_i Q_j Q_k \ldots$ in the expansion (6.98) are cancelled by those of $\partial Q_i \partial Q_j \partial Q_k \ldots$ in the corresponding derivative. Consequently, we obtain the *symmetry* selection rule that $\Gamma(\psi'_{vib}) \otimes \Gamma(\boldsymbol{\mu}) \otimes \Gamma(\psi_{vib})$ must contain the totally symmetric IR—*regardless* of which terms dominate in the Taylor series expansion of $\boldsymbol{\mu}$. For *fundamental* transitions from the (totally symmetric) vibrational ground state to levels with one quantum in mode Q_i and zero quanta on all others, $\Gamma(\psi'_{vib}) \otimes (\boldsymbol{\mu}) \otimes \Gamma(\psi_{vib}) = \Gamma(Q_j) \otimes \Gamma(\boldsymbol{\mu})$. Hence the E1 selection rule for fundamental transitions is that $\Gamma(Q_j) = \Gamma(\boldsymbol{\mu})$; that is, Q_j must transform as a vector component. As two examples of this, we cite BF_3 and acetylene. In BF_3, the normal mode symmetries are a_1', a_2'', and $2e'$ (the prefix 2 denotes that this molecule has two pairs of degenerate e' vibrations). In the D_{3h} point group, (x, y) and z transform as e' and a_2'', respectively—so that the $2e'$ and a_2'' vibrations exhibit E1-allowed fundamentals, but the totally symmetric a_1' breathing mode does not. In acetylene, we showed that the seven vibrational modes are $2\sigma_g^+$, σ_u^+, π_u, and π_g. The vector components transform as σ_u^+ and π_u in $D_{\infty h}$, so only the σ_u^+ and π_u vibrations have E1-allowed fundamentals in acetylene. These group theoretical rules ensure that fundamentals are observed only in normal modes in which the molecule's electric dipole moment oscillates.

The selection rules on Δv can be extracted by applying the second quantization formalism to Eq. 6.98. In particular, the first-order terms proportional to Q_i permit transitions with $\Delta v_i = \pm 1$, the second-order terms in Q_iQ_j are responsible for the overtone and combination bands with $\Delta(v_i + v_j) = 0$ or ± 2, and so on. The symmetry selection rule must simultaneously be satisfied. It is well known to students in organic chemistry that overtone and combination bands are frequently prominent in infrared spectra, and so the second- and higher order terms in Eq. 6.98 are not negligible.

CO_2 presents a good example of vibrational selection rules in polyatomics. It possesses three normal mode species (Fig. 6.10): a σ_g^+ symmetric stretch with frequency $\nu_1 \sim 1390\,\text{cm}^{-1}$, a doubly degenerate π_u bend with frequency $\nu_2 \sim 667\,\text{cm}^{-1}$, and a σ_u^+ asymmetric stretch with frequency $\nu_3 \sim 2280\,\text{cm}^{-1}$. The lowest few vibrational levels in CO_2 are shown in Fig. 6.11. Each level is labeled with the number of quanta $(v_1\, v_2\, v_3)$ in each mode; the vibrational state symmetries are also given. The E1 allowed transitions among these levels are shown by the solid connecting lines. Those transitions originating from the (0 1 0) level are *hot bands* which will have appreciable intensity in a CO_2 sample at 300 K ($kT = 208\,\text{cm}^{-1}$), because this level lies only $667\,\text{cm}^{-1}$ above the vibrationless level (0 0 0).

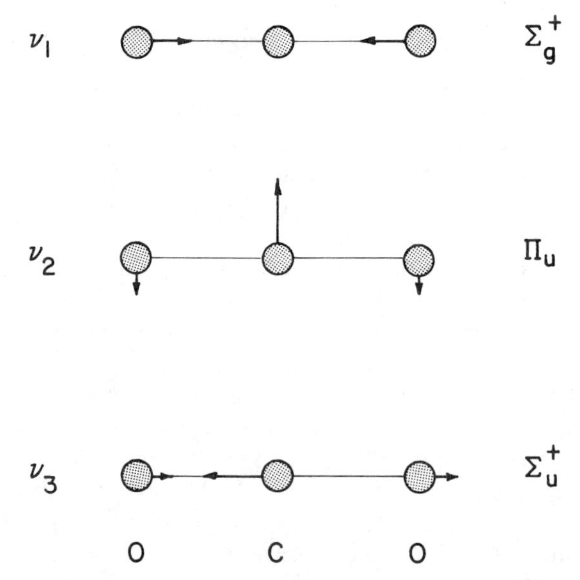

Figure 6.10 Normal modes in $D_{\infty h}$ CO_2. Note that the relative displacements in the Σ_g^+ mode are a consequence of the mode symmetry; those in the Π_u mode may be found from the requirement that the center of mass is undisplaced in any vibrational mode.

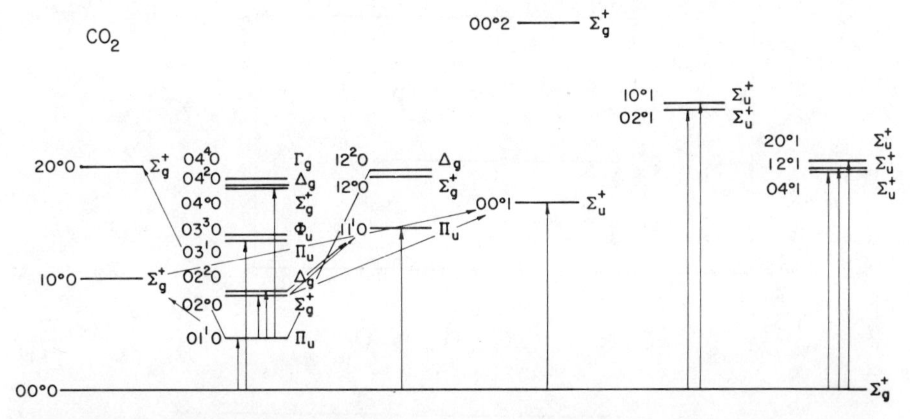

Figure 6.11 Energy level diagram for low-lying vibrational states in CO_2. Levels are labeled with the vibrational quantum numbers $v_1v_2^lv_3$. Some of the observed infrared transitions are indicated by arrows. Adapted from G. Herzberg, *Infrared and Raman Spectra of Polyatomic Molecules*, Van Nostrand, Princeton, NJ, 1945.

The intense 10.6-μm infrared laser transitions in CO_2 are due to the $(0\,0\,1) \rightarrow (0\,2\,0)$ and $(0\,0\,1) \rightarrow (1\,0\,0)$ transitions. Since these are both $\sigma_u^+ \rightarrow \sigma_g^+$ transitions, they are symmetry-allowed (and z-polarized, because z transforms as σ_u^+). It is interesting that even though these two E1 transitions arise from third- and second-order terms, respectively, in the expansion of $\boldsymbol{\mu}$, efficient lasing has been achieved in both of them.

Degenerate vibrational modes give rise to *vibrational angular momentum* in polyatomic molecules. In the case of CO_2, the degenerate π_u normal modes Q_2 and Q_2' shown in Fig. 6.12 exhibit nuclear motion in mutually perpendicular planes containing the molecular axis. In cylindrical coordinates, the nuclear positions during vibration may be specified by the coordinates z, r, and ϕ. Since the cylindrically symmetric vibrational potential energy function is independent of ϕ, the vibrational wavefunctions must exhibit the ϕ-dependence $\exp(\pm il\phi)$ with l an integer. Such a wave function describes vibrational motion in which the nuclei exhibit a constant angular momentum $l\hbar$ about the z axis. It may be shown [8] that when v_2 quanta are distributed between the π_u modes, the allowed values of l are $v_2, v_2 - 2, \ldots, 0$ or 1. In the $(0\,1\,0)$ level, for example, the vibrational angular momentum quantum number becomes $l = v_2 = 1$; the vibrational wave function then becomes proportional to $\exp(\pm i\phi)$. The vibrational motion in state $(0\,1\,0)$ therefore cannot be confined to one of the coordinates Q_2 or Q_2' (which exhibit zero angular momentum about the z axis). It is classically described by a linear combination of Q_2 and Q_2' in which the nuclei follow closed trajectories in a plane normal to the z axis (Fig. 6.13). In the $(v_1v_2^lv_3)$ level notation of Fig. 6.11, the l quantum number is included as a superscript to the number of quanta v_2 in the π_u mode. The l specification is superfluous if the vibrational state symmetry is known: levels with σ, π, δ, ...

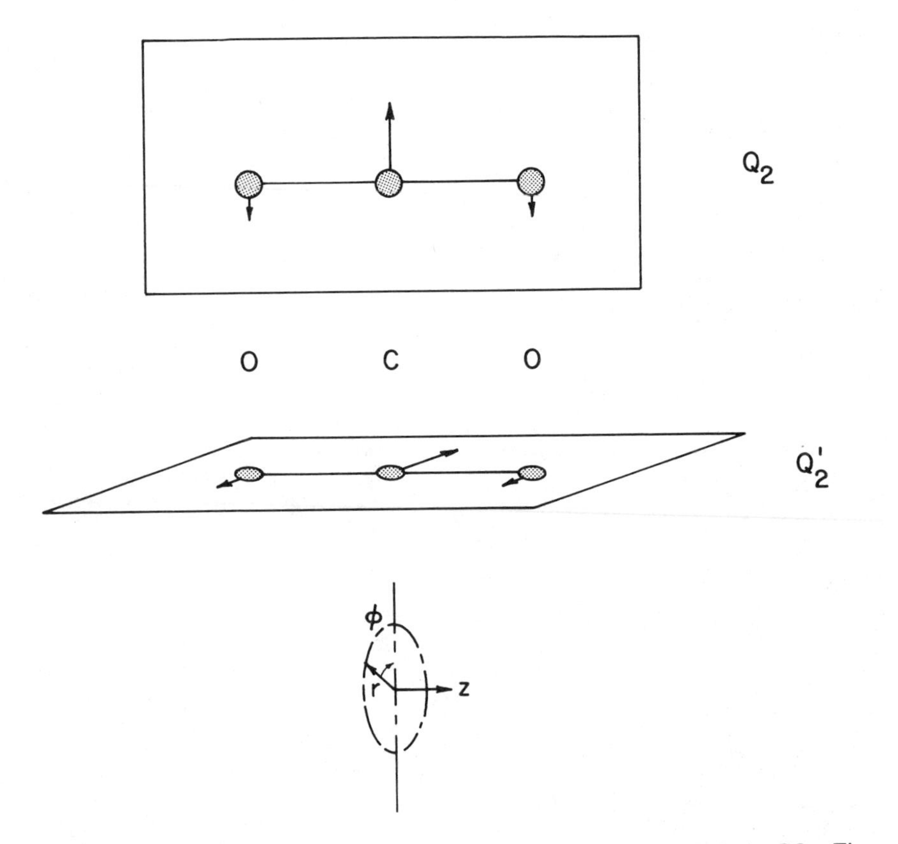

Figure 6.12 The doubly degenerate π_u bending modes Q_2 and Q_2' in CO_2. The nuclear positions in CO_2 vibrations may be expressed in terms of the cylindrical coordinates z, r, and ϕ.

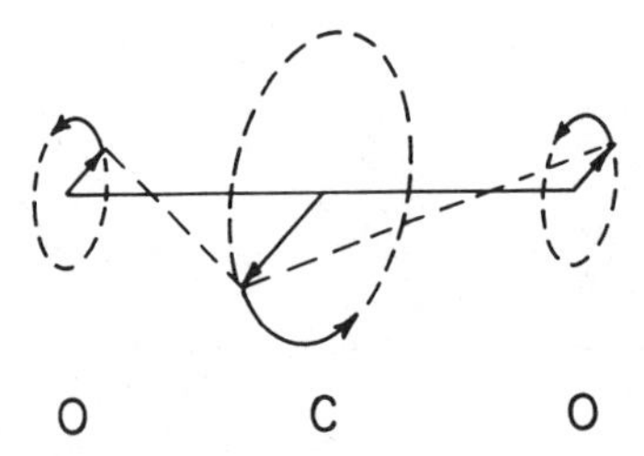

Figure 6.13 Nuclear trajectories in a CO_2 vibrational state with nonzero vibrational angular momentum quantum number l.

symmetry exhibit $l = 0, 1, 2, \ldots$, respectively. Hence, the symmetric product representations we described earlier (in connection with obtaining the symmetries of states with multiply excited degenerate modes) embody the rules for composition of vibrational angular momenta. The foregoing discussion has tacitly assumed that vibration–rotation coupling is negligible; under such coupling the vibrational angular momentum may no longer be a constant of the motion. Vibrational angular momenta also occur in degenerate vibrations of nonlinear polyatomics, where the nuclear displacements may trace ellipses, circles, or lines in a plane perpendicular to the axis of highest symmetry [6].

6.6 ROTATIONAL FINE STRUCTURE OF VIBRATIONAL BANDS

As in diatomics, vibrational transitions in polyatomic molecules are inevitably accompanied by rotational fine structure. In *linear* molecules, the vibrational and rotational selection rules in vibration–rotation spectra are closely analogous to the electronic and rotational selection rules, respectively, in diatomic electronic band spectra. When applied to a $D_{\infty h}$ molecule, the general symmetry arguments of the previous Section lead to the E1 selection rules

$$\Delta l = 0, \pm 1$$

$$\sigma^+ \not\leftrightarrow \sigma^-$$

$$g \not\leftrightarrow g, \quad u \not\leftrightarrow u$$

for vibrational transitions in linear molecules. Here l is the vibrational angular momentum quantum number. The selection rule on Δl is reminiscent of the condition $\Delta\Lambda = 0, \pm 1$ for E1-allowed electronic transitions in diatomics. (The g/u labels are dropped in the case of $C_{\infty v}$ molecules.) For transitions between two σ-type vibrational levels ($l = 0 \rightarrow l' = 0$), the vibrational transition moment is polarized along the molecular axis, and the transition is called a *parallel* transition. The rotational selection rule in parallel vibrational transitions is $\Delta J = \pm 1$; that is, only the P and R branches occur. For transitions in which $\Delta l = \pm 1$ ($\sigma \leftrightarrow \pi$, $\pi \leftrightarrow \delta$, etc.), the transition moment lies perpendicular to the molecular axis, and the transition is termed a *perpendicular* transition. In such a transition, the Q branch also becomes allowed. The frequencies of the P-, Q-, and R-branch lines in vibration–rotation spectra are given by formulas analogous to Eqs. 4.76; $\bar{\nu}_0$ represents the vibrational energy change in the transition. All three rotational branches appear in $\Delta l = 0$ vibrational bands with $l \neq 0$ (e.g., $\pi \leftrightarrow \pi$ transitions). These rotational selection rules are all identical to those that apply to rotational fine structure in diatomic electronic transitions, if l is replaced by Λ in the discussion above (Section 4.6).

In *symmetric top* molecules, the rotational selection rules depend on the relative orientations of the figure axis (Chapter 5) and the vibrational transition moment. When these are *parallel*, the E1 rotational selection rules are [8]

$$\begin{aligned} \Delta K &= 0 \\ \Delta J &= 0, \pm 1 \end{aligned} \tag{6.99}$$

when $K \neq 0$. When K is zero, the transition $\Delta J = 0$ is forbidden. When the transition moment is *perpendicular* to the figure axis, one obtains the contrasting selection rules [8]

$$\begin{aligned} \Delta K &= \pm 1 \\ \Delta J &= 0, \pm 1 \end{aligned} \tag{6.100}$$

Analysis of the rotational fine structure in vibration–rotation spectra thus offers potential for deducing the direction of the transition moment (and thus the vibrational symmetry species) of a vibrational band. If the transition moment has components parallel and normal to the figure axis, then both $\Delta K = 0$ and $\Delta K = \pm 1$ transitions will be observed.

This variety in rotational selection rules, coupled with our natural endowment of molecules with diverse rotational constants, leads to wide variations in the rotational fine structure exhibited by symmetric and near-symmetric tops. For definiteness, we consider a prolate symmetric top whose rotational energy levels are given in Eq. 5.26. Rotational lines will be found at the frequencies

$$\bar{\nu} = \bar{\nu}_0 + B'J'(J' + 1) + (A' - B')K'^2 - [B''J''(J'' + 1) + 1) + (A'' - B'')K''^2] \tag{6.101}$$

where $\bar{\nu}_0$ is the frequency of the pure vibrational transition and (A', B'), (A'', B'') are the rotational constants of the upper and lower vibrational states. According to Eq. 6.101, the rotational structure can be regarded as a superimposition of sets of diatomic P, Q, and R rotational branches (corresponding to $\Delta J = -1, 0,$ and $+1$, respectively) centered at origins with the K-dependent frequencies

$$\bar{\nu} = \bar{\nu}_0 + (A' - B')K'^2 - (A'' - B'')K''^2 \tag{6.102}$$

For parallel bands ($K' = K'' \equiv K$) the origin positions are

$$\bar{\nu} = \bar{\nu}_0 + [(A' - B') - (A'' - B'')]K^2 \tag{6.103}$$

whereas for perpendicular transitions ($K' = K'' \pm 1 \equiv K \pm 1$) they become

$$\bar{\nu} = \bar{\nu}_0 + [(A' - B') - (A'' - B'')]K^2 \pm 2(A' - B')K + A' - B' \tag{6.104}$$

The initial and final rotational levels responsible for a given transition may be specified by writing P, Q, and R as a superscript to denote $\Delta K = -1, 0,$ or $+1$, and by supplying the numerical value of K'' as a subscript. Hence the symbol

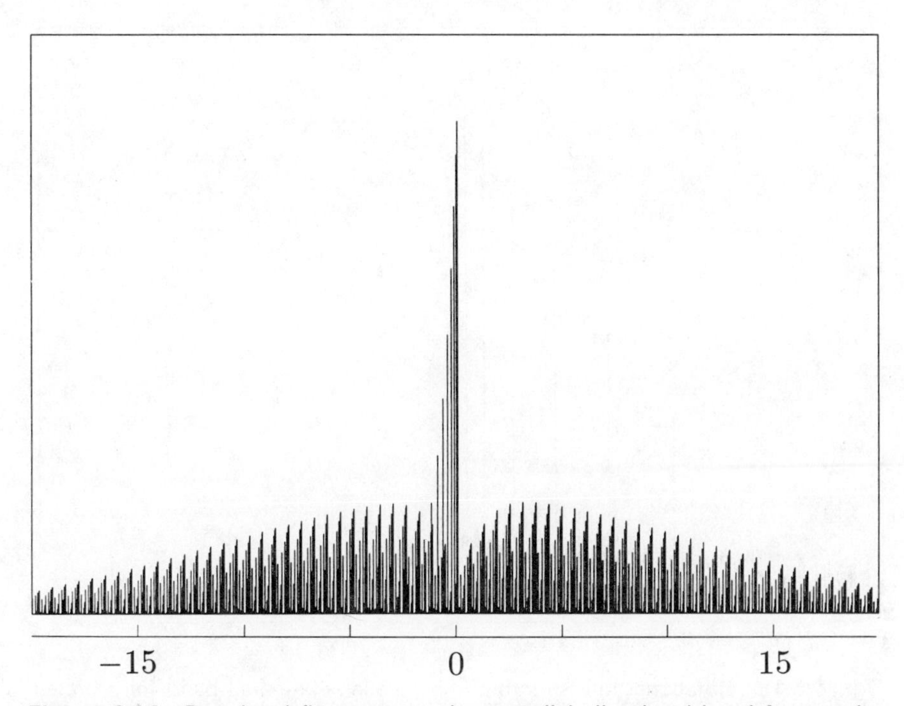

Figure 6.14 Rotational fine structure in a parallel vibrational band for a prolate symmetric top in which $(A' - B') - (A'' - B'')$ is small: $A'' = 5.28\ \text{cm}^{-1}$, $A' = 5.26\ \text{cm}^{-1}$, and $B'' = B' = 0.307\ \text{cm}^{-1}$. The origin positions are closely spaced, and the spectrum resembles the vibration—rotation spectrum of a diatomic molecule. Horizontal energy scale is in cm^{-1}.

${}^{P}Q_2(3)$ represents a $\Delta J = 0$, $\Delta K = -1$ transition from $J'' = 3$, $K'' = 2$ to $J' = 3$, $K' = 1$. In a *parallel* transition, the origin positions (6.103) will frequently depend weakly on K, because the rotational constants are nearly the same in the upper and lower vibrational states. In such a case the rotational structure will resemble that in Fig. 6.14, which is reminiscent of the HCl vibration–rotation spectrum of Fig. 3.3. Since the positions of the ${}^{Q}Q_K(J)$ lines in a parallel transition vary little with K and J when $A' \simeq A''$ and $B' \simeq B''$ (Eq. 6.101), considerable intensity is concentrated near the frequency of the pure vibrational transition. Figure 6.15 illustrates the rotational structure in a parallel transition in which $[(A' - B') - (A'' - B'')]$ is appreciable; the origin positions are now well separated. In many prolate tops (e.g., CH_3Cl), the rotational constant A about the figure axis is much larger than B. The positions of the ${}^{Q}Q_K(J)$ lines then tend to depend strongly on K, but weakly on J (Eq. 6.101), so that the spectrum in Fig. 6.15 is dominated by a series of bunched Q-branch lines. In *perpendicular* transitions, the subband origins depend strongly on K by virtue of the $\pm 2(A' - B')K$ term in Eq. 6.104. The rotational structure then exhibits dense groups of ${}^{P}Q_K$ and ${}^{R}Q_K$ lines, because the position of any ${}^{P,R}Q_K(J)$ lines varies little with J.

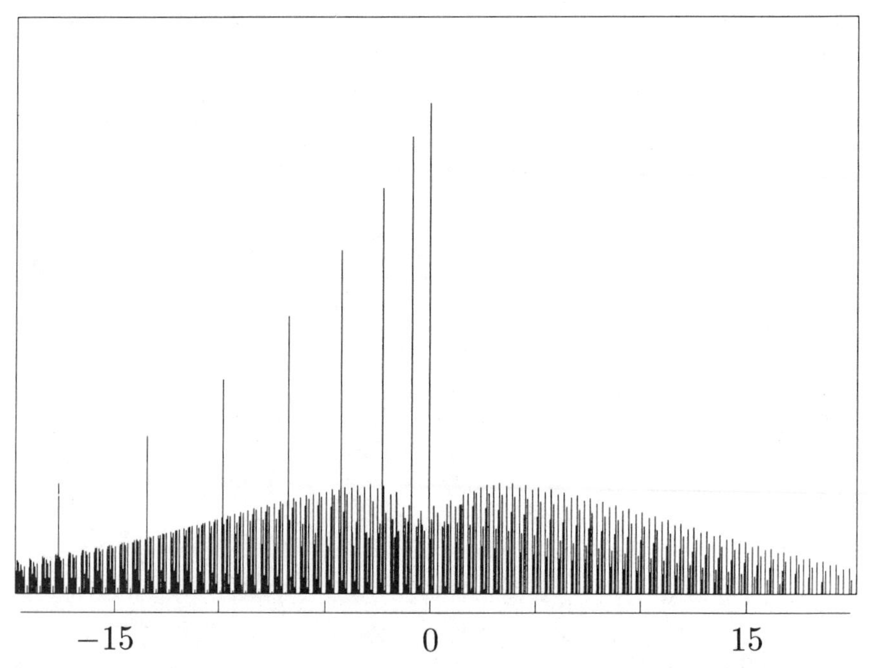

Figure 6.15 Rotational fine structure in a parallel vibrational band for a prolate symmetric top in which $(A' - B') - (A'' - B'')$ is substantial: $A'' = 5.28 \text{ cm}^{-1}$, $A' = 5.00 \text{ cm}^{-1}$, and $B'' = B' = 0.307 \text{ cm}^{-1}$. The origin positions here become well dispersed. Horizontal energy scale is in cm^{-1}.

These spectra serve to illustrate the sensitivity of rotational fine structure to the transition moment orientations and rotational constants. In practice, individual rotational lines cannot be resolved in most infrared vibration–rotation spectra, because the rotational constants are too small. In spectra such as that in Fig. 6.14, the bunched groups of Q-branch lines frequently materialize as single intense bands, while the more sparse P and R branches form weak continua. Rotational structures are frequently analyzed by comparing them with computer-generated spectra derived from assumed rotational constants and selection rules. By weighting the rotational line intensities with appropriate Boltzmann factors (cf. Eq. 3.28) and assigning each rotational line a frequency width commensurate with the known instrument resolution, realistic simulations of experimental spectra are possible if the rotational constants and selection rules are properly adjusted.

6.7 BREAKDOWN OF THE NORMAL MODE APPROXIMATION

Our development of the normal mode description of polyatomic vibrations in Sections 6.1–6.4 rested on the assumption that the potential energy function (6.4) is harmonic in the nuclear coordinates. As in diatomics, this assumption breaks down for sufficiently large vibrational energies in real polyatomic molecules. When several quanta of excitation are placed into a normal mode, it begins to redistribute its energy to other normal modes. Such behavior is guaranteed by the cubic and higher order terms in the vibrational potential, since their presence rules out the possibility of finding $3N - 6$ independently oscillating normal coordinates that obey the uncoupled differential equations (6.49).

At high vibrational energies, there is compelling evidence [9] that the nuclear motion cannot be even approximately described in terms of normal coordinates. A case in point is the thermal dissociation of benzene, $C_6H_6 \xrightarrow{\Delta} C_6H_5 + H$, where a hydrogen atom is created by selective stretching of a single C–H bond. Such motion is inconsistent with the normal mode descriptions of the six benzene vibrations composed primarily of C–H stretches (Fig. 6.16), in which the stretching amplitudes must be identical in at least two of the C–H bonds by symmetry. The infrared–visible absorption spectrum of benzene exhibits a prominent series of overtone bands at frequencies that closely obey the equation [10]

$$\nu = (x_1 + x_2)v - x_2 v^2 \qquad v = 1, 2, \ldots, 8 \tag{6.105}$$

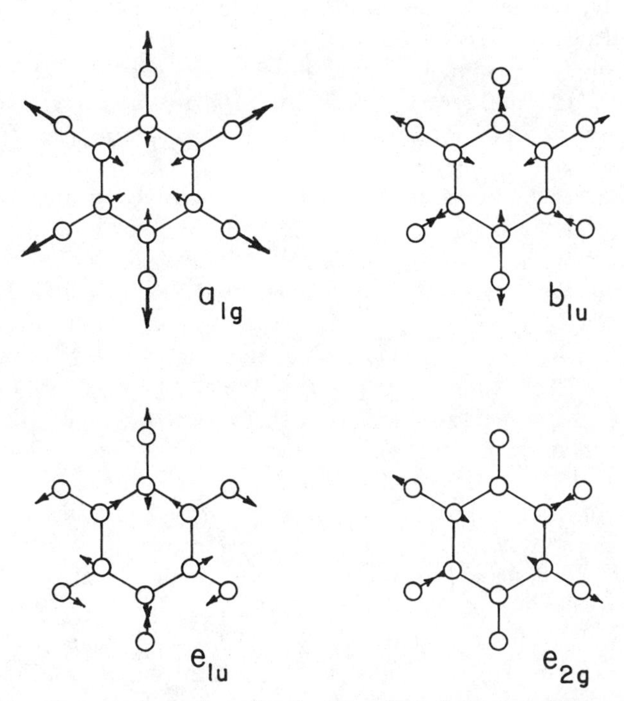

Figure 6.16 Benzene normal modes dominated by C–H stretching motions. Since there are six C–H bonds, there are six such modes: one each of a_{1g} and b_{1u} symmetry, and two degenerate pairs of e_{1u} and e_{2g} symmetry.

with $x_1 = 3153\,\mathrm{cm}^{-1}$ and $x_2 = 58.4\,\mathrm{cm}^{-1}$. This overtone spectrum is strikingly similar to the vibrational spectrum of the isolated CH radical, and has been assigned to an anharmonic *local* stretching mode confined to a *single* C–H bond. The observed frequencies (6.105) are consistent with transitions from $v = 0$ to $v = 1, 2, \ldots, 8$ in a one-dimensional oscillator subject to the quartic potential

$$V(q) = a_0 + a_2 q^2 + a_3 q^3 + a_4 q^4 \tag{6.106}$$

If one treats the anharmonicities $a_3 q^3$ and $a_4 q^4$ to second and first order respectively in a harmonic oscillator basis. Here q is the C–H bond displacement coordinate, and the expansion coefficients in the potential are related to the overtone spectrum parameters by

$$\begin{aligned} x_1 &= a_2 \\ x_2 &= 6a_4 - 30a_3^2/a_4 \end{aligned} \tag{6.107}$$

Additional evidence for the validity of the local mode description in vibrationally excited states is furnished by overtone spectra obtained using highly sensitive thermal lensing spectroscopy techniques [11] in several other aromatic hydrocarbons. The sixth overtone band of the C–H stretching mode is found at virtually the same frequency in benzene ($16{,}480\,\mathrm{cm}^{-1}$), naphthalene ($16{,}440\,\mathrm{cm}^{-1}$), and anthracene ($16{,}470\,\mathrm{cm}^{-1}$). This sameness is difficult to rationalize in a normal mode description, in which the nature of the parent hydrocarbon skeleton is expected to influence the allowed frequencies of the collective nuclear motions. Furthermore, the observed width of the C–H stretching vibrational band is the same ($\sim 360\,\mathrm{cm}^{-1}$) for both the e_{1u} fundamental in benzene (cf. Fig. 6.16) and for the sixth overtone. If one distributes six quanta among the C–H stretching normal modes in Fig. 6.16, one obtains 462 distinct levels. Using the symmetry classification techniques outlined in Section 6.4, it may be shown that 150 of these will exhibit the overall a_{2u} or e_{1u} vibrational level symmetry required in the D_{6h} point group for observation of an E1 overtone transition from the a_{1g} ground state. (These are 75 doubly degenerate states of e_{1u} symmetry.) Hence one would expect a noticably broader vibrational band in the sixth overtone than in the fundamental, in consequence of the far greater variety of E1-accessible states generated using six quanta, if the normal mode approximation were accurate at these energies. Such a picture is not supported by the spectroscopic evidence [11].

In the local mode treatment [11] of the C–H stretching vibrations in benzene, the six bonds oscillate independently with energies (cf. Eq. 6.105)

$$E_{\mathrm{vib}}(v_i) = -1562 + 3153(v_i + \tfrac{1}{2}) - 58.4(v_i + \tfrac{1}{2})^2 \qquad i = 1, \ldots, 6 \tag{6.108}$$

The total vibrational energy residing in the C–H vibrations is

$$E_{\mathrm{vib}} = \sum_{i=1}^{6} E_{\mathrm{vib}}(v_i) \tag{6.109}$$

and the corresponding vibrational states $|v_1 \; v_2 \; \cdots \; v_6\rangle$ are products of one-dimensional anharmonic oscillator states, which are eigenfunctions of a Hamiltonian with the potential function (6.106). For a given total number

$$\mathrm{v} = \sum_{i=1}^{6} v_i \tag{6.110}$$

of vibrational quanta, the possible vibrational states may be divided into classes of degenerate states. One such class is the nondegenerate class (1 1 1 1 1 1), in which each local mode contains one quantum. An example of a degenerate class is (4, 2), in which one local mode has four quanta and the other two are placed together in any of the other six. The degeneracy of this mode is 30 [11]. The classes, energies, and degeneracies of all 462 C–H local mode states in benzene are shown in Fig. 6.17. The product vibrational wavefunctions within each class

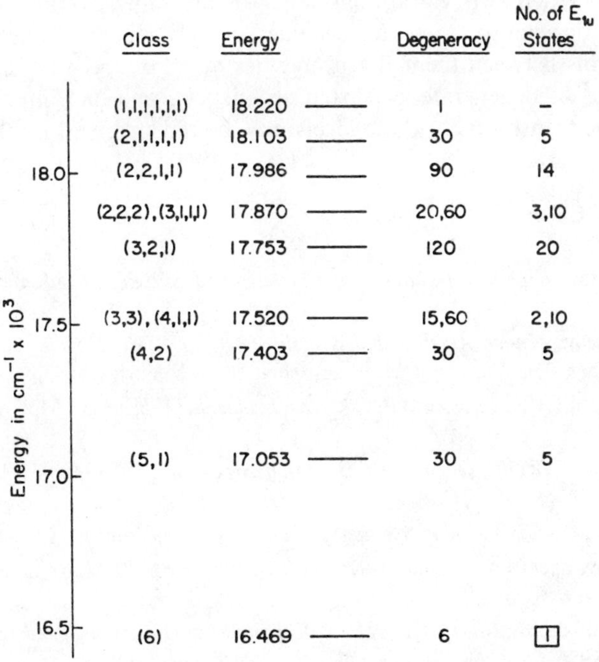

Class	Energy	Degeneracy	No. of E_{1u} States
(1,1,1,1,1,1)	18.220	1	–
(2,1,1,1,1)	18.103	30	5
(2,2,1,1)	17.986	90	14
(2,2,2),(3,1,1,1)	17.870	20,60	3,10
(3,2,1)	17.753	120	20
(3,3),(4,1,1)	17.520	15,60	2,10
(4,2)	17.403	30	5
(5,1)	17.053	30	5
(6)	16.469	6	1

Energy in $\mathrm{cm}^{-1} \times 10^3$ (axis: 16.5, 17.0, 17.5, 18.0)

Figure 6.17 Classes, energies, and degeneracies of the C–H local mode vibrational states in benzene. Reproduced with permission from R. L. Swofford, M. E. Long, and A. C. Albrecht, *J. Chem. Phys.* **65**: 187 (1976).

form a basis for irreducible representations of D_{6h}; one may construct linear combinations of these functions transforming as a_{1g}, a_{2g}, b_{1u}, b_{2u}, e_{2g}, or e_{1u}. In this way, one finds that there exist 150 linearly independent combinations of anharmonic local mode states with e_{1u} overall symmetry, as shown in Fig. 6.17. (This counting of state symmetries if, of course, independent of whether the normal or local mode formulation is used.) It may be shown [11] that in the local mode formulation, E1 transitions are possible only when *one* of the v_i changes, and the remaining v_j are unaffected (i.e., combination transitions among local mode states are forbidden). Hence, for $v = 6$ the only e_{1u} state accessible from the a_{1g} ground state is the one belonging to the (6) class in which all six quanta reside in one of the C–H bonds. This is why the absorption band of the sixth overtone in benzene is no broader than that in the C–H stretching fundamental. The whole question of whether vibrational motion in polyatomics is more appropriately described in the normal mode or local mode formulation has fundamental implications for vibrational spectroscopy, intramolecular vibrational redistribution (IVR), and dissociation. It is also important in radiationless relaxation processes such as internal conversion and intersystem crossing (Chapter 7).

Another manifestation of vibrational anharmonicity occurs in *Fermi resonance* [8]. When two vibrational states of the same overall symmetry are accidentally degenerate, they can become strongly mixed by the anharmonic coupling terms between them. Their energies may be repelled considerably (in the language of degenerate perturbation theory), and the intensities of the spectroscopic transitions to these levels may be redistributed by the mixing.

REFERENCES

1. J. B. Marion, *Classical Dynamics of Particles and Systems*, Academic, New York, 1965.
2. E. D. Nering, *Linear Algebra and Matrix Theory*, Wiley, New York, 1963; F. R. Gantmacher, *The Theory of Matrices*, Chelsea, New York, 1960.
3. E. B. Wilson, J. C. Decius, and P. C. Cross, *Molecular Vibrations*, McGraw-Hill, New York, 1955.
4. M. Tinkham, *Group Theory and Quantum Mechanics*, McGraw-Hill, New York, 1964.
5. D. S. Schonland, *Molecular Symmetry*, Van Nostrand, London, 1965.
6. G. W. King, *Spectroscopy and Molecular Structure*, Holt, Rinehart, & Winston, New York, 1964.
7. C. D. H. Chisholm, *Group Theoretical Techniques in Quantum Chemistry*, Academic, London, 1976.
8. G. Herzberg, *Molecular Spectra and Molecular Structure, II. Infrared and Raman Spectra of Polyatomic Molecules*, Van Nostrand, Princeton, NJ, 1945.
9. P. Avouris, W. M. Gelbart, and M. A. El-Sayed, *Chem. Rev.* **77**: 793 (1977).

10. J. W. Ellis, *Phys. Rev.* **32**: 906 (1928); **33**: 27 (1929); *Trans. Faraday Soc.* **25**: 888 (1924).

11. R. L. Swofford, M. E. Long, and A. C. Albrecht, *J. Chem. Phys.* **65**: 179 (1976). See also W. Siebrand, *J. Chem. Phys.* **44**: 4055 (1966); W. Siebrand and D. F. Williams, *J. Chem. Phys.* **49**: 1860 (1968); B. R. Henry and W. Siebrand, *J. Chem. Phys.* **49**: 5369 (1968); R. Wallace, *Chem. Phys.* **11**: 189 (1975).

PROBLEMS

1. The force constants of the H–C and C–N bonds in linear HCN are 5.8×10^5 and 17.9×10^5 dyne/cm, respectively. Use the treatment of the linear ABC molecule in Section 6.1 to predict the HCN stretching frequencies in cm^{-1}. Compare these with the actual stretching frequencies, 2062 and 3312 cm^{-1}, and comment on the validity of the harmonic approximation to the vibrational potential.

2. The frequencies of the stretching fundamentals of linear CS_2 are 657 and 1523 cm^{-1}. Calculate the C–S bond force constant in two different ways. Are the resulting values consistent? Why or why not? Which of these fundamentals is E1 infrared-active?

3. Determine the symmetry species of the normal modes in SF_6 (O_h), P_4 (T_d), and $C_5H_5^-$ (D_{5h}). Which of these normal modes have E1-allowed fundamentals?

4. Consider a planar, T-shaped molecule of C_{2v} symmetry (ClF_3 has approximately this geometry). The pertinent coordinate systems are shown below.

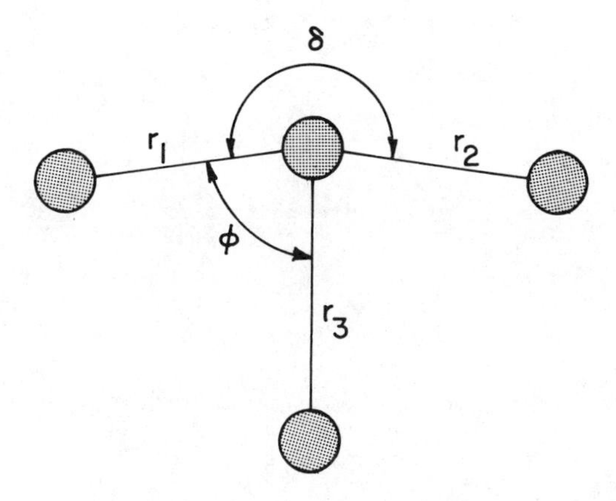

(a) How many vibrational modes will involve only nuclear displacements in the molecular plane, and according to what irreducible representations must they transform?

(b) We may form the symmetry coordinates

$$S_1 = (r_1 + r_2)/2$$
$$S_2 = r_3$$
$$S_3 = \phi$$
$$S_4 = (r_1 - r_2)/2$$
$$S_5 = \delta$$

from the internal coordinates r_1, r_2, r_3, ϕ, and δ. According to which irreducible representation does each of these transform? Assuming the potential energy function

$$2V = k_1(r_1^2 + r_2^2) + k_2 r_3^2 + k_3\phi^2 + k_4\delta^2$$

determine the **F** matrix.

(c) Obtain the **G** matrix for the in-plane vibrations, form the matrix **F · G**, and determine the value(s) of any vibrational frequencies that can be obtained without solving quadratic or higher order equations for λ.

5. Consider a hypothetical square-planar A_4 molecule of D_{4h} symmetry.

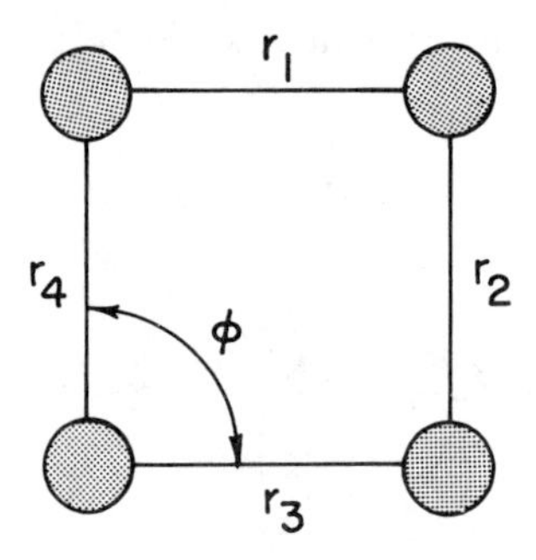

(a) How many in-plane vibrational modes does A_4 have, and what are their symmetry species?

(b) As symmetry coordinates for the in-plane vibrations, we may take

$$S_1 = (r_1 + r_2 + r_3 + r_4)/4$$
$$S_2 = (r_1 - r_2 + r_3 - r_4)/4$$
$$S_3 = (r_1 - r_3)/2$$
$$S_4 = (r_2 - r_4)/2$$
$$S_5 = \phi$$

where the internal displacement coordinates r_1, r_2, r_3, r_4, ϕ are defined in the accompanying figure (ϕ is actually the displacement of the bond angle from its equilibrium value of 90°). What irreducible representation of D_{4h} does each of these belong to? Assuming the potential

$$2V = k_1(r_1^2 + r_2^2 + r_3^2 + r_4^2) + k_2\phi^2$$

set up the **F** matrix for the in-plane vibrations.

(c) Obtain the **G** and **F · G** matrices for the in-plane vibrations, and determine the A_4 vibrational frequencies in terms of the force constants, the equilibrium bond length and the nuclear mass.

7

ELECTRONIC SPECTROSCOPY OF POLYATOMIC MOLECULES

Many of the ideas that are essential to understanding polyatomic electronic spectra have already been developed in the three preceding chapters. As in diatomics, the Born-Oppenheimer separation between electronic and nuclear motions is a useful organizing principle for treating electronic transitions in polyatomics. Vibrational band intensities in polyatomic electronic spectra are frequently (but not always) governed by Franck-Condon factors in the vibrational modes. The rotational fine structure in gas-phase electronic transitions parallels that in polyatomic vibration–rotation spectra (Section 6.6), except that the rotational selection rules in symmetric and asymmetric tops now depend on the relative orientations of the *electronic* transition moment and the principal axes. Analyses of rotational contours in polyatomic band spectra thus provide valuable clues about the symmetry and assignment of the electronic states involved.

Polyatomic band spectra still abound in features that have no antecedents in diatomic spectra. Polyatomic spectra are often far more *congested* (in the sense that they exhibit many more vibrational bands per frequency interval), because the number of vibrational modes scales with molecular size as $3N - 6$. A thermal gas sample of naphthalene ($C_{10}H_8$) cannot be selectively pumped into a single vibrational level in its lowest excited singlet S_1 state at 300 K, because the rotational fine structure at this temperature merges the closely spaced vibrational bands into a barely resolved continuum. A qualitatively new phenomenon arises from the presence of nontotally symmetric modes in polyatomics. Such modes can cause *vibronic coupling* between electronic states belonging to different symmetry species, allowing electronic transitions which would otherwise be E1-forbidden to gain appreciable E1 intensity. This coupling is responsible for the "first allowed" electronic transition in SO_2, the well-known 2600-Å $S_1 \leftarrow S_0$ band system in benzene, and the rich S_1 state photochemistry of the carbonyl group in aldehydes and ketones. Predictions of vibrational band intensities in such transitions require quantitative theories of vibronic coupling. In contrast, many other electronic band spectra arise from intrinsically E1-allowed transitions in which vibrational band intensities are straightforwardly given by products of Franck-Condon factors and squared electronic transition moments (cf. Eq. 4.51). Examples of these are the "second allowed" transition in SO_2 and the $S_1 \leftarrow S_0$ spectrum of aniline, $C_6H_5NH_2$.

In isolated polyatomic molecules of sufficient size, electronically excited states decay nonradiatively and irreversibly into states with lower electronic energy. (Since such a process is necessarily isoenergetic in an isolated molecule, the electronic energy difference is converted into excess vibrational energy.) Such spontaneous *radiationless relaxation* processes, unknown in collisionless diatomics, pervade the photophysics of molecules with $\gtrsim 4$ atoms. Their discovery prompted fundamental questions about the nature of quantum mechanical stationary states in molecules with dense vibrational level structure, and their investigation became one of the most active research areas in chemical physics during the 1960s and 1970s.

Discussions of polyatomic band spectra in a text of this scope can cover only a small fraction of the molecular types that have been explored in this vast field. We begin by treating electronic transitions in triatomic molecules, which are of interest to environmental scientists (viz. NO_2, O_3) and astrophysicists. The electronic band spectrum of SO_2 is considered in detail and presents us with a prototype example of vibronic coupling. We then deal with several aromatic hydrocarbons: aniline, naphthalene, and benzene. These chemically similar molecules exhibit sharply contrasting $S_1 \leftarrow S_0$ spectra arising from transitions from their ground states to their lowest excited singlet states, and serve to illuminate the sensitivity of band spectra to symmetry, vibronic coupling, and geometry changes accompanying transitions. This chapter concludes by developing quantitative theories for vibronic coupling and radiationless relaxation in polyatomics.

7.1 TRIATOMIC MOLECULES

As in diatomics (Section 4.3), the molecular orbitals in polyatomic molecules may be expressed as linear combinations of atomic orbitals (AOs) centered on the nuclei. A minimal basis set of AOs contains all of the AOs that are occupied in the separated atoms [1, 2]. In the bent ozone molecule, for example, the separated O atoms have the ground state configuration $(1s)^2(2s)^2(2p)^4$. The minimal basis set for ground-state O_3 therefore consists of the 1*s*, 2*s*, and three 2*p* orbitals centered on each of the oxygen nuclei (Fig. 7.1). The orientations selected for the 2*p* AOs in the basis set are of course arbitrary (aside from the constraint that basis AOs centered on any nucleus must be linearly independ-

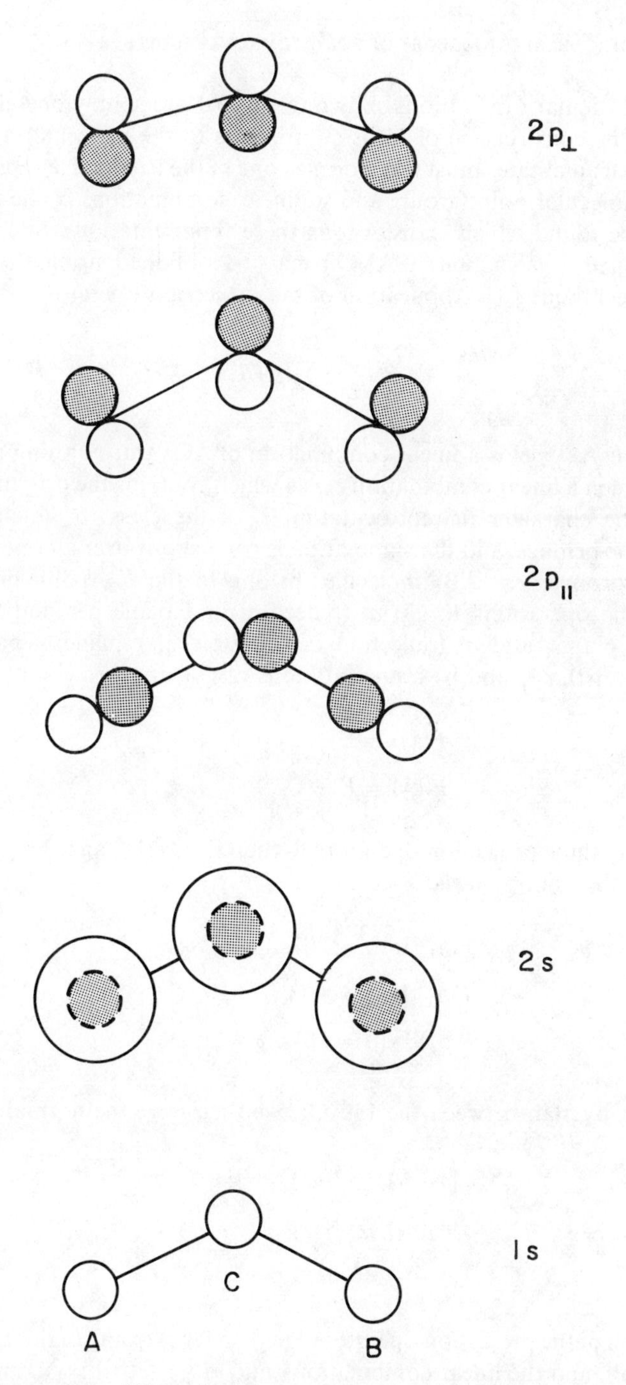

Figure 7.1 Minimal basis set of atomic orbitals (AOs) in O_3.

ent); the particular orientations shown in Fig. 7.1 are simply one choice which facilitates the construction of MOs appropriate to the molecular symmetry.

The electronic states must transform as one of the irreducible representations Γ of the molecular point group, and so linear combinations of the AOs in Fig. 7.1 must be found which transform as these representations. Such *symmetry-adapted linear combinations* (SALCs) may be obtained using the projection operator technique [3]. Application of the projection operator

$$\hat{P}(\Gamma) = \sum_R \chi^\Gamma(R)\hat{R} \tag{7.1}$$

to any basis AO yields a linear combination of AOs transforming as the IR Γ, provided such a linear combination exists which contains the original AO. Here $\chi^\Gamma(R)$ is the character in representation Γ of the class to which the group operation $\hat{R}$ belongs, and the summation is carried out over all operations $\hat{R}$ in the point group. Bent AB_2 molecules belong to the C_{2v} point group, which contains the operations E, C_2, σ_v (reflection in a plane perpendicular to the molecular plane), and σ_v' (reflection in the molecular plane). The projection operators for the a_1 and b_2 representations of C_{2v} are then

$$\begin{aligned}\hat{P}(a_1) &= \hat{E} + \hat{C}_2 + \hat{\sigma}_v + \hat{\sigma}_v' \\ \hat{P}(b_2) &= \hat{E} - \hat{C}_2 - \hat{\sigma}_v + \hat{\sigma}_v'\end{aligned} \tag{7.2}$$

By applying these projection operators to the 1s AOs $|1s_A\rangle$ and $|1s_C\rangle$ in Fig. 7.1, we obtain the unnormalized SALCs

$$\begin{aligned}\hat{P}(a_1)|1s_A\rangle &= 2(|1s_A\rangle + |1s_B\rangle) \\ \hat{P}(a_1)|1s_C\rangle &= 4|1s_C\rangle \\ \hat{P}(b_2)|1s_A\rangle &= 2(|1s_A\rangle - |1s_B\rangle)\end{aligned} \tag{7.3}$$

Neglecting overlap between the 1s AOs, we then have the normalized SALCs

$$\begin{aligned}|\sigma_1(a_1)\rangle &= (|1s_A\rangle + |1s_B\rangle)/\sqrt{2} \\ |\sigma_2(a_1)\rangle &= |1s_C\rangle \\ |\sigma_1(b_2)\rangle &= (|1s_A\rangle - |1s_B\rangle)/\sqrt{2}\end{aligned} \tag{7.4}$$

Application of the projection operators $\hat{P}(a_2)$ or $\hat{P}(b_1)$ to any of the 1s AOs yields a null result, and the linear combinations $\hat{P}(a_1)|1s_B\rangle$, $\hat{P}(b_2)|1s_B\rangle$, and $\hat{P}(b_2)|1s_C\rangle$ all either vanish or reproduce the unnormalized SALCs (7.3). Hence, only three linearly independent SALCs are generated from the three 1s basis AOs, as expected. In an LCAO–MO–SCF calculation, the two SALCs of a_1 symmetry will mix to yield the lowest two MOs of a_1 symmetry, $|1a_1\rangle$ and $|2a_1\rangle$ (Fig. 7.2).

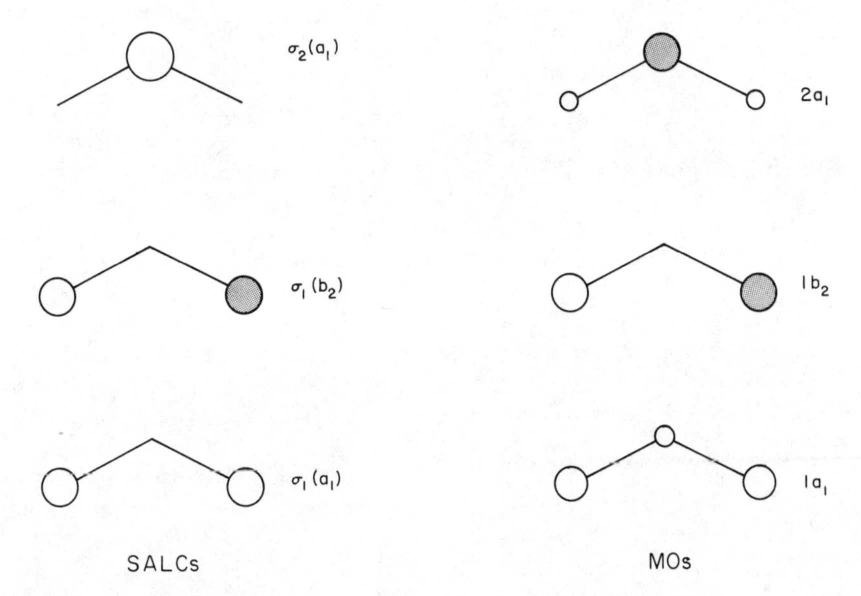

Figure 7.2 Symmetry-adapted linear combinations (SALCs) and molecular orbitals (MOs) generated from the 1s AOs in a bent AB_2 molecule.

The extent of this mixing will be small in triatomics composed of atoms with valence electrons in their 2*s* or higher energy AOs, since the overlap between the inner-shell 1*s* AOs in these molecules will be insignificant. The rest of the minimal-basis SALCs can similarly be generated using the projection operator technique on the 2*s* and 2*p* AOs. Mixing between SALCs having similar energy and belonging to the same symmetry species then yields valence MOs with the qualitative nodal patterns shown on the right side of Fig. 7.3.

When the bond angle ϕ in a bent AB_2 molecule is increased toward 180°, these nodal patterns must be preserved. As the molecule approaches linear geometry, the resulting orbital symmetries in $D_{\infty h}$ may be found by noting that the symmetry operations C_2, σ_v, and σ_v' in C_{2v} correspond to the classes C_2', σ_h, and σ_v, respectively, in the linear point group. The correlations between AB_2 orbital symmetries in C_{2v} and $D_{\infty h}$ may then be worked out by requiring that the characters of the corresponding classes of operations be identical in both irreducible representations. (Some of the characters for σ_h, which are not ordinarily listed in $D_{\infty h}$ character tables, are $+1$ (σ_g^+ and σ_g^-), -2 (π_g), -1 (σ_u^+ and σ_u^-), and $+2$ (π_u).) This yields the correlations

C_{2v}	E	$C_2(C_2')$	$\sigma_v(\sigma_h)$	$\sigma_v'(\sigma_v)$	$D_{\infty h}$
a_1	1	1	1	1	σ_g^+
b_2	1	-1	-1	1	σ_u^+
$a_1 + b_1$	2	0	2	0	π_u
$a_2 + b_2$	2	0	-2	0	π_g

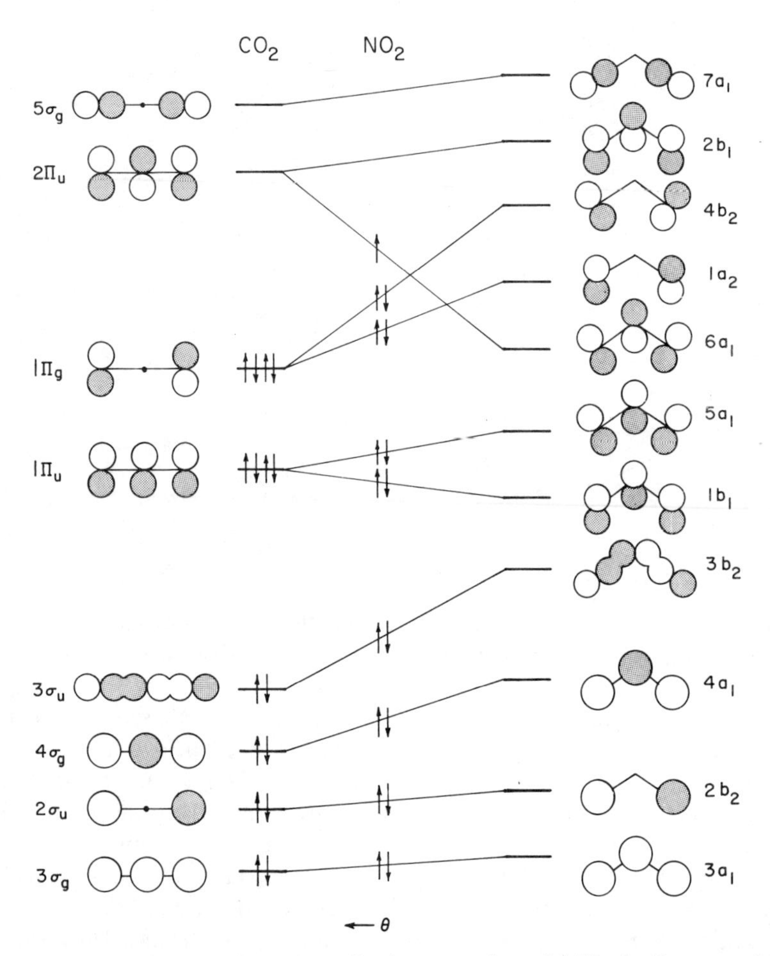

Figure 7.3 Nodal patterns and qualitative energies of MOs in linear and bent triatomic molecules according to Walsh [4]. The lowest three MOs, composed of 1s AOs on the constituent atoms (Fig. 7.2), participate little in the chemical bonding and are excluded. Horizontal coordinate is the bending angle θ. Orbital occupancies are shown for linear CO_2 ($\theta = 180°$) and for bent NO_2 ($\theta = 134°$). Irreducible representations are given for the MOs in $D_{\infty h}$ and C_{2v} point groups at left and right, respectively.

and these are reflected in the orbital correlation diagram in Fig. 7.3. Note that when a linear molecule becomes bent, its doubly degenerate π_u orbitals in $D_{\infty h}$ are split into pairs of nondegenerate a_1 and b_1 orbitals in C_{2v}.

Orbital correlations alone cannot rationalize the electronic structure in triatomic molecules; one needs to know how the MO energies are ordered and how they are influenced by the bond angle ϕ. In a remarkably prescient series of papers published in the early 1950s (long before accurate wave functions became available for polyatomic molecules), Walsh [4] developed semiempirical rules for predicting geometry effects on orbital energies in small polyatomics. Walsh's rules for triatomic AB_2 are incorporated in the correlation diagram in Fig. 7.3, where the orbital energies and bond angle are qualitatively plotted along the vertical and horizontal axes, respectively. Most of the orbital energies are seen to increase slightly when linear AB_2 becomes bent. The $2\pi_u$ orbitals in linear AB_2 present a key anomaly: one of them correlates with the $6a_1$ orbital, whose energy falls violently as the bond angle is reduced.

The electron configurations in AB_2 may now be constructed using the Aufbau prescription of placing electrons in successive MOs according to the Pauli principle. The first six electrons go into the nonvalence orbitals $1a_1$, $2a_1$, and $1b_2$ (Fig. 7.2). The remaining electrons are placed in the valence orbitals which are shown in the correlation diagram (Fig. 7.3). The CO_2 molecule, which has 16 valence electrons, exhibits the ground-state configuration

$$\ldots (3\sigma_g)^2(2\sigma_u)^2(4\sigma_g)^2(3\sigma_u)^2(1\pi_u)^4(1\pi_g)^4 \qquad {}^1\Sigma_g^+$$

Since the majority of the occupied orbital energies are minimized at $\phi = 180°$, CO_2 has linear geometry. In NO_2, which has 17 valence electrons, the bent geometry becomes favored because the "extra" electron goes into the $6a_1$ orbital whose energy drops sharply when the bond angle is decreased. The electron configuration of NO_2 then becomes

$$\ldots (3a_1)^2(2b_2)^2(4a_1)^2(3b_2)^2(1b_1)^2(5a_1)^2(1a_2)^2(4b_2)^2(6a_1)^1 \qquad {}^2A_1$$

In accordance with these predictions, ground-state triatomics with 16 or fewer valence electrons (CO_2, CS_2, OCS, N_2O) are experimentally found to be linear, while those with more than 16 valence electrons (NO_2, O_3, SO_2) are bent.

The low-resolution absorption spectrum of SO_2 is shown in Fig. 7.4. The intense band system lying between 1900 and 2300 Å is called the "second allowed" band. It exhibits a decadic molar absorption coefficient $\varepsilon \sim 3000$ L mol^{-1} cm^{-1} at maximum, which is characteristic of a strongly E1-allowed electronic transition in triatomics. (Absorption coefficients are defined in Appendix D.) The less prominent "first allowed" band system between 2400 and 3400 Å shows ε values on the order of 300 L mol^{-1} cm^{-1}. In the "forbidden" band between 3400 and 4000 Å, the absorption coefficients of $\sim$0.1 L mol^{-1} cm^{-1} are typical of those found in spin-forbidden transitions. The valence structure in SO_2 is isoelectronic with that in O_3, and we will use the orbital nomenclature in Fig. 7.3 to discuss electronic transitions in SO_2. In so doing, we should bear in mind that the "$6a_1$" orbital in Fig. 7.3, for example, is not actually the sixth lowest-energy a_1 MO in SO_2: This molecule (unlike O_3) has additional inner-shell a_1 MOs arising from SALCs of $2s$ and $2p$ AOs centered on the sulfur atom.

The ground state in SO_2 has the closed-shell configuration

$$\ldots (1a_2)^2(4b_2)^2(6a_1)^2 \qquad {}^1A_1$$

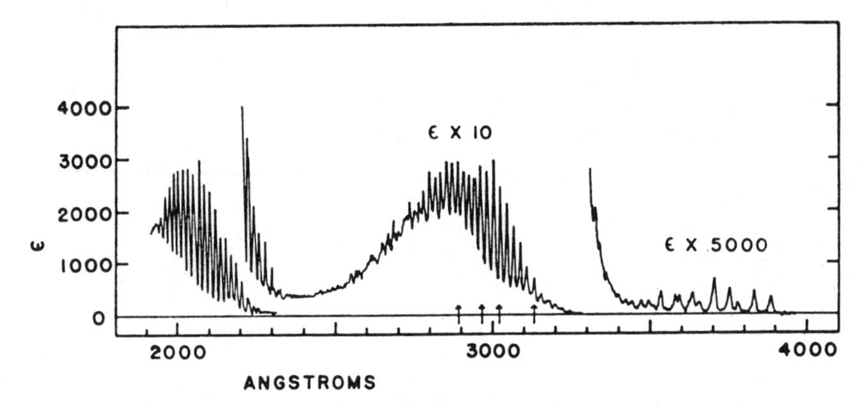

Figure 7.4 Absorption spectrum of SO_2 gas from 1900 to 4000 Å. The strong "second allowed" band system appears between 1900 and 2300 Å; the "first allowed" band occurs between 2400 and 3400 Å; and the very weak "forbidden band" lies between 3400 and 4000 Å. Reproduced by permission from S. J. Strickler and D. B. Howell, *J. Chem. Phys.* **49**: 1948 (1968).

and is a totally symmetric singlet state. According to Walsh's rules (Fig. 7.3), some of the lowest-lying excited states in SO_2 should be

$$\ldots (1a_2)^2(4b_2)^2(6a_1)^1(2b_1)^1 \qquad {}^1B_1, {}^3B_1$$

$$\ldots (1a_2)^2(4b_2)^1(6a_1)^2(2b_1)^1 \qquad 1A_2, {}^3A_2$$

$$\ldots (1a_2)^1(4b_2)^2(6a_1)^2(2b_1)^1 \qquad {}^1B_2, {}^3B_2$$

Each of these open-shell configurations gives rise to a singlet and a triplet state; the triplet state in each configuration exhibits the lower energy due to Hund's rule. The overall electronic state symmetries are given by the direct products of irreducible representations for singly occupied MOs (e.g., $b_2 \otimes b_1 = A_2$). The *second allowed* band between 1900 and 2300 Å is due to the ${}^1B_2 \leftarrow {}^1A_1$ transition, which promotes an electron from the $1a_2$ orbital to the $2b_1$ orbital. This transition is group-theoretically E1-allowed ($A_1 \otimes B_2 = B_2$) and y-polarized. This is consistent with an analysis of rotational fine structure in this band system [5], which indicates that the electronic transition is polarized in the molecular plane. According to Fig. 7.3, this transition removes an electron from an essentially nonbonding π-type orbital ($1a_2$) and places it into an antibonding π^* orbital ($2b_1$). This should weaken the S–O bond in the ${}^1B_2$ state relative to the ${}^1A_1$ ground state, and thus endow the ${}^1B_2$ state with longer bonds. This is in fact what happens: The S–O bond lengths in the ${}^1A_1$ and ${}^1B_2$ states are 1.432 and 1.560 Å, respectively. Furthermore, the $1a_2$ orbital energetically favors larger bond angles (its correlation curve minimizes at $\phi = 180°$), whereas the $2b_1$ orbital energy varies more weakly with ϕ. Hence the ${}^1B_2 \leftarrow {}^1A_1$ transition should produce an excited state with a smaller equilibrium bond angle. This is borne out by the experimental bond angles in the ${}^1A_1$ and ${}^1B_2$ states (119.5° and

104.3°, respectively). Bent SO_2 exhibits three vibrational modes: an a_1 symmetric stretch (ν_1), an a_1 bending mode (ν_2) and a b_2 asymmetric stretch (ν_3). Since the equilibrium bond angles are so different in the $^1A_1$ and $^1B_2$ states, the resulting displacement along normal coordinate Q_2 between the minima in the respective potential energy surfaces causes the Franck-Condon factors $|\langle v_1' v_2' v_3' | v_1'' v_2'' v_3'' \rangle|^2$ to assume appreciable magnitudes for *many* values of $\Delta v_2 = v_2' - v_2''$. This is responsible for the *progression* of nearly equally spaced vibrational bands which is observed on the long-wavelength side of the second allowed system in Fig. 7.4. These arise principally from transitions from $v_2'' = 0$ in the $^1A_1$ state to $v_2' = 0, 1, 2, 3, \ldots$ in the $^1B_2$ state. Such progressions are generally associated with significant changes in equilibrium geometry between the upper and lower electronic states.

The assignment of the *first allowed* band system was controversial for many years [5]. It is now known that the low-energy portion of this system (where the spectrum maintains a nearly regular band spacing in a progression between ~2800 and 3400 Å in Fig. 7.4) is due to the $^1A_2 \leftarrow {}^1A_1$ transition. Since no vector component transforms as A_2 in C_{2v}, this transition is E1 symmetry-forbidden. It is observed anyway (though with lower intensity than the second allowed $^1B_2 \leftarrow {}^1A_1$ transition), due to a breakdown in the Born-Oppenheimer approximation. We recall in Section 3.1 that motion in two electronic states $|\psi_k\rangle$ and $|\psi_{k'}\rangle$ in a diatomic molecule may be coupled by a nuclear kinetic energy term proportional to $\langle \psi_k | \partial / \partial R | \psi_{k'} \rangle$. Generalizing this treatment to triatomic SO_2, one finds that nuclear kinetic energy coupling is possible between the $^1A_2$ state and some other excited singlet state (which we provisionally call ^1B), provided that

$$\langle {}^1B | \partial / \partial Q_k | {}^1A_2 \rangle \neq 0 \tag{7.5}$$

Here Q_k is one of the SO_2 vibrational mode coordinates. If Q_k were one of the a_1 normal modes, the ^1B state would have to be another $^1A_2$ state to render the integrand in Eq. 7.5 totally symmetric as required for a nonvanishing matrix element. If mode Q_k is the b_2 asymmetric stretching mode, however, the integrand will transform as $\Gamma(^1B) \otimes b_2 \otimes A_2$. This becomes totally symmetric if the electronic state ^1B has $^1B_1$ symmetry. As it happens, we have already shown that Walsh's rules predict the existence of a low-lying $^1B_1$ excited state. The close energy separation between this $^1B_1$ state and the $^1A_2$ state then makes for substantial kinetic energy coupling between these states by the nontotally symmetric b_2 vibrational mode. As a result, the $^1A_2$ state no longer has purely $^1A_2$ character, but contains an admixture of $^1B_1$ character as well. Since a $^1B_1 \leftarrow {}^1A_1$ electronic transition is E1 symmetry-allowed (and polarized perpendicular to the molecular plane), this $^1B_1$ admixture to the $^1A_2$ state renders the otherwise forbidden $^1A_2 \leftarrow {}^1A_1$ transition partly allowed. This coupling of different electronic states by nuclear motion in nontotally symmetric modes is an example of *vibronic coupling*, a phenomenon widely observed in polyatomic band spectra.

In the Walsh picture, the $^1A_2 \leftarrow {}^1A_1$ transition excites an electron from the $4b_2$ orbital to the $2b_1$ orbital (Fig. 7.3). Such a transition from a nonbonding to an antibonding π^* orbital should produce an increased bond length, as is observed (1.53 versus 1.432 Å). In analogy to the $^1B_2 \leftarrow {}^1A_1$ transition, it also decreases the bond angle as predicted (from 119.5° to 99°). The question arises as to where the $^1B_1$ state, from which the $^1A_1 \leftarrow {}^1A_1$ transition presumably borrows its intensity via vibronic coupling, can be found in the absorption spectrum. This $^1B_1$ state is not conspicuous in the SO_2 spectrum in Fig. 7.4; it may contribute a weak continuum to the high-energy side (~2400–2800 Å) of the first allowed system.

The *forbidden* band system (3400–4000 Å) arises from a $^3B_1 \leftarrow {}^1A_1$ spin-forbidden transition. (This $^3B_1$ state is the triplet counterpart to the aforementioned $^1B_1$ state.) In this transition, an electron jumps from the $6a_1$ orbital to the $2b_1$ orbital, orbitals that favor smaller and larger bond angles, respectively (Fig. 7.3). Thus Walsh's rules predict that SO_2 in the $^3B_1$ state will exhibit a larger bond angle than in the ground state, as indeed it does (126.1° as compared to 119.5°).

Investigators have searched in vain [5] for absorption bands due to the two remaining low-lying triplet states ($^3A_2$ and $^3B_2$) whose existence is predicted by Walsh's ordering of orbital energies. Electronic structure calculations have predicted that the $^3B_2$ state should absorb in the region around 6000 Å, and that the $^3A_2$ state should produce a band system at 3400–3900 Å. Perturbations in the $^3B_1 \leftarrow {}^1A_1$ spectrum have been attributed to the presence of a nearby $^3B_2$ state.

7.2 AROMATIC HYDROCARBONS

The lowest energy electronic transitions in homocyclic aromatic hydrocarbons occur at near-ultraviolet, visible, and infrared wavelengths from 2500 Å out to beyond 7000 Å. They involve excitations of electrons in delocalized π-type MOs, which are composed principally of carbon $2p$ orbitals oriented perpendicular to the aromatic plane. The remaining minimal-basis carbon valence orbitals (the $2s$ orbitals and the $2p$ orbitals oriented in the molecular plane) are utilized to form in-plane σ-type MOs directed along the chemical bonds. Excitations of electrons in σ-type MOs to unoccupied MOs require far higher photon energies (in the vacuum ultraviolet), and are not considered in this Section.

A well-studied aromatic molecule is aniline, $C_6H_5NH_2$, which differs from benzene in that one of the six hydrogens has been replaced by the amino group —NH_2. To a first approximation, the π MOs may be formed from linear combinations of the six out-of-plane carbon $2p$ AOs. In benzene, these MOs would have to transform as irreducible representations of the D_{6h} point group (see below). It is reasonable to expect that amino group substitution will perturb the π-electron system in aniline, however, reducing its effective point group symmetry to C_{2v}. (In fact, the plane containing the three atoms in the amino

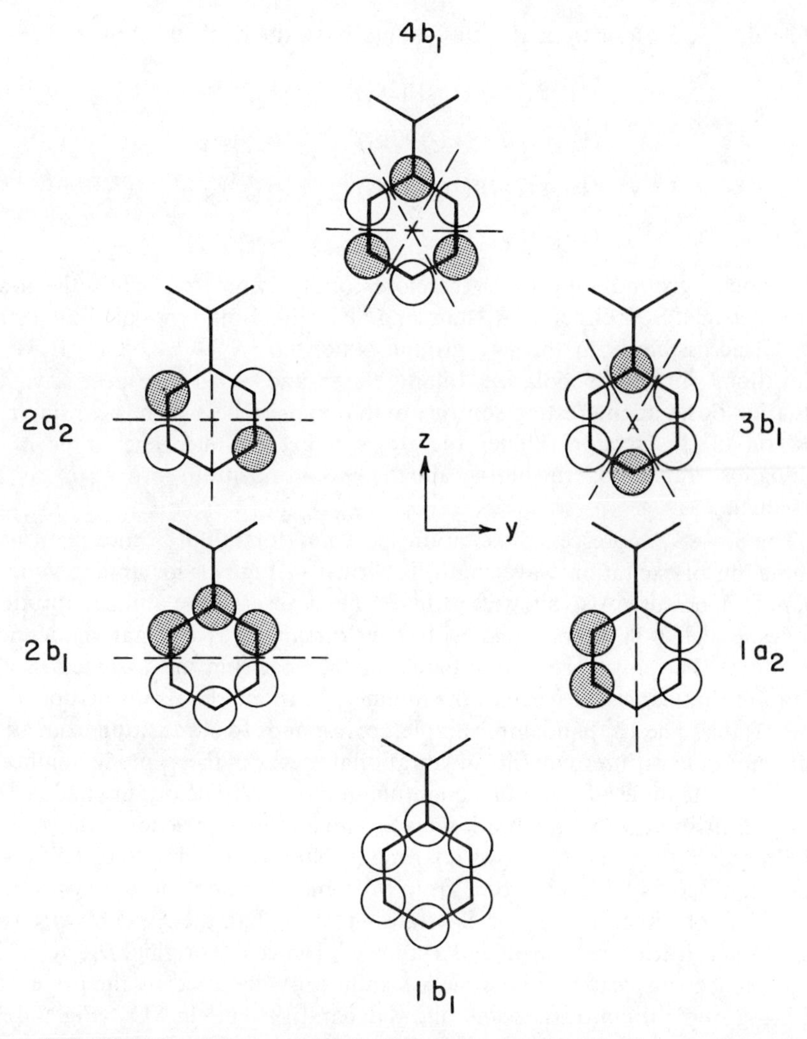

Figure 7.5 Nodal patterns and irreducible representations in C_{2v} of π orbitals in aniline, $C_6H_5NH_2$.

group is inclined from the aromatic plane by 39° in ground-state aniline, so that the environment experienced by the π-electron system does not even show C_{2v} symmetry. This point will be ignored in our treatment.) The nodal patterns for the six lowest-energy π MOs are shown in Fig. 7.5. This diagram also gives the alignments of the x, y, and z axes which are standard in discussions of excited-state symmetries in aniline. Since each carbon atom contributes one electron to the π system, ground-state aniline will have the closed-shell π-electron configuration

$$(1b_1)^2(2b_1)^2(1a_2)^2 \qquad {}^1A_1$$

while the lowest four excited states should have the configurations

$$(1b_1)^2(2b_1)^2(1a_2)^1(2a_2)^1 \qquad {}^1A_1,\ {}^3A_1$$
$$(1b_1)^2(2b_1)^2(1a_2)^1(3b_1)^1 \qquad {}^1B_2,\ {}^3B_2$$
$$(1b_1)^2(2b_1)^1(1a_2)^2(2a_2)^1 \qquad {}^1B_2,\ {}^3B_2$$
$$(1b_1)^2(2b_1)^1(1a_2)^2(3b_1)^1 \qquad {}^1A_1,\ {}^3A_1$$

The singlet ground state S_0 is therefore totally symmetric, while the lowest excited singlet S_1 is either a ${}^1A_1$ state or a ${}^1B_2$ state. Both types of excited states are E1-accessible from the ${}^1A_1$ ground state: the ${}^1A_1 \leftarrow {}^1A_1$ and ${}^1B_2 \leftarrow {}^1A_1$ transitions should be polarized along the z and y axes, respectively. This situation poses an interesting contrast to that in benzene, which is considered at the end of this section. Under the D_{6h} symmetry of benzene, some of the analogous transitions (including the $S_1 \leftarrow S_0$ transition) prove to be E1-forbidden.

The $S_1 \leftarrow S_0$ fluorescence excitation spectrum (total fluorescence intensity as a function of excitation wavelength) is shown in Fig. 7.6 for aniline vapor at room temperature. Also shown in Fig. 7.6 are several of the aniline vibrational modes (labeled in accordance with the Varsanyi [6] normal mode nomenclature). The most prominent bands in the spectrum are assigned using a notation that concisely specifies the aniline vibrational levels in the upper and lower states. The 12^1_0 band, for example, corresponds to a transition from an S_0-state molecule with zero quanta of vibrational energy in the a_1 mode number 12, to an S_1-state molecule with one quantum in mode 12. The origin band (arising from a transition between vibrationless S_0 and S_1 states) is denoted the 0^0_0 band. The very fact that Fig. 7.6 shows a strong 0^0_0 band—implying that the $S_1 \leftarrow S_0$ transition occurs without succor from vibronic coupling through vibrational excitation of either the S_1 or S_0 state—means that the $S_1 \leftarrow S_0$ electronic transition is intrinsically strongly E1-allowed. Hence the original D_{6h} symmetry of the π-electron system in benzene is significantly distorted by the presence of amino group. In contrast to the allowed band systems in SO_2 (Fig. 7.4), no regular progressions of nearly equally spaced bands occur in the aniline excitation spectrum. For that matter, the great majority of the aniline bands arise from transitions in which no vibrational quantum number changes by more than 1 unit. Unlike the allowed transitions in SO_2, then, the $S_1 \leftarrow S_0$ aniline transition is not accompanied by large geometry changes along its totally symmetric modes.

To determine the symmetry of the S_1 state, Christofferson et al. [7] analyzed the rotational structure of several of the bands. The appearance of the 0^0_0 absorption band under high resolution is shown in the top portion of Fig. 7.7. In a molecule as large as aniline, such a contour comprises some 30,000 rotational transitions, and so there is little hope for resolving individual lines. Instead, the contour is compared with a computer simulation that calculates the asymmetric rotor energy level differences, weights the intensities of allowed rotational

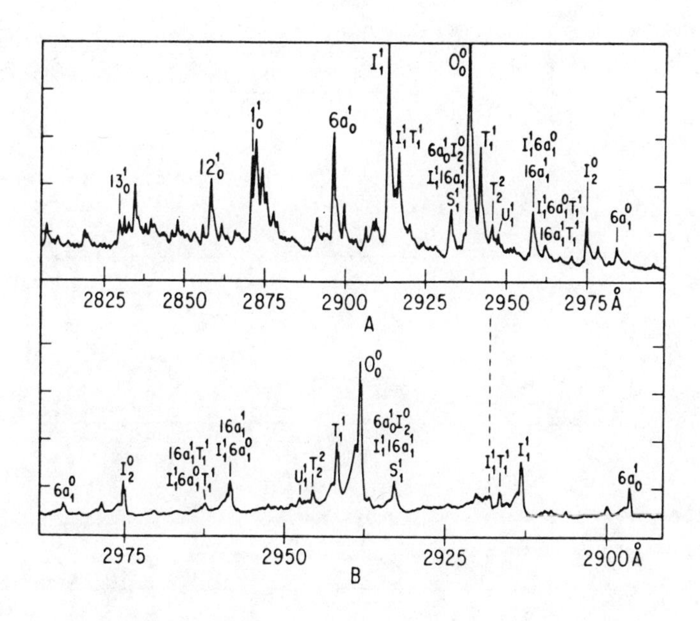

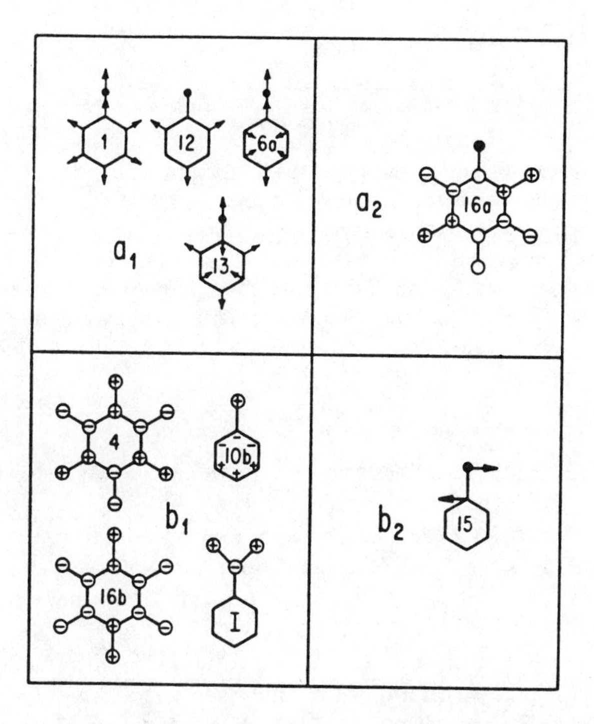

Figure 7.6 Fluorescence excitation spectra of aniline vapor: (a) 0.3-torr pressure, 0.32 Å resolution; (b) 0.24-torr pressure, 0.08-Å resolution. Also shown are several normal vibrations that are active in the fluorescence excitation spectra. Used with permission from D. Chernoff and S. A. Rice, *J. Chem. Phys.* **70**: 2521 (1979).

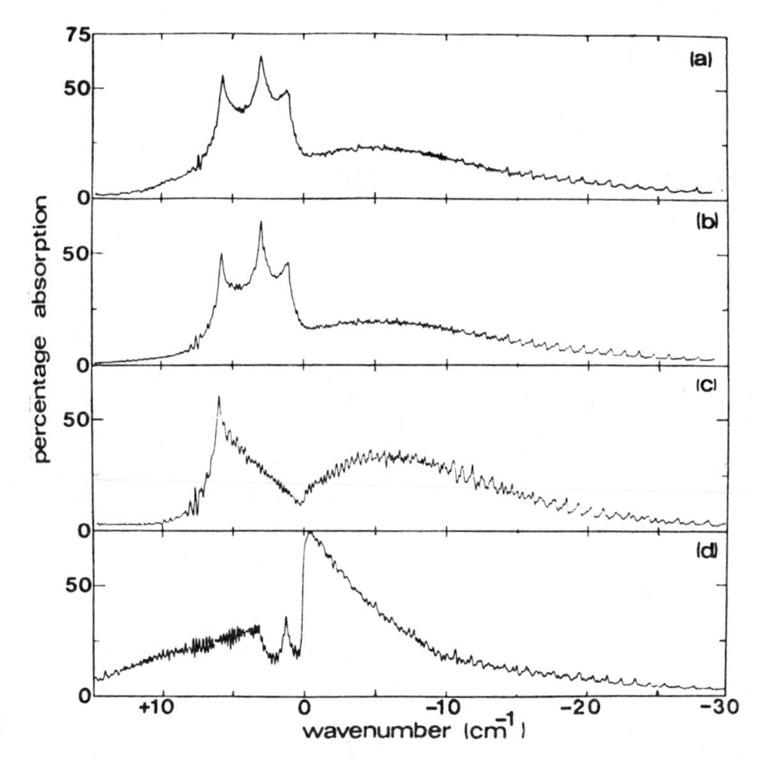

Figure 7.7 Rotational fine structure of the 0^0_0 absorption band in aniline vapor. The experimental rotational contour is shown at top; the theoretical simulations labeled (b), (c), and (d) were generated using identical rotational constants, but assumed $S_1 \leftarrow S_0$ transition moments polarized along the *b, a,* and *c* principal rotational axes, respectively. This work established that the $S_1 \leftarrow S_0$ transition in aniline is polarized along the *b* axis (Fig. 7.8), or equivalently, *y*-polarized (cf. Fig. 7.5). Reproduced by permission from J. Christofferen, J. M. Hollas, and G. H. Kirby, *Mol. Phys.* **16**, 441 (1969).

transitions with the degeneracies and Boltzmann factors of the lower levels, and superimposes the lines (broadened to replicate the known experimental resolution) to generate a theoretical spectrum. The rotational constants A'', B'', C'' of S_0 state aniline were already known from microwave spectroscopy. The S_1 state rotational constants A', B', C' may be varied to optimize the theoretical spectrum to fit to the experimental contour. A crucial point here is that the contour shape is extremely sensitive to the rotational selection rules, which in turn hinge on the relative orientations of the $S_1 \leftarrow S_0$ transition moment and the principal rotational axes *a*, *b*, and *c* (Fig. 7.8). We have already seen in Section 6.6 how the rotational structure in vibration–rotation spectra is influenced by the alignment of the vibrational transition moment with respect to the rotational axes. The three theoretical contours in Fig. 7.8 were generated using

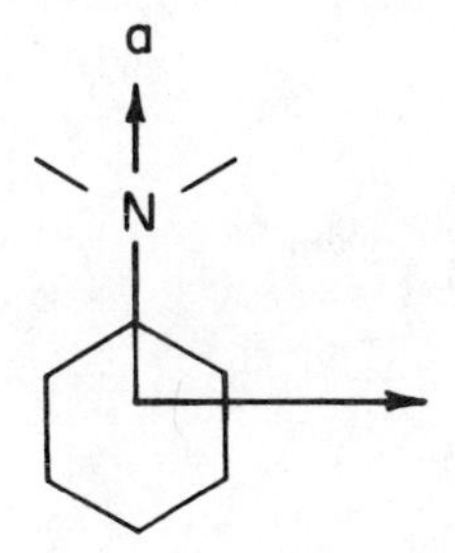

Figure 7.8 Orientations of principal rotational axes in aniline. The *c* axis is perpendicular to the paper.

identical upper state rotational constants, but assuming different directions for the $S_1 \leftarrow S_0$ transition moments. It is clear that the observed contour is only consistent with an electronic transition polarized along the *b* principal rotational axis, which is equivalent to the *y* axis in Fig. 7.5. Hence the S_1 state is a $^1B_2$ rather than $^1A_1$ state. This example illustrates the power of rotational contour analysis for assigning electronic transitions.

Naphthalene, $C_{10}H_8$, is a planar molecule belonging to the D_{2h} point group. Ten π-type SALCs may be derived by application of the D_{2h} projection operators to the out-of-plane carbon $2p$ orbitals; the qualitative nodal patterns of the resulting π MOs are shown in Fig. 7.9. We use the Pariser coordinate system, in which the in-plane *x* and *y* axes are aligned with the long and short axes in naphthalene; the *z* axis is perpendicular to the molecular plane. Since naphthalene has ten electrons (one donated by each carbon) in its π orbitals, its ground state S_0 will have the closed-shell configuration

$$(1b_{1u})^2(1b_{2g})^2(1b_{3g})^2(1a_u)^2(2b_{1u})^2 \qquad {}^1A_g$$

The lowest few excited singlet states are then expected to be

$$\ldots(1a_u)^2(2b_{1u})^1(2b_{2g})^1 \qquad {}^1B_{3u}$$

$$\cdots(1a_u)^2(2b_{1u})^1(2b_{3g})^1 \qquad {}^1B_{2u}$$

$$\cdots(1a_u)^1(2b_{1u})^2(2b_{2g})^1 \qquad {}^1B_{2u}$$

The $S_1 \leftarrow S_0$ transition to the lowest excited singlet state in napthalene is the $^1B_{3u} \leftarrow {}^1A_g$ transition. Since the ^1A_g state is totally symmetric and the vector *x* transforms as B_{3u} in D_{2h}, this transition should presumably be E1 symmetry-allowed and polarized along the long axis. The next higher spin-allowed transition ($S_2 \leftarrow S_0$) is a symmetry-allowed $^1B_{2u} \leftarrow {}^1A_g$ transition. The vector *y* transforms as B_{2u}, so this transition should be polarized along the short axis. The $S_2 \leftarrow S_0$ transition is in fact responsible for an intense electronic band system in the near ultraviolet.

In Fig. 7.10, $S_1 \leftarrow S_0$ fluorescence excitation spectra are contrasted for room-temperature collisionless naphthalene (at a pressure of $\sim 5 \times 10^{-5}$ torr) and for

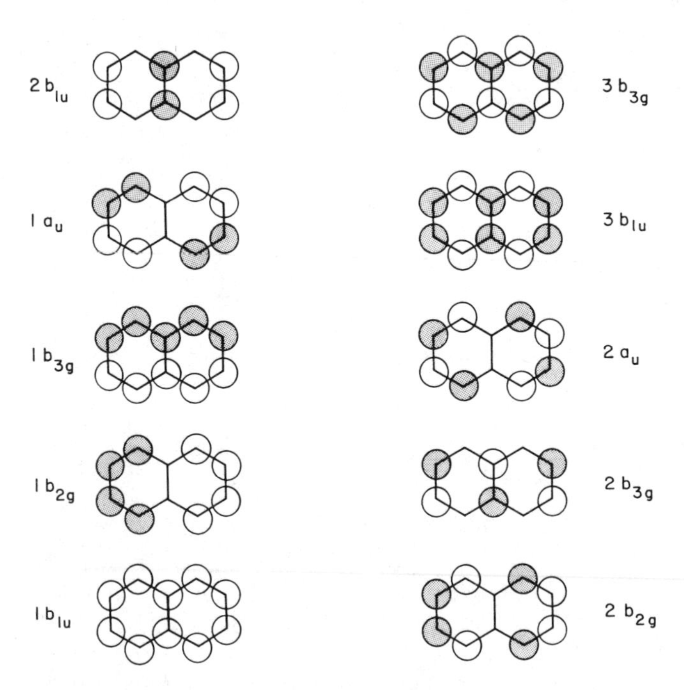

Figure 7.9 Nodal patterns and irreducible representations in D_{2h} of π orbitals in naphthalene, $C_{10}H_8$.

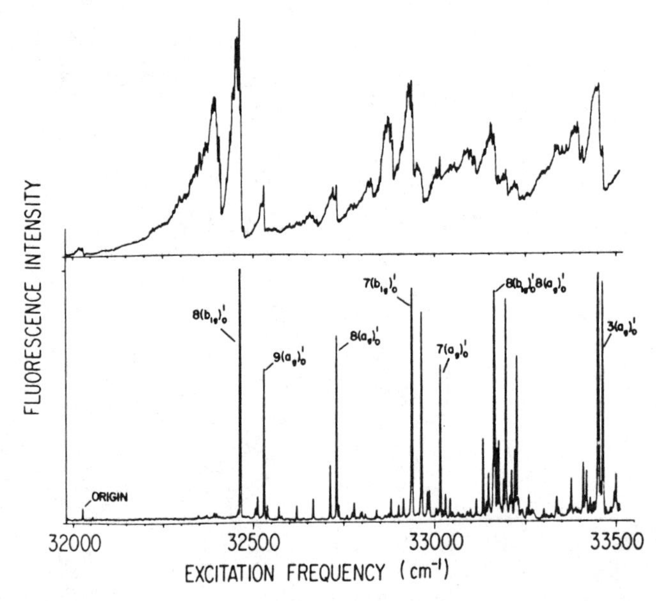

Figure 7.10 Fluorescence excitation spectra of naphthalene: (a) in room-temperature vapor at 5×10^{-5} torr; (b) in supersonic jet, 0.07-torr hydrocarbon in 4 atm helium. The $8(b_{1g})^1_0$ peak in the jet spectrum extends considerably offscale. Used with permission from S. Behlen, D. McDonald, V. Sethuraman, and S. A. Rice, *J. Chem. Phys.* **75**: 5685 (1981).

naphthalene in a supersonic jet. The "bulb" spectrum is broadened by rotational profiles at room temperature. The supersonic jet spectrum differs radically from the bulb spectrum because the effective rotational and vibrational temperatures in the jet are less than ~ 10 K. The rotational cooling sharpens the vibrational bands, while the vibrational cooling removes hot bands caused by transitions from vibrationally excited S_0 molecules. The jet spectrum is clearly more useful for making spectral assignments and for studying the S_1 state vibrational structure.

It can be shown using the group theoretical methods in Chapter 6 that naphthalene possesses nine a_g, eight b_{1g}, three b_{2g}, four b_{3g}, four a_u, four b_{1u}, eight b_{2u}, and eight b_{3u} normal modes. Each normal mode is denoted with a numerical prefix indicating its rank in frequency among modes of the same symmetry. For example, mode $8(b_{1g})$ is the b_{1g} mode having the eighth highest (i.e., the lowest) frequency among b_{1g} modes. This notation is employed in the vibrational band assignments in Fig. 7.10: The band $7(a_g)_0^1$ arises from a transition between an S_0 molecule with zero vibrational quanta in mode $7(a_g)$ and an S_1 molecule with one quantum in mode $7(a_g)$. A distinguishing feature of the naphthalene $S_1 \leftarrow S_0$ spectrum is the weakness of the 0_0^0 transition between vibrationless S_1 and S_0 states (this transition is labeled "origin" in Fig. 7.10). Even though the $S_1 \leftarrow S_0$ transition is E1 symmetry-allowed in naphthalene, it so happens that the pertinent electronic transition moment $\mathbf{M}_e$ (Section 4.4) is numerically much smaller than expected for a strong E1 transition. (In the oscillator strength language which will be developed in Chapter 8, the $S_1 \leftarrow S_0$ transition exhibits $f \sim 0.001$; the strongly allowed $S_2 \leftarrow S_0$ transition that occurs at higher energies exhibits $f \sim 1.0$.) The presence of b_{1g} vibrational modes allows vibronic coupling between the ${}^1B_{3u}$ S_1 state and the nearby ${}^1B_{2u}$ S_2 state (since $B_{3u} \otimes b_{1g} \otimes B_{2u} = A_g$). The ${}^1B_{3u}\,S_1$ state then gains some of the ${}^1B_{2u}$ character of the S_2 state, and consequently borrows some of the intensity of the strong $S_2 \leftarrow S_0$ transition. This vibronic coupling accounts for the $8(b_{1g})_0^1$, $7(b_{1g})_0^1$, and to some extent the $8(b_{1g})_0^1 8(a_g)_0^1$ bands, all of which are an order of magnitude more intense than the origin band at 32,018.5 cm^{-1}. Since most of the intensity in these bands is due to the ${}^1B_{2u}$ admixture into the ${}^1B_{3u}$ S_1 state, these bands are polarized along the short axis (rather than along the long axis as would be expected for an intrinsically strong ${}^1B_{3u} \leftarrow {}^1A_g$ $S_1 \leftarrow S_0$ transition). It will be shown in the following section that the vibrational selection rule in these *vibronically induced* transitions is $\Delta v = \pm 1, \pm 3, \ldots$ in the normal mode responsible for the vibronic coupling. This is why one observes transitions such as $8(b_{1g})_0^1$, but not $8(b_{1g})_0^2$.

The bands designated $9(a_g)_0^1$, $8(a_g)_0^1$, $7(a_g)_0^1$, and $3(a_g)_0^1$ are also vibronically induced bands, occasioned by vibronic coupling of the ${}^1B_{3u}$ S_1 state and some higher energy ${}^1B_{3u}$ S_n state by the respective totally symmetric modes ($B_{3u} \otimes a_g \otimes B_{3u} = A_g$). While this mixing of S_n into S_1 does not modify the B_{3u} character in S_1 levels having excitation in a_g modes, it will enlarge the $S_1 \leftarrow S_0$ transition moment if the $S_n \leftarrow S_0$ transition is more intense than the intrinsic $S_1 \leftarrow S_0$ transition in the absence of vibronic coupling. These bands in the

$S_1 \leftarrow S_0$ spectrum should exhibit the long-axis polarization of the intrinsic $S_n \leftarrow S_0$ ($^1B_{3u} \leftarrow {}^1A_g$) transition. All of these band polarizations have been confirmed by obtaining absorption spectra of napthalene doped as an impurity into host crystals, where the napthalene orientations relative to the crystal axes are independently known.

Another notable feature in the napthalene $S_1 \leftarrow S_0$ spectrum is the lack of progression formation—the absence of intense bands like $8(b_{1g})_0^1\ (a_g)_0^n$ with $n > 1$. This indicates that large geometry changes do not occur along totally symmetry modes in the $S_1 \leftarrow S_0$ transitions. In this respect, napthalene is similar to aniline and contrasts with SO_2.

The benzene molecule C_6H_6 exhibits D_{6h} symmetry. The nodal patterns of the six π-type SALCs formed from linear combinations of out-of-plane carbon $2p$ AOs are shown in Fig. 7.11. With six π electrons, benzene has the totally

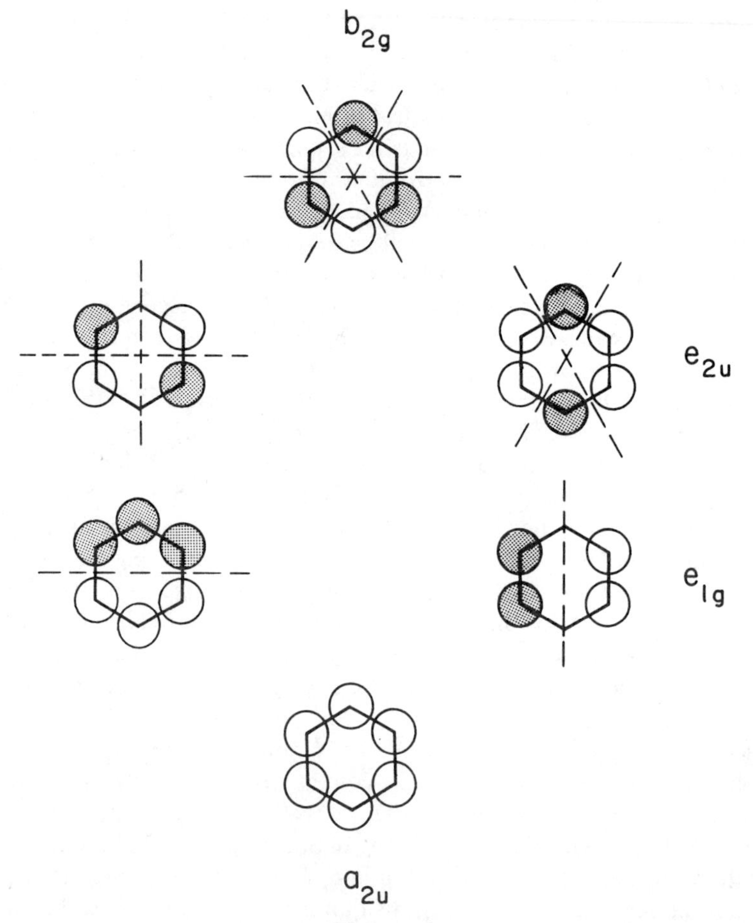

Figure 7.11 Nodal patterns and irreducible representations in D_{6h} of π orbitals in benzene, C_6H_6.

symmetric A_{1g} ground-state configuration $(a_{2u})^2(e_{1g})^4$. Its lowest lying excited states arise from the configuration $(a_{2u})^2(e_{1g})^3(e_{2u})^1$; they have the overall electronic symmetries $e_{1g} \otimes e_{2u} = B_{1u} \oplus B_{2u} \oplus E_{1u}$. The intrinsic (i.e., not vibronically induced) electronic transitions $^1B_{1u} \leftarrow {}^1A_{1g}$ and $^1B_{2u} \leftarrow {}^1A_{1g}$ are symmetry-forbidden. The $^1E_{2u} \leftarrow {}^1A_{1g}$ transition is E1-allowed, and has been assigned to an intense ($f \sim 1.0$) band system at 1850 Å. The lowest spin-allowed transition $S_1 \leftarrow S_0$ in benzene is associated with a much weaker system ($f \sim 0.001$) at 2600 Å; its spectrum is shown in Fig. 7.12. This must be either the

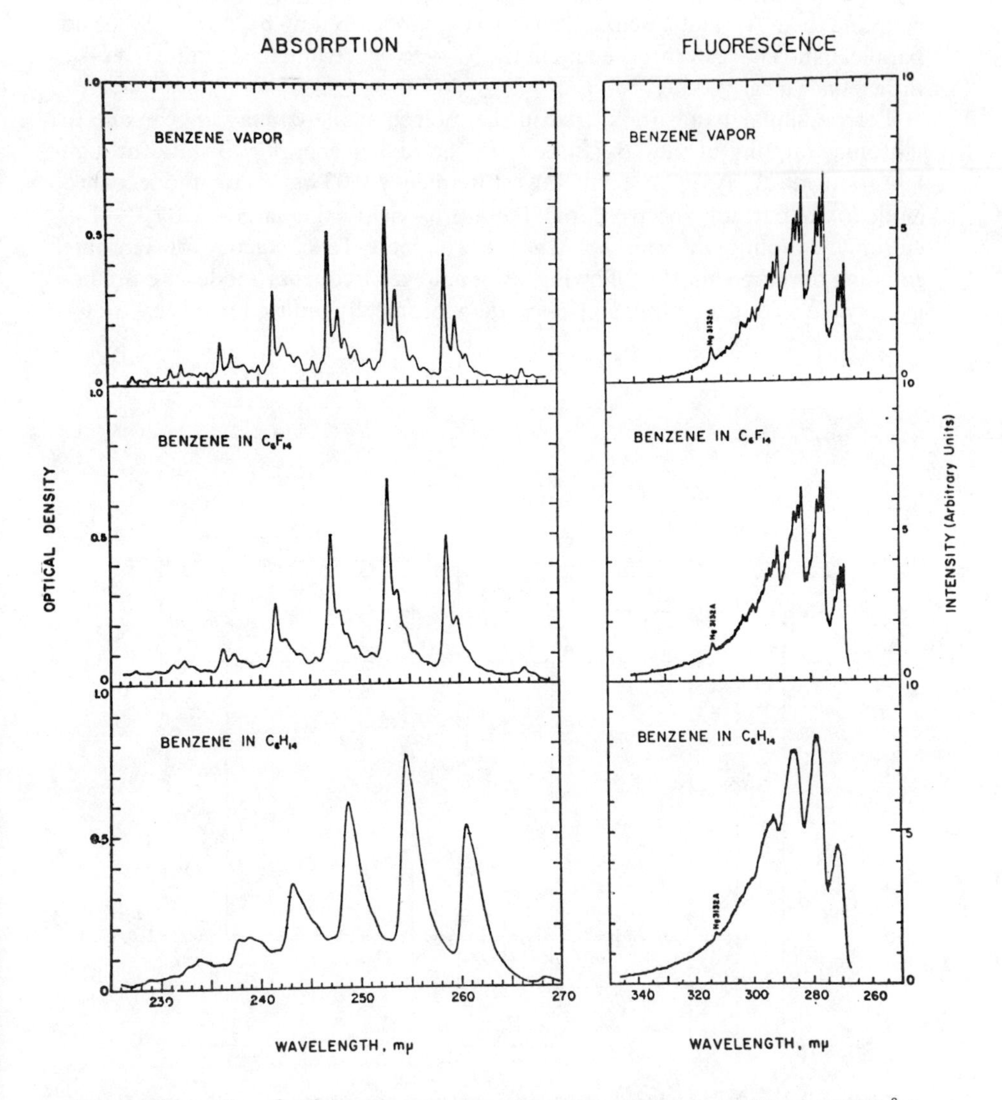

Figure 7.12 $S_1 \leftarrow S_0$ absorption spectra of benzene between 2300 and 2700 Å in the vapor, in C_6F_{14}, and in C_6H_{14}. Used with permission from C. W. Lawson, F. Hirayama, and S. Lipsky, *J. Chem. Phys.* **51**: 1595 (1961).

$^1B_{1u} \leftarrow {}^1A_{1g}$ or the $^1B_{2u} \leftarrow {}^1A_{1g}$ transition, with E1 intensity gained through vibronic coupling.

The 30 normal vibrations available in benzene include two a_{1g}, one a_{2g}, two b_{2g}, one e_{1g}, four e_{2g}, one a_{2u}, two b_{1u}, two b_{2u}, three e_{1u}, and two e_{2u} modes. For a vibronically induced E1 transition from S_0 to S_1, the latter state must be coupled to either an E_{1u} state (which transforms as x, y) or an A_{2u} state (which transforms as z). In the latter case, the $S_1 \leftarrow S_0$ absorption spectrum would exhibit *parallel* bands polarized along the z axis. For S_1 states having B_{1u} and B_{2u} symmetry, respectively, b_{2g} and b_{1g} modes would be required to effect coupling to an A_{2u} state. Benzene has two b_{2g} modes, but no b_{1g} modes. Since no parallel bands have ever turned up in the $S_1 \leftarrow S_0$ spectrum, the S_1 state is a $^1B_{2u}$ rather than a $^1B_{1u}$ state.

Perpendicular bands (polarized in the molecular plane) may appear due to vibronic coupling of the $^1B_{2u}$ state with the higher energy E_{1u} state through vibrations of e_{2g} symmetry. The lowest frequency (605 cm^{-1}) e_{2g} mode, called mode 6 in current spectroscopic literature, yields the largest $^1B_{2u} \leftrightarrow {}^1E_{1u}$ vibronic coupling in benzene. (In the Herzberg-Teller theory of vibronic coupling developed in the following section, lower-frequency modes are predicted to yield stronger coupling.) Hence, the vibronically induced bands 6^1_0, 6^0_1, 6^2_1,

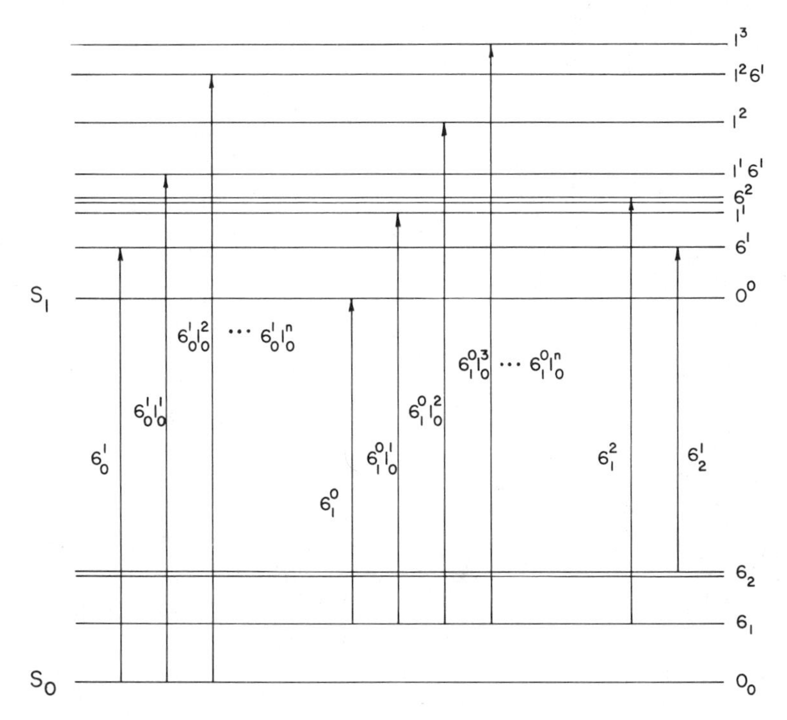

Figure 7.13 Energy level diagram for vibronically induced $S_1 \leftarrow S_0$ transitions in benzene.

6^1_2 are prominent in the $S_1 \leftarrow S_0$ spectrum; the last three are hot bands which occur in bulb spectra. The distinctively regular progression in the benzene spectrum of Fig. 7.12 appears because the geometry change associated with the $S_1 \leftarrow S_0$ transition has a large component along the totally symmetric "breathing" mode (which is called mode 1). For this reason, one observes, in addition to the transitions 6^1_0, 6^0_1, 6^2_1, 6^1_2, the transitions $6^1_0 1^n_0$, $6^0_1 1^n_0$, $6^2_1 1^n_0$, $6^1_2 1^n_0$ with $n \geqslant 1$ (Fig. 7.13).

7.3 QUANTITATIVE THEORIES OF VIBRONIC COUPLING

In our qualitative discussions of vibronic coupling in Sections 7.1 and 7.2, we did not develop any method for predicting vibrational band intensities or vibrational selection rules in electronic band spectra in which vibronic coupling is important. Vibronic coupling generally arises from interactions between nuclear and electronic motions [8]. The Herzberg-Teller (HT) theory of vibronic coupling deals with the ramifications of the vibrational coordinate dependence of the electronic transition moment; it thus examines one aspect of coupling between nuclear and electronic motions. The Born-Oppenheimer (BO) vibronic coupling theory is concerned with interactions arising from the nuclear kinetic energy operator $\hat{T}(Q) = -(\hbar^2/2)\sum \partial^2/\partial Q_i^2$; these interactions are analogous to those described for diatomic molecules in Section 3.1. Earlier in this chapter, we rationalized vibronic coupling in the $^1A_2$ state of SO_2 in terms of such nuclear kinetic energy coupling. The HT theory has accounted for vibronic band intensity distributions to a first approximation in many organic molecules; the BO theory has been invoked to improve on the quantitative predictions of HT theory.

Let the initial and final states in an E1 optical transition be represented by the Born-Oppenheimer states

$$|\Psi_{mv}(q, Q)\rangle = |\psi_m(q, Q)\chi_v^m(Q)\rangle \tag{7.6}$$

and

$$|\Psi_{nw}(q, Q)\rangle = |\psi_n(q, Q)\chi_w^n(Q)\rangle \tag{7.7}$$

where q and Q represent the electronic and normal coordinates, $|\psi_m\rangle$ and $|\psi_n\rangle$ are the initial and final electronic states, and $|\chi_v^m\rangle$ and $|\chi_w^n\rangle$ are the respective vibrational states. In analogy to Eq. 4.49 for diatomics, the transition moment integral is given by

$$\boldsymbol{\mu}_{mn,vw} = \langle \chi_v^m | \mathbf{M}_e^{m \to n} | \chi_w^n \rangle \tag{7.8}$$

where $\mathbf{M}_e^{m \to n}$ is the electronic transition moment

$$\mathbf{M}_e^{m \to n} = \langle \psi_m | \boldsymbol{\mu}_{\mathrm{el}} | \psi_n \rangle \tag{7.9}$$

In the Herzberg-Teller theory, $\mathbf{M}_e^{m\to n}$ (whose notation we now truncate to $\mathbf{M}_{mn}$) is expanded in a Taylor series in some normal coordinate Q_k about its equilibrium value $Q_k = 0$,

$$\mathbf{M}_{mn}(Q) = \mathbf{M}_{mn}(0) + \left(\frac{\partial \mathbf{M}_{mn}}{\partial Q_k}\right)_0 Q_k + \cdots \tag{7.10}$$

This renders the E1 transition moment equal to

$$\boldsymbol{\mu}_{mn,vw} = \mathbf{M}_{mn}(0)\langle \chi_v^m | \chi_w^n \rangle + \left(\frac{\partial \mathbf{M}_{mn}}{\partial Q_k}\right)_0 \langle \chi_v^m | Q_k | \chi_w^n \rangle + \cdots \tag{7.11}$$

$\mathbf{M}_{mn}(0)$ is nonzero only if the electronic transition is intrinsically E1-allowed (as in the aniline $S_1 \leftarrow S_0$ system or in the second allowed system of SO_2). It is small for the naphthalene $S_1 \leftarrow S_0$ system, and it vanishes for the $S_1 \leftarrow S_0$ transition in an isolated benzene molecule. In the latter two molecules, intense vibrational bands can thus arise only from the first-order term (or in principle from higher-order terms) in Eq. 7.11. HT theory retains only the first-order term in Eq. 7.11. If Q_k is a nontotally symmetric mode having the same frequency and equilibrium value in both electronic states m and n, the Franck-Condon amplitude $\langle \chi_v^m | \chi_w^n \rangle$ vanishes by symmetry unless the number of quanta in mode Q_k changes by $\Delta v = 0, \ \pm 2, \ \pm 4, \ldots$. However, $\langle \chi_v^m | Q_k | \chi_w^n \rangle$ vanishes unless $\Delta v = \pm 1, \pm 3, \ldots$ If Q_k is totally symmetric, however, both terms in Eq. 7.11 may simultaneously be nonzero, allowing interferences to occur between the zeroth-order (intrinsic) and first-order (vibronically induced) components of $\boldsymbol{\mu}_{mn,vw}$.

The vibronically induced component, which is proportional to $(\partial \mathbf{M}_{mn}/\partial Q_k)_0$, controls the vibrational band intensity according to HT theory when the intrinsic electronic transition is forbidden. Since the electronic dipole moment operator $\boldsymbol{\mu}_{el}$ depends only on the electronic coordinates (Eq. 4.47), we have

$$\left(\frac{\partial \mathbf{M}_{mn}}{\partial Q_k}\right)_0 = \langle \psi_m | \boldsymbol{\mu}_{el} \left| \frac{\partial \psi_n}{\partial Q_k} \right\rangle + \left\langle \frac{\partial \psi_m}{\partial Q_k} \right| \boldsymbol{\mu}_{el} | \psi_n \rangle$$

$$\equiv \sum_i \left[\langle \psi_m | \boldsymbol{\mu}_{el} | \psi_i \rangle \langle \psi_i | \frac{\partial \psi_n}{\partial Q_k} \right\rangle + \left\langle \frac{\partial \psi_m}{\partial Q_k} | \psi_i \rangle \langle \psi_i | \boldsymbol{\mu}_{el} | \psi_n \rangle \right] \tag{7.12}$$

This leads to an explicit expression for the vibronically induced component of the transition moment according to Herzberg-Teller theory,

$$\boldsymbol{\mu}_{mn,vw}^{\mathrm{HT}} = \sum_i \left[\langle \psi_m | \boldsymbol{\mu}_{el} | \psi_i \rangle \langle \psi_i | \frac{\partial}{\partial Q_k} | \psi_n \rangle - \langle \psi_m | \frac{\partial}{\partial Q_k} | \psi_i \rangle \langle \psi_i | \boldsymbol{\mu}_{el} | \psi_n \rangle \right] \langle \chi_v^m | Q_k | \chi_w^n \rangle \tag{7.13}$$

The first set of terms on the right side of this expression may be physically interpreted as follows. The upper electronic state $|\psi_n\rangle$ in the transition $nw \leftarrow mv$

can be mixed by vibration in normal coordinate Q_k with some other electronic state $|\psi_i\rangle$ to an extent that is proportional to $\langle\psi_i|\partial/\partial Q_k|\psi_n\rangle$. The $nw \leftarrow mv$ transition moment, which would be zero in the absence of vibronic coupling if the intrinsic transition is forbidden ($\mathbf{M}_{mn} = 0$), can then *borrow intensity* from the $i \leftarrow m$ electronic transition provided that $\langle\psi_m|\boldsymbol{\mu}_{el}|\psi_i\rangle$ is nonzero. Since $\partial/\partial Q_k$ transforms as Q_k under molecular point group operations, the matrix element $\langle\psi_i|\partial/\partial Q_k|\psi_n\rangle$ required for intensity-borrowing is nonvanishing only if the direct product $\Gamma(\psi_i) \otimes \Gamma(Q_k) \otimes \Gamma(\psi_n)$ contains the totally symmetric irreducible representation. This symmetry requirement is identical to the one we used in our more qualitative discussions of vibronic coupling in SO_2, naphthalene and benzene. We may interpret Eq. 7.13 less formally by saying that when state $|\psi_n\rangle$ is vibronically coupled to some other state $|\psi_i\rangle$, it becomes replaced by the mixed state $|\psi_n\rangle + \alpha|\psi_i\rangle$, where α is a small number proportional to $\langle\psi_i|\partial/\partial Q_k|\psi_n\rangle$. The transition moment $\boldsymbol{\mu}_{mn} = \langle\psi_m\chi_v^m|\boldsymbol{\mu}_{el}|\psi_n\chi_w^n\rangle$ is then superseded in the Herzberg-Teller picture by

$$\boldsymbol{\mu}_{mn} = \langle\psi_m\chi_v^m|\boldsymbol{\mu}_{el}|\psi_n\chi_w^n\rangle + \alpha\langle\psi_m\chi_v^m|\boldsymbol{\mu}_{el}|\psi_i\chi_w^n\rangle \tag{7.14}$$

which says that if the intrinsic $nw \leftarrow mv$ transition is forbidden ($\langle\psi_m\chi_v^m|\boldsymbol{\mu}_{el}|\psi_n\chi_w^n\rangle = 0$), it can still occur if the $i \leftarrow m$ transition is allowed ($\langle\psi_m\chi_v^m|\boldsymbol{\mu}_{el}|\psi_i\chi_w^n\rangle \neq 0$) and if states $|\psi_n\rangle$ and $|\psi_i\rangle$ are vibronically coupled ($\alpha \neq 0$).

The second set of terms in the summation of Eq. 7.13 represents vibronic coupling of the *initial* state $|\psi_m\rangle$ with other states $|\psi_i\rangle$ through vibration in mode Q_k. The transition $nw \leftarrow mv$ then borrows intensity from the electronic transitions $i \leftarrow n$. In most situations of interest, $|\psi_m\rangle$ and $|\psi_n\rangle$ are ground and excited states, respectively, in an absorptive transition. Other electronic states $|\psi_i\rangle$ tend to lie closer in energy to $|\psi_n\rangle$ than to $|\psi_m\rangle$, and so workers have frequently assumed that the "intensity-lending" states $|\psi_i\rangle$ are much more strongly coupled to the excited state $|\psi_n\rangle$ than to the ground state $|\psi_m\rangle$. This approximation (which ignores the second set of terms) has been challenged, however [9]. It may be shown that

$$\left(\frac{\partial|\psi_n\rangle}{\partial Q_k}\right)_0 = \sum_{i\neq n} \frac{\langle\psi_i|\partial H_0/\partial Q_k|\psi_n\rangle}{E_i - E_n}|\psi_i\rangle \tag{7.15}$$

where $\hat{H}_0$ is the electronic Hamiltonian and the E_i, E_n are electronic-state energies. For a nontotally symmetric, harmonic mode with identical frequencies in both electronic states, we have the selection rule $\Delta v = \pm 1$ in the vibrational integral $\langle\psi_v^m|Q_k|\chi_w^n\rangle$. Use of second quantization then shows that

$$\langle\chi_v^m|Q_k|\chi_w^n\rangle = \left(\frac{\hbar}{2\mu_k\omega_k}\right)^{1/2} \sqrt{[v_k, w_k]} \sum_{j\neq k}^{3N-6} \langle\chi_v^m(Q_j)|\chi_w^n(Q_j)\rangle \tag{7.16}$$

where μ_k and ω_k are the reduced mass and frequency in mode k, and $[v_k, w_k]$ is the larger of the two quantum numbers of vibrational states v and w in normal coordinate Q_k. Combining Eqs. 7.15 and 7.16 yields the conventional expression

$$\boldsymbol{\mu}_{mn,vw}^{\mathrm{HT}} = \sum_{i \neq n} \frac{\langle \psi_m | \boldsymbol{\mu}_{\mathrm{el}} | \psi_i \rangle \langle \psi_i | \partial H_0 / \partial Q_k | \psi_n \rangle}{E_i - E_n} \times \left(\frac{\hbar}{2\mu_k \omega_k} \right)^{1/2} \sqrt{[v_k, w_k]} \prod_{j \neq k}^{3N-6} \langle \chi_v^m(Q_j) | \chi_w^n(Q_j) \rangle \tag{7.17}$$

for the vibronically induced transition moment in the Herzberg-Teller theory. (The last factor in Eqs. 7.16 and 7.17 is simply a product of Franck-Condon amplitudes in all modes other than mode Q_k.) Herzberg-Teller coupling appears to account fairly well for the relative intensities of the vibronically induced bands in the naphthalene $S_1 \leftarrow S_0$ spectrum. Equation 7.17 predicts that vibronically induced band intensities will vary as $1/\omega_k$ with the frequency of the nontotally symmetric mode. This is in accord with the observation that lower-frequency modes of appropriate symmetry tend to be more active in vibronically induced spectra.

Vibronic coupling through the nuclear kinetic energy operator $\hat{T}(Q)$ rather than through Q-dependence in the electronic transition moment $\mathbf{M}_{mn}$ can be treated in a manner that parallels our discussion of the Born-Oppenheimer approximation in diatomics (Section 3.1). The Born-Oppenheimer theory of vibronic coupling predicts that the induced transition moment will be

$$\boldsymbol{\mu}_{mn,vw}^{\mathrm{BO}} = \sum_i \left(\frac{\hbar\omega_k}{E_i - E_n} \langle \psi_m | \boldsymbol{\mu}_{\mathrm{el}} | \psi_i \rangle \langle \psi_i | \frac{\partial}{\partial Q_k} | \psi_n \rangle + \frac{\hbar\omega_k}{E_i - E_m} \langle \psi_m | \frac{\partial}{\partial Q_k} | \psi_i \rangle \langle \psi_i | \boldsymbol{\mu}_{\mathrm{el}} | \psi_n \rangle \right) \langle \chi_v^m | Q_k | \chi_w^n \rangle \tag{7.18}$$

If vibronic coupling of the lower electronic state $|\psi_m\rangle$ to higher states $|\psi_i\rangle$ is ignored (by setting $\langle \psi_m | \partial/\partial Q_k | \psi_i \rangle = 0$), comparison of Eqs. 7.13 and 7.18 immediately shows that

$$\left| \frac{\boldsymbol{\mu}_{nm,vw}^{\mathrm{BO}}}{\boldsymbol{\mu}_{mn,vw}^{\mathrm{HT}}} \right| = \hbar\omega_k / (E_i - E_n) \tag{7.19}$$

if the coupling is dominated by one of the higher states $|\psi_i\rangle$. The vibrational spacing $\hbar\omega_k$ in mode Q_k is frequently small compared to the energy gap $(E_i - E_n)$ between coupled electronic states. For example, $|\psi_i\rangle$ and $\psi_n\rangle$ can represent the $S_1(^1B_{3u})$ and $S_2(^1B_{2u})$ states that are separated by $\sim 3800\ \mathrm{cm}^{-1}$ in naphthalene; they are vibronically coupled by b_{1g} modes with $h\omega_k = 512$ and $944\ \mathrm{cm}^{-1}$ in the jet spectrum shown in Fig. 7.10. Hence, BO coupling has frequently been assumed to be insignificant relative to HT coupling, a presumption that has been questioned by Orlandi and Siebrand [9]. It is necessary to invoke BO as well as HT coupling to reproduce the details of the naphthalene $S_1 \leftarrow S_0$ band intensity distribution, because the intensity borrowing and lending states (S_2 and S_1) are unusually close together in this molecule.

We finally comment on the vibrational selection rules for vibronically induced transitions. The intensity of a vibrational band which occurs through vibronic coupling in normal coordinate Q_k is proportional to $|\langle\chi_v^m|Q_k|\chi_w^n\rangle|^2$ in both the Herzberg-Teller and Born-Oppenheimer theories. When mode k is nontotally symmetric, the symmetry selection rule in that mode will be $\Delta v = \pm 1, \pm 3, \ldots$ if the equilibrium position is undisplaced along Q_k in the electronic transition. If mode k is also harmonic, with similar frequencies in both electronic states, the more restrictive selection rule $\Delta v = \pm 1$ applies. In such a case, the band intensity becomes proportional to $\hbar[v_k, w_k]/2\mu_k\omega_k$ times a product of Franck-Condon factors over all vibrational modes other than mode k according to Eq. 7.16. For this reason, one observes the vibronically induced transitions $6_0^1, 6_1^0, 6_1^2, 6_2^1$ in the $S_1 \leftarrow S_0$ spectrum of benzene, but not 6_1^1, 6_0^2, or 6_0^4. One similarly obtains the vibronically induced bands $8(b_{1g})_0^1$ and $7(b_{1g})_0^1$ in the naphthalene $S_1 \leftarrow S_0$ spectrum, but not bands like $8(b_{1g})_0^2$. In the first allowed band progression of SO_2 (in which the upper state $^1A_2$ is vibronically coupled to a higher $^1B_1$ electronic state by asymmetric stretching mode 2, which has b_2 symmetry), the selection rules permit the transitions $1_0^n 2_0^1$ and $1_0^n 2_0^3$ with various n, but not $1_0^n 2_0^0$ or $1_0^n 2_0^2$.

7.4 RADIATIONLESS RELAXATION IN ISOLATED POLYATOMICS

When an isolated (collision-free) molecule is prepared in an excited vibronic level, its probable subsequent fate depends fundamentally on whether the molecule is "small" or "large" (the criterion for "largeness" will be developed in this section). If one excites the $B^1\Pi_u$ state in $v = 5$ of Na_2 at sufficiently gas low pressures, that pumped level will decay almost exclusively by emitting a fluorescence photon—and one can be confident that essentially *all* of the fluorescence in a system of Na_2 molecules so excited will be emitted specifically by $v = 5$ in the $B^1\Pi_u$ state. However, a pumped vibrational level (say 6^1) in collision-free S_1 benzene (which is assuredly a "large" molecule in the context of radiationless relaxation theory) has access to many decay routes that do not involve emission of a photon. These radiationless decay routes include conversion of a 6^1 S_1 molecule into a vibrationally excited S_0-state isoenergetic with the 6^1 S_1 state (internal conversion), and conversion into some vibrational level of T_1 (lowest triplet state) benzene with a total energy that closely matches that of 6^1 S_1 benzene (intersystem crossing). Internal conversion and intersystem crossing are generic terms for spin-allowed and spin-forbidden nonradiative electronic-state changes, respectively.

Since the molecule is isolated, these radiationless processes must be energy-conserving. Since they are energy-conserving, it might seem a priori that they should be reversible. For example, T_1 benzene formed by intersystem crossing

(ISC) from 6^1S_1 benzene should be capable of reverting back to the initial state. The processes are in fact *irreversible*, even in *isolated* large molecules. (In nondilute gases and in solution, these radiationless decay processes would appear irreversible in any case, because collisions would rapidly remove the excess vibrational energy from the vibrationally hot S_0 or T_1 molecules formed in IC or ISC, rendering them energetically incapable of recreating a 6^1S_1 molecule. The point we are making here is that the energy-conserving non-radiative decay processes themselves are irreversible in collision-free large molecules.) Unlike diatomics, large molecules can thus spontaneously decay nonradiatively into electronic states other than the pumped state, and emit luminescence from those states. 6^1S_1 benzene can emit a fluorescence photon, or it can undergo ISC to some vibrational level within the T_1 manifold of levels (which may then phosphoresce), or it may undergo IC to the S_0 state (Fig. 7.14). The peculiarities of large-molecule photophysics and their contrasts to small-molecule behavior occupied the attention of a number of foremost theoreticians during the 1960s and 1970s [10].

The general problem of isolated-molecule nonradiative relaxation may be stated as follows. A Born-Oppenheimer molecular state $|\psi_s\rangle$ in electronic state manifold A is prepared in a molecule by photon absorption from some lower

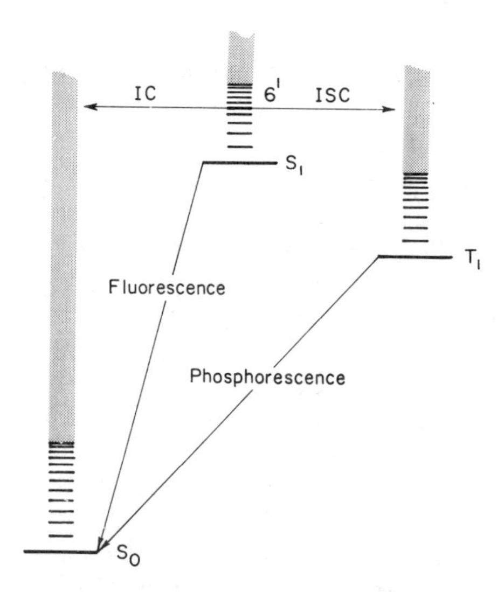

Figure 7.14 Possible radiationless processes following creation of $6^1\ S_0$ benzene, one of the vibronic levels which is EI-accessible from ground-state benzene (cf. Fig. 7.13). This level may undergo internal conversion (IC) to an isoenergetic, vibrationally hot S_0 molecule, or it may undergo intersystem crossing (ISC) to an isoenergetic level in triplet state T_1. The $T_1 \rightarrow S_0$ phosphorescence transition can be monitored for experimental evidence of ISC. Time-dependent $S_1 \rightarrow S_0$ fluorescence decay furnishes a probe for depopulation of S_1 through radiative (fluorescence) and nonradiative (IC, ISC) decay.

Born-Oppenheimer state $|\psi_g\rangle$ as shown in Fig. 7.15. Electronic state B possesses a manifold of Born-Oppenheimer states $|\psi_i\rangle$ with energies E_i which span the energy region of the state $|\psi_s\rangle$ having energy E_s. One wishes to calculate the rate of decay of the initially excited level $|\psi_s\rangle$ into the $|\psi_i\rangle$ manifold due to perturbations like nonadiabatic and spin–orbit coupling, and also to determine the conditions under which such decay will be irreversible. The Born-Oppenheimer states and Hamiltonian are analogous to those we used for diatomics at the beginning of Chapter 3. The total Hamiltonian

$$\hat{H} = \hat{T}(q) + \hat{T}(Q) + U(q, Q) + V(Q) + \hat{H}_{so} \tag{7.20}$$

contains the electronic kinetic energy, nuclear kinetic energy, electronic potential energy, nuclear repulsion, and spin–orbit operators, respectively, which are functions of the electronic and/or nuclear coordinates q and Q. The reduced Schrödinger equation for electronic motion is

$$[\hat{T}(q) + U(q, Q)]|\psi_l(q, Q)\rangle = E_l(Q)|\psi_l(q, Q)\rangle \tag{7.21}$$

where the $|\psi_l\rangle$ are fixed-nuclei electronic wave functions and the $E_l(Q)$ are potential energy surfaces for nuclear motion in the electronic states $|\psi_l\rangle$. The total wave function is the superimposition of Born-Oppenheimer states

$$|\psi(q, Q)\rangle = \sum_l |\psi_l(q, Q)\rangle|\chi_l(Q)\rangle \tag{7.22}$$

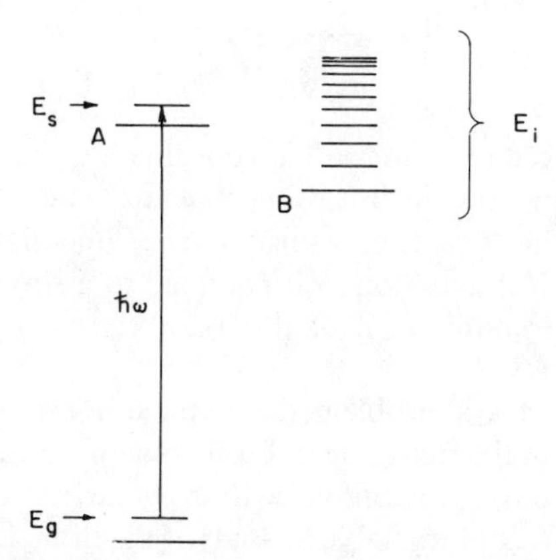

Figure 7.15 General problem for nonradiative decay of an excited Born-Oppenheimer state with energy E_s in electronic state A, prepared by photon excitation of a level with energy E_g in the electronic ground state. The prepared state $|\psi_s\rangle$ is connected by perturbations (spin–orbit coupling, nonradiative coupling, etc.) to a set of Born–Oppenheimer states $|\psi_i\rangle$ with energies E_i in electronic state B. The states $|\psi_i\rangle$ are not accessible by El transitions from the ground state.

and satisfies the total Schrödihger equation

$$\hat{H}|\psi(q, Q)\rangle = W|\psi(q, Q)\rangle \tag{7.23}$$

This requires that

$$[\hat{T}(q) + \hat{T}(Q) + U(q, Q) + V(Q) + H_{so} - W] \sum_l |\psi_l(q, Q)\chi_l(Q)\rangle = 0 \tag{7.24}$$

In analogy to what was done in Section 3.1, we multiply this equation by $\langle\psi_m(q, Q)|$, and use the facts that

$$\begin{aligned} \langle\psi_m|\hat{T}(q) + U(q, Q)|\psi_l\rangle &= E_m\delta_{ml} \\ \langle\psi_m|V(Q)|\psi_l\rangle &= V(Q)\delta_{ml} \\ \langle\psi_m|W|\psi_l\rangle &= W\delta_{ml} \end{aligned} \tag{7.25}$$

to obtain the coupled equations

$$\begin{aligned} &[\hat{T}(Q) + E_m(Q) + V(Q) + \langle\psi_m|\hat{T}(Q)|\psi_m\rangle + \langle\psi_m|\hat{H}_{so}|\psi_m\rangle - W]|\chi_m(Q)\rangle \\ &\quad = -\sum_{l \neq m} \langle\psi_m|\hat{T}(Q)|\psi_l\rangle|\chi_l(Q)\rangle \\ &\quad\quad + \sum_{l \neq m} \left[\begin{array}{c} 2\sum_k \dfrac{\hbar^2}{2\mu_k} \langle\psi_m|\dfrac{\partial}{\partial Q_k}|\psi_l\rangle \dfrac{\partial|\chi_l\rangle}{\partial Q_k} \\ -\langle\psi_m|\hat{H}_{so}|\psi_l\rangle|\chi_l(Q)\rangle \end{array} \right] \end{aligned} \tag{7.26}$$

where μ_k is the reduced mass in normal coordinate Q_k and $|\chi_m\rangle$ and $|\chi_l\rangle$ are the vibrational wave functions in Born-Oppenheimer states $|\psi_m\rangle$ and $|\psi_l\rangle$. When the right side of this equation vanishes (i.e., nonadiabatic and spin–orbit coupling are negligible), the motion is confined to a single Born-Oppenheimer state ($|\psi_m\rangle$ in this example). All of this parallels what we demonstrated for diatomics in Section 3.1.

Returning to the general problem, the pumped BO state $|\psi_s\rangle$ may be coupled by terms like those on the right side of Eq. 7.26 with varying strengths to a large number of levels in the $|\psi_i\rangle$ manifold with irregularly spaced energies E_i. Since this general problem is not tractable to analytic solution, Bixon and Jortner [11] studied an idealized system in which the energies of the $|\psi_i\rangle$ manifold are equally spaced,

$$E_i = E_s - \alpha + i\varepsilon \qquad i = 0, \pm 1, \pm 2, \ldots \tag{7.27}$$

where E_s is the energy of BO state $|\psi_s\rangle$ and α, ε are real constants. They also made the simplifying assumption that the coupling matrix element

$$\langle \psi_s | \hat{H} | \psi_i \rangle = v \tag{7.28}$$

has the same value regardless of the final level i, so that all BO levels $|\psi_i\rangle$ have identical nonadiabatic couplings to $|\psi_s\rangle$. The density of final states ρ_i is $1/\varepsilon$, a constant independent of E_i, since the latter energies are equally spaced (Fig. 7.16). It was also assumed that

$$\begin{aligned} \langle \psi_s | \hat{H} | \psi_s \rangle &= E_s \qquad & \langle \psi_g | \hat{H} | \psi_s \rangle &= 0 \\ \langle \psi_i | \hat{H} | \psi_i \rangle &= E_i \qquad & \langle \psi_g | \hat{H} | \psi_i \rangle &= 0 \\ \langle \psi_i | \hat{H} | \psi_{i'} \rangle &= 0 \end{aligned} \tag{7.29}$$

so that nonadiabatic coupling occurs only between $|\psi_s\rangle$ and the states in the $|\psi_i\rangle$ manifold. In the presence of the perturbations (i.e., nonadiabatic and spin–orbit coupling) which are included in the total Hamiltonian $\hat{H}$, the BO states $|\psi_s\rangle$ and the $|\psi_i\rangle$ will become modified into the mixed states

$$|\Psi_n\rangle = a_n |\psi_s\rangle + \sum_i b_i^n |\psi_i\rangle \tag{7.30}$$

which satisfy

$$\hat{H} |\Psi_n\rangle = E_n |\Psi_n\rangle \tag{7.31}$$

It is important to differentiate here between the energies E_s, E_i (which are eigenvalues of the BO Hamiltonian $\hat{T}(q) + U(q, Q)$) and the energies E_n (which

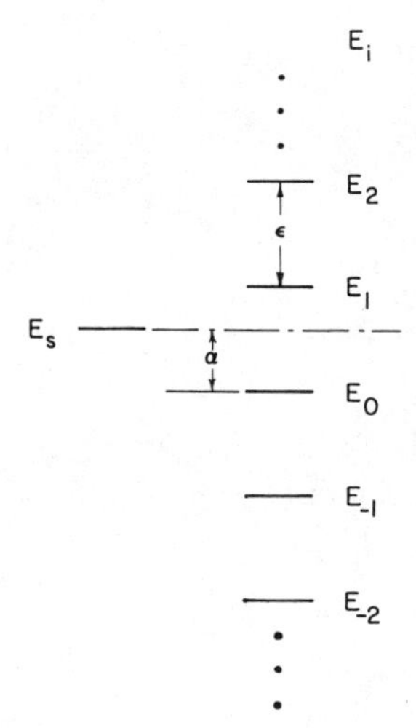

Figure 7.16 Simplified level scheme studied by Bixon and Jortner. The levels E_i in electronic state B are equally separated with spacing ϵ; the level E_0 is offset from the photon-excited level E_s by an arbitrary energy α. The coupling matrix element $\langle \psi_s | \hat{H} | \psi_i \rangle = v$ is assumed to be the same for all states $|\psi_i\rangle$.

are eigenvalues of the complete Hamiltonian $\hat{H}$ including nonadiabatic and spin–orbit coupling). The expansion coefficients in Eq. 7.30 must satisfy

$$\mathbf{H}\begin{bmatrix} a_n \\ \vdots \\ b_i^n \\ \vdots \end{bmatrix} = E_n \begin{bmatrix} a_n \\ \vdots \\ b_i^n \\ \vdots \end{bmatrix} \tag{7.32}$$

Using the matrix elements of $\hat{H}$ assigned by Bixon and Jortner in Eqs. 7.28 and 7.29, this is equivalent to saying that

$$\begin{bmatrix} E_s & v & v & \cdots \\ v & \ddots & & 0 \\ v & & E_i & \\ \vdots & 0 & & \ddots \end{bmatrix}\begin{bmatrix} a_n \\ \vdots \\ b_i^n \\ \vdots \end{bmatrix} = E_n \begin{bmatrix} a_n \\ \vdots \\ b_i^n \\ \vdots \end{bmatrix} \tag{7.33}$$

where the only nonzero off-diagonal elements in the Hamiltonian matrix are the H_{si}. Expanding the matrix equation (7.33) leads to

$$(E_s - E_n)a_n + v \sum_i b_i^n = 0 \tag{7.34}$$

$$(E_i - E_n)b_i^n + va_n = 0 \tag{7.35}$$

From the second of these equations, we have

$$b_i^n = \frac{-va_n}{E_i - E_n} = \frac{-va_n}{E_s - \alpha + i\varepsilon - E_n} \tag{7.36}$$

so that

$$(E_s - E_n)a_n = \sum_{i=-\infty}^{\infty} \frac{v^2 a_n}{E_s - \alpha + i\varepsilon - E_n} \tag{7.37}$$

and

$$E_s - E_n = -v^2 \sum_{i=-\infty}^{\infty} (E_n - E_s + \alpha - i\varepsilon)^{-1} \tag{7.38}$$

This is an equation that can be solved for the perturbed-state energies E_n in terms of E_s, the coupling v, and the constants α and ε. It can be shown [11] that Eq. 7.38 is mathematically equivalent to the statement that

$$E_s - E_n = \frac{\pi v^2}{\varepsilon} \cot\left[\left(\frac{\pi}{\varepsilon}\right)(E_n - E_s + \alpha)\right] \tag{7.39}$$

which may be solved graphically for the perturbed energies E_n (Fig. 7.17). It is clear from inspection of the graph that there will be one new eigenvalue E_n of the perturbed Hamiltonian between each pair of BO eigenvalues E_i (Fig. 7.18). For large $|i|$ ($E_i \ll$ or $\gg E_s$) E_n will differ little from E_i, so that the E_n are little shifted from the E_i for states $|\psi_i\rangle$ which are far off-resonance. Normalization of the perturbed states $|\psi_n\rangle$ in Eq. 7.30 requires that

$$a_n^2 + \sum_{i=-\infty}^{\infty} (b_i^n)^2 = 1$$

$$= a_n^2 + v^2 \sum_{i=-\infty}^{\infty} \frac{a_n^2}{(E_s - \alpha + i\varepsilon - E_n)^2} \tag{7.40}$$

or

$$a_n^2 = \left(1 + v^2 \sum_{-\infty}^{\infty} \frac{1}{(E_s - \alpha + i\varepsilon - E_n)^2}\right)^{-1}$$

$$\equiv \frac{v^2}{(E_n - E_s)^2 + v^2 + \left(\dfrac{\pi v^2}{\varepsilon}\right)^2} \tag{7.41}$$

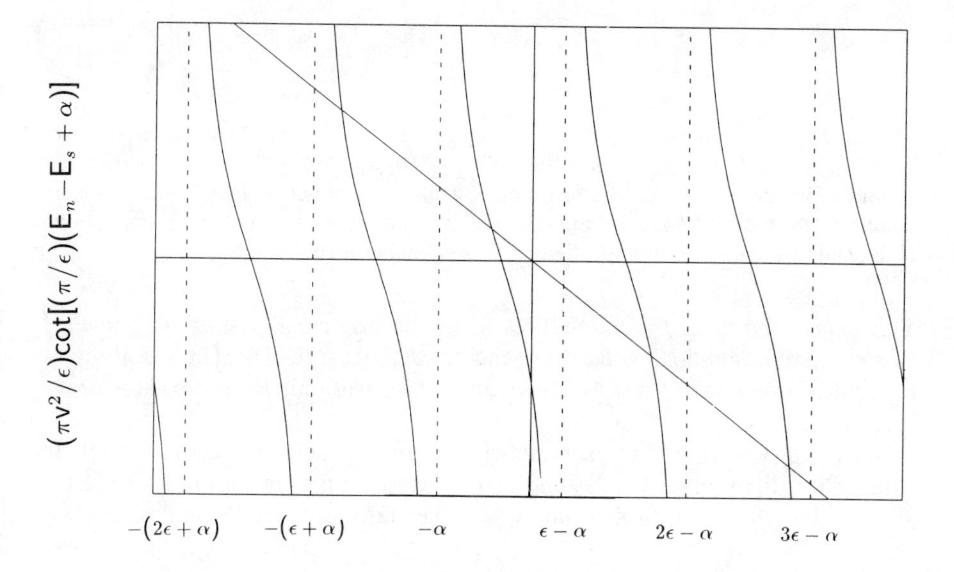

Figure 7.17 Graphical solution of the equation $(E_s - E_n) = (\pi v^2/\epsilon) \times \cot[(\pi/\epsilon)(E_n - E_s + \alpha)]$. The eigenvalues are given by intersections of the straight line with the periodic cotangent function.

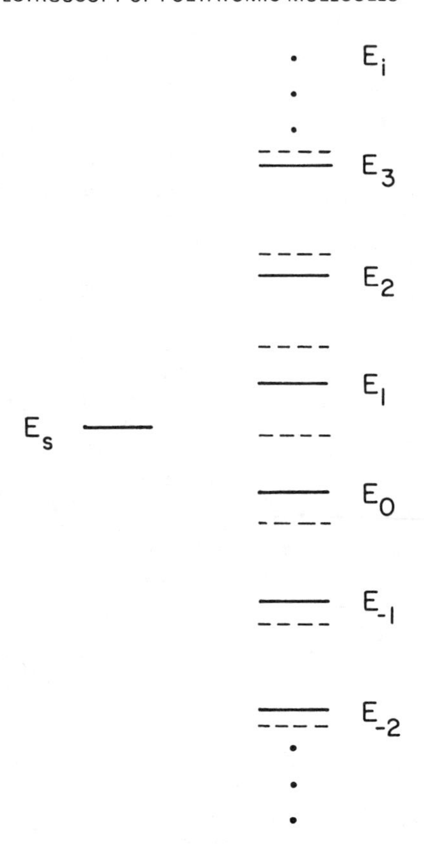

Figure 7.18 Eigenvalues E_s and E_i ($i = 0, \pm 1, \pm 2, \ldots$) of the Born-Oppenheimer Hamiltonian, solid lines; eigenvalues E_n ($n = 0, \pm 1, \pm 2, \ldots$) of the perturbed Hamiltonian, dashed lines. The latter eigenvalues are obtained from the graphical solutions in Fig. 7.17. Note the general property that there is one perturbed eigenvalue between every pair of Born-Oppenheimer levels.

This expression gives the admixtures a_n^2 of the original BO state $|\psi_s\rangle$ in the various mixed states $|\psi_n\rangle$ which have energies E_n. Because of the $(E_n - E_s)^2$ term in the denominator of Eq. 7.41, these admixtures will only be appreciable near resonance.

We are now prepared to examine what happens when a molecule is excited with a short light pulse. We assume that the transition from $|\psi_g\rangle$ to $|\psi_s\rangle$ is E1-allowed, but that transitions from $|\psi_g\rangle$ to the states $|\psi_i\rangle$ are forbidden, i.e.,

$$\langle\psi_s|\boldsymbol{\mu}|\psi_g\rangle \neq 0$$
$$\langle\psi_i|\boldsymbol{\mu}|\psi_g\rangle = 0 \tag{7.42}$$

This will be the case, for example, when $|\psi_s\rangle$ is an excited singlet state vibrational level that is E1-connected to the ground state, and when the $|\psi_i\rangle$ are vibrational levels belonging to one of the triplet electronic states. Excitation by a photon of appropriate energy will then momentarily produce the pure BO state $|\psi_s\rangle$, which may be expanded in terms of the mixed states as

$$|\psi_s\rangle = \sum_n |\psi_n\rangle\langle\psi_n|\psi_s\rangle \equiv \sum a_n|\psi_n\rangle \tag{7.43}$$

at the time $t = 0$ when excitation occurs. At later times this prepared state will evolve as

$$\begin{aligned} |\psi(t)\rangle &= \sum_n a_n e^{-iE_n t/\hbar}|\psi_n\rangle \\ &= \sum_n a_n e^{-iE_n t/\hbar}\left(a_n|\psi_s\rangle + \sum_i b_i^n|\psi_i\rangle\right) \end{aligned} \tag{7.44}$$

The probability of finding the molecule in BO state $|\psi_s\rangle$ after excitation will be

$$\begin{aligned} |\langle\psi_s|\psi(t)\rangle|^2 &\equiv P_s(t) \\ &= \left|\sum_n \langle\psi_s|a_n e^{-iE_n t/\hbar}|a_n\psi_s + \sum b_i^n\psi_i\rangle\right|^2 \\ &= \left|\sum_n a_n^2 e^{-iE_n t/\hbar}\right|^2 \\ &= \left|\sum_n \frac{v^2}{(E_n - E_s)^2 + v^2 + \left(\dfrac{\pi v^2}{\varepsilon}\right)^2} e^{-iE_n t/\hbar}\right|^2 \end{aligned} \tag{7.45}$$

This is difficult to evaluate a priori, because the mixed state energies E_n are not given by analytic expressions (Fig. 7.17). We therefore coarsen our approximation by assuming that they are given by

$$E_n \simeq E_s + n\varepsilon \qquad n = 0, \pm 1, \pm 2, \ldots \tag{7.46}$$

(This is not totally unreasonable, because there will be an E_n level between every pair of E_i levels, will and so the two sets of levels have similar average spacings in Fig. 7.18). Letting

$$\Delta^2 \equiv v^2 + \left(\frac{\pi v^2}{\varepsilon}\right)^2 \tag{7.47}$$

we have the time-dependent probability

$$P_s(t) = \left|\sum_n \frac{v^2 e^{in\varepsilon t/\hbar}}{n^2\varepsilon^2 + \Delta^2}\right|^2 \tag{7.48}$$

that the molecule will be found in state $|\psi_s\rangle$ after excitation. In the limits $v \gg \varepsilon$ (or $v\rho_i \gg 1$) and $t \ll \hbar/\varepsilon$, this sum may be replaced by the integral [10]

$$P_s(t) = \left| \int_{-\infty}^{\infty} dn \, \frac{v^2}{n^2\varepsilon^2 + \Delta^2} \cos(\varepsilon n t/\hbar) \right|^2$$
$$= e^{-2\pi v^2 t/\varepsilon\hbar} \equiv e^{-k_{NR}t} \qquad (7.49)$$

Hence when $v \gg \varepsilon$ and $t \ll \hbar/\varepsilon$, the excited BO state $|\psi_s\rangle$ undergoes irreversible first-order decay with a rate constant

$$k_{NR} = 2\pi v^2/\varepsilon\hbar \qquad (7.50)$$

which is proportional to the square of the nonadiabatic coupling matrix element and to the density of final states $\rho_i = 1/\varepsilon$. If, however, we drop the assumptions $v \gg \varepsilon$ and $t \ll \hbar/\varepsilon$, the sum for $P_s(t)$ in Eq. 7.48 must be evaluated explicitly. This was done by Gelbart et al. for several model systems [10], and we show their results for $v = 0.5\varepsilon$ in Fig. 7.19. $P_s(t)$ initially decays exponentially, becoming

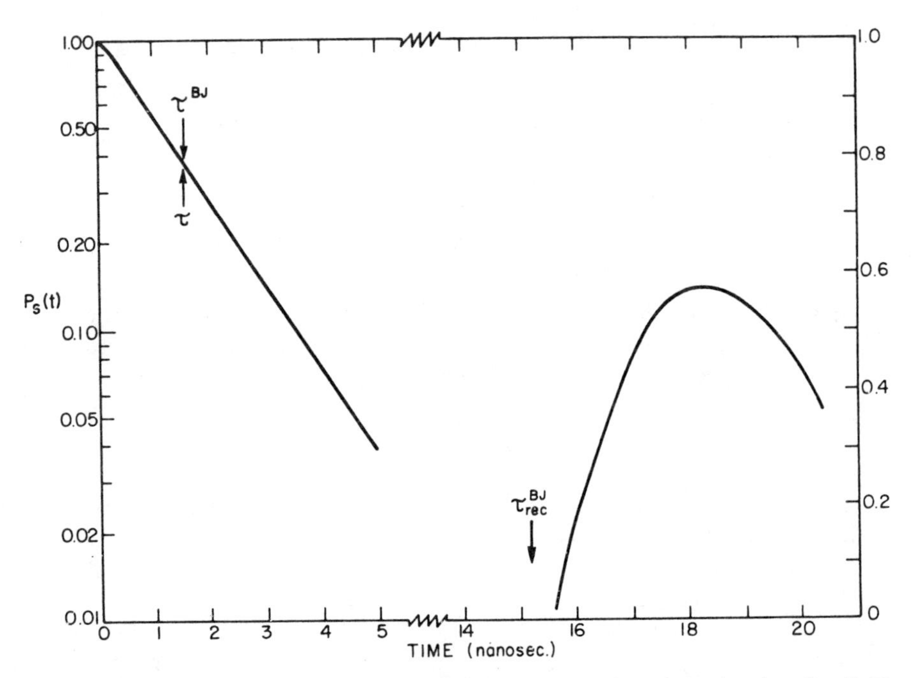

Figure 7.19 The time-dependent decay function $P_s(t)$ evaluated using Eq. 7.48 with $v = 0.5\epsilon$. $P_s(t)$ exhibits exponential decay (manifested by a straight line in this semilog plot) at early times; a partial *recurrence* occurs at $t \gtrsim 2\pi\hbar/\epsilon \equiv \tau_{rec}^{BJ}$. The energy spacing ϵ was chosen to render the exponential lifetime $\tau = 1/k_{NR}$ between 1 and 2 *ns* for realism. Reproduced by permission from P. Avouris, W. M. Gelbart, and M. A. El-Sayed, *Chem. Rev.* **77**: 793 (1977).

very small for $t \gg 1/k_{NR}$. However, it then begins to build up again to a large value at the *recurrence time* $\tau_{rec} = 2\pi\hbar/\varepsilon$, since Eq. 7.48 contains a superposition of periodic terms proportional to $\exp(in\varepsilon t/\hbar)$. In real molecules, the population of pumped state $|\psi_s\rangle$ is depleted by spontaneous emission (fluorescence, phosphorescence) long before this recurrence time is reached. For example, we can consider the radiationless decay of vibrationless S_1 benzene into T_1 benzene (ISC). The density of T_1 vibrational states isoenergetic with the 0° S_1 state is $\rho \sim 3 \times 10^5$ cm (i.e., 3×10^5 levels per cm^{-1}) in benzene, and this number can be identified with $1/\varepsilon$ in our discussion. The corresponding recurrence time is then $\tau_{rec} = 2\pi\hbar/\varepsilon \sim 10^{-5}$ s, which should be compared with the $\sim 10^{-8}$ s fluorescence lifetime in benzene. The radiative decay timescale in benzene preempts that of recurrence by several orders of magnitude.

One of the nominal criteria for the validity of the integral approximation in Eq. 7.49 was $v \gg \varepsilon$. The foregoing discussion shows that this is too strict, since the calculations of Gelbart et al. prove that irreversible decay can be obtained for $v = 0.5\varepsilon$. A more realistic criterion is $v/\varepsilon \gtrsim 1$, or $v\rho \gtrsim 1$ in molecules with an energy-dependent density of final states.

The problem of estimating densities of vibrational states ρ is a large one that we will only touch on here. For single harmonic oscillator with uniform energy spacing $h\nu$, ρ is of course $1/h\nu$. The number of ways of placing n vibrational quanta in q identical oscillators (total energy $E = nh\nu$) is [12]

$$W(E) = (n + q - 1)!/(q - 1)!n! \tag{7.51}$$

The total number of states in such a system with vibrational energies between 0 and E is

$$G(E) = \sum_{i=0}^{n} \frac{(i + q - 1)!}{i!(q - 1)!} = \frac{(n + q)!}{n!q!} \tag{7.52}$$

and the density of states $\rho(E)$ at vibrational energy E is obtained from

$$\rho(E) = dG(E)/dE \tag{7.53}$$

For a system of three such oscillators, the following table describes the behavior of $G(E)$ with increasing $n = E/h\nu$:

n	$G(E)$
0	1
1	4
2	10
3	20
4	35

$G(E)$ and $\rho(E)$ can become incredibly large at moderate vibrational energies (e.g., 5000 cm^{-1}) in molecules of respectable size (benzene has 30 vibrational modes). Approximations are then required to evaluate them, particularly when the normal modes have a range of frequencies [13].

The criterion that $v\rho \gtrsim 1$ now implies that a critical number of vibrations is required to make an irreversible radiationless relaxation process possible. We may consider a hypothetical case in which the energy gap between the vibrationless electronic states is $\sim$1 eV (8066 cm^{-1}); that is, state $|\psi_s\rangle$ decays into a set of final states $|\psi_i\rangle$ which have $\sim$1 eV of excess vibrational energy. For IC and ISC, typical values of the coupling v may be taken to be $\sim 10^{-1}$ and 10^{-4} cm^{-1}, respectively [11]. A table of products $v\rho$ calculated by Bixon and Jortner for nonlinear molecules with N atoms in which all $(3N - 6)$ vibrational modes oscillate with frequency 1000 cm^{-1} is shown below; the densities of states ρ were evaluated according to the method of Haarhoff [13]:

N	$\rho\,(E = 1\text{ eV})$	$v\rho$ (IC)	$v\rho$ (ISC)
3	0.06 cm	6×10^{-3}	6×10^{-6}
4	4	0.4	4×10^{-4}
5	50	5	5×10^{-3}
10	4×10^{5}	4×10^{4}	40

Hence, internal conversion is typically expected to occur in molecules with $\geqslant$4 atoms, and intersystem crossing sets in when $N \gtrsim 10$. These are rough guidelines for the "large-molecule" regime in which nonradiative relaxation is prevalent in isolated molecules. It includes all aromatic molecules (the smallest common one of which is benzene); formaldehyde and larger molecules with the carbonyl chromophore; and all laser dyes such as rhodamines, oxazines and commarins (Chapter 9).

REFERENCES

1. H. F. Schaefer III, *The Electronic Structure of Atoms and Molecules: A Survey of Rigorous Quantum Mechanical Results*, Addison-Wesley, Reading, MA, 1972.
2. W. H. Flygare, *Molecular Structure and Dynamics*, Prentice-Hall, Englewood Cliffs, NJ, 1978.
3. D. S. Schonland, *Molecular Symmetry*, Van Nostrand, London, 1965.
4. A. D. Walsh, *J. Chem. Soc.*, 2260, 2266, 2288, 2296, 2301, 2306, 2318, 2321, 2325, 2330 (1953).
5. J. Heicklen, N. Kelly, and K. Partymiller, *Rev. Chem. Intermed.* **3**: 315 (1980).
6. G. Varsanyi, *Assignments for Vibrational Spectra of Seven Hundred Benzene Derivatives*, Wiley, New York, 1974.
7. J. Christoffersen, J. M. Hollas, and G. H. Kirby, *Mol. Phys.* **16**, 441 (1969).
8. G. Fischer, *Vibronic Coupling*, Academic, London, 1984.

9. G. Orlandi and W. Siebrand, *J. Chem. Phys.* **58**: 4513 (1973).
10. P. Avouris, W. M. Gelbart, and M. A. El-Sayed, *Chem. Rev.* **77**: 793 (1977).
11. M. Bixon and J. Jortner, *J. Chem. Phys.* **48**: 715 (1968).
12. W. Forst, *Chem. Rev.* **71**: 339 (1971).
13. P. C. Haarhoff, *Mol. Phys.* **7**: 101 (1963).

PROBLEMS

1. The $S_1 \leftarrow S_0$ band system of formaldehyde (CH_2O) has been thoroughly studied, and it has been established that the vibrationless S_1 and S_0 states have ${}^1A_2$ and ${}^1A_1$ symmetry, respectively, in C_{2v}. The $S_1 \leftarrow S_0$ absorption spectrum and descriptions of normal vibrations in the S_1 and S_0 states are shown in Figure P7.1 and in the list below.

Normal mode		Symmetry	$\tilde{X}^1A_1$	$\tilde{A}^1A_2$
ν_1	C—H symmetric stretch	a_1	2766.4	2847
ν_2	C=O stretch	a_1	1746.1	1173
ν_3	H—C—H bend	a_1	1500.6	1290
ν_4	Out-of-plane wag	b_1	1167.3	124.6
ν_5	C—H asymmetric stretch	b_2	2843.4	2968
ν_6	In-plane wag	b_2	1251.2	904

(a) Is the intrinsic $S_1 \leftarrow S_0$ transition E1-allowed in formaldehyde?
(b) Assuming that the exhibited bands gain intensity by Herzberg-Teller coupling between S_1 and higher excited singlets S_n, what are the symmetries of the electronic states S_n? What polarizations do these bands exhibit?
(c) What information about the relative geometries of the S_1 and S_0 states can be inferred from the progressions in this spectrum?

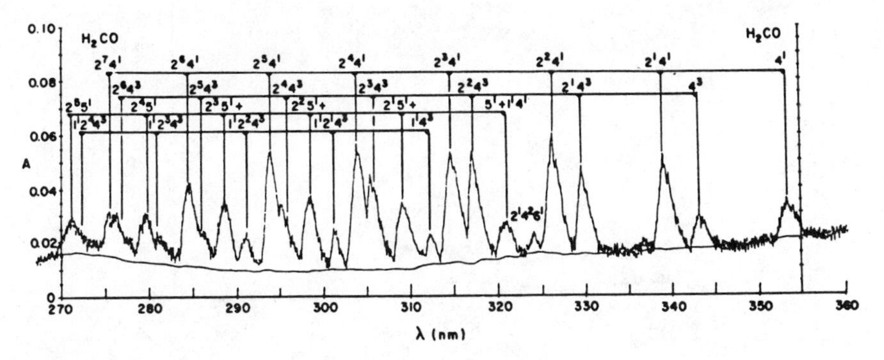

Figure P7.1 Reproduced with permission from E. K. C. Lee, *Adv. Photochem.* **12**: 18 (1980).

2. In *s*-triazine (a D_{3h} molecule), the three lowest excited singlet states are predicted to be closely spaced in energy with symmetries $^1A_1''$, $^1A_2''$, and $^1E''$. An *s*-triazine crystal absorption spectrum (taken with unpolarized light) is shown in Figure P7.2. The weak 0–0 band is E1 symmetry-forbidden, and appears because the D_{3h} symmetry of *s*-triazine is slightly distorted by the crystal environment. From analysis of the polarized crystal absorption spectra, the following fundamentals are found in the $S_1 \leftarrow S_0$ spectrum:

Normal mode	Symmetry	Polarization
6	e'	$\parallel (z)$
4	a_2''	$\perp (x, y)$
5	a_2''	$\perp (x, y)$
10	e''	$\perp (x, y)$
16	e''	$\perp (x, y)$

(a) Assuming that these fundamentals gain intensity through Herzberg-Teller coupling, deduce the symmetry of the lowest excited singlet state. Show that this choice is consistent with *all* pertinent data given in this problem.

(b) Mode 12 in *s*-triazine has a_1' symmetry. By what mechanism (other than environment symmetry-breaking) can the 12_0^1 band appear in this crystal absorption spectrum?

(c) What symmetries of electronically excited states are vibronically coupled to the S_1 state in the 6^1, 4^1, 5^1, 10^1, and 16^1 vibrational levels? Considering this, why does the 6_0^1 band exhibit such large intensity?

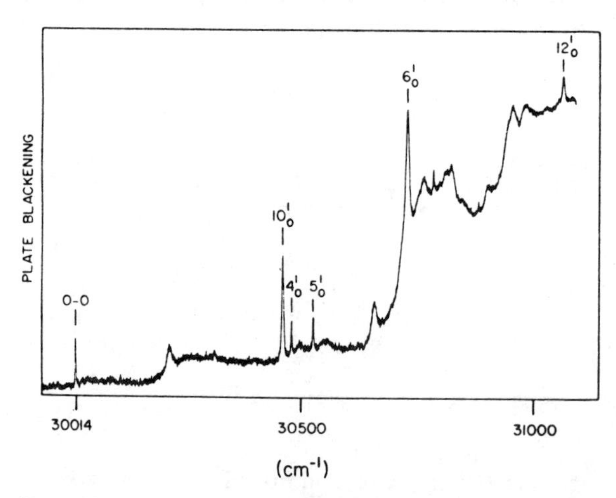

Figure P7.2 Reproduced with permission from N. J. Kruse and G. J. Small, *J. Chem. Phys.* **56**: 2987 (1972).

3. Polarized $S_1 \leftarrow S_0$ absorption spectra of tetracene in a transparent ordered host crystal at 4.2 K are shown in Figure P7.3. The upper spectrum was obtained using light polarized along the tetracene y axis, and the lower spectrum was obtained using z-polarized light. The y and z axes lie in the molecular plane along the short and long axes of tetracene, respectively. The tetracene S_0 state has ^1A_g symmetry in D_{2h}.

(a) Most of the assigned bands in the y-polarized spectrum arise from fundamentals and combinations in a_g modes: the 308-cm^{-1} band is $12(a_g)_0^1$, the 308 + 609-cm^{-1} band is $12(a_g)_0^1\ 11(a_g)_0^1$, etc. What is the symmetry of the electronic state S_1?

(b) The 479-, 1166-, and 1506-cm^{-1} bands in the z-polarized spectrum are fundamentals arising from vibronic coupling between S_1 and a higher lying $^1B_{1u}$ state. What is the symmetry of the normal modes responsible for this vibronic coupling?

(c) Use Herzberg-Teller theory to develop an algebraic expression for the ratio of intensities for the 479- and 1166-cm^{-1} bands in the z-polarized spectrum. What structural information about tetracene would be needed to test the theory against these data?

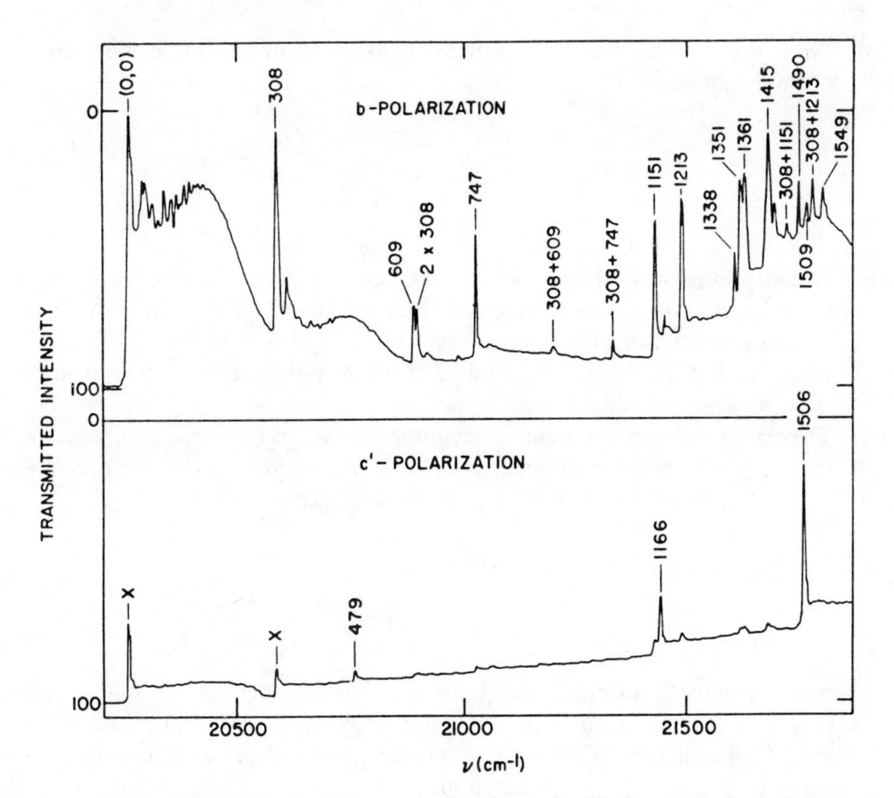

Figure P7.3 Reproduced with permission from G. Fischer and G. J. Small, *J. Chem. Phys.* **56**: 5937 (1972).

4. One way to evaluate energy-dependent vibrational densities of states $\rho(E)$ is to invert the classical vibrational partition function

$$Q = \int_0^\infty \rho(E)e^{-E/kT} \equiv \int_0^\infty \rho(E)e^{-Es}\,dE = Q(s)$$

The desired density of states is the inverse Laplace transform of $Q(s)$,

$$\rho(E) = \mathscr{L}^{-1}[Q(s)]$$

For a harmonic oscillator, the classical partition function is

$$Q(s) = \lim_{T \to \infty} [1/(1 - \exp(-h\nu/kT))] = kT/h\nu = 1/sh\nu$$

and for ν independent oscillators with frequencies ν_i it is

$$Q(s) = (kT)\nu \bigg/ \prod_i^{\nu} h\nu_i = s^{-\nu} \bigg/ \prod_i^{\nu} h\nu_i$$

(a) Show that the classical vibrational density of states in a molecule with ν normal modes is

$$\rho(E) = \frac{E^{\nu-1}}{(\nu-1)!\prod_i h\nu_i}$$

(b) Obtain numerical values of $\rho(E)$ at $E = 8000\,\text{cm}^{-1}$ for CO_2 (whose vibrational frequencies are given in Section 6.5) and for ethylene, whose 12 nondegenerate vibrational frequencies are 825, 943, 950, 995, 1050, 1342, 1443, 1623, 2990, 3019, 3106, and $3272\,\text{cm}^{-1}$. What circumstances lead to large densities of states for given E?

(c) The classical density of states is accurate only when E is large compared to the zero-point vibrational energy $E_0 = \frac{1}{2}h\Sigma\nu_i$. A more accurate expression for $\rho(E)$ at lower energies is the *semiclassical* expression

$$\rho(E) = \frac{(E + E_0)^{\nu-1}}{(\nu-1)!\prod_i h\nu_i}$$

which results from replacing E in the classical density of states with $(E + E_0)$. Reevaluate $\rho(E)$ for CO_2 and ethylene at $8000\,\text{cm}^{-1}$ using the semiclassical density of states. At what energies do the two approximations agree to within 10% in these two molecules?

5. In the isolated benzene molecule, the ${}^3B_{1u}T_1$ state lies 9200 cm^{-1} below the ${}^1B_{2u}$ S_1 state. Assume that T_1 benzene has the following vibrational frequencies in cm^{-1}:

3060	a_{1g}	850	e_{1g}
990	a_{1g}	3000	e_{1u}
1200	a_{2g}	1500	e_{1u}
650	a_{2u}	1050	e_{1u}
3050	b_{1u}	3000	e_{2g}
1000	b_{1u}	1500	e_{2g}
1500	b_{2g}	1200	e_{2g}
550	b_{2g}	600	e_{2g}
1900	b_{2g}	1200	e_{2u}
1145	b_{2u}	400	e_{2u}

(a) Calculate the semiclassical density of states $\rho(E)$ in T_1 benzene at the energy of the vibrationless S_1 state. (Be sure to consider the vibrational mode degeneracies.)

(b) If the value of the spin–orbit coupling matrix element v between S_1 and T_1 is on the order of 10^{-4} cm^{-1}, is irreversible $T_1 \leftarrow S_1$ intersystem crossing likely to occur in benzene?

6. The lowest excited triplet state (${}^3A_2$) in formaldehyde lies some 3500 cm^{-1} below the S_1 origin. Assuming that the vibrational frequencies in T_1 are the same as in S_1 (Problem 7.1), evaluate the semiclassical density of states in T_1 at the energy of the vibrationless S_1 state. If the relevant spin–orbit coupling matrix element v is 10^{-4} cm^{-1}, what will be the first-order rate constant k_{NR} in s^{-1} for $T_1 \leftarrow S_1$ intersystem crossing?

8

SPECTRAL LINESHAPES AND OSCILLATOR STRENGTHS

We have tacitly assumed in the preceding seven chapters that spectroscopic transitions occur at sharp, well-defined frequencies, leading to absorption lines with zero frequency width. In reality, numerous mechanisms (including lifetime broadening, Doppler broadening, and collisional broadening) endow experimental absorption lines with finite widths and characteristic lineshape functions. We develop a framework in the first part of this chapter for predicting these lineshape functions in gases, where they assume a particularly simple form. In condensed media, the absorption lineshapes (e.g., in solid naphthalene or Nd^{3+} : glass) are controlled by electron–phonon interactions, thermal broadening, and broadening that arises from the fact that not all molecules experience identical environments within the medium (inhomogeneous broadening). Such condensed-phase broadening mechanisms lie beyond the scope of this chapter.

Our discussion of Einstein coefficients and oscillator strengths in Sections 8.3 and 8.4 yields fundamental relationships among absorption coefficients, luminescence lifetimes, and probabilities for stimulated emission. The latter process is responsible for light amplification by stimulated emission (laser action), and these relationships figure prominently in the derivation of lasing criteria in Chapter 9.

8.1 ELECTRIC DIPOLE CORRELATION FUNCTIONS

According to the formalism we developed in Chapter 1, the probability amplitude for an E1 one-photon transition from state $|k\rangle$ to state $|m\rangle$ is

$$c_m^{(1)}(t) = \frac{1}{i\hbar} \int_0^t e^{i\omega_{mk}t_1} \langle m|W(t_1)|k\rangle dt_1$$

$$= \frac{\omega_{mk}}{\hbar} \langle m|e\mathbf{r} \cdot \mathbf{A}_0|k\rangle \int_0^t e^{i(\omega_{mk}-\omega)t_1} dt_1 \tag{8.1}$$

for a molecule that is exposed to a vector potential $\mathbf{A}(\mathbf{r}, t) = \mathbf{A}_0 \exp[i(\mathbf{k} \cdot \mathbf{r} - \omega t)]$ turned on at time $t = 0$. Since the time integral in this equation is

$$\int_0^t e^{i(\omega_{mk}-\omega)t_1} dt_1 = \frac{e^{i(\omega_{mk}-\omega)t} - 1}{i(\omega_{mk} - \omega)} \tag{8.2}$$

the probability that the transition will occur between times 0 and t becomes

$$\begin{aligned} |c_m^{(1)}(t)|^2 &= \frac{2\omega_{km}^2}{\hbar^2} |\langle m|\boldsymbol{\mu} \cdot \mathbf{A}_0|k\rangle|^2 \frac{1 - \cos[(\omega_{mk} - \omega)t]}{(\omega_{mk} - \omega)^2} \\ &= \frac{2\omega_{km}^2}{\hbar^2} |\langle m|\boldsymbol{\mu} \cdot \mathbf{A}_0|k\rangle|^2 \frac{2 \sin^2[(\omega_{mk} - \omega)t/2]}{(\omega_{mk} - \omega)^2} \end{aligned} \tag{8.3}$$

In the limit of long times $t \gg \omega_{mk}^{-1}$, the ω-dependent factor in Eq. 8.3 approaches a constant times the Dirac delta function,

$$\frac{2 \sin^2[(\omega_{mk} - \omega)t/2]}{(\omega_{mk} - \omega)^2} \rightarrow \pi t \delta(\omega_{mk} - \omega) \tag{8.4}$$

(The proportionality factor πt is required here to ensure that the delta function is normalized to unity.) The external electric field $\mathbf{E} = \mathbf{E}_0 \exp[i(\mathbf{k} \cdot \mathbf{r} - \omega t)]$ is related to the vector potential in the Coulomb gauge by $\mathbf{E} = -\partial \mathbf{A}/\partial t$. Noting that the delta function (8.4) will constrain ω to equal ω_{mk} in Eq. 8.3, we may write

$$|c_m^{(1)}(t)|^2 = \frac{2\pi}{\hbar^2} t |\langle m|\boldsymbol{\mu} \cdot \mathbf{E}_0|k\rangle|^2 \delta(\omega_{mk} - \omega) \tag{8.5}$$

The transition probability is clearly proportional to the time duration t for which the external field is applied. It is therefore meaningful to define the transition probability per unit time,

$$P_{k \rightarrow m} = \frac{2\pi}{\hbar^2} |\langle m|\boldsymbol{\mu} \cdot \mathbf{E}_0|k\rangle|^2 \delta(\omega_{mk} - \omega) \tag{8.6}$$

This expression coincides with the well-known *Golden Rule* formulation [1] of the molecular transition probability under the external perturbation $W = -\boldsymbol{\mu} \cdot \mathbf{E}_0$.

We now wish to generalize this expression to a system of molecules at thermal equilibrium. Let p_k and p_m be the probabilities that a molecule will be found in state k and in state m, respectively. The rate of *energy loss* from the radiation field due to E1 one-photon absorption/emission processes will be [3]

$$
\begin{aligned}
-\dot{E} &= \sum_{km} \hbar\omega_{mk} P_{k\to m} \\
&= \frac{2\pi}{\hbar} \sum_{km} \omega_{mk}(p_k - p_m)|\langle m|\mathbf{E}_0 \cdot \boldsymbol{\mu}|k\rangle|^2 \delta(\omega_{mk} - \omega) \\
&= \frac{2\pi}{\hbar} \sum_{km} \omega_{mk} p_k(1 - e^{-\hbar\omega_{mk}/kT})|\langle m|\mathbf{E}_0 \cdot \boldsymbol{\mu}|k\rangle|^2 \delta(\omega_{mk} - \omega)
\end{aligned}
\tag{8.7}
$$

The average energy density stored in the electromagnetic field is $E = \varepsilon_0 \mathbf{E}_0/2$ [2]. Since the delta function in Eq. 8.7 requires ω to equal ω_{mk} in each term of the summation (8.7), the optical absorption coefficient will be proportional to

$$
\varepsilon(\omega) = \frac{-\dot{E}}{\omega \bar{E}} = \frac{4\pi}{\varepsilon_0 \hbar}(1 - e^{-\hbar\omega/kT}) \sum_{km} p_k |\langle m|\hat{E} \cdot \boldsymbol{\mu}|k\rangle|^2 \delta(\omega_{mk} - \omega) \tag{8.8}
$$

where $\hat{E}$ is a unit vector directed along the electric field. Since we are now concentrating on the shape (rather than intensity) of the absorption lines, we define the *lineshape function*

$$
I(\omega) = \frac{3\varepsilon_0 \hbar \varepsilon(\omega)}{4\pi(1 - e^{-\hbar\omega/kT})} = 3 \sum_{km} p_k |\langle m|\hat{E} \cdot \boldsymbol{\mu}|k\rangle|^2 \delta(\omega_{mk} - \omega) \tag{8.9}
$$

In view of the integral representation (1.113) of the Dirac delta function this is equivalent to

$$
\begin{aligned}
I(\omega) &= \frac{3}{2\pi} \sum_{km} p_k \langle k|\hat{E} \cdot \boldsymbol{\mu}|m\rangle\langle m|\hat{E} \cdot \boldsymbol{\mu}|k\rangle \int_{-\infty}^{\infty} dt\, e^{i[(E_m - E_k)\hbar - \omega]t} \\
&= \frac{3}{2\pi} \int_{-\infty}^{\infty} dt\, e^{-i\omega t} \sum_{km} p_k \langle k|\hat{E} \cdot \boldsymbol{\mu}|m\rangle\langle m|e^{iE_m t/\hbar}(\hat{E} \cdot \boldsymbol{\mu})e^{-iE_k t/\hbar}|k\rangle
\end{aligned}
\tag{8.10}
$$

In the Schrödinger representation of the latter matrix element in (8.10), the molecular states are regarded as time-dependent basis functions $\exp(-iE_k t/\hbar)|k\rangle$ and $\exp(-iE_m t/\hbar)|m\rangle$, and the operator $\dot{E} \cdot \boldsymbol{\mu}$ is considered to be time-independent. For present purposes, it is more illuminating to use the Heisenberg representation, in which the molecular states are the time-independent basis functions $|k\rangle$, $|m\rangle$ and the operator is viewed as time-dependent. Since $\hat{H}|i\rangle = E_i|i\rangle$ for each of the molecular states $|i\rangle$, we have

$$
I(\omega) = \frac{2}{2\pi} \int_{-\infty}^{\infty} dt e^{-i\omega t} \sum_{km} p_k \langle k|\hat{E} \cdot \boldsymbol{\mu}|m\rangle\langle m|e^{i\hat{H}t/\hbar}\hat{E} \cdot \boldsymbol{\mu} e^{-i\hat{H}t/\hbar}|k\rangle \tag{8.11}
$$

and so the time-dependent electric dipole moment operator in the Heisenberg representation is

$$\boldsymbol{\mu}(t) = e^{i\hat{H}t/\hbar}\boldsymbol{\mu}e^{-i\hat{H}t/\hbar} \tag{8.12}$$

In terms of this, the lineshape function becomes

$$I(\omega) = \frac{3}{2\pi}\int_{-\infty}^{\infty} dt\, e^{-i\omega t} \sum_k p_k \langle k|\hat{E}\cdot\boldsymbol{\mu}(0)\hat{E}\cdot\boldsymbol{\mu}(t)|k\rangle \tag{8.13}$$

Performing the summation in (8.13) with the Boltzmann weighting factors p_k amounts to evaluating an ensemble average. Using the subscript zero to denote the ensemble average of a matrix element, we then have

$$I(\omega) = \frac{3}{2\pi}\int_{-\infty}^{\infty} dt\, e^{-i\omega t}\langle k|\hat{E}\cdot\boldsymbol{\mu}(0)\hat{E}\cdot\boldsymbol{\mu}(t)|k\rangle_0 \tag{8.14}$$

Since the orientation of the unit vector $\hat{E}$ is arbitrary in an isotropic sample, and since

$$\begin{aligned}\langle k|\mu_x(0)\mu_x(t)|k\rangle_0 &= \langle k|\mu_y(0)\mu_y(t)|k\rangle_0 = \langle k|\mu_z(0)\mu_z(t)|k\rangle_0 \\ &= \tfrac{1}{3}\langle k|\boldsymbol{\mu}(0)\cdot\boldsymbol{\mu}(t)|k\rangle_0\end{aligned} \tag{8.15}$$

our lineshape function assumes the final form [3]

$$I(\omega) = \frac{1}{2\pi}\int_{-\infty}^{\infty} dt\, e^{-i\omega t}\langle\boldsymbol{\mu}(0)\cdot\boldsymbol{\mu}(t)\rangle_0 \tag{8.16}$$

$I(\omega)$ is then given by the Fourier transform of the *electric dipole correlation* function $\langle\boldsymbol{\mu}(0)\cdot\boldsymbol{\mu}(t)\rangle_0$. For a single molecule, the quantity $\boldsymbol{\mu}(0)\cdot\boldsymbol{\mu}(t)$ gives the projection of its dipole moment at time t along its initial direction at time $t = 0$. In a collection of molexules, the *total* dipole moment must be used in $\boldsymbol{\mu}(t)$, so that the correlation function will generally contain cross terms between dipole moment operators belonging to different molecules. At high concentrations (e.g., in pure polar liquids), where orientational motion between neighboring molecules may be highly correlated, $\langle\boldsymbol{\mu}(0)\cdot\boldsymbol{\mu}(t)\rangle_0$ therefore cannot be interpreted in terms of reorientation of a single molecule. Such cross terms are unimportant in gases and in dilute solutions of polar molecules in nonpolar solvents (where the motions of neighboring polar molecules are essentially uncoupled), and in these systems the dipole correlation function gives a quantitative measure of how a single molecule loses its orientational memory as a result of collisions or other perturbations. The correlation function typically approaches zero at long times in liquids in gases, as the molecular orientations become randomized through stochastic processes (e.g., angular momentum changes in bimolecular collisions).

8.2 LIFETIME BROADENING

We next evaluate the lineshape function (8.16) for two concrete situations in gases. (The complexity of molecular motions in liquids precludes computation of their dipole correlation functions in a text of this scope.) In the first situation, we imagine that we are examining lineshapes in the far-infrared spectrum of a collision-free, rotating polar molecule. Its dipole moment $\boldsymbol{\mu}_0$ is assumed to rotate classically without interruption with angular frequency ω_0 about an axis normal to $\boldsymbol{\mu}_0$. In a dilute gas, we would then have

$$\begin{aligned}\langle \boldsymbol{\mu}(0)\cdot\boldsymbol{\mu}(t)\rangle_0 &= \mu_0^2 \cos \omega_0 t \\ &= \frac{\mu_0^2}{2}(e^{i\omega_0 t} + e^{-i\omega_0 t})\end{aligned} \tag{8.17}$$

for $-\infty < t < +\infty$. The relevant lineshape function is then

$$\begin{aligned}I(\omega) &= \frac{\mu_0^2}{4\pi}\int_{-\infty}^{\infty} dt\, e^{-i\omega t}(e^{i\omega_0 t} + e^{-i\omega_0 t}) \\ &= \frac{\mu_0^2}{2}[\delta(\omega - \omega_0) + \delta(\omega + \omega_0)]\end{aligned} \tag{8.18}$$

These two delta functions correspond to absorption and emission of radiation at frequency ω_0, respectively, with spectral lineshapes exhibiting zero full width at half maximum (fwhm). Such uninterrupted molecular rotation, in which the dipole correlation function (8.17) maintains perfect sinusoidal coherence for an indefinite period of time, produces no broadening in the lineshape function $I(\omega)$.

Suppose now that we have the more realistic correlation function

$$\begin{aligned}\langle \boldsymbol{\mu}(0)\cdot\boldsymbol{\mu}(t)\rangle_0 &= 0 & t < 0 \\ &= \mu_0^2 e^{-\gamma t/2}\cos \omega_0 t & t > 0\end{aligned} \tag{8.19}$$

This would represent a molecule rotating classically with frequency ω_0, with a transition moment that decays exponentially with $1/e$ lifetime $\tau = 2/\gamma$ following excitation at time $t = 0$. Such decay may occur via radiationless transition, spontaneous emission (Section 8.4), or collisional deactivation in the excited state. The corresponding (real) lineshape function is

$$\begin{aligned}I(\omega) &= \frac{\mu_0^2}{4\pi}\,\mathrm{Re}\int_0^{\infty} e^{-i\omega t}e^{-\gamma t/2}(e^{i\omega_0 t} + e^{-i\omega_0 t})dt \\ &= \frac{\mu_0^2}{4\pi}\left[\frac{\gamma/2}{(\omega - \omega_0)^2 + \gamma^2/4} + \frac{\gamma/2}{(\omega + \omega_0)^2 + \gamma^2/4}\right]\end{aligned} \tag{8.20}$$

Like Eq. 8.18, this expression exhibits terms corresponding to absorption and emission, respectively. The normalized absorption lineshape function

$$P_L(\omega) = \frac{\gamma}{2\pi} \frac{1}{(\omega - \omega_0)^2 + \gamma^2/4} \tag{8.21}$$

satisfies the condition

$$\int_{-\infty}^{\infty} P_L(\omega) d\omega = 1 \tag{8.22}$$

$P_L(\omega)$ is called the *Lorentzian lineshape function.* Its fwhm is equal to γ, and is inversely proportional to the lifetime $\tau = 2/\gamma$. It approaches zero as $\omega \to \pm\infty$, and maximizes at $\omega = \omega_0$ (Fig. 8.1). Physically, γ itself will have several components in any real absorption line, arising from spontaneous emission (fluorescence or phosphorescence), nonradiative excited-state decay (intersystem crossing, internal conversion, photochemistry), collisional deactivation, etc.:

$$\gamma = \gamma_{\text{rad}} + \gamma_{\text{nonrad}} + \gamma_{\text{coll}} + \cdots \tag{8.23}$$

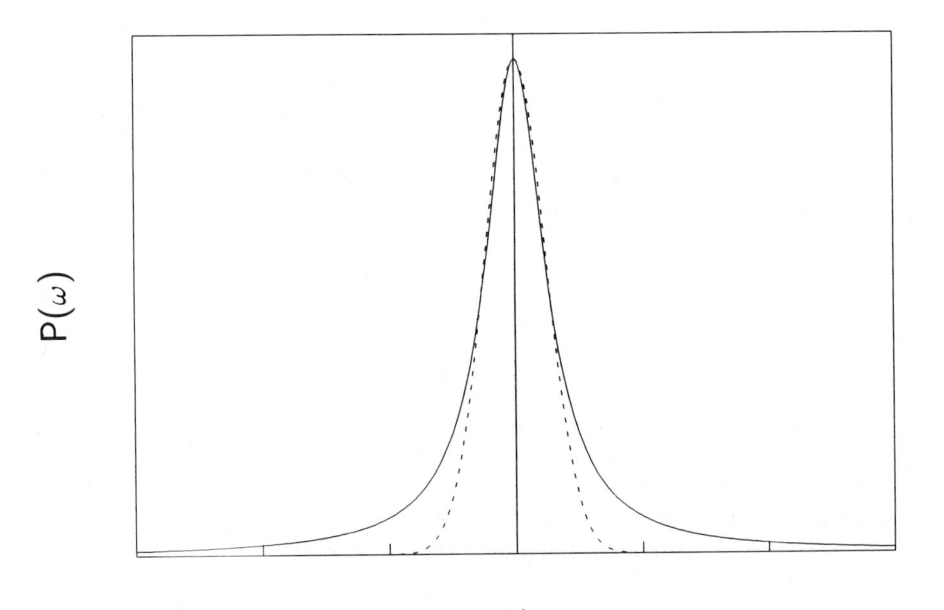

Figure 8.1 Lorentzian and Gaussian profiles $P_L(\omega)$ and $P_G(\omega)$ (solid and dashed curves, respectively) with the same peak height and the same fwhm. The two profiles are very similar for $|\omega| \lesssim$ fwhm/2, but the Lorentizian profile falls off much more slowly than the Gaussian profile at large $|\omega|$.

or

$$1/\tau = 1/\tau_{\text{rad}} + 1/\tau_{\text{nonrad}} + 1/\tau_{\text{coll}} + \cdots \tag{8.24}$$

Lifetime broadening is ubiquitous, because all excited states decay. The observed absorption lineshape is seldom Lorentzian, however, since other mechanisms than lifetime broadening usually dominate the actual lineshape.

8.3 DOPPLER BROADENING AND VOIGT PROFILES

As a result of the Doppler effect, a molecule traveling with velocity component v_z along the propagation axis of an incident light beam will experience a light frequency that is shifted from that experienced by a stationary molecule by

$$\Delta\omega = \omega - \omega_0 = -\omega_0(v_z/c) \tag{8.25}$$

Here ω_0 is the frequency experienced by the stationary molecule. In a thermal gas sample at temperature T, the one-dimensional Boltzmann velocity distribution in v_z will be

$$P(v_z) = P_0 e^{-mv_z^2/2kT} \tag{8.26}$$

where m is the molecular mass. (This equation should not be confused with the *three-dimensional* velocity distribution, which is proportional to $v^2 \exp(-mv^2/2kT)$.) Using $v_z = c(\omega - \omega_0)/\omega_0$, we obtain

$$P_G(\omega) = P_0 \exp[-m(\omega - \omega_0)^2 c^2/2\omega_0^2 kT] \tag{8.27}$$

This is the *Gaussian* lineshape function, which arises from Doppler broadening. Like the Lorentzian function $P_L(\omega)$, it maximizes at $\omega = \omega_0$, and approaches zero as $\omega \to \pm\infty$. The fwhm of $P_G(\omega)$ is

$$\text{fwhm} = \left(\frac{2\omega_0}{c}\right)\left(\frac{2kT \ln 2}{m}\right)^{1/2} \tag{8.28}$$

so that the Gaussian lineshape broadens with temperature as $T^{1/2}$. The Lorentzian and Gaussian lineshapes are contrasted in Fig. 8.1.

The Lorentzian and Gaussian lineshapes physically differ in one important respect. All molecules in a homogeneous gas (excepting specialized situations where the gas is subjected to a nonuniform external field) in a given excited state have an identical probability per unit time of decaying either radiatively or nonradiatively. All molecules in such a sample thus contribute equally to $P_L(\omega)$ at all frequencies ω. This is an example of *homogeneous* broadening. In Doppler broadening, it is clear from Eqs. 8.26 and 8.27 that slowly moving molecules contribute more to the center of the Gaussian profile $P_G(\omega)$, while faster molecules account for the wings where $\omega \gg \omega_0$ or $\omega \ll \omega_0$. Doppler broadening is thus an example of *inhomogeneous* broadening.

Both lifetime and Doppler broadening are always present in gas samples. If these two are the only important broadening mechanisms, the resulting profile is given by the convolution

$$P_V(\omega) = \int_{-\infty}^{\infty} P_G(\omega - \omega')P_L(\omega')d\omega' \tag{8.29}$$

of the Gaussian and Lorentzian profiles (Fig. 8.2). This function, called the *Voigt* lineshape function, is nonanalytic (i.e., inexpressible in closed form), and it has been extensively tabulated for analysis of gas sample absorption lineshapes. With our current access to interactive computers, such tabulations are rapidly becoming unnecessary.

Doppler broadening in gases can readily be eliminated by manipulation of experimental conditions. One way to achieve this is to do spectroscopy on supersonic gas jets, in which the translational velocity distribution can be made to resemble a delta function along the jet direction (i.e., the velocity distribution

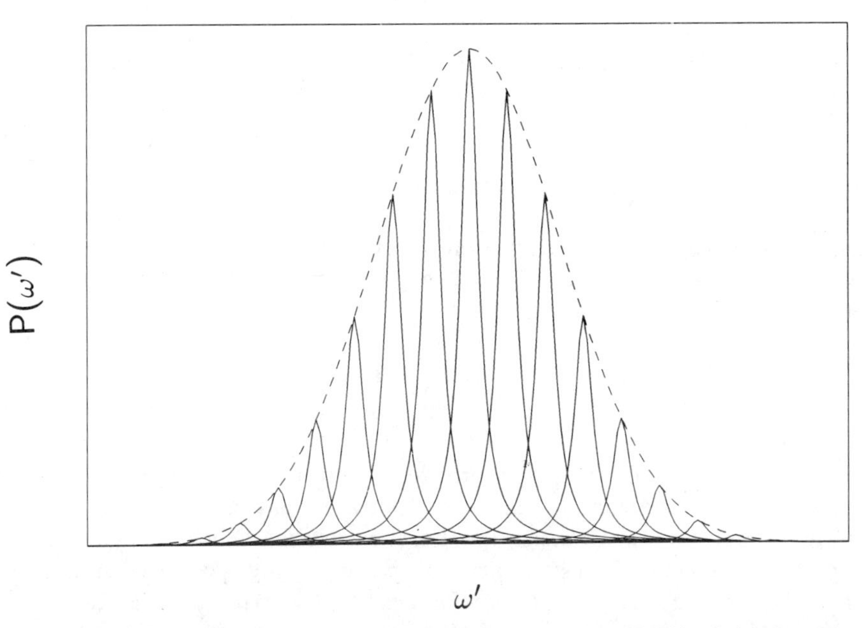

Figure 8.2 The Voigt profile, formed by the convolution (8.29) of the Gaussian profile with the Lorentzian profile, effectively sums Lorentian profiles (solid curves) centered at all frequencies ω', weighted by the value of the Gaussian profile (dashed curve) at those frequencies.

is highly non-Boltzmann), while the velocity distribution transverse to the jet direction is characterized by low temperatures $T \lesssim 20\,\text{K}$. At such low temperatures, Eq. 8.28 yields an absorption line fwhm that is far narrower than that observed in room-temperature gases. Supersonic jet spectroscopy has therefore afforded experimentalists an opportunity to obtain highly detailed, well-resolved vibrational structure in large-molecule electronic band spectra (cf. Fig. 7.10). An alternative way of obtaining Doppler-free spectra (applicable to *thermal* gases) is to do two-photon spectroscopy with counterpropagating laser beams; this will be discussed in the section on two-photon absorption in Chapter 10.

Experimentalists must always contend with *instrumental* broadening as well, since spectrometers and excitation lasers never operate with zero bandpass or output bandwidth. The resolution of a grating spectrometer (defined as $R = \lambda/\Delta\lambda$, with λ and $\Delta\lambda$ equal respectively to the operating wavelength and the wavelength bandwidth passed by the instrument) is ideally inversely proportional to the width of the instrument's exit slit. Any real instrument suffers from aberrations that lower R from its ideal value: coma, spherical aberrations, and chromatic aberrations are some of the artifacts that can contribute to instrumental broadening in the less well-designed instruments. Additional problems arise in grating spectrometers with mechanically ruled gratings, because the groove spacing in such gratings cannot be made absolutely uniform; these gratings inevitably diffract more than one wavelength at a time into any given direction. This drawback has been greatly reduced by the introduction of holographic gratings, in which the diffraction grooves are automatically formed with uniform spacing upon exposure to an interference pattern of two coherent laser beams with well-defined wavelength.

8.4 EINSTEIN COEFFICIENTS

The Einstein coefficients prove to be useful for understanding the relationships among the probabilities for spontaneous emission, stimulated emission, and absorption. They are thus valuable for understanding the criteria for achieving laser action, where the competition between spontaneous and stimulated emission in the laser medium is crucial. The Einstein coefficients also lead to important insights into the relationships between the absorption and fluorescence properties of molecules, relationships that are often taken for granted in the chemical physics literature.

We begin our discussion with an ensemble of identical two-level systems in which the upper and lower state populations are N_2 and N_1, respectively. The energy levels are spaced by $\Delta E = h\nu$, and the systems are at thermal equilibrium with a radiation energy distribution over light frequencies ν given by $\rho(\nu)$. It is assumed that only three mechanisms exist for transferring systems between levels 1 and 2: one-photon absorption, spontaneous emission (radiation of a single photon), and stimulated emission (Fig. 8.3). In the latter process, a photon

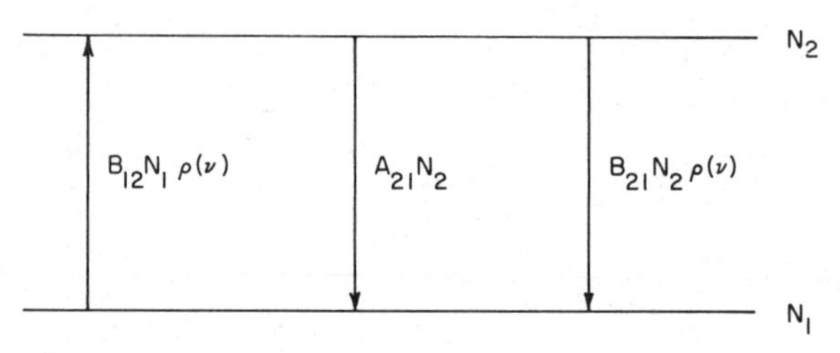

Figure 8.3 Optical processes in a two-level system with level populations N_1 and N_2. The rates of photon absorption, spontaneous emission, and stimulated emission are $B_{12}N_1\rho(\nu)$, $A_{21}N_2$, and $B_{21}N_2\rho(\nu)$, respectively.

with frequency ν (which matches the energy level difference ΔE) and given polarization is incident upon a system in upper level 2. It causes the system to relax to level 1, giving up a photon of identical energy, polarization, and propagation. (Since two identical, coherent photons emerge from one incident photon, stimulated emission forms the basis for laser amplification in Chapter 9.) The number of photons absorbed per unit time will be proportional to both N_1 and $\rho(\nu)$, and we define the Einstein absorption coefficient B_{12} by setting this photon absorption rate equal to $B_{12}N_1\rho(\nu)$. The number of photons emitted per unit time will be equal to $N_2A_{21} + N_2B_{21}\rho(\nu)$, where we have now implicitly defined the Einstein coefficients B_{21} for stimulated emission and A_{21} for spontaneous emission (whose rate is necessarily independent of $\rho(\nu)$). The absorption and emission rates must balance at thermal equilibrium, so that

$$N_2(B_{21}\rho(\nu) + A_{21}) = N_1B_{12}\rho(\nu) \tag{8.30}$$

We may solve this for the radiation energy distribution

$$\rho(\nu) = \frac{N_2A_{21}}{N_1B_{12} - N_2B_{21}} \tag{8.31}$$

However, $\rho(\nu)$ also be given by the Planck blackbody distribution at thermal equilibrium [4]

$$\rho(\nu) = \frac{8\pi h\nu^3}{c^3}\frac{1}{e^{h\nu/kT} - 1} \tag{8.32}$$

We also know that N_2 and N_1 are related at equilibrium by [4]

$$N_2/N_1 = (g_2/g_1)e^{-h\nu/kT} \tag{8.33}$$

where g_1 and g_2 are the statistical weights of levels 1 and 2. It is easy to show that Eqs. 8.31 through 8.33 are mutually consistent only if

$$B_{21} = (g_1/g_2)B_{12} \tag{8.34}$$

and

$$A_{21} = \frac{8\pi h\nu^3}{c^3} B_{21} \tag{8.35}$$

Hence, the Einstein coefficients for absorption, spontaneous emission, and stimulated emission are all simply related. The ν^3 factor that enters in the spontaneous emission coefficient A_{21} (Eq. 8.35) has had historical importance in the development of lasers, since it implies that spontaneous emission competes more effectively with stimulated emission at higher frequencies. High-frequency lasers have therefore been more difficult to construct. This is one of the reasons why X-ray lasers have only recently been built, and why the first laser was an ammonia maser operating on a microwave umbrella-inversion vibration rather than a visible laser.

It may be shown [5] that in the E1 approximation the Einstein absorption coefficient for the two-level system is given by

$$B_{12} = \frac{8\pi^3}{3h^2} |\langle 1|\boldsymbol{\mu}|2\rangle|^2 \tag{8.36}$$

This leads immediately to explicit expressions for the two Einstein emission coefficients,

$$B_{21} = \frac{8\pi^3}{3h^2}\left(\frac{g_1}{g_2}\right) |\langle 1|\boldsymbol{\mu}|2\rangle|^2 \tag{8.37}$$

and

$$A_{21} = \frac{64\pi^4\nu^3}{3hc^3}\left(\frac{g_1}{g_2}\right) |\langle 1|\boldsymbol{\mu}|2\rangle|^2 \tag{8.38}$$

8.5 OSCILLATOR STRENGTHS

We now exploit the relationships among Einstein coefficients to obtain relationships between the electronic absorption and emission properties in polyatomics. For simplicity, we assume that an absorptive transition originates from the ground vibrational state 0 in electronic state 1, and terminates with vibrational state n in electronic state u. (These vibrational level designations are shorthand labels for collective vibrational states involving all of the normal modes.) The Einstein coefficient for this absorption process is

$$B_{10\to un} = \frac{8\pi^3}{3h^2} |\langle 10|\boldsymbol{\mu}|un\rangle|^2 \tag{8.39}$$

which in the Born-Oppenheimer approximation (not good for vibronically induced transitions!) is

$$B_{10\to un} = \frac{8\pi^3}{3h^2} |\langle 1|\boldsymbol{\mu}_{el}|u\rangle|^2 |\langle 0|n\rangle|^2$$

$$= \frac{8\pi^3}{3h^2} |\mathbf{M}_e^{1\to u}|^2 |\langle 0|n\rangle|^2 \tag{8.40}$$

Here $|\langle 0|n\rangle|^2$ is the Franck-Condon factor for the $0 \to n$ vibrational band intensity in the electronic spectrum, and $\mathbf{M}_e^{1\to u}$ is the electronic transition moment. If one sums Eq. 8.40 over all of the upper state vibrational levels n reached from the vibrationless lower state, one obtains

$$\sum_n B_{10\to un} = \frac{8\pi^3}{3h^2} |\mathbf{M}_e^{1\to u}|^2 \sum_n \langle 0|n\rangle\langle n|0\rangle$$

$$= \frac{8\pi^3}{3h^2} |\mathbf{M}_e^{1\to u}|^2 \tag{8.41}$$

which depends only on the electronic transition moment and is *independent* of the upper state vibrational structure. This sum is therefore a measure of the allowedness of the electronic transition. To describe the latter, it is conventional to use the dimensionless *oscillator strength* [6]

$$f_{1\to u} \equiv \frac{8\pi^2 m_e}{3he^2} \bar{\nu}_{1u} |\mathbf{M}_e^{1\to u}|^2 \tag{8.42}$$

where m_e is the electron mass, e is the electron charge, and $\bar{\nu}_{1u}$ is the mean frequency of the electronic transition. Defined in this way, the sum of oscillator strengths $\sum_u f_{1\to u}$ for electronic transitions from electronic state 1 to all other electronic states is supposed to equal unity, but in practice this depends on how $\bar{\nu}_{1u}$ is specified. As a rule, $f_{1\to u}$ on the order of unity is associated with a strongly allowed E1 transition in aromatic hydrocarbons, while E1-forbidden transitions carry $f_{1\to u} \lesssim 10^{-2}$.

Next we examine the fluorescence properties embodied in the spontaneous emission coefficient. In a two-level system with low radiation density (i.e., negligible stimulated emission and negligible pumping of level 2 by absorption from level 1), we have

$$-\frac{dN_2}{dt} = N_2 A_{21} \tag{8.43}$$

This has the time-dependent solution

$$N_2(t) = N_2(0)e^{-A_{21}t}$$

$$\equiv N_2(0)e^{-t/\tau_{rad}} \tag{8.44}$$

meaning the upper level decays by spontaneous emission (fluorescence in the case of a spin-allowed transition) with an exponential lifetime $\tau_{\rm rad} = 1/A_{21}$. When we consider the analogous case of a polyatomic fluorescing from its vibrationless electronic state u down to a *manifold* of vibrational levels m in electronic state 1, the radiative lifetime is given instead by

$$1/\tau_{\rm rad} = \sum_m A_{u0\to 1m} \tag{8.45}$$

In the Born-Oppenheimer approximation, this becomes

$$1/\tau_{\rm rad} = \frac{64\pi^4}{3hc^3}(g_1/g_2)|\langle u|\boldsymbol{\mu}_{\rm el}|1\rangle|^2 \sum_m \nu^3_{u0\to 1m}|\langle 0|m\rangle|^2 \tag{8.46}$$

which simplifies into

$$1/\tau_{\rm rad} = \frac{64\pi^4}{3hc^3}(g_1/g_2)\overline{|M_e^{1\to u}|^2\nu^3_{u\to 1}} \tag{8.47}$$

if the value of ν^3 is assumed constant over the fluorescence spectrum. At this level of approximation, it is clear that $\tau_{\rm rad}$ is *independent* of the upper state vibrational level. (To see this, assume that upper vibrational level l rather than level 0 emits, replacing the right side of Eq. 8.45 with $\sum_m A_{ul\to 1m}$. The l-dependence of $1/\tau_{\rm rad}$ then drops out by the time the analog of Eq. 8.47 is reached.) As we pointed out at the end of Section 8.2, $\tau_{\rm rad}$ is related to the observed lifetime τ by $1/\tau = 1/\tau_{\rm rad} + 1/\tau_{\rm nonrad} + 1/\tau_{\rm coll} + \cdots$, so that the observed lifetime equals $\tau_{\rm rad}$ only if the other excited-state deactivation pathways have negligible rates. The fluorescence quantum yield of the excited state u is defined as

$$Q_{\rm F} = \gamma_{\rm rad}/(\gamma_{\rm rad} + \gamma_{\rm nonrad} + \gamma_{\rm coll} + \cdots) \tag{8.48}$$

and approaches unity if $\gamma_{\rm rad} \gg \gamma_{\rm nonrad} + \gamma_{\rm coll} + \cdots$. Since $\gamma_{\rm rad}$ is nearly independent of the fluorescing vibrational level l in the Born-Oppenheimer approximation, a strong l-dependence in $Q_{\rm F}$ implies that there are vibrational level-dependent intersystem crossing, internal conversion, photochemical, or collisional deactivation processes present if the Born-Oppenheimer approximation holds in Eq. 8.46.

We next combine Eqs. 8.41 and 8.47 to find a relationship between the absorption spectrum and the fluorescence lifetime. In particular, the equation

$$1/\tau_{\rm rad} = \frac{8\pi h\overline{\nu^3}_{u\to 1}}{c^3}(g_1/g_2)\sum_n B_{10\to un} \tag{8.49}$$

implies that the fluorescence lifetime for the $u \to 1$ fluorescence transition may be predicted by summing the Einstein absorption coefficients for all of the $0 \to n$

vibrational bands in the $1 \rightarrow u$ electronic absorption spectrum. In many practical situations (i.e., outside of supersonic jets) the measured absorption spectrum is not a series of discrete vibrational lines, but a continuous spectrum (inhomogeneously broadened in a solution, for example). A working formula that has been evolved for singlet–singlet transitions in such cases is [6]

$$1/\tau_{\text{rad}} = \frac{8\pi \cdot 2303 n_f^3}{Nc^2 n_a} \langle \nu_f^{-3} \rangle^{-1} \int \frac{\varepsilon(\nu)}{\nu} d\nu \tag{8.50}$$

where n_a and n_f are the medium (solvent) refractive index at the absorption and fluorescence wavelengths, ν is frequency in s^{-1}, τ_{rad} is in s, and $\varepsilon(\nu)$ is the molar absorption coefficient in L/mol·cm. Equation (8.50) has been verified for a number of rigid molecules by Strickler and coworkers [7]. In nonrigid molecules having different equilibrium geometries in electronic states 1 and u, $|\mathbf{M}_{\text{el}}^{1 \rightarrow u}|^2$ (which depends on the electronic wave functions) will not have the same value in both Eqs. 8.46 and 8.41, so that Eq. 8.50 will not be valid.

The strongest commonly observed absorption bands in organic molecules are exhibited by laser dyes such as rhodamine 6G, a xanthene dye with a rigid chromophore which exhibits $\varepsilon_{\max} \sim 10^5$ L/mol·cm at 5300 Å (Chapter 9). Its fluorescence lifetime τ is in the neighborhood of 5 ns (depending on solvent), and is dominated by τ_{rad} because Q_F is nearly unity under conditions in which stimulated emission is suppressed. At the other extreme, $\varepsilon_{\max}$ for I_2 in an inert solvent like cyclohexane is about 7×10^2 L/mol·cm for the spin-forbidden $X^1\Sigma_g^+ \rightarrow B^3\Pi_{0u}$ transition. In accordance with Eq. 8.50, τ_{rad} for this transition is in excess of 10 μs. Little I_2 $B^3\Pi_{0u} \rightarrow X^1\Sigma_g^+$ emission can be seen in solution, however ($Q_F < 10^{-4}$), because collisions of $B^3\Pi_{0u}$ I_2 with the solvent induces rapid predissociation into I atoms on a time scale of $\sim$15 ps.

REFERENCES

1. K. Gottfried, *Quantum Mechanics*, Vol. 1, W. A. Benjamin, New York, 1966.
2. M. H. Nayfeh and M. K. Brussel, *Electricity and Magnetism*, Wiley, New York, 1985.
3. R. G. Gordon, *Adv. Magn. Reson.* **3**: 1 (1968).
4. F. Reif, *Fundamentals of Statistical and Thermal Physics*, McGraw-Hill, New York, 1965.
5. E. U. Condon and G. H. Shortley, *The Theory of Atomic Spectra*, Cambridge Univ. Press, Cambridge, England, 1935.
6. J. B. Birks, *Photophysics of Aromatic Molecules*, Wiley-Interscience, London, 1970.
7. S. J. Strickler and R. A. Berg, *J. Chem. Phys.* **37**: 814 (1962).

PROBLEMS

1. The E1 transition moment $\langle 2p|\boldsymbol{\mu}|1s\rangle$ for a $1s \rightarrow 2p$ transition in H can be evaluated analytically using the hydrogenic wave functions.

(a) Calculate the Einstein coefficients B_{12}, A_{21}, and B_{21} for this transition. Include units.

(b) For what radiation energy density $\rho(\nu)$ at the transition frequency will stimulated emission from a $2p$ level become competitive with spontaneous emission?

(c) Determine the radiative lifetime of a $2p$ state in hydrogen.

2. The absorption and fluorescence spectra of the organic dye rhodamine 6G in ethanol are shown in Figure P8.2. Use reasonable approximations to determine τ_{rad} and the oscillator strength to within 20% for the $S_1 \leftarrow S_0$ electronic transition. (Assume that the $S_1 \leftarrow S_0$ system is confined between 400 and 600 nm; the absorption bands at wavelengths shorter than 400 nm arise from transitions to higher singlet states.)

3. Calculate the expected fwhm of the R(0) line in the HCl vibration–rotation spectrum of Fig. 3.3 if the linewidth is dominated by Dopper broadening. Assume that the gas temperature is 300 K. Is the observed broadening primarily Doppler broadening?

4. An absorption spectrophotometer is operated with a resolution of 1 cm^{-1} in the near ultraviolet. What excited-state lifetimes will yield Lorentzian lineshape functions with fwhm larger than this?

5. Figure P8.5 shows the normalized dipole correlation functions $\langle \boldsymbol{\mu}(0) \cdot \boldsymbol{\mu}(t) \rangle / \langle \boldsymbol{\mu}^2(0) \rangle$ obtained by taking the inverse Fourier transform of near-

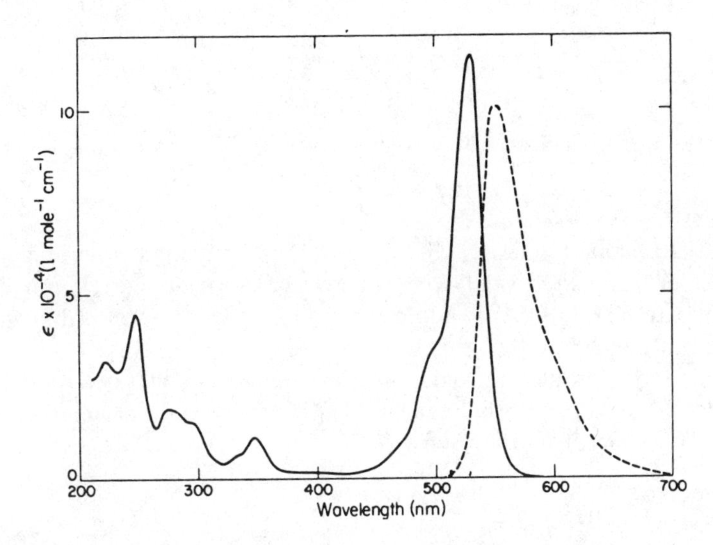

Figure P8.2 Reproduced with permission from K. H. Drexhage, in *Topics in Applied Physics,* Vol. 1, *Dye Lasers,* F. P. Schäfer (Ed.), Springer-Verlag, New York, 1973, p.168.

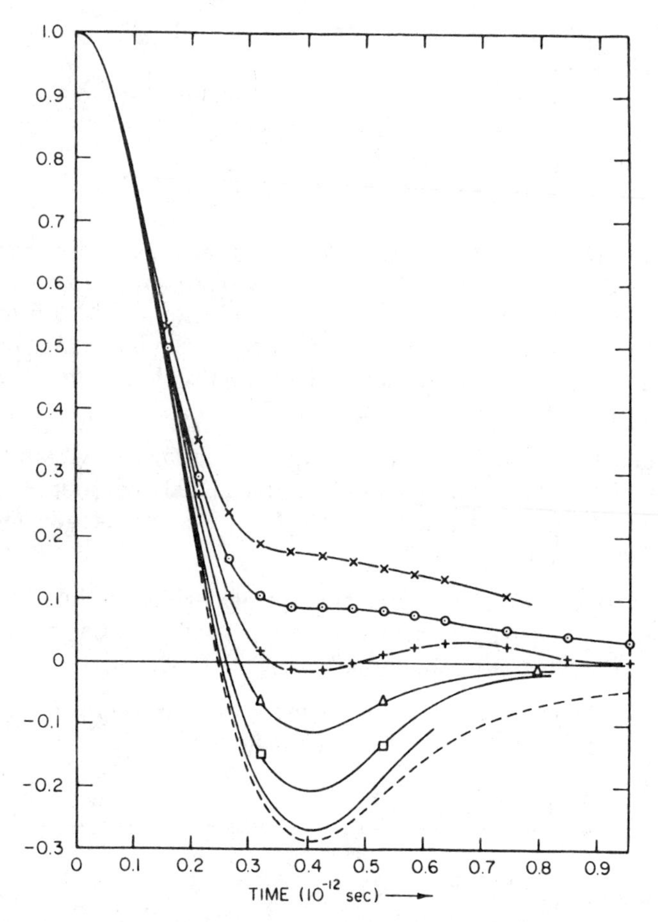

Figure P8.5 Reproduced with permission from R. G. Gordon, *Adv. Magn. Reson.* **3**; 1 (1968).

infrared absorption lines of CO in several contrasting environments [3]. Account for the qualitative differences between the various correlation functions (assume that the solutions are dilute enough to render cross terms in $\langle \boldsymbol{\mu}(0) \cdot \boldsymbol{\mu}(t) \rangle$ insignificant). Why is the correlation function for "free" CO (i.e., isolated molecules) not a simple sinusoidal function of time, like that of a set of classical dipoles rotating at a uniform frequency ω_0? How would the correlation function for free HCl differ from that of free CO?

9

LASERS

Since their introduction to spectroscopy laboratories in the 1960s, lasers have revolutionized the field to the point where most current spectroscopy would be unrecognizable to observers from the pre-laser era. This stems partly from the fact that tunable lasers can be made to operate with intense output which is highly monochromatic (output bandwidths $\Delta\bar{\nu} \lesssim 10^{-3}\,\mathrm{cm}^{-1}$ are now available using specialized ring lasers), so that high-resolution spectroscopy which was once barely feasible can now be done routinely. The fact that lasers can easily generate picosecond and subpicosecond light pulses with selectable wavelength and repetition rate has brought time-resolved spectroscopy securely into the time scale of vibrational motions and fast photochemistry. The truly novel laser-based spectroscopies, however, have evolved primarily from the ability of lasers to generate such extraordinarily high power densities that the nonlinear terms in the Dyson expansion of $c_m(t)$ in Eq. 1.96, which are barely noticeable with classical light sources, become prominent in laser-excited molecules. Processes like two-photon absorption, second- and third-harmonic generation, four-wave mixing, stimulated Raman scattering, self-focusing and self-phase modulation were all discovered in laser laboratories. Ordinary Raman scattering was well established in classical spectroscopy before the 1950s, but required cumbersomely large excitation lamps; laser excitation has enormously expanded its resolution and productivity. We discuss the characteristics of lasers in this chapter, and we deal with several topics in nonlinear optics in the last two chapters of this book.

9.1 POPULATION INVERSIONS AND LASING CRITERIA

To establish the conditions required for laser action, we consider a slab of a laser medium containing a large number of two-level species. A light beam initially with intensity I_0 will have the intensity

$$I = I_0 e^{-\alpha l} \tag{9.1}$$

after passing a distance l through the medium if the latter obeys Beer's law. According to Eq. 9.1, the absorption coefficient α is

$$\alpha = -\frac{1}{I}\frac{dI}{dl} \tag{9.2}$$

so that the change in beam intensity after passing through a distance dl is

$$-dI = \alpha I\, dl \tag{9.3}$$

We now rephrase Eq. 9.3 in terms of the microscopic dynamics of the two-level system. If $W_{1\to 2}$ and $W_{2\to 1}$ are the probabilities for upward and downward transitions per unit time in a two-level system, the change in beam intensity will be

$$-dI = h\nu(W_{1\to 2}N_1 - W_{2\to 1}N_2)dt \tag{9.4}$$

where $dt = n\,dl/c$ is the time required for a photon to traverse distance dl through a medium with refractive index n. Using the Einstein coefficients and two-level notation introduced in Chapter 8, this becomes

$$-dI = h\nu(B_{12}N_1 - B_{21}N_2)I \cdot n\,dl/c \tag{9.5}$$

From Eq. 9.2, we may now obtain an expression for the absorption coefficient of the laser medium,

$$\alpha = -\frac{1}{I}\frac{dI}{dl} = \frac{nh\nu}{c}(B_{12}N_1 - B_{21}N_2) \tag{9.6}$$

Note that we include no term in A_{21} (for spontaneous emission) in Eqs. 9.4 through 9.6: Fluorescence is so widely dispersed over all propagation directions that its contribution to the coherent beam directed along the propagation axis of the incident light can be ignored. Using the relationships among the three Einstein coefficients, the absorption coefficient of the laser medium is finally [1]

$$\alpha = \frac{c^2 A_{21}}{8\pi\nu^2 n^2}\left(\frac{g_2}{g_1}\right)\left(N_1 - \frac{g_1}{g_2}N_2\right) \tag{9.7}$$

This absorption coefficient can have either sign a priori, with the consequences tabulated below:

	Sign of α	Result
$N_1 < \frac{g_1}{g_2} N_2$	<0	Amplification, $I > I_0$
$N_1 > \frac{g_1}{g_2} N_2$	>0	Absorption, $I < I_0$
$N_1 = \frac{g_1}{g_2} N_2$	$=0$	Threshold, $I = I_0$

At thermal equilibrium, $N_2 = (g_2/g_1)N_1 \exp(-h\nu/kT)$, meaning that one inevitably has

$$\alpha = \frac{c^2 A_{21}}{8\pi\nu^2 n^2}\left(\frac{g_2}{g_1}\right) N_1[1 - \exp(-h\nu/kT)] > 0 \tag{9.8}$$

for a collection of two-level systems in a Boltzmann distribution with finite, positive temperature. Hence, no system at thermal equilibrium can lase; to get the required "negative absorption coefficient," N_2 must be artificially increased by external pumping of the upper level through optical or other means.

To set up a prototype laser cavity (Fig. 9.1), we place a laser medium of length L between parallel mirrors with light intensity reflectivities $r_1 \leqslant 1$ and $r_2 \leqslant 1$ at the optical frequency $\nu = \Delta E/h$ corresponding to the transition between the upper and lower levels of the two-level system. Upon traveling twice through the gain medium with absorption coefficient α in a single round-trip pass through the cavity, an incident light beam with intensity I_0 will emerge with intensity $I_0 \exp(-2\alpha L)$. At the same time, it will be reduced by the factor $r_1 r_2 \equiv \exp(-2\gamma)$ as the beam strikes each reflector once in a round-trip pass; this defines a cavity *loss coefficient* $\gamma = -\frac{1}{2}\ln(r_1 r_2)$. For lasing to occur, the round-trip gains must

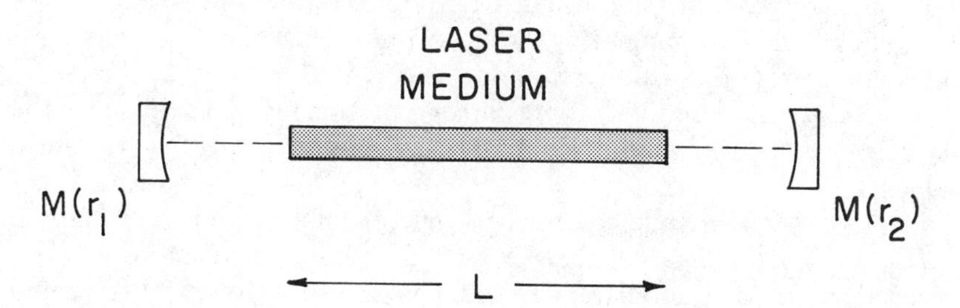

Figure 9.1 A laser cavity. The laser gain medium has physical length L; the end reflecting mirrors M have light intensity reflectivities r_1 and r_2.

exceed the cavity losses, or

$$I_0 e^{-2\alpha L} e^{-2\gamma} > I_0 \tag{9.9}$$

This requires that the negative absorption coefficient $\alpha(\nu)$ at the lasing frequency ν satisfy

$$-\alpha(\nu) > \gamma/L \tag{9.10}$$

In view of Eq. 9.7, this implies the condition

$$\frac{-c^2 A_{21}}{8\pi \nu^2 n^2}\left(\frac{g_2}{g_1}\right)\left(N_1 - \frac{g_1}{g_2} N_2\right) > \frac{\gamma}{L} \tag{9.11}$$

or

$$\left(\frac{g_1}{g_2}\right) N_2 - N_1 > \frac{8\pi n^2 \nu^2}{c^2 A_{21}}\left(\frac{g_1}{g_2}\right)\frac{\gamma}{L} \tag{9.12}$$

As a final detail, we allow for broadening (e.g., Doppler broadening, which determines the lasing lineshape in He/Ne lasers) of the two-level transition. Assume that the spectral lineshape function is $g(\nu)$, normalized so that $\int g(\nu)d\nu = 1$. Then the negative absorption coefficient in Eq. 9.7 at the lineshape center frequency ν_0 should be multiplied by $g(\nu_0)$, which will be inversely proportional to the lineshape fwhm $\Delta\nu$ since $g(\nu)$ is normalized; the product $g(\nu_0)\Delta\nu$ will be about unity. This implies that the lasing criterion in Eq. 9.12 should be replaced by [1]

$$\left(\frac{g_1}{g_2}\right) N_2 - N_1 \gtrsim \frac{8\pi n^2 \nu^2}{c^2 A_{21}}\left(\frac{g_1}{g_2}\right)\frac{\gamma \Delta\nu}{L} \tag{9.13}$$

The factors favoring lasing are now easily identified. The right side of inequality (9.13) can be minimized by selecting a strongly allowed E1 transition (large A_{21}), by using a large laser medium length L (in principle), by using sharp transition bandwidths $\Delta\nu$, and by minimizing the cavity loss coefficient γ. (In practice, γ has contributions from diffraction, scattering, and reflection losses at the medium boundaries as well as from the cavity mirror reflection losses mentioned earlier.) The upper state population N_2 must be kept large enough to maintain the inequality.

Not all of these requirements are symbiotic. In an argon ion (Ar^+) laser, the upper lasing level population N_2 is created in a plasma in which Ar atoms are ionized and excited by bombardment with hot electrons. However, the resulting high temperatures markedly increase the transition bandwidth $\Delta\nu$ through Doppler broadening, raising the lasing threshold. As another example, a laser at the Lawrence Livermore Laboratory amplifies 1.06-μm pulses through one of the world's longest Nd^{3+}:YAG gain media—but this fact can incur problems, because the intense laser pulses themselves can cause the gain medium to become lenslike (with nonuniform refractive index), leading to pulse self-focusing and gain medium damage over such distances.

In the next section, we describe how these lasing criteria are met and maintained in two real laser types, the He/Ne laser and the dye laser.

9.2 THE He/Ne AND DYE LASERS

Discovered by A. Javan and coworkers in 1960 at Bell Laboratories, the He/Ne laser is the most commonly used laser for alignment and holographic purposes. The upper and lower laser levels in this system are pairs of highly excited ($\gtrsim$ 150,000 cm^{-1}) electronic states in the Ne atom. Lasing has been achieved between many such pairs: The 6328 Å line is the one most commonly used, but the system can lase at 3.39, 1.12, 1.21, 1.16, and 1.19 μm, and many other lines stemming from the complicated multiplet structure of rare gas atom excited states.

When a mixture of He and Ne is subjected to a suitably energetic electric discharge, the following energy transfer processes take place:

1. $He + e^- \rightarrow He^* + e^-$

   $He^* + Ne \rightarrow He + Ne^* \; (N_2)$

   $\downarrow h\nu\text{(laser)}$

   $Ne^* \; (N_1)$

2. $Ne + e^- \rightarrow Ne^* \; (N_2) + e^-$

   $\downarrow h\nu\text{(laser)}$

   $Ne^* \; (N_1)$

(The asterisks denote electronic excitation.) In the first mechanism, electronically excited Ne* produced in the discharge transfers its excitation collisionally to Ne, which subsequently lases down to a lower excited level. In the second mechanism, excited Ne* is produced directly by the electric discharge. The latter mechanism can cause lasing at several lines in pure Ne; however, the addition of He greatly increases the lasing efficiency for the following reasons. He gas subjected to an electric discharge is initially pumped to a wide range of excited states, which can then rapidly relax radiatively to a succession of lower lying levels by E1 transitions (Fig. 9.2). The latter process, known as cascading, proceeds according to the selection rules $\Delta l = \pm 1$, $\Delta j = 0, \pm 1$ for the excited electron. A large fraction of the plasma-excited He atoms accumulate this way in the 2^1S and 2^3S states (both with the $(1s)^1(2s)^1$ configuration) of He. Once

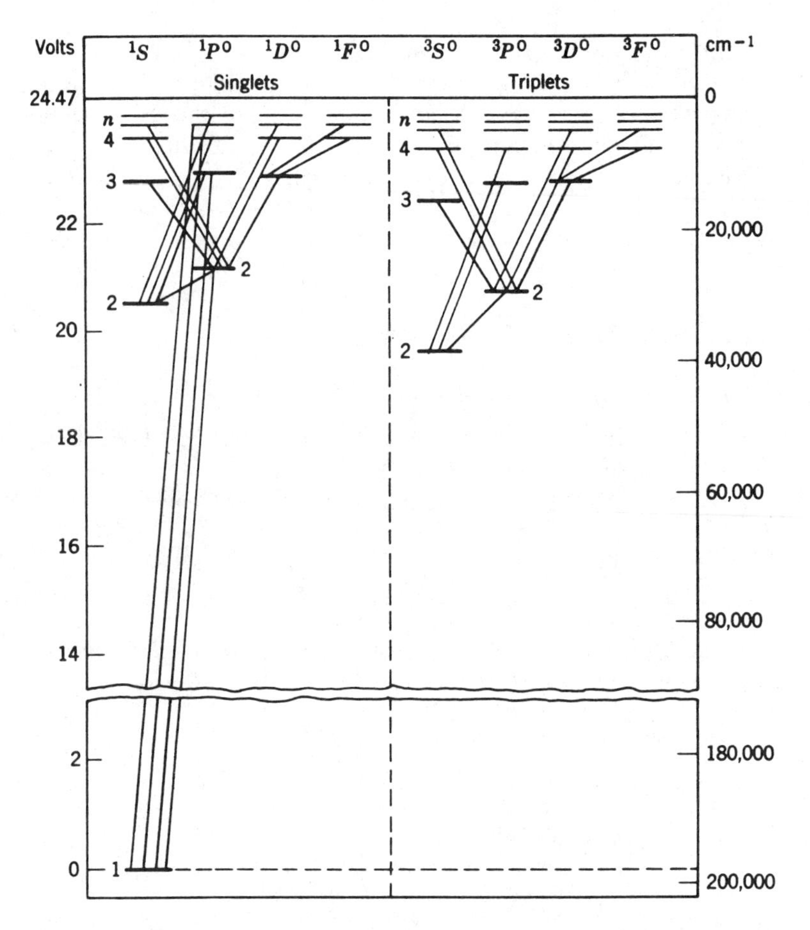

Figure 9.2 Energy level diagram for He. Since the 2^1S and 2^3S levels are E1-metastable, populations accumulate by cascading toward these two levels. Reproduced by permission from B. Lengyel, *Lasers, 2d ed., Wiley-Interscience, New York, 1971.*

populated, these levels are E1-metastable because neither one is E1-connected to any lower level (only the $(1s)^2$ 1^1S ground state lies below the 2^3S state). These levels do not lase in He, because (aside from the lack of any E1-connected lower-lying level) the 1^1S ground state invariably has a larger population N_1 than either 2^1S or 2^3S. Hence, 2^1S and 2^3S He atoms typically retain their excitation (at suitably adjusted He and Ne pressures) until they collide with a Ne atom, creating Ne*. The efficiency of this collisional excitation transfer is abetted by the near-resonance of the 2^1S and 2^3S energy levels with several of the $\cdots(2p)^5(5s)$ and $\cdots(2p)^5(4s)$ excited levels in Ne, respectively (Fig. 9.3).

The multiplet structure in Ne* arises from a *pair-coupling* (not Russell-Saunders) angular momentum coupling scheme, which is obeyed in rare-gas atom excited states. The ground and excited states of Ne exhibit the electron

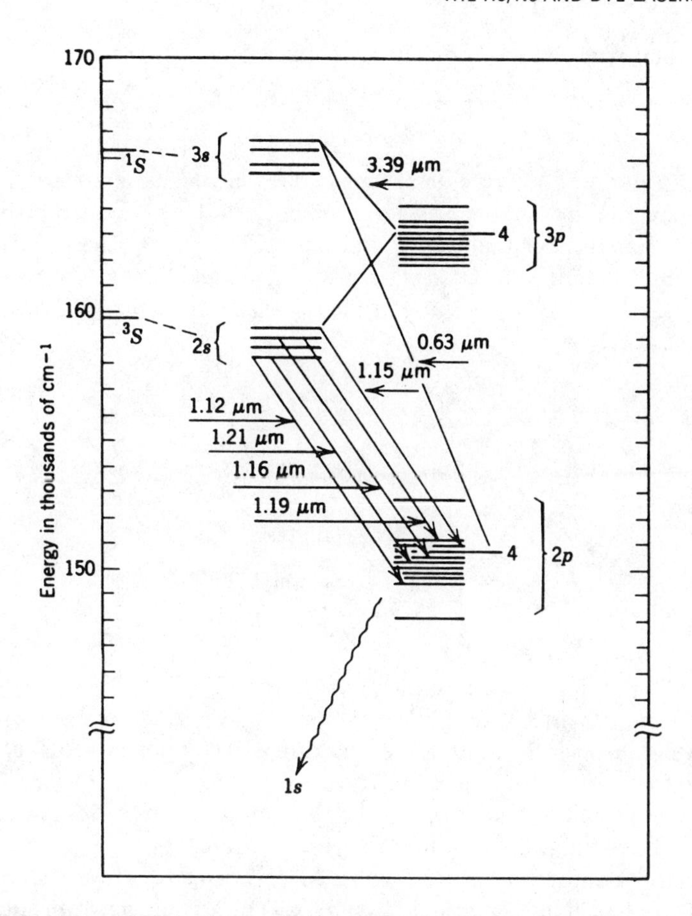

Figure 9.3 Energy level diagram for Ne excited states in the energy region of the metastable He 2^1S and 2^3S levels, which are shown at extreme left. The strongest laser transitions are indicated with their wavelengths. Adapted from B. Lengyel, *Lasers*, 2d ed., Wiley-Interscience, New York, 1971; used with permission.

configurations $\cdots(2p)^6$ and $\cdots(2p)^5(nl)^1$, respectively. In the latter configurations, the angular momenta of the $2p$ "core" electrons form the resultants $\mathbf{L}_c$, $\mathbf{S}_c$ of their orbital and spin angular momenta, which couple to form the core angular momentum $\mathbf{J}_c$. Clearly $L_c = 1$ and $S_c = \frac{1}{2}$ (since these are the quantum numbers of the missing electron in the otherwise-filled $(2p)^5$ core shell), and so the possible J_c values are $\frac{1}{2}$ and $\frac{3}{2}$. The excited "valence" electron $(nl)^1$, whose angular momenta are comparatively weakly coupled to the core angular momenta because its orbital is considerably more voluminous than those of the core electrons, exhibits $L_v = l$ and $S_v = \frac{1}{2}$. The angular momenta $\mathbf{J}_c$ and $\mathbf{L}_v$ then couple to form a resultant

$$\mathbf{K} = \mathbf{J}_c + \mathbf{L}_v \tag{9.14}$$

and the total Ne* angular momentum is given by

$$\mathbf{J} = \mathbf{K} + \mathbf{S}_v \tag{9.15}$$

The Racah notation for rare gas excited states neatly summarizes all of these pertinent angular momenta in the term symbol $nl[K]_J$. As two examples of Ne* multiplets, we consider the $3s[K]_J$ and the $np[K]_J$ states. The former are the lowest-lying $\cdots(2p)^5(3s)^1$ excited states with $L_v = 0$ and $S_v = \frac{1}{2}$. Since one always has $L_c = 1$, $S_c = \frac{1}{2}$, and $J_c = \frac{1}{2}$ or $\frac{3}{2}$ in the rare-gas atoms, the possible K values are from Eq. 9.14

$$K = J_c + L_v, \ldots, |J_c - L_v| = J_c = \tfrac{1}{2}, \tfrac{3}{2} \tag{9.16}$$

since $L_v = 0$ in the $3s[K]_J$ multiplet. Then the possible J values are [Eq. 9.15]

$$\begin{aligned} J &= K + S_v, \ldots, |K - S_v| \\ &= \begin{cases} 2, 1 & \text{when } K = \frac{3}{2} \\ 1, 0 & \text{when } K = \frac{1}{2} \end{cases} \end{aligned} \tag{9.17}$$

The $3s[K]_J$ multiplet therefore has four sublevels: $3s[\frac{1}{2}]_0$, $3s[\frac{1}{2}]_1$, $3s[\frac{3}{2}]_1$, and $3s[\frac{3}{2}]_2$; the higher $ns[K]_J$ multiplets exhibit similar sets of four sublevels. In a similar vein, the $np[K]_J$ multiplets each exhibit 10 components, because $L_c = 1$ can combine with $J_c = \frac{1}{2}, \frac{3}{2}$ to form $K = \frac{1}{2}, \frac{3}{2}, \frac{5}{2}; \frac{1}{2}, \frac{3}{2}$.

With these multiplet splittings, a large number of Ne* laser transitions is possible. The common 6328-Å line arises from a $5s[K]_J \rightarrow 3p[K]_J$ E1 transition, the 3.39-μm infrared line arises from a $5s[K]_J \rightarrow 4p[K]_J$ transition, and the series of infrared lines between 1.12 and 1.19 μm arises from transitions from various $4s[K]_J$ levels to lower-energy $3p[K]_J$ levels. Most laser applications in spectroscopy require monochromatic output, and the undesired laser lines are easily suppressed in favor of the selected one. One can make the cavity loss coefficient $\gamma = -\frac{1}{2}\ln(r_1 r_2)$ considerably larger for the infrared wavelengths than for the 6328-Å wavelength, for example, by using laser cavity mirrors with coatings exhibiting larger $r_1 r_2$ at 6328 Å than in the infrared. One may alternatively insert a prism or other dispersive element into the cavity so that only the desired wavelength will strike both cavity mirrors at normal incidence in a self-consistent, recycling trajectory (Fig. 9.4).

One important property of the above Ne* excited states which favors their involvement in lasing is their radiative lifetimes. To maintain favorable N_2 and N_1 in Eq. 9.13 and achieve continuous laser action, the upper state population N_2 should have a high probability for stimulated emission prior to deexcitation by other pathways, and the lower state population N_1 should be depleted rapidly after the laser transition to prevent its buildup. Thus, the upper and lower levels ideally should have relatively long and short radiative lifetimes, respectively. This condition is met in each of the He/Ne transitions we have

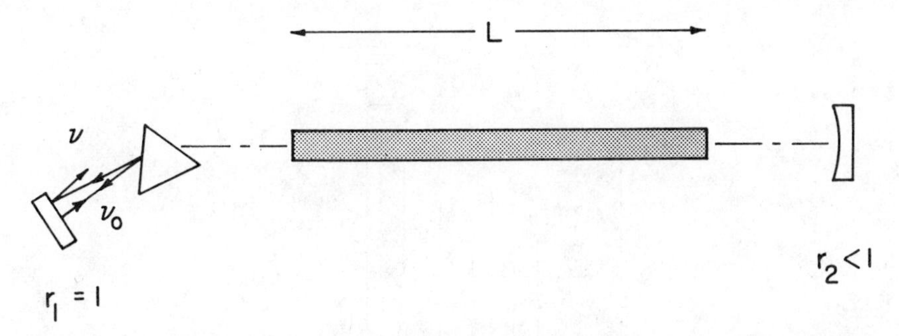

Figure 9.4 Laser cavity with intracavity tuning prism. Laser oscillation at the selected frequency ν_0 is refracted by the prism in a direction normal to the reflector with $r_1 = 1$, and retraces its trajectory through the gain medium. Oscillation at other frequencies ν strikes the reflector at nonnormal incidence, and leaves the cavity.

described, since the $ns[K]_J$ and $np[K]_J$ levels have radiative lifetimes of $\sim 10^{-7}$ and $\sim 10^{-8}$ s, respectively. This condition is not met in the well known N_2 laser, which oscillates with high efficiency in a series of $C^3\Pi_u \rightarrow B^3\Pi_g$ transitions between 3370.44 and 3371.44 Å. Since the upper state lifetime is shorter than the $B^3\Pi_g$ state lifetime, the laser action terminates in about 20 ns, and N_2 laser operation is restricted to the pulsed mode.

Prior to the discovery of organic dye lasers in 1966, laser action had already been demonstrated in several hundred gases (e.g., He/Ne, CO, N_2) and solids (e.g., ruby, Nd^{3+}:glass, Nd^{3+}:YAG). The truly novel property of dye lasers was their tunability—the fact that their output wavelengths could be varied over a broad range (tens of nm) by adjustment of the dye concentration and/or resonator conditions. While dyes may lase in vapors and solids as well as in liquid solutions, the latter rapidly became the media of choice owing to their economy and ease of handling. Population inversions in early dye lasers were frequently achieved by broadband excitation with xenon gas-filled flashlamps. Dye lasers are now more commonly monochromatically pumped using another laser (typically a 5145-Å argon ion laser or a 5320-Å beam from a frequency-doubled Nd^{3+}:YAG laser).

The processes that are critical to lasing in an organic dye [2] are summarized in Fig. 9.5. Absorption of a visible photon creates a vibrationally excited S_1 molecule. In aqueous or alcoholic solution, the excess vibrational energy is lost within several picoseconds to the medium, leaving an S_1 molecule whose vibrational energy distribution is thermally equilibrated. Laser action may then occur, terminating in S_0 molecules with varying degrees of vibrational excitation. To obtain $S_1 \leftarrow S_0$ energy gaps appropriate for visible laser transitions, one must resort to molecules with extended π electron systems (alternating single and double bonds) which are larger than those in any of the ultraviolet-absorbing molecules we considered in Chapter 7. A typical size is exhibited by the rhodamine 6G cation (Problem 8.2), which has 64 atoms and therefore 186

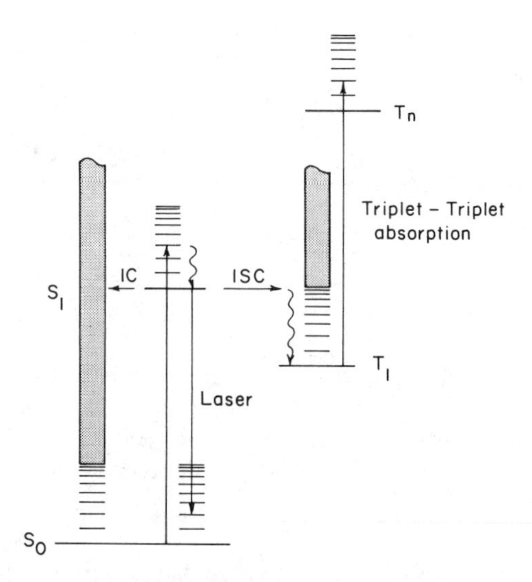

Figure 9.5 Energy levels for an organic dye laser. IC and ISC denote internal conversion and intersystem crossing, respectively (Chapter 7). Wavy arrows indicate vibrational relaxation. S_0 and S_1 are singlet electronic states, while T_1 and T_n are triplet states.

vibrational modes. In view of the resulting potential for spectral congestion, it is no surprise that this dye's $S_1 \leftarrow S_0$ absorption and fluorescence spectra are vibrationally unresolved continua. It is this feature that endows a dye laser with its most conspicuous asset, continuous tunability.

Two of the processes which detract from lasing, $S_1 \rightarrow S_0$ internal conversion (IC) and $S_1 \rightarrow T_1$ intersystem crossing (ISC), are shown in Fig. 9.5. IC depletes the S_1 population, reducing the laser gain. The consequences of ISC are often more serious. Since the T_1 state is metastable (the $T_1 \rightarrow S_0$ phosphorescence transition is spin-forbidden), molecules accumulate in T_1 if the ISC quantum yield is substantial. Many dyes exhibit large $T_1 \rightarrow T_n$ absorption coefficients at the wavelengths of the $S_1 \rightarrow S_0$ laser transition, due to transitions from T_1 to some higher triplet state T_n. Such triplet–triplet absorption can dramatically reduce the laser gain. Since rapid IC and/or ISC occur in the vast majority of organic compounds (Section 7.4), a survey of some one thousand dyes that were commercially available in 1969 gleaned only four that proved useful in dye lasers [2]. Most of the dyes that are currently used fall into three structural classes: xanthenes, oxazines, and coumarins.

It has been empirically established [2] that the presence of torsional modes (intramolecular rotations about bonds) accelerates $S_1 \rightarrow S_0$ IC in dyes. For example, the dye phenolphthalein has essentially zero fluorescence yield, while fluorescein has 90% yield (Fig. 9.6): The torsional modes in phenolphthalein are frozen out by the presence of the oxygen bridge in fluorescein. In the absence of such torsional modes, most of the $S_1 \leftarrow S_0$ electronic energy difference in IC is

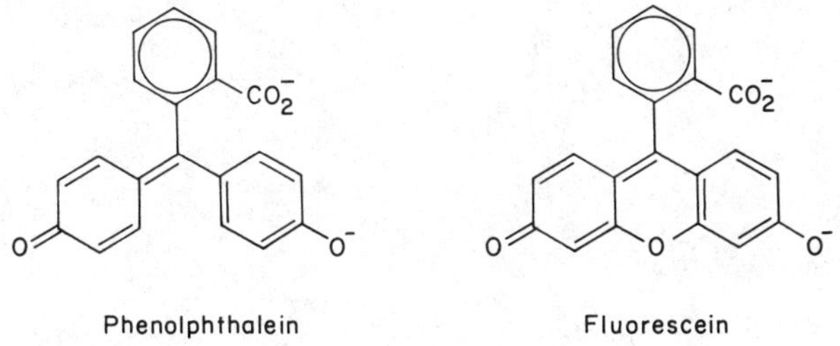

Phenolphthalein

Fluorescein

Figure 9.6.

absorbed as S_0 state vibrational energy in high-frequency C–H stretching modes. (It may be shown that the IC probability is larger if this energy difference is deposited into a smaller number of vibrational quanta in the lower electronic state [3]. For a fixed energy gap, this implies that higher frequency "accepting" modes are more effective in promoting IC.) For this reason, the presence of more H atoms attached to the chromophore will reduce the fluorescence yield (and also the lasing efficiency). Dyes frequently show larger fluorescence yields in deuterated solvents, because proton exchange with the solvent replaces the C–H modes with lower frequency C–D modes via the isotope effect.

Intersystem crossing will obviously be accelerated by heavy-atom substitutients, and no efficient laser dye contains them. The four bromine atoms in eosin (Fig. 9.7) increase its $S_1 \rightarrow T_1$ ISC yield to 76%, as compared to the 3% value observed in fluorescein. Solvents containing heavy atoms (e.g., CBr_4) also contribute to T_1 buildup in laser dyes, and are avoided.

Drexhage [2] has formulated a remarkable rule that relates the ISC rates in dyes to the topologies of their π-electron structures. If the ring atoms that contribute to the π-electron structure form a closed loop of adjacent sites, the molecule exhibits a higher ISC yield than if the loop is broken by the presence of a ring atom uninvolved the π-electron system. This rule can be rationalized semiclassically by noting that the orbital angular momentum in dyes with such closed loops will produce larger spin–orbit coupling and enhanced $S_1 \rightarrow T_1$ yields. As an example, acridine dyes (Fig. 9.8) have not materialized as a class of laser dyes. The two resonance structures shown in this figure both exhibit the

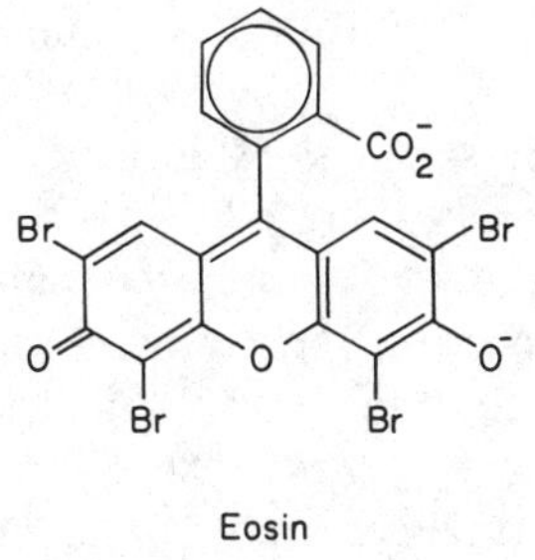

Eosin

Figure 9.7.

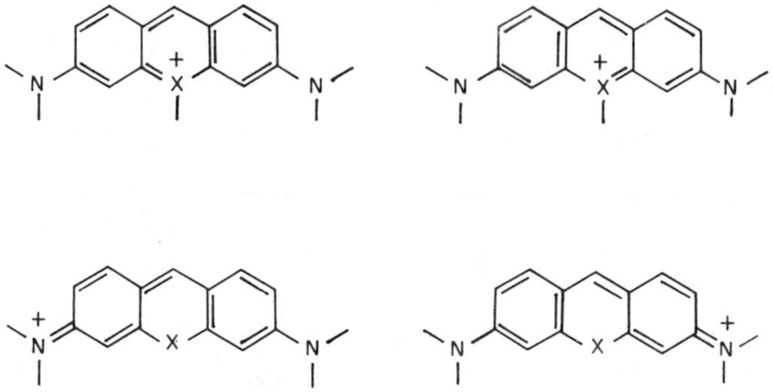

Figure 9.8 Resonance structures contributing to π electron densities in acridine dyes ($X = N$) and rhodamine dyes ($X = O$). The resonance structures containing $=X^+<$ are dominant in acridines, but are less favorable in rhodamines.

ammonium group $=N^+<$. Both structures have comparable weighting in the π-electron structure of acridines, so that considerable π-electron density resides on the bridging N atom. In the important rhodamine laser dyes, the bridging group $=N^+<$ in the second resonance structure is replaced by the energetically less favored oxonium group $=O^+-$. Appreciably less π-electron density is then found on the bridging atom in rhodamines than in acridines, with the result that the latter show far higher $S_1 \rightarrow T_1$ yields. Likewise, carbazine dyes (in which the bridging atom is a tetrahedral C atom, Fig. 9.9) exhibit very low triplet yields.

Many laser dyes (xanthenes and oxazines) are employed as cations in polar solvents, and the S_1 state in such dyes may be quenched by electron transfer from the negatively charged counterion. The quenching rate decreases with the identity of the counterion as $I^- > Br^- > Cl^- > ClO_4^-$. Perchlorate is therefore the counterion of choice, particularly at higher concentrations in nonpolar solvents, where rapid diffusion occurs between the dye and counterion.

In the xanthene dyes (Fig. 9.9), the absorption and lasing wavelengths are sensitive to the substituents on the xanthene chromophore, with the result that a continuous range of lasing wavelengths is accessible between ~540 and ~650 nm through appropriate choice of dye. (The carboxyphenyl group common to all rhodamine dyes is not part of the rigid xanthene chromophore, and it has little effect on its lasing properties.) Longer lasing wavelengths (630–750 nm) are available using oxazine dyes, in which the $=\overset{|}{C}-$ bridge group in xanthenes is supplanted by $=N-$. Since the $S_1 \leftarrow S_0$ energy gap is smaller in the oxazines than in the xanthenes, IC is more problematic in the oxazines, and the use of deuterated solvents improves their lasing efficiency. In both the xanthene and oxazine dyes, the $S_1 \leftarrow S_0$ transition moment is polarized along the long molecular axis. By symmetry, neither the S_1 nor S_0 state in these dye chromophores exhibits a permanent dipole moment.

In contrast, the coumarin dyes, while relatively nonpolar in the ground state,

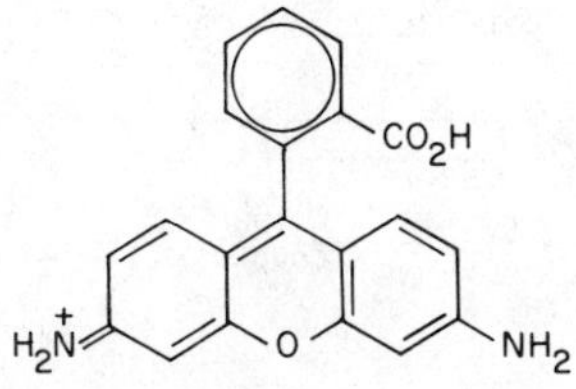

Rhodamine 110 (560)

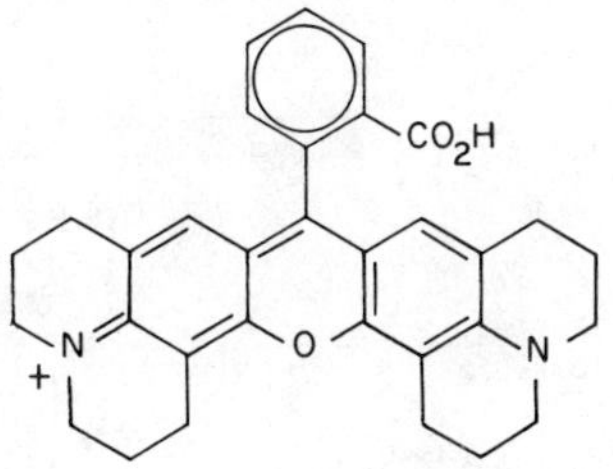

Rhodamine 101 (640)

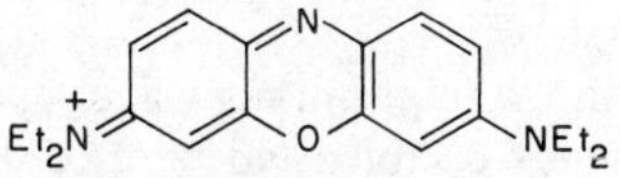

Oxazine 1

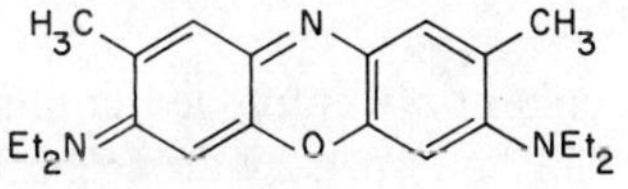

Oxazine 4

Figure 9.9 Typical xanthene and oxazine dyes.

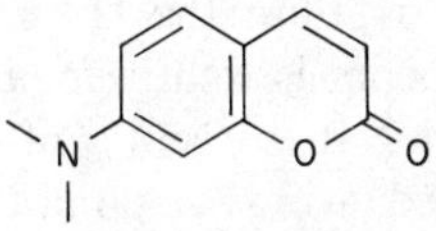

S_0

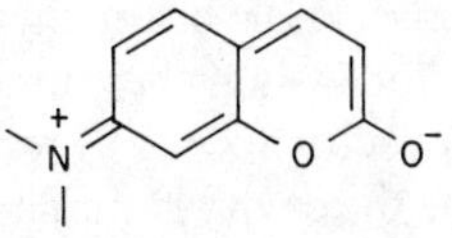

S_1

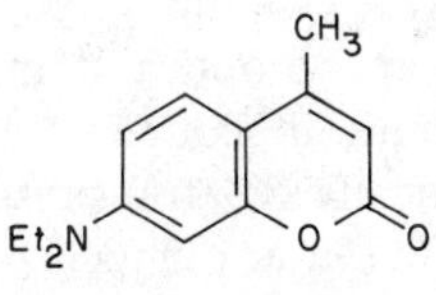

Coumarin 1

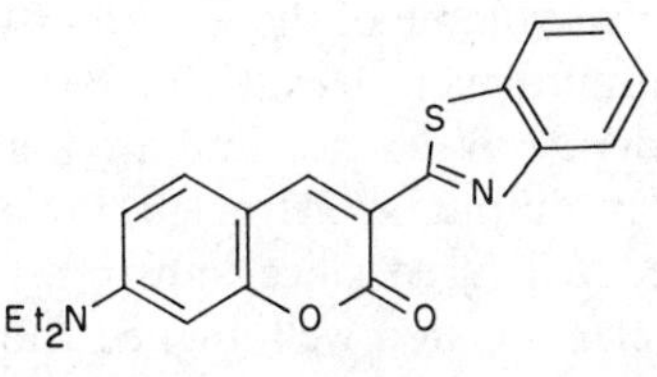

Coumarin 6

Figure 9.10 Dominant resonance structures for S_0 and S_1 electronic states in coumarin dyes, and two typical coumarins.

are highly polar in the S_1 state (Fig. 9.10). Excitation of a coumarin S_1 state in a polar solvent (e.g., methanol) then produces a sudden increase in the molecule's permanent dipole moment, followed by rapid solvent dipole reorganization [2]. This lowers the S_1 state energy relative to that in nonpolar solvents, and decreases the $S_1 \leftarrow S_0$ energy separation; the result is a large wavelength shift (Stokes shift) between the absorption and fluorescence maxima. For example, the absorption and lasing maxima of coumarin 1 in methanol are at 373 and 460 nm. Owing to their smaller chromophore size, the coumarins lase at shorter wavelengths (440 to 540 nm) than the xanthenes and oxazines.

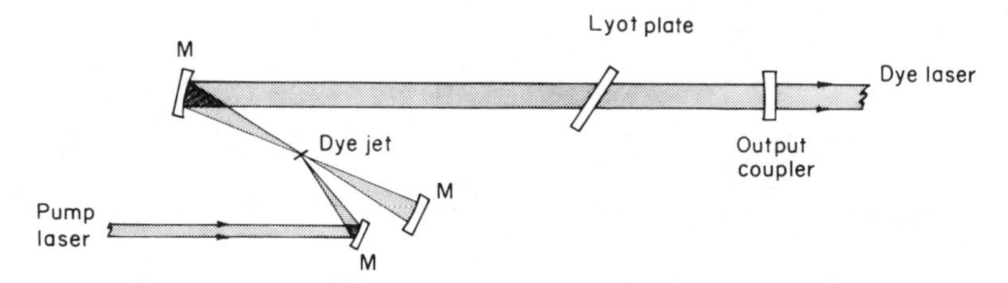

Figure 9.11 Folded-cavity configuration for dye laser pumped by an argon ion or frequency-doubled Nd^{3+}:YAG laser. The optics denoted *M* are >99.9% reflecting at the incident wavelength. The dye jet is flowed along an axis perpendicular to the plane of the paper. The Lyot plate provides wavelength tunability (see text).

A commonly employed configuration for a cw (continuous-wave) dye laser externally pumped by an argon ion or frequency-doubled Nd^{3+}:YAG laser is shown in Fig. 9.11. A rhodamine dye solution in ethylene glycol (a viscous solvent) is flowed in a jet stream ~0.25 nm thick at ~7 m/s. The use of such a jet obviates the thermal inhomogeneities, window damage, and extensive dye photochemical degradation that would attend exposure of a stationary dye solution to an intense laser beam in a cell. To attain the large population inversions requisite for lasing, the pump laser beam is tightly focused using a spherical concave reflector to ~0.01 mm diameter at the center of the jet. The dye laser beam inside the cavity is similarly focused, using two confocal spherical reflectors (i.e., reflectors whose focal points coincide at the center of the jet) to provide good spatial overlap between the amplified dye laser beam and the small pumped volume of the jet. All of these spherical surfaces are coated with specialized multilayer dielectric materials, which are essentially 100% reflecting at the incident wavelengths and are highly resistant to optical damage.

The laser output is extracted through the output coupler, a multilayer dielectric-coated fused silica substrate with typically 5% transmission at the lasing wavelength. In a well-aligned rhodamine 6G laser pumped by a 5145-Å cw argon ion laser, 25–30% of the optical pump power can be converted into dye laser output. Coarse wavelength tuning can be effected by incorporating a prism into the dye cavity (Fig. 9.4), or by including a diffraction grating as one of the cavity elements. The current tuning element of choice is a Lyot plate, aligned with its surface normal at the Brewster angle θ_B [4] with respect to the laser propagation axis as shown in Fig. 9.11. A dye laser beam polarized in the plane of the Lyot plate (which is made of birefringent single-crystal quartz) and incident on the plate at θ_B will be partly reflected off the plate and partly refracted into the plate. (This polarization, termed the *S* polarization, corresponds to polarization along an axis perpendicular to the paper in Fig. 9.11.) Laser light with the orthogonal polarization (*P* polarization) incident at θ_B will be totally refracted through the plate. Since a typical laser photon experiences many round-trip passes in the cavity prior to exiting through the output coupler, the presence of these Brewster-angle surfaces therefore strongly favors

P polarization in the laser beam. The crucial property of the Lyot plate is that it rotates the polarization of all wavelengths traversing the plate, with the exception of those wavelengths λ obeying

$$d(n_o - n_e) = n\lambda \qquad n = 1, 2, \ldots \tag{9.18}$$

Here n_o, n_e are the ordinary and extraordinary refractive indices [4] of the birefringent plate, and d is the plate thickness. Since a laser beam that is consistently P-polarized during each cavity pass is strongly favored, the Lyot plate effectively suppresses lasing at wavelengths that fail to satisfy condition (9.18). When the plate is rotated about its normal, the wavelengths that fulfill Eq. 9.18 become shifted. Hence a Lyot plate (or a stack of two or three Lyot plates with different thickness) can afford continuous tuning of the laser output across the entire dye gain bandwidth.

9.3 AXIAL MODE STRUCTURE AND SINGLE MODE SELECTION

Provided the loss coefficient γ is accurately given (taking into account all cavity loss mechanisms), lasing will occur a priori for all frequencies for which inequality (9.10) is satisfied. The resulting lasing bandwidth, shown graphically in Fig. 9.12, extends over all frequencies for which the gain coefficient $[-\alpha(\nu)]$ exceeds γ/L. A new constraint is now posed by the physical boundary condition that the laser oscillation must have nodes at the surfaces of the cavity end mirrors. This means that an integral number of laser half-wavelengths must fit

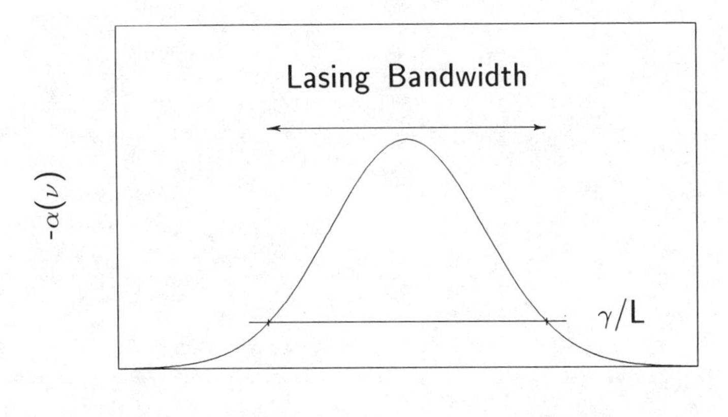

Figure 9.12 Dye laser gain curve $-\alpha(\nu)$ versus laser frequency ν. Lasing occurs for frequencies satisfying $-\alpha(\nu) > \gamma/L$.

into the cavity's *optical* path length, so that

$$\frac{n\lambda}{2} = L' \equiv \int n(s)ds \qquad n = 1, 2, 3, \ldots \tag{9.19}$$

Here $n(s)$ is the refractive index along the light propagation path in the cavity, and the line integral is taken over a single traversal of the cavity (one-half round trip). For a laser medium with refractive index n and physical length L_1 situated in air between cavity mirrors separated by physical length $L_2 > L_1$, the optical path length is $L' = nL_1 + (L_2 - L_1)$. Equation 9.19 can be rearranged using $\lambda = c/v$ to give the allowed *longitudinal* (or *axial*) *mode* frequencies [1]

$$v_n = nc/2L' \qquad n = 1, 2, 3, \ldots \tag{9.20}$$

These axial mode frequencies are equally spaced with separation $\Delta v \equiv v_{n+1} - v_n = c/2L' = 1/T$, where T is the time required for one cavity round-trip. In the absence of tuning elements, the actual lasing frequencies will be those which simultaneously satisfy both Eqs. 9.10 and 9.20, and all axial modes with frequencies v for which $-\alpha(v) > \gamma/L$ will lase (Fig. 9.13). The shape of the gain curve $-\alpha(v)$ depends on the operative line-broadening mechanism: Doppler broadening in He/Ne or Ar^+ lasers, lattice broadening in solid-state Nd^{3+}:YAG lasers, and so forth.

The axial mode frequencies themselves are not infinitely sharp. The widths of the individual modes depend on how long a typical laser photon stays inside the cavity. In consequence of lifetime broadening, we have

$$\Delta E\, \Delta t \geqslant h/2 \tag{9.21}$$

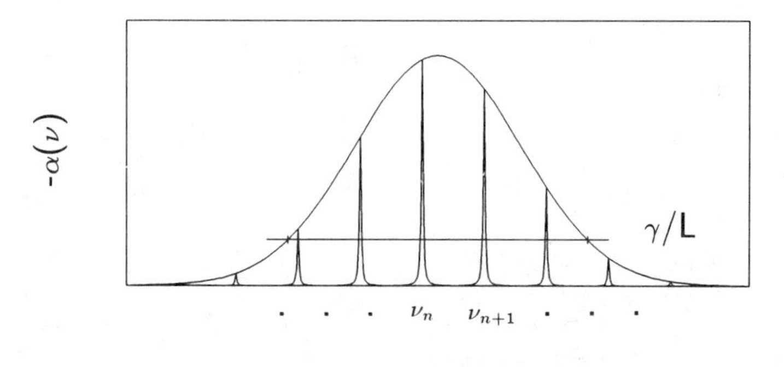

Figure 9.13 Dye laser gain curve $-\alpha(v)$, with axial mode frequencies $v_n = nc/2L'$ superimposed. Lasing occurs in axial modes v_n for which $-\alpha(v_n) > \gamma/L$. Axial mode profiles are represented as Lorentzian profiles for realism, since they are broadened by finite photon lifetimes in the cavity.

and so we can deduce that since $\Delta E = h\,\Delta\nu = hc\,\Delta\bar{\nu}$ for photons,

$$t_p \Delta\nu \geqslant 1/4\pi \tag{9.22}$$

where $t_p = \Delta t$ is the photon lifetime inside the cavity. The photon lifetime t_p is controlled by the cavity loss coefficient γ. Cavity losses tend to deplete the cavity photon population by a factor of $\exp(-\gamma)$ per cavity pass; by the definition of t_p, this factor can be equated with $\exp(-T/2t_p)$, where $T = 2L'/c$ is the round-trip time. (One round-trip equals two cavity passes.) Hence the photon lifetime is

$$t_p = T/2\gamma = L'/c\gamma \tag{9.23}$$

According to Eq. 9.22, the order of magnitude of $\Delta\bar{\nu}$ is then

$$\Delta\bar{\nu} \gtrsim 1/4\pi c t_p = \gamma/4\pi L' \tag{9.24}$$

for the width of individual axial modes in cm^{-1}.

As an example, we consider a laser with 1-m optical path length and cavity end reflectors with reflectivities $r_1 = 1.00, r_2 = 0.95$ (these are typical values for a rear (high) reflector and an output coupler reflector respectively in a practical laser). If cavity reflector losses dominate the loss coefficient,

$$\gamma = -\tfrac{1}{2}\ln[(1)(0.95)] = +0.025 \tag{9.25}$$

and so the individual mode width will be the order of $\Delta\bar{\nu} \gtrsim 2.5 \times 10^{-5}\ cm^{-1}$.

This suggests a potential for compressing the intense laser output into a remarkably narrow (by classical spectroscopy standards!) bandwidth of $< 10^{-4}\ cm^{-1}$, if the laser oscillation can be limited to just one of these axial modes. In fact, this is readily achieved by incorporating an etalon, in effect an additional pair of partially mirrored parallel reflectors separated by the etalon length L_e, into the cavity (Fig. 9.14). All lasing frequencies must then simultaneously satisfy two sets of axial mode boundary conditions

$$\begin{aligned} \nu_n &= nc/2L' \qquad & n &= 1, 2, 3, \ldots \\ \nu_{n'} &= n'c/2L_e \qquad & n' &= 1, 2, 3, \ldots \end{aligned} \tag{9.26}$$

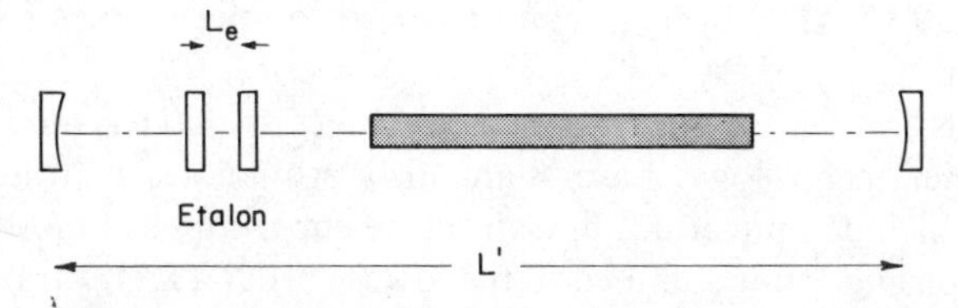

Figure 9.14 Laser cavity of optical path length L' augmented with an etalon, whose partially reflecting surfaces are separated by a precisely variable distance L_e.

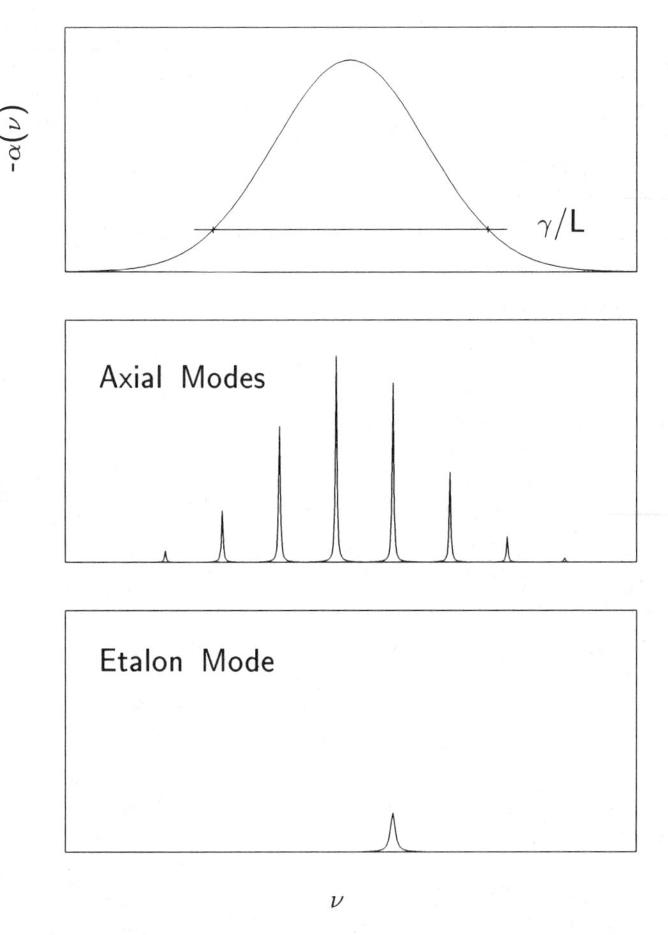

Figure 9.15 Laser gain curve $-\alpha(\nu)$, axial mode frequencies $\nu_n = nc/2L'$, and an etalon mode frequency $\nu_{n'} = n'c/2L_e$. Since $L_e \ll L'$, other etalon modes are too widely spaced from $\nu_{n'}$ to appear in this frequency range. Lasing occurs selectively at the axial mode which overlaps the etalon mode. Tuning the etalon separation L_e controls the axial mode selected.

The etalon spacing L_e can be adjusted in practice so that only one axial cavity mode within the gain bandwidth $-\alpha(\nu) > \gamma/L$ satisfies both of Eqs. 9.18, as shown in Fig. 9.15. Moreover, L_e can be varied to select successive cavity axial modes at will.

This discussion barely touches on the topics of cavity modes and high-resolution laser technology. Laser light also propagates in transverse cavity modes [5], which can introduce fine structure superimposed on the axial mode frequencies if lasing is not confined to the lowest-order TEM_{00} transverse mode. Ring dye lasers are currently the most widely used frequency-stabilized high-resolution lasers.

9.4 MODE-LOCKING AND ULTRASHORT LASER PULSES

We now concentrate on the time dependence of laser oscillation in a cavity whose gain curve encompasses a large number of axial modes. Each axial mode amplitude will oscillate with a time dependence of the form $\exp[i\omega_n(t - x/c) + i\phi_n]$, with circular frequency

$$\omega_n = 2\pi\nu_n = \pi nc/L' \qquad n = 1, 2, 3, \ldots \tag{9.27}$$

ϕ_n is the phase of oscillation for mode n. The total amplitude of laser oscillation will then behave as

$$\mathbf{E}(t) = \sum_n \mathbf{E}_n e^{i\omega_n(t - x/c) + i\phi_n} \tag{9.28}$$

where the amplitude factors $\mathbf{E}_n$ reflect the weighting of the laser gain curve at the axial mode frequencies ω_n. Equation 9.28 describes the time dependence of a randomly spaced sequence of light pulses, since $\mathbf{E}(t)$ is a superimposition of different frequency components added together with random phases ϕ_n.

We now consider what happens when all of the axial modes are forced to oscillate at the *same* phase, say ϕ_0. For simplicity, we assume that $(2k + 1)$ modes oscillate with identical amplitude $\mathbf{E}_0$, and that the equally spaced mode frequencies run from $\omega = \omega_0 - k\,\Delta\omega$ to $\omega = \omega_0 + k\,\Delta\omega$, with $\Delta\omega = 2\pi\,\Delta\nu = \pi c/L' = 2\pi/T$. The total oscillation amplitude then simplifies into

$$\begin{aligned}\mathbf{E}(t) &= \mathbf{E}_0 e^{i\phi_0} \sum_{n=-k}^{k} e^{i(\omega_0 + n\Delta\omega)(t - x/c)} \\ &= \mathbf{E}_0 e^{i[\omega_0(t - x/c) + \phi_0]} \frac{\sin[(k + \frac{1}{2})\Delta\omega(t - x/c)]}{\sin[\frac{1}{2}\Delta\omega(t - x/c)]}\end{aligned} \tag{9.29}$$

The intensity of the associated light wave is then

$$|\mathbf{E}(t)|^2 = |\mathbf{E}_0|^2 \frac{\sin^2[(k + \frac{1}{2})\Delta\omega(t - x/c)]}{\sin^2[\frac{1}{2}\Delta\omega(t - x/c)]} \tag{9.30}$$

This function is periodic, with period T equal to the cavity round-trip time (Fig. 9.16); it corresponds physically to the fact that *one* light pulse is propagating back and forth inside the cavity at all times. Since part of this pulse is transmitted outside the cavity everytime it strikes the output coupler reflector, the laser output consists of a train of pulses equally spaced in time by T. The zeros in $|\mathbf{E}(t)|^2$ on either side of the primary pulse peaks are separated by the duration $2T/(2k + 1)$, which gives an upper bound estimate of the laser pulse width τ_p. For a 1-m optical path length cavity, the round-trip time T is $2L'/c = 6.67$ ns. If 9 axial modes are forced to oscillate in phase with equal

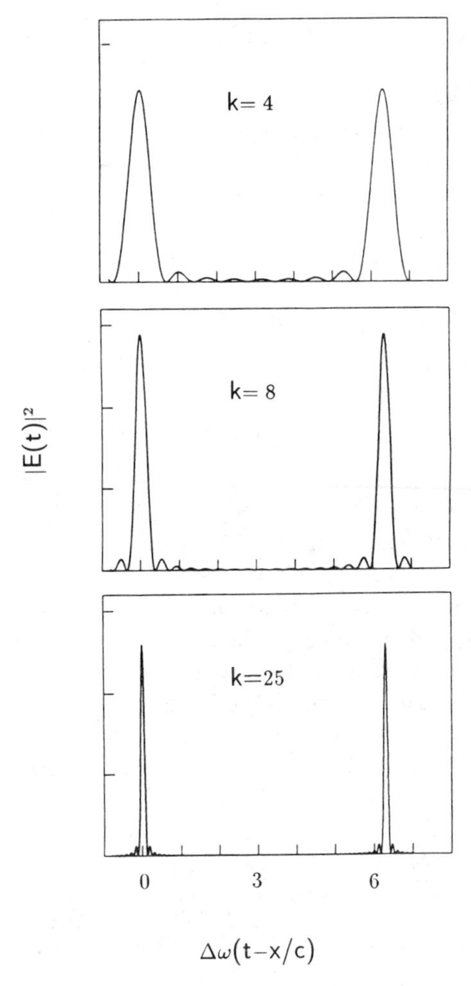

Figure 9.16 Graphs of the periodic time-dependent function $|\mathbf{E}(t)|^2$ for $k=4$, 8, and 25, corresponding to $(2k+1) = 9$, 17, and 51 axial modes locked in the same phase with equal amplitude. Note the laser pulse sharpening which occurs as the number of locked modes increases.

amplitude, the pulse widths are on the order of 1.5 ns; for 51 phase-locked modes, the pulse widths would be reduced to 260 ps. This trend is illustrated in Fig. 9.16. Since the pulse widths depend on the number of locked modes $2k + 1$ via $\tau_p \sim 2T/(2k + 1)$, generation of extremely short laser pulses has become the province of mode-locked solid-state and dye lasers, whose broad gain bandwidths can permit simultaneous phase-locked oscillation in thousands of axial modes.

Mode-locking does not occur spontaneously in a simple laser cavity. Either it must be actively driven by a cavity element which introduces cavity losses with a period of exactly $T/2$ (i.e., one-half the optical round-trip time), or it must be passively induced by an intracavity nonlinear absorber [6] which discourages lasing at phases other than some phase ϕ_0 at which strong lasing initially occurs. In an active acousto-optic [7] mode-locker, coarse wavelength selection of the laser output is provided by an interactivity triangular prism placed near the rear mirror (Fig. 9.4). A thin layer of piezoelectric material is deposited on one of the triangular faces. An oscillating voltage applied to the piezoelectric creates a mechanical stress, which is transmitted to the prism in the form of a standing longitudinal acoustic wave if the driven wave frequency matches a prism resonance frequency. The mechanical rarefactions and compressions induced in the prism by the standing acoustic wave produce a spatial alternation in refractive index, creating a transient diffraction grating which deflects the laser beam from its cavity path when the voltage is applied to the piezoelectric. By proper synchronization of the applied voltage frequency with the cavity round-trip time, the longitudinal modes can thus be forced to oscillate in phase. Acousto-optic mode lockers are commonly used in argon ion lasers, which can operate in several strong visible lines (notably 4880 and 5145 Å). The 5145-Å line is generally selected with the tuning prism; Doppler broadening of this line in the argon plasma tube allows some 40 axial modes to lase under the gain bandwidth curve in a 1-m cavity. Active mode-locking of such a laser typically produces laser pulses with ~ 500 ps fwhm.

It is beyond the scope of the present chapter to review the technology and capabilities of mode-locked lasers. The currently favored systems for picosecond pulse generation are mode-locked Nd^{3+}:YAG lasers (which afford 1.06-μm pulses ~ 15 ps fwhm and may be wavelength-converted by using their 5320-Å second-harmonic pulses to pump tunable dye lasers), although dye lasers pumped by mode-locked Ar^+ lasers are still widely used. Pulses from Nd^{3+}:YAG-based systems have been compressed to less than 1 ps fwhm in optical fibers. The very shortest pulses now reported have been generated in passively mode-locked Ar^+-pumped colliding-pulse-mode ring dye lasers, which have yielded pulses as short as 40 fs wide (1 fs $= 10^{-15}$ s).

REFERENCES

1. B. A. Lengyel, *Lasers*, 2d ed., Wiley-Interscience, New York, 1971.
2. K. H. Drexhage, in Dye Lasers, *Springer-Verlag Topics in Applied Physics*, Vol. 1, *Dye Lasers*, F. P. Schäfer (Ed.), Springer-Verlag, Berlin, 1973.
3. P. Avouris, W. M. Gelbart, and M. A. El-Sayed, *Chem. Rev.* **77**: 793 (1977).
4. E. Hecht and A. Zajac, *Optics*, Addison-Wesley, Reading, MA, 1976; M. V. Klein, *Optics*, Wiley, New York, 1970; M. Born and E. Wolf, *Principles of Optics*, Pergamon, Oxford, 1970.
5. A. E. Siegman, *An Introduction to Lasers and Masers*, McGraw-Hill, New York, 1971.
6. D. J. Bradley, in *Springer-Verlag Topics in Applied Physics*, Vol. 18, *Ultrashort Light Pulses*, S. L. Shapiro (Ed.), Springer-Verlag, Berlin, 1977.
7. A. Yariv, *Quantum Electronics*, 2d ed., Wiley, New York, 1975.

PROBLEMS

1. As an exercise in evaluating criteria for lasing in an idealized system, consider the $3^2P_{3/2} \rightarrow 3^2S_{1/2}$ transition in a Na atom. The radiative lifetime of the $3^2P_{3/2}$ levels is 5×10^{-8} s; the photon energy for the transition is 16,978 cm^{-1}. It is proposed to explore the possibility of lasing in a uniform 10-cm cavity bounded by end reflectors with $r_1 = 1.00$ and $r_2 = 0.98$. Assume that the translational temperature in the Na vapor is 300 K, and that no cavity losses other than transmission losses at the end reflectors are operative.

(a) Determine the population inversion $(g_1/g_2)N_2 - N_1$ required for lasing in this system; include units. How is the answer changed if the translational temperature is increased to 600 K?
(b) What are the most fundamental problems that limit the practicality of such a laser?

2. Several compounds from the limitless roster of organic species that cannot serve as useful laser dyes are listed below. For each of these, indicate the most important physical reason(s) why the molecule is an unsuitable laser dye candidate. Consider only the $S_1 \rightarrow S_0$ transitions.

(a) Naphthalene
(b) Aniline
(c) Rosamine 4
(d) Dithiofluorescein
(e) Acridine
(f) Iodoanthracene

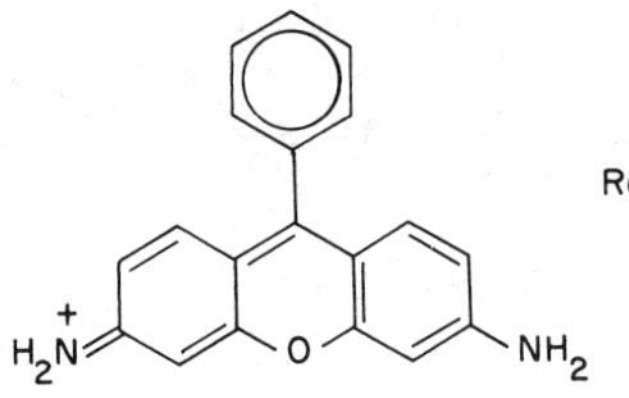

Rosamine 4

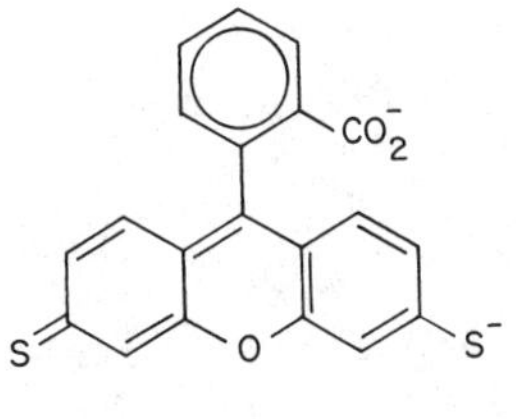

Dithiofluorescein

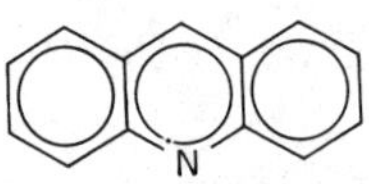

Acridine

3. A dye laser 0.5 m long is operated in a single axial mode with end reflectors characterized by $r_1 = 1.00$, $r_2 = 0.95$. The single-mode output bandwidth is $10^{-3}\,\text{cm}^{-1}$. Is this bandwidth limited by end reflector losses? What effect would doubling the cavity length L' have on the output bandwidth if the cavity losses are dominated by r_2? If the cavity losses are uniformly distributed along L' (e.g., through diffraction losses?)

4. An etalon is used for single-mode selection in a 1-m rhodamine 6G laser. If the dye gain bandwidth is commensurate with the width of the rhodamine 6G fluorescence spectrum shown in Problem 8.2, what etalon separations L_e and etalon surface reflectivities would ensure that only one axial mode is selected at any time?

5. A 0.75-m solid-state Nd^{3+}:YAG laser is acousto-optically mode-locked to yield ultrashort pulses centered at $\bar{\nu}_L = 9416\ \text{cm}^{-1}$.

(a) Assuming that the lasing bandwidth function is given by

$$|\alpha(\bar{\nu}) + \gamma/L'| = C, \qquad |\bar{\nu} - \bar{\nu}_L| \leq 0.5\,\text{cm}^{-1}$$
$$= 0, \qquad |\bar{\nu} - \bar{\nu}_L| > 0.5\,\text{cm}^{-1}$$

where C is a positive constant, how many axial modes will lase? What pulse duration will result from perfect mode-locking in this laser? What will be the time separation between adjacent pulses?

(b) Suppose now that the lasing bandwidth function is given by a Gaussian function of $\bar{\nu}$, centered at $\bar{\nu}_L$ with an fwhm of $1\ \text{cm}^{-1}$. How are the answers in part (a) qualitatively charged?

10

TWO-PHOTON PROCESSES

Up to now, we have been primarily concerned with one-photon absorption and emission processes, whose probability amplitudes are given by the first-order term

$$c_m^{(1)}(t) = \frac{1}{i\hbar} \int_{t_0}^{t} e^{-i\omega_{km}t_1} \langle m|W(t_1)|k\rangle dt_1 \tag{10.1}$$

in the time-ordered perturbation expansion (1.96). We have seen that evaluation of the time integral (10.1) in the cw limit $t_0 \to -\infty, t \to +\infty$ leads to a statement of the one-photon Ritz combination principle $E_m - E_k = \hbar\omega$, where ω is the circular frequency of the applied radiation field (Eq. 1.112). The discussions of oscillator strengths and radiative lifetimes in Chapter 8 proceeded from the assumption that one-photon processes accounted for all spectroscopic transitions of interest.

Many radiative transitions cannot be treated under the framework of one-photon processes. Raman transitions (which are two-photon processes) were discovered by Raman and Krishnan in 1928; evidence for two-photon absorption and more exotic multiphoton phenomena accumulated rapidly after the introduction of lasers in the 1960s. Some of the characteristics of two-photon processes are illustrated by the Raman spectra of *p*-difluorobenzene (Fig. 10.1). These spectra were generated by exposing the pure liquid or vapor to a nearly monochromatic cw beam from either a He/Ne or an argon ion laser, and analyzing the wavelengths of light scattered by the sample at a right angle from the laser beam. They are plotted as scattered light intensity versus the difference $\omega - \omega'$ between incident and scattered frequencies. *p*-Difluorobenzene exhibits

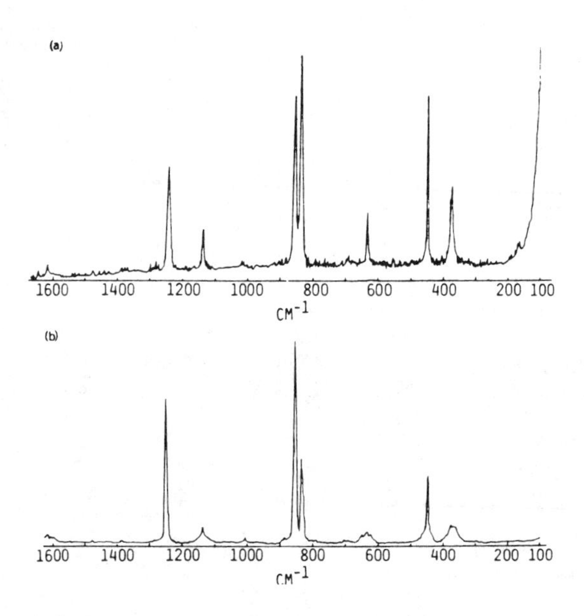

Figure 10.1 Raman spectra of *p*-difluorobenzene (a) pure liquid and (b) vapor, recorded as light intensity $I(\omega')$ scattered at frequency ω' versus the difference $(\omega - \omega')$ between incident and scattered frequencies. The spectra excited using an argon ion laser (4880 Å) and a He/Ne laser (6328 Å) are nearly identical. Used with permission from R. L. Zimmerman and T. M. Dunn, *J. Mol. Spectrosc.* **110**; 312 (1985).

an $S_1 \leftarrow S_0$ electronic spectrum with an origin band at 2713.5 Å in the near ultraviolet, and is practically transparent at the visible He/Ne and Ar^+ laser wavelengths (6328 and 4880 Å, respectively). Photons at the scattered frequencies ω' are produced essentially *instantaneously* (within $\lesssim 1$ fs) upon disappearance of incident photons at frequency ω. Consequently, this process cannot be interpreted as a sequence of one-photon absorption and emission steps. A one-photon absorptive transition with an oscillator strength of ~ 1 in the UV-visible would populate an excited state with a radiative lifetime on the order of ns (Chapter 8). In a *p*-difluorobenzene molecule subjected to a visible laser, the emergence of photon ω' would typically be delayed by a far longer time if it followed the (extremely weak) one-photon absorption process at 6328 or 4880 Å.

The frequencies $\omega - \omega'$ of the Raman lines in Fig. 10.1 prove to be independent of the excitation laser frequency ω, and analysis shows that they are equal to vibrational energy level separations in S_0 *p*-difluorobenzene. This is an example of the energy conservation law $\hbar(\omega - \omega') = E_m - E_k$ in Raman spectroscopy: An incident photon with energy $\hbar\omega$ interacts with the molecule, a transition occurs from level $|k\rangle$ to level $|m\rangle$, and a scattered photon emerges with a shifted energy $\hbar\omega'$ that compensates for the energy gained or lost by the molecule (Fig. 10.2). When $\omega > \omega'$, the process is called a Stokes Raman

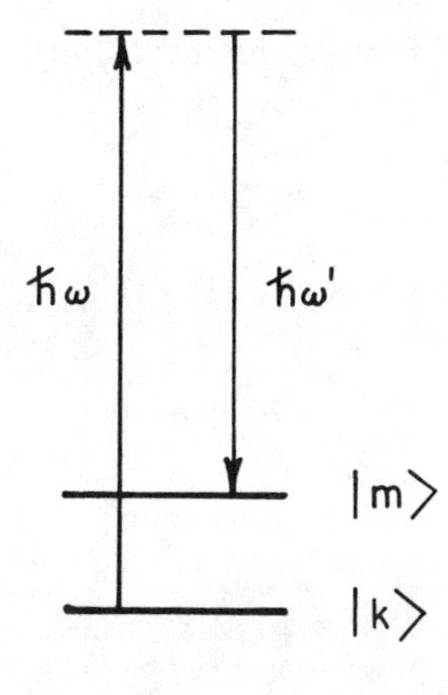

Figure 10.2 Energy level diagram for Raman scattering. A photon is incident at frequency ω and a photon is scattered at frequency ω'; the energy difference $\hbar(\omega - \omega')$ matches a molecular level separation $E_m - E_k$. The dashed line corresponds to a *virtual state,* which need not coincide with any eigenstate of the molecule (Section 10.1).

transition; when $\omega < \omega'$, it is an anti-Stokes transition. We will see that Stokes transitions are generally stronger than anti-Stokes transitions, and only the Stokes portions of the *p*-difluorobenzene spectra are reported in Fig. 10.1. Detailed study of these spectra reveals that some of the Raman lines (e.g., the lines at 3084, 859, 636, and 376 cm^{-1} in the liquid spectrum) are fundamentals in vibrations of a_g, b_{2g}, and b_{3g} symmetry in the D_{2h} point group. Such fundamentals are symmetry-forbidden in one-photon vibrational spectroscopy (Chapter 6). This illustrates the value of vibrational Raman spectroscopy for characterizing vibrational modes that are spectrally dark in the infrared. Raman spectra have also been used to probe rotational and (less frequently) electronic structure.

The other important two-photon process is two-photon absorption (TPA), in which two photons are simultaneously absorbed and a molecule is promoted from some state $|k\rangle$ to a higher-energy state $|m\rangle$. The selection rules in TPA are different from those in one-photon absorption, and TPA has proved fruitful in identifying electronic states that are inaccessible to conventional electronic spectroscopy.

10.1 THEORY OF TWO-PHOTON PROCESSES

The probabilities of two-photon $|k\rangle \to |m\rangle$ transitions are controlled by the second-order coefficients

$$c_m^{(2)}(t) = \frac{1}{(i\hbar)^2} \sum_n \int_{t_0}^{t} e^{-i\omega_{nm}t_1} \langle m|W(t_1)|n\rangle dt_1 \times \int_{t_0}^{t_1} e^{-i\omega_{kn}t_2} \langle n|W(t_2)|k\rangle dt_2 \tag{10.2}$$

from the Dyson expansion (1.96). We may allow for the presence of two different radiation fields with vector potentials $\mathbf{A}_1(\mathbf{r}, t)$ and $\mathbf{A}_2(\mathbf{r}, t)$ in the Coulomb gauge by setting

$$W(t_1) = \frac{i\hbar q}{mc} \mathbf{A}_2(\mathbf{r}, t_1)\cdot\nabla$$

$$W(t_2) = \frac{i\hbar q}{mc} \mathbf{A}_1(\mathbf{r}, t_2)\cdot\nabla \tag{10.3}$$

If one lets

$$\begin{aligned}\mathbf{A}_1(\mathbf{r}, t_2) &= \mathbf{A}_1 \cos(\mathbf{k}_1\cdot\mathbf{r} - \omega_1 t_2)\\ &= \frac{\mathbf{A}_1}{2}\{\exp(i(\mathbf{k}_1\cdot\mathbf{r} - \omega_1 t_2)] + \exp[-i(\mathbf{k}_1\cdot\mathbf{r} - \omega_1 t_2)]\}\\ &\equiv \frac{\mathbf{A}_1}{2}(\mathbf{r})\exp(-i\omega_1 t_2) + \frac{\mathbf{A}_1}{2}(-\mathbf{r})\exp(i\omega_1 t_2)\end{aligned} \tag{10.4}$$

and similarly treats $\mathbf{A}_2(\mathbf{r}, t_1)$, we have

$$\begin{aligned}\langle n|W(t_2)|k\rangle &= \frac{i\hbar q}{2mc}\langle n|\mathbf{A}_1(\mathbf{r})\cdot\nabla|k\rangle e^{-i\omega_1 t_2} + \frac{i\hbar q}{2mc}\langle n|\mathbf{A}_1(-\mathbf{r})\cdot\nabla|k\rangle e^{i\omega_1 t_2}\\ &\equiv \frac{i\hbar q}{2mc}(\alpha_{nk}e^{-i\omega_1 t_2} + \bar{\alpha}_{nk}e^{i\omega_1 t_2})\end{aligned} \tag{10.5}$$

and

$$\begin{aligned}\langle m|W(t_1)|n\rangle &= \frac{i\hbar q}{2mc}\langle m|\mathbf{A}_2(\mathbf{r})\cdot\nabla|n\rangle e^{-i\omega_2 t_1} + \frac{i\hbar q}{2mc}\langle m|\mathbf{A}_2(-\mathbf{r})\cdot\nabla|n\rangle e^{+i\omega_2 t_1}\\ &\equiv \frac{i\hbar q}{2mc}(\alpha_{mn}e^{-i\omega_2 t_1} + \bar{\alpha}_{mn}e^{i\omega_2 t_1})\end{aligned} \tag{10.6}$$

We finally obtain

$$\begin{aligned}c_m^{(2)}(t) = \frac{q^2}{4m^2c^2}\sum_n \int_{t_0}^{t} e^{-i\omega_{nm}t_1}(\alpha_{mn}e^{-i\omega_2 t_1} + \bar{\alpha}_{mn}e^{i\omega_2 t_1})dt_1\\ \times \int_{t_0}^{t_1} e^{-i\omega_{kn}t_2}(\alpha_{nk}e^{-i\omega_1 t_2} + \bar{\alpha}_{nk}e^{i\omega_1 t_2})dt_2\end{aligned} \tag{10.7}$$

as the second-order contribution to $c_m(t)$. This summation contains four cross terms for each n. Their interpretations will become clear as we develop Eq. 10.7

farther, and we list them for reference below:

Term	Process
$\alpha_{mn}\alpha_{nk}$	Two-photon absorption
$\bar{\alpha}_{mn}\alpha_{nk}$	Raman (Stokes if $\omega_1 > \omega_2$)
$\alpha_{mn}\bar{\alpha}_{nk}$	Raman (anti-Stokes if $\omega_1 < \omega_2$)
$\bar{\alpha}_{mn}\bar{\alpha}_{nk}$	Two-photon emission

These processes can also be visualized in the same order using qualitative energy level diagrams in Fig. 10.3. The dashed lines in this figure denote *virtual states*, which are not generally true eigenstates of the molecular Hamiltonian unless one of the radiation field frequencies ω_1, ω_2 happens to be tuned to one of the molecular energy level differences. All of these two-photon processes are effectively instantaneous, and the virtual states do not exhibit measurable lifetimes. A second way [1] of visualizing these processes, which appears to be cumbersome for displaying these (relatively) simple second-order phenomena but which proves to be valuable in sorting out still higher order processes like second-harmonic generation (Chapter 11), is to use time-ordered graphs (Fig. 10.4). The time coordinate in these graphs is vertical, pointing upwards. The photons are represented by wavy lines. The vertical lines, which are divided into segments labeled k, n, and m, identify the molecular states that are occupied at various times; the center portions of these lines denote the time intervals during which the molecule is in the virtual state labeled n. The state of the system at any time t can thus be inferred by noting which portion of the vertical line and which (wavy) photon line(s) intersect the horizontal line representing time t. In the first diagram (corresponding to two-photon absorption), there are two photons, $(\mathbf{k}_1, \omega_1)$ and $(\mathbf{k}_2, \omega_2)$, and the molecule is in the initial state $|k\rangle$ at time t_a. By time

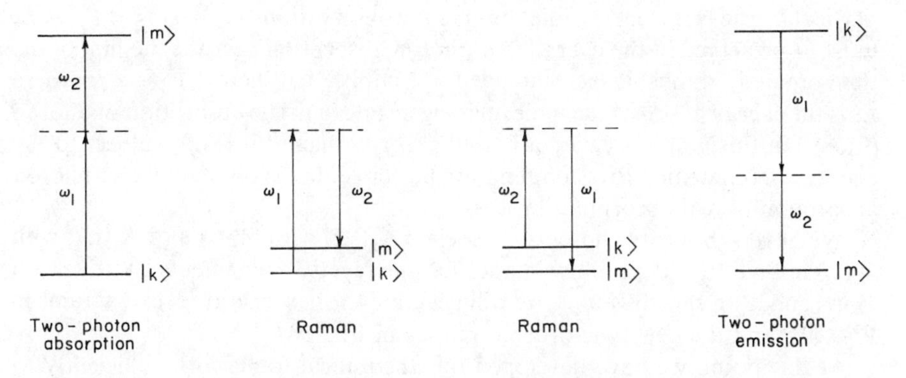

Figure 10.3 Energy level diagrams representing the four contributions to $c_m^{(2)}(t)$ when the perturbation matrix elements are given by Eqs. 10.5 and 10.6.

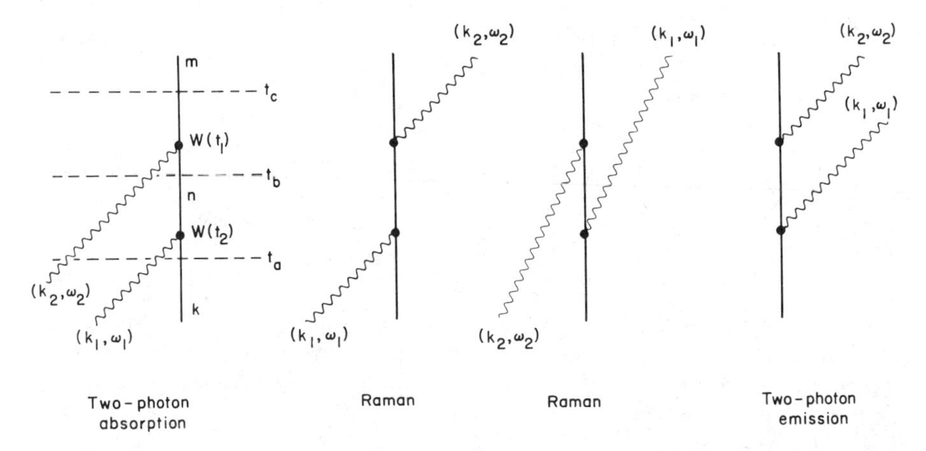

Figure 10.4 Time-ordered graphs corresponding to the four contributions to $c_m^{(2)}(t)$ when the perturbation matrix elements are given by Eqs. 10.5 and 10.6. These are shown in the same order as the energy level diagrams in Fig. 10.3.

t_b, the molecule has undergone a transition to virtual state $|n\rangle$ by virtue of the radiation-molecule interaction $W(t_2)$, and only the $(\mathbf{k}_2, \omega_2)$ photon remains unabsorbed. At time t_c the molecule has reached its final state $|m\rangle$ as a result of the interaction between virtual state $|n\rangle$ and the $(\mathbf{k}_2, \omega_2)$ photon via coupling by the $W(t_1)$ term. The intersections of the photon lines with the vertical lines, which are labeled with interaction Hamiltonian terms like $W(t_1)$ or $W(t_2)$, are called interaction vertices.

The role implied by these time-ordered graphs for the virtual states called "$|n\rangle$" should not be taken too literally. In the treatment that follows, these virtual states are in effect expanded in infinite series of true molecular eigenstates $|n\rangle$, and no virtual state in any of the processes will coincide with any single, particular true $|n\rangle$. Hence, while the energy conservation $\hbar\omega_1 + \hbar\omega_2 = E_m - E_k$ must be preserved in the overall two-photon absorption process, the first of the time-ordered graphs is *not* intended to imply that $\hbar\omega_1 = E_n - E_k$, where E_n is the energy of some true molecular eigenstate $|n\rangle$. The absorption of photon $(\mathbf{k}_1, \omega_1)$ in this graph is called a *virtual* absorption, and it is not subject to the energy level-matching Ritz combination principle that is obeyed by one-photon absorption (a real absorption process).

We have arbitrarily chosen to associate $\mathbf{A}_2(\mathbf{r}, t)$ with $W(t_1)$ and $\mathbf{A}_1(\mathbf{r}, t)$ with $W(t_2)$ in Eqs. 10.3. If we allow in addition the reverse assignments [$\mathbf{A}_1(\mathbf{r}, t)$ with $W(t_1)$ and $\mathbf{A}_2(\mathbf{r}, t)$ with $W(t_2)$], we will generate the new energy level diagrams in Fig. 10.5 and the new time-ordered graphs in Fig. 10.6.

At this point, we have developed our theoretical framework sufficiently to deal explicitly with TPA and Raman spectroscopy. Spontaneous two-photon emission (which is depicted by the last of each set of time-ordered graphs in Figs.

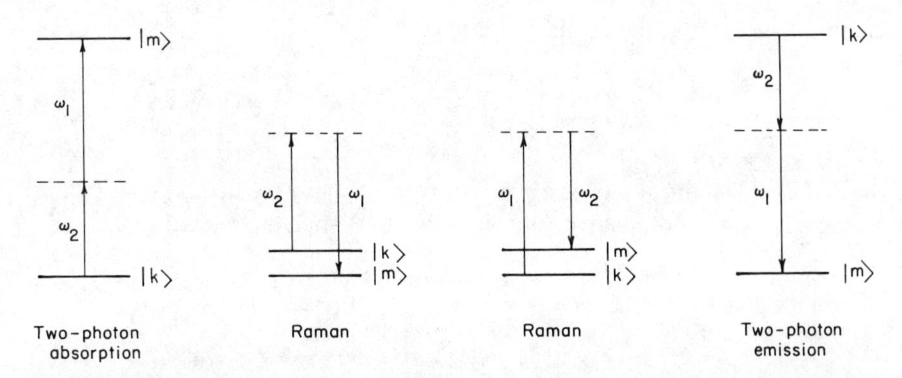

Figure 10.5 Energy level diagrams for four additional contributions to $c_m^{(2)}(t)$, generated by associating $\mathbf{A}_1$ with $W(t_1)$ and $\mathbf{A}_2$ with $W(t_2)$.

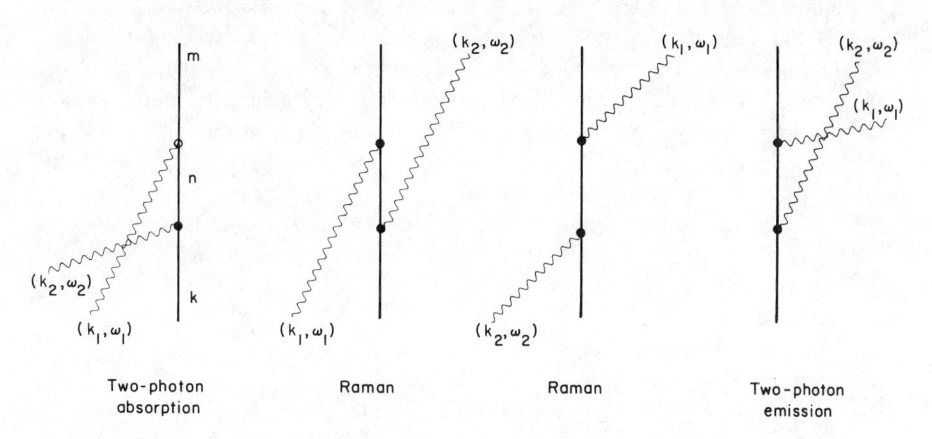

Figure 10.6 Time-ordered graphs corresponding to the energy level diagrams in Fig. 10.5.

10.4 and 10.6) exhibits transition rates far smaller than those of E1-allowed one-photon emission [1], and has not been detected. It is likely to contribute to decay in astrophysical systems in which one-photon decay is E1-forbidden.

10.2 TWO-PHOTON ABSORPTION

We now develop the terms pertinent to TPA in Eq. 10.7. They become

$$c_m^{\mathrm{TPA}}(t) = \frac{q^2}{4m^2c^2} \sum_n \alpha_{mn}\alpha_{nk} \int_{t_0}^{t} dt_1 e^{-i(\omega_{nm}+\omega_2)t_1} \int_{t_0}^{t_1} dt_2 e^{-i(\omega_{kn}+\omega_1)t_2} \tag{10.8}$$

Setting $t_0 = -\infty$ in the cw limit, we have

$$c_m^{\mathrm{TPA}}(t) = \frac{q^2}{4m^2c^2} \sum_n \alpha_{mn}\alpha_{nk} \int_{-\infty}^{t} dt_1 e^{-i(\omega_{nm}+\omega_2)t_1} \int_{-\infty}^{t_1} dt_2 e^{-i(\omega_{kn}+\omega_1)t_2} \quad (10.9)$$

To make the last integral on the right converge, we may replace ω_{kn} by $(\omega_{kn} + i\varepsilon)$, where ε is small and positive, and then let $\varepsilon \to 0$ after the integration:

$$\begin{aligned}\int_{-\infty}^{t_1} dt_2 e^{-i(\omega_{kn}+\omega_1+i\varepsilon)t_2} &= \left.\frac{-e^{-i(\omega_{kn}+\omega_1+i\varepsilon)t_2}}{i(\omega_{kn}+\omega_1+i\varepsilon)}\right|_{-\infty}^{t_1} \\ &= \frac{-e^{-i(\omega_{kn}+\omega_1+i\varepsilon)t_1}}{i(\omega_{kn}+\omega_1+i\varepsilon)} \xrightarrow[\varepsilon\to 0]{} \frac{-e^{-i(\omega_{kn}+\omega_1)t_1}}{i(\omega_{kn}+\omega_1)} \end{aligned} \quad (10.10)$$

This is more than just a mathematical artifice. This substitution is tantamount to replacing the energy E_n by $(E_n - i\hbar\varepsilon)$, so that the intermediate state $|n\rangle$ exhibits the time dependence $\exp(-iE_n t/\hbar - \varepsilon t)$ and hence physically decays with lifetime $1/2\varepsilon$. The constant ε can be identified with $\gamma/4$, where γ is the Lorentzian linewidth (Chapter 8). Such linewidths are generally much smaller than level energies E_n, so that dropping ε at the end of the integration yields good approximations to $c_m^{(2)}(t)$ in Eq. 10.10. Next, we have

$$c_m^{\mathrm{TPA}}(t) = -\frac{q^2}{4m^2c^2} \sum_n \alpha_{mn}\alpha_{nk} \int_{-\infty}^{t} dt_1 e^{-i(\omega_{km}+\omega_1+\omega_2)t_1} \frac{1}{i(\omega_{kn}+\omega_1)} \quad (10.11)$$

and so in the cw limit

$$\begin{aligned} c_m^{\mathrm{TPA}}(\infty) &= -\frac{2\pi q^2}{4im^2c^2} \sum_n \frac{\alpha_{mn}\alpha_{nk}}{(\omega_{kn}+\omega_1)} \delta(\omega_{km}+\omega_1+\omega_2) \\ &= \frac{-2\pi q^2}{4im^2c^2} \sum_n \frac{\langle m|\mathbf{A}_2(\mathbf{r})\cdot\nabla|n\rangle\langle n|\mathbf{A}_1(\mathbf{r})\cdot\nabla|k\rangle}{\omega_{kn}+\omega_1} \delta(\omega_{km}+\omega_1+\omega_2) \end{aligned} \quad (10.12)$$

In the E1 approximation (Chapter 1), this is equivalent to

$$c_m^{\mathrm{TPA}}(\infty) \propto \sum_n \frac{\mathbf{E}_2\cdot\langle m|\boldsymbol{\mu}|n\rangle\langle n|\boldsymbol{\mu}|k\rangle\cdot\mathbf{E}_1}{\omega_{kn}+\omega_1} \delta(\omega_{km}+\omega_1+\omega_2) \quad (10.13)$$

where $\mathbf{E}_1$ and $\mathbf{E}_2$ are the electric vectors of the incident light waves $(\mathbf{k}_1, \omega_1)$ and $(\mathbf{k}_2, \omega_2)$. Since the roles of vector potentials $\mathbf{A}_1$ and $\mathbf{A}_2$ can be reversed in TPA and $c_m(t)$ must exhibit symmetry reflecting this fact, we finally conclude that

$$c_m^{\mathrm{TPA}}(\infty) \propto \sum_n \left(\frac{\mathbf{E}_2\cdot\langle m|\boldsymbol{\mu}|n\rangle\langle n|\boldsymbol{\mu}|k\rangle\cdot\mathbf{E}_1}{\omega_{kn}+\omega_1} + \frac{\mathbf{E}_1\cdot\langle m|\boldsymbol{\mu}|n\rangle\langle n|\boldsymbol{\mu}|k\rangle\cdot\mathbf{E}_2}{\omega_{kn}+\omega_2} \right) \times \delta(\omega_{km}+\omega_1+\omega_2) \quad (10.14)$$

Since the delta function in Eqs. 10.13–10.14 is proportional to $\delta[E_k - E_m + \hbar(\omega_1 + \omega_2)]$, it yields the obvious energy-conserving criterion $(E_m - E_k) = \hbar(\omega_1 + \omega_2)$ relevant to TPA. It is thus clear that the terms included in Eq. 10.8 are the ones associated with TPA, and furthermore that the particular physical processes connected with the other terms in Eq. 10.7 may be identified by examining the signs of the time-dependent exponential arguments.

Equation 10.14 describes the TPA transition amplitude for a molecule subjected to two light beams with arbitrary electric field vectors and propagation vectors. A particularly useful application of TPA in gas phase spectroscopy employs two counterpropagating laser beams with $\mathbf{k}_1 \cdot \mathbf{k}_2 = -|\mathbf{k}_1||\mathbf{k}_2|$. In this case, a molecule traveling with velocity $\mathbf{v}_x$ parallel to $\mathbf{k}_2$ will experience Doppler shifts

$$\frac{\omega_1 - \omega_1^0}{\omega_1^0} = v_x/c \qquad \frac{\omega_2 - \omega_2^0}{\omega_2^0} = -v_x/c \tag{10.15}$$

in the frequencies ω_1, ω_2 relative to the frequencies ω_1^0, ω_2^0 experienced by a molecule at rest (Fig. 10.7). The total energy absorbed in a transition involving photons $(\mathbf{k}_1, \omega_1)$ and $(\mathbf{k}_2, \omega_2)$ will then be proportional to

$$\begin{aligned}\omega_{12} &= \omega_1 + \omega_2 = \omega_1^0 + \omega_2^0 + \frac{v_x}{c}(\omega_1^0 - \omega_2^0) \\ &\equiv \omega_{12}^0 + \frac{v_x}{c}(\omega_1^0 - \omega_2^0)\end{aligned} \tag{10.16}$$

or

$$\omega_{12} - \omega_{12}^0 = \frac{v_x}{c}(\omega_1^0 - \omega_2^0) \tag{10.17}$$

The Gaussian absorption profile that results from Doppler broadening of the TPA transition probability as a function of ω_{12} will then be

$$P(\omega_{12}) = P_0 e^{-mv_x^2/2kT} = P_0 e^{-mc^2(\omega_{12} - \omega_{12}^0)^2/2kT(\omega_1^0 - \omega_2^0)^2} \tag{10.18}$$

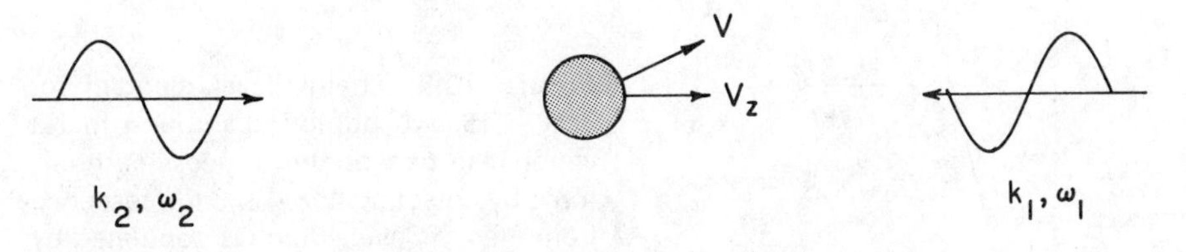

Figure 10.7 Two-photon absorption in a molecule subjected to counterpropagating light beams $(\mathbf{k}_1, \omega_1)$ and $(\mathbf{k}_2, \omega_2)$ directed along the x axis.

This profile exhibits a full width at half-maximum

$$\text{fwhm} = \frac{2|\omega_1^0 - \omega_2^0|}{c}\left(\frac{2kT\ln 2}{m}\right)^{1/2} \tag{10.19}$$

where m is the molecular mass. If photons of identical frequency are used ($\omega_1^0 = \omega_2^0$), the Doppler broadening cancels between the counterpropagating photons, and one nominally obtains zero fwhm. (In practice, one will still observe lifetime and possibly other residual broadening effects.) Doppler-free TPA spectroscopy is the most practical means of obtaining high-resolution absorption spectra in thermal gases.

The $3^2S \rightarrow 5^2S$ transition in Na vapor provided the now-famous prototype system for observing Doppler-free TPA [2, 3]. The one-photon $3^2S \rightarrow 5^2S$ transition, which would occur at 301.11 nm, is E1-forbidden ($\Delta l = 0$). If two counterpropagating photons with identical wavelength $\lambda = 602.23$ nm are used, we have $\omega_1 = \omega_2 \equiv \omega$ and $\mathbf{k}_2 = -\mathbf{k}_1$ (Fig. 10.8). The leading contributions to the $3^2S \rightarrow 5^2S$ TPA transition amplitude will then be

$$\begin{aligned} c_m^{\text{TPA}}(\infty) &\propto \frac{\mathbf{E}_2\cdot\langle 5^2S|\boldsymbol{\mu}|3^2P\rangle\langle 3^2P|\boldsymbol{\mu}|3^2S\rangle\cdot\mathbf{E}_1 + \mathbf{E}_1\cdot\langle 5^2S|\boldsymbol{\mu}|3^2P\rangle\langle 3^2P|\boldsymbol{\mu}|3^2S\rangle\cdot\mathbf{E}_2}{E_{3P} - E_{3S} - \hbar\omega} \\ &+ \frac{\mathbf{E}_2\cdot\langle 5^2S|\boldsymbol{\mu}|4^2P\rangle\langle 4^2P|\boldsymbol{\mu}|3^2S\rangle\cdot\mathbf{E}_1 + \mathbf{E}_1\cdot\langle 5^2S|\boldsymbol{\mu}|4^2P\rangle\langle 4^2P|\boldsymbol{\mu}|3^2S\rangle\cdot\mathbf{E}_2}{E_{4P} - E_{3S} - \hbar\omega} \\ &+ \frac{\mathbf{E}_2\cdot\langle 5^2S|\boldsymbol{\mu}|5^2P\rangle\langle 5^2P|\boldsymbol{\mu}|3^2S\rangle\cdot\mathbf{E}_1 + \mathbf{E}_1\cdot\langle 5^2S|\boldsymbol{\mu}|5^2P\rangle\langle 5^2P|\boldsymbol{\mu}|3^2S\rangle\cdot\mathbf{E}_2}{E_{5P} - E_{3S} - \hbar\omega} \\ &+ \cdots \end{aligned} \tag{10.20}$$

The intermediate states $|n\rangle$ are restricted to the m^2P states (with $m \geqslant 3$) by the E1 selection rule $\Delta l = \pm 1$ in each of the matrix elements of $\boldsymbol{\mu}$. Contributions

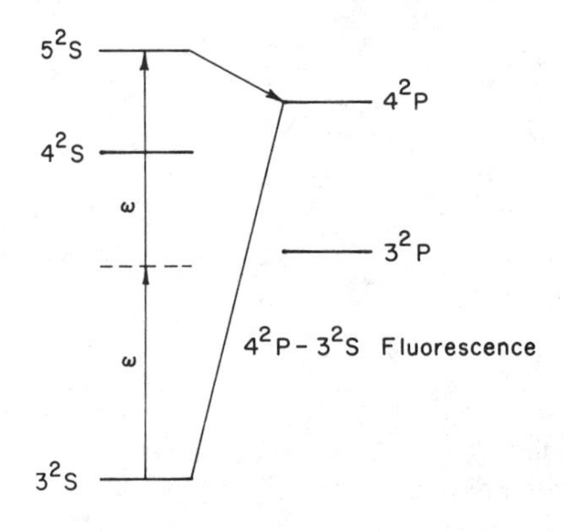

Figure 10.8 Energy level diagram for $3^2S \rightarrow 5^2S$ two-photon absorption in Na vapor. The two-photon process is monitored by detecting $4^2P \rightarrow 3^2S$ fluorescence from the 4^2P level, which is populated by cascading from 5^2S atoms created by two-photon absorption.

from the 6^2P, 7^2P states, etc. will be smaller than those in Eq. 10.20, because the energy denominator ($E_{mP} - E_{3S} - \hbar\omega$) increases with m. In one of the earliest Na vapor TPA experiments, a N_2 laser-pumped rhodamine B dye laser provided linearly polarized pulses at 602.23 nm. These laser pulses were passed through a thermal Na vapor cell, and then reflected backward by a mirror, causing them to collide inside the cell with later pulses passing through the cell for the first time. The $3^2S \rightarrow 5^2S$ TPA transition was detected by monitoring $4^2P \rightarrow 3^2S$ fluorescence from the 4^2P Na atoms, generated by cascading from the 5^2S atoms created by TPA; this particular fluorescence transition in Na occurs at a visible wavelength that is easily monitored by conventional phototubes.

The elimination of Doppler broadening in this experiment allows the clear observation of *hyperfine structure* that arises from the interaction of electronic and nuclear angular momenta. The total atomic angular momentum is

$$\mathbf{F} = \mathbf{L} + \mathbf{S} + \mathbf{I} \qquad (10.21)$$

where $\mathbf{I}$ is the nuclear spin angular momentum. In ^2S states of ^{23}Na, $L = 0$, $I = \frac{3}{2}$ and $S = \frac{1}{2}$, so that the possible F values are

$$F = I + S, \ldots, |I - S| = 2, 1 \qquad (10.22)$$

in both the 3^2S and 5^2S states. The splitting between the $F = 1, 2$ sublevels is larger in the 3^2S than the 5^2S state (as might be expected because the 5^2S orbital is more diffuse and has less electron probability density near the nucleus). It can be shown that the selection rule on ΔF is $\Delta F = 0$ in TPA [2]; thus *two* transitions will be observed ($F = 1 \rightarrow 1$ and $F = 2 \rightarrow 2$ as shown in Fig. 10.9) at

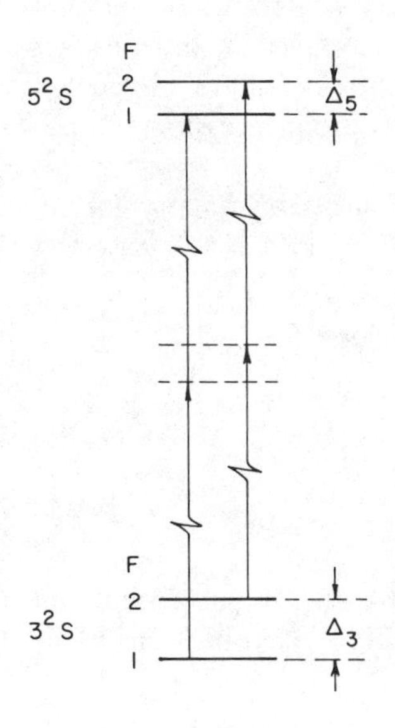

Figure 10.9 Detailed energy level diagram for $3^2S \rightarrow 5^2S$ two-photon absorption in Na, showing splitting of the n^2S levels into hyperfine components with $F = 1, 2$. Dashed lines indicate virtual states.

single-photon frequencies separated by

$$2\,\Delta\omega = \Delta_3 - \Delta_5 \quad \text{or} \quad \Delta\omega = \tfrac{1}{2}(\Delta_3 - \Delta_5) = 2.6 \times 10^{-2}\ \text{cm}^{-1} \tag{10.23}$$

(the factor of 2 is required here because this is a two-photon transition). The actual TPA spectrum obtained this way is shown in Fig. 10.10. The hyperfine components exhibit approximately a 5:3 intensity ratio, because 5 and 3 are the degeneracies of the $F = 2$ and 1 sublevels. The resolved hyperfine peaks result from Doppler-free TPA of photons travelling in opposite directions. The broad background in which these Doppler-free peaks are superimposed arises from TPA of pairs of photons traveling in the *same* direction, since nothing in this apparatus can present TPA of copropagating photons. This background can be removed using circularly polarized photons, however (Fig. 10.11).

It is instructive to touch briefly on the TPA spectroscopy of benzene [4] since we have discussed its one-photon absorption spectroscopy in Chapter 7. For TPA from the $^1A_{1g}$ benzene ground state to some final vibronic state f,

$$c_m^{\text{TPA}}(\infty) \propto \sum_n \left(\frac{\mathbf{E}_2 \cdot \langle f|\boldsymbol{\mu}|n\rangle\langle n|\boldsymbol{\mu}|\text{A}_{1g}\rangle \cdot \mathbf{E}_1}{(E_n - E_{\text{A}_{1g}} - \hbar\omega_1)} + \frac{\mathbf{E}_1 \langle f|\boldsymbol{\mu}|n\rangle\langle n|\boldsymbol{\mu}|\text{A}_{1g}\rangle \cdot \mathbf{E}_2}{(E_n - E_{\text{A}_{1g}} - \hbar\omega_2)} \right) \tag{10.24}$$

For E1-allowed TPA, it is then necessary that both

$$\Gamma(n) \otimes \Gamma(\boldsymbol{\mu}) \otimes \text{A}_{1g}$$
$$\Gamma(f) \otimes \Gamma(\boldsymbol{\mu}) \otimes \Gamma(n)$$

simultaneously contain A_{1g} for some intermediate state $|n\rangle$. Since (x, y) and z transform as E_{1u} and A_{2u} in D_{6h}, respectively, the intermediate states $|n\rangle$ must have E_{1u} or A_{2u} symmetry. Consequently, the allowed symmetries of the final vibronic states $|f\rangle$ are A_{1g}, A_{2g}, E_{1g}, and E_{2g}. This exemplifies the obvious fact that the selection rules in TPA are anti-Laporte in centrosymmetric molecules, and that TPA can be used to study excited states that are inaccessible to one-photon absorption from the ground state.

While the $^1B_{2u}$ S_1 state in benzene has inappropriate symmetry for an intrinsically E1-allowed $S_1 \leftarrow S_0$ TPA transition, $S_1 \leftarrow S_0$ TPA is still observed due to vibronic coupling. Since

$$\text{B}_{2u} \otimes \begin{bmatrix} \text{b}_{2u} \\ \text{b}_{1u} \\ \text{e}_{2u} \\ \text{e}_{1u} \end{bmatrix} = \begin{bmatrix} \text{A}_{1g} \\ \text{A}_{2g} \\ \text{E}_{1g} \\ \text{E}_{2g} \end{bmatrix} \tag{10.25}$$

in D_{6h}, there are four symmetries of normal modes (b_{2u}, b_{1u}, e_{2u}, e_{1u}) that can serve as promoting modes in vibronically induced $S_1 \leftarrow S_0$ transitions. This

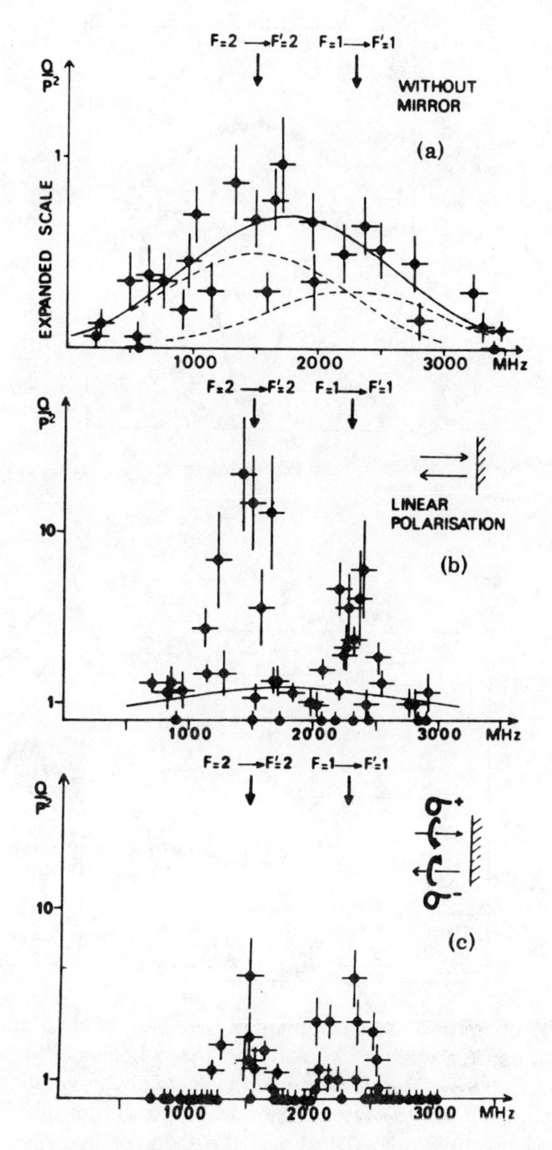

Figure 10.10 Two-photon absorption profiles in Na vapor, obtained using linearly and circularly polarized light beams. Profile (a) was obtained using only copropagating beams. Profiles (b) and (c) were obtained using counterpropagating beams that were linearly and circularly polarized, respectively. Reproduced by permission from F. Biraben, B. Cagnac, and G. Grynberg, *Phys. Rev. Lett.* **32**; 643 (1974).

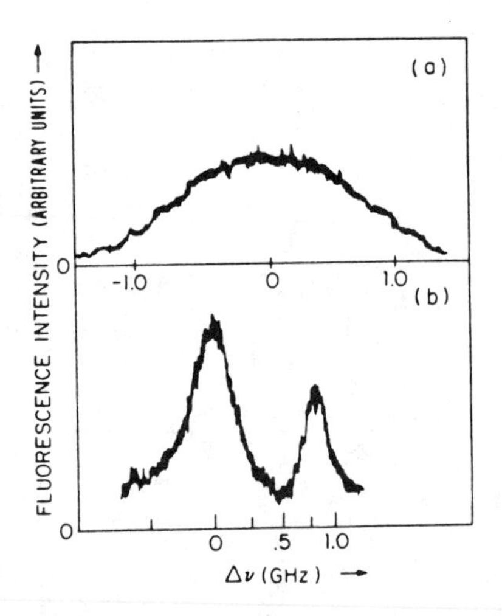

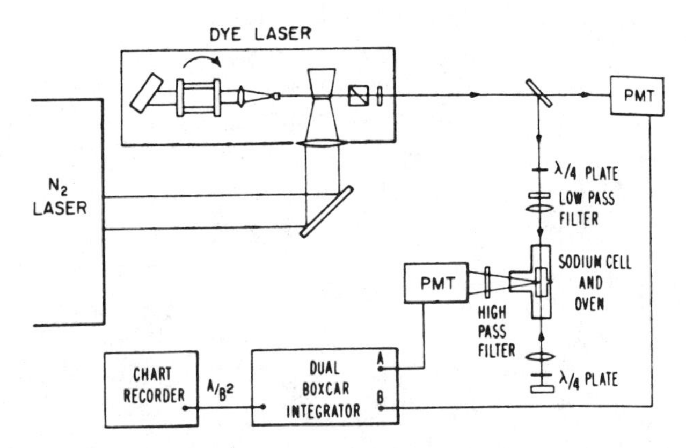

Figure 10.11 Apparatus for measurement of two-photon absorption profiles in Na vapor using counterpropagating circularly polarized beams. PMT denotes photomultiplier tube. The dye laser is wavelength-scanned by rotating an intracavity Fabry-Perot etalon. Profile (a) was obtained by two-photon absorption from one linearly polarized beam. Profile (b) shows the Doppler-free $F = 1 \rightarrow 1$ and $2 \rightarrow 2$ hyperfine peaks, obtained using counterpropagating circularly polarized beams. Used with permission from M. D. Levenson and N. Bloembergen, *Phys. Rev. Lett.* **32**, 645 (1974). Note that this work arrived in a dead heat with that of Biraben et al., Fig. 10.10.

situation contrasts with that in the one-photon spectrum, where vibrational modes of e_{2g} symmetry are required for the intensity-borrowing (Section 7.2).

10.3 RAMAN SPECTROSCOPY

In ordinary Raman scattering, we are concerned with the two-photon process whereby photon $(\mathbf{k}_1, \omega_1)$ is annihilated, photon $(\mathbf{k}_2, \omega_2)$ is created, and the molecule undergoes a transition from state $|k\rangle$ to state $|m\rangle$. Energy conservation requires that $\hbar(\omega_1 - \omega_2) = E_m - E_k$. The possible time-ordered graphs satisfying these conditions are shown in Fig. 10.12. These two graphs should not be regarded as physically distinct in the sense that the first graph depicts "absorption" of photon $(\mathbf{k}_1, \omega_1)$ followed by "emission" of photon $(\mathbf{k}_2, \omega_2)$, while the second graph depicts these events occurring in reverse sequence. They are simply a bookkeeping method for keeping track of different terms in the perturbation expression (10.7). (For that matter, the portions of the time lines labeled "n" in Fig. 10.12 are unresolvably short, and the interaction vertices labeled "$W(t_1)$" and "$W(t_2)$" coincide for practical purposes.) Using the terms associated with these two diagrams in Eq. 10.7, it is straightforward to show by a procedure similar to that carried out for TPA in Section 10.2 that the second-order probability amplitude for Raman scattering in the cw limit is

$$c_m^{\mathrm{R}}(\infty) \propto \sum_n \left(\frac{\mathbf{E}_2 \cdot \langle m|\boldsymbol{\mu}|n\rangle\langle n|\boldsymbol{\mu}|k\rangle \cdot \mathbf{E}_1}{\omega_{kn} + \omega_1} + \frac{\mathbf{E}_1 \cdot \langle m|\boldsymbol{\mu}|n\rangle\langle n|\boldsymbol{\mu}|k\rangle \cdot \mathbf{E}_2}{\omega_{kn} - \omega_2} \right)$$
$$\times\ \delta(\omega_{km} + \omega_1 - \omega_2) \tag{10.26}$$

For a given intermediate state $|n\rangle$, the first and second right-hand terms in Eq. 10.26 correspond to the first and second time-ordered graphs in Fig. 10.12, respectively. When $\omega_1 > \omega_2$, the scattered radiation frequency ω_2 is said to be

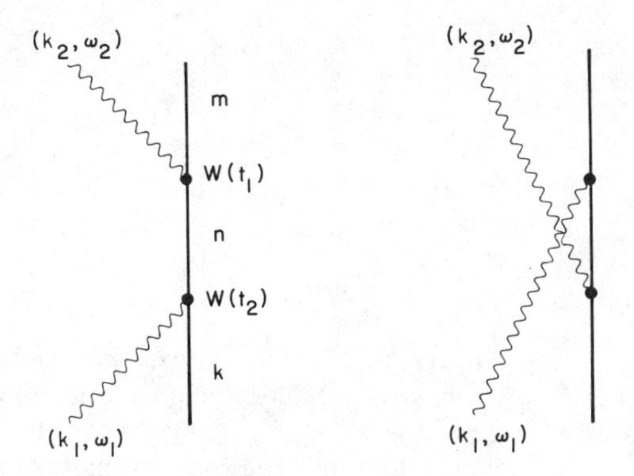

Figure 10.12 Time-ordered graphs for the Raman process with incident frequency ω_1 and scattered frequency ω_2.

Stokes-shifted from the incident frequency ω_1; anti-Stokes scattering is obtained when $\omega_1 < \omega_2$. The energy difference $\hbar(\omega_2 - \omega_1)$ normally matches either a molecular vibrational–rotational or rotational level difference, and the incident frequency ω_1 is usually some readily generated visible frequency (e.g., an Ar^+ or He/Ne laser line) in conventional Raman spectroscopy. In such cases $|\omega_2 - \omega_1|/\omega_1 \ll 1$. We may specialize Eq. 10.26 to chemical applications of Raman spectroscopy [1] by using Born-Oppenheimer states for the molecular zeroth-order states,

$$\begin{aligned}
|k\rangle &= |\psi_{\rm el}^0(q, Q)\chi_{0v}(Q)\rangle \\
|n\rangle &= |\psi_{\rm el}^n(q, Q)\chi_{nv''}(Q)\rangle \\
|m\rangle &= |\psi_{\rm el}^0(q, Q)\chi_{0v'}(Q)\rangle
\end{aligned} \qquad (10.27)$$

This implies that the initial and final states $|k\rangle$ and $|m\rangle$ are different vibrational levels within the same (normally the ground) electronic state, and that the intermediate states $|n\rangle$, in terms of which the virtual states are expanded, are vibrational levels in electronically excited manifolds (Fig. 10.13). Using this notation, the Raman transition amplitude becomes

$$c_m^{\rm R}(\infty) \propto \sum_{n,v''} \left(\frac{\mathbf{E}_2 \cdot \langle \chi_{0v'} | \boldsymbol{\mu}_{0n} | \chi_{nv''} \rangle \langle \chi_{nv''} | \boldsymbol{\mu}_{n0} | \chi_{0v} \rangle \cdot \mathbf{E}_1}{\omega_{0n} + \omega_{0v,nv''} + \omega_1} + \frac{\mathbf{E}_1 \cdot \langle \chi_{0v'} | \boldsymbol{\mu}_{0n} | \chi_{nv''} \rangle \langle \chi_{nv''} | \boldsymbol{\mu}_{n0} | \chi_{0v} \rangle \cdot \mathbf{E}_2}{\omega_{0n} + \omega_{0v,nv''} - \omega_2} \right) \qquad (10.28)$$

where

$$\boldsymbol{\mu}_{0n} = \langle \psi_{\rm el}^0 | \boldsymbol{\mu}_{\rm el} | \psi_{\rm el}^n \rangle \equiv \boldsymbol{\mu}_{0n}(Q) \qquad (10.29)$$

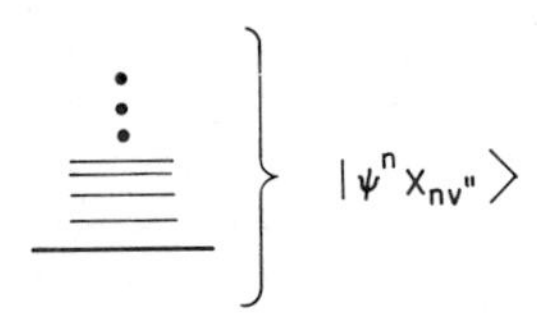

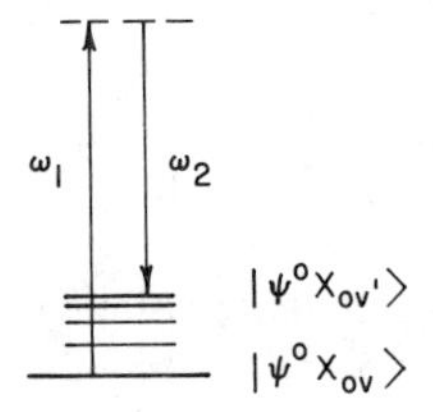

Figure 10.13 Energy level scheme for chemical applications of ordinary Raman scattering. The first allowed electronic state $|\psi^n\rangle$ lies well above the energy $\hbar\omega_1$ of the incident photon; the energy separations $\hbar(\omega_1 - \omega_2)$ correspond to rotational/vibrational energies in the ground electronic state $|\psi^0\rangle$.

is a transition moment function for the electronic transition $|n\rangle \leftarrow |0\rangle$. In analogy to the R-dependent diatomic transition moment function $\mathbf{M}_e(R)$ in Eq. 4.49, $\boldsymbol{\mu}_{0n}(Q)$ depends on the vibrational coordinates Q in polyatomics. The quantity $\hbar\omega_{0n} = (E_0 - E_n)$ is the electronic energy difference between states $|k\rangle$ and $|n\rangle$, and $\hbar\omega_{0v',nv''}$ is the vibrational energy difference between states $|k\rangle$ and $|n\rangle$. In a conventional Raman experiment where $|\omega_2 - \omega_1|/\omega_1 \ll 1$ and $|\omega_{0v',nv''}| \ll |\omega_{0n}|$ (Fig. 10.13), one may use the approximation $\omega_2 \simeq \omega_1 \equiv \omega$ and neglect the vibrational energy difference terms $\omega_{0v',nv''}$ in the energy denominators, with the result that the Raman transition amplitude takes on the symmetrical form

$$c_m^{\mathrm{R}}(\infty) \propto \sum_{n,v''} \left(\frac{\mathbf{E}_2 \cdot \langle \chi_{0v'} | \boldsymbol{\mu}_{0n} | \chi_{nv''} \rangle \langle \chi_{nv''} | \boldsymbol{\mu}_{n0} | \chi_{0v} \rangle \cdot \mathbf{E}_1}{\omega_{0n} + \omega} + \frac{\mathbf{E}_1 \cdot \langle \chi_{0v'} | \boldsymbol{\mu}_{0n} | \chi_{nv''} \rangle \langle \chi_{nv''} | \boldsymbol{\mu}_{n0} | \chi_{0v} \rangle \cdot \mathbf{E}_2}{\omega_{0n} - \omega} \right)$$

$$\doteq \langle \chi_{0v'} | \sum_n \frac{(\mathbf{E}_2 \cdot \boldsymbol{\mu}_{0n})(\boldsymbol{\mu}_{n0} \cdot \mathbf{E}_1)}{\omega_{0n} + \omega} + \frac{(\mathbf{E}_1 \cdot \boldsymbol{\mu}_{0n})(\boldsymbol{\mu}_{n0} \cdot \mathbf{E}_2)}{\omega_{0n} - \omega} | \chi_{0v} \rangle \qquad (10.30)$$

The *polarizability tensor* $\boldsymbol{\alpha}(\omega)$ for a molecule in electronic state $|0\rangle$ subjected to a sinusoidal electric field with circular frequency ω has components

$$\alpha_{ij}(\omega) = \sum_n \left(\frac{(\boldsymbol{\mu}_{0n})_i (\boldsymbol{\mu}_{n0})_j}{\hbar\omega_{0n} + \hbar\omega} + \frac{(\boldsymbol{\mu}_{0n})_j (\boldsymbol{\mu}_{n0})_i}{\hbar\omega_{0n} - \hbar\omega} \right) \qquad (10.31)$$

This bears a family resemblance to expression (1.35), which was derived for the polarizability of a molecule subjected to a *static* electric field ($\omega = 0$). Equation 10.31 is a generalization of Eq. 1.35 for the *dynamic* polarizability in the presence of applied fields with nonzero frequency ω. (The dynamic polarizability is reduced when the applied frequency is increased, because the electronic motions in molecules cannot react instantaneously to rapid changes in the external electric field.) A comparison of Eqs. 10.30 and 10.31 then shows that

$$c_m^{\mathrm{R}}(\infty) \propto \mathbf{E}_2 \cdot \langle \chi_{0v'} | \boldsymbol{\alpha}(\omega) | \chi_{0v} \rangle \cdot \mathbf{E}_1 \qquad (10.32)$$

which is to say that the transition amplitude is proportional to the matrix element of the frequency-dependent molecular electric polarizability taken between the initial and final vibrational states. The polarizability tensor $\boldsymbol{\alpha}(\omega)$ generally depends on Q as well, through the Q-dependence of $\boldsymbol{\mu}_{0n}$ in Eq. 10.29—otherwise, Eq. 10.32 tells us that no vibrational Raman transitions would take place between the orthonormal Born-Oppenheimer states $|k\rangle$ and $|m\rangle$.

We may obtain working selection rules for vibrational Raman transitions by expanding the molecular polarizability tensor in the normal coordinates Q_i

about the molecular equilibrium geometry,

$$\boldsymbol{\alpha} = \boldsymbol{\alpha}_0 + \sum_i \left(\frac{\partial \boldsymbol{\alpha}}{\partial Q_i}\right)_0 Q_i + \frac{1}{2} \sum_{ij} \left(\frac{\partial^2 \boldsymbol{\alpha}}{\partial Q_i \partial Q_j}\right)_0 Q_i Q_j + \cdots \tag{10.33}$$

so that

$$\langle \chi_{0v'} | \boldsymbol{\alpha} | \chi_{0v} \rangle = \boldsymbol{\alpha}_0 \delta_{v'v} + \sum_i \langle \chi_{0v'} | \left(\frac{\partial \boldsymbol{\alpha}}{\partial Q_i}\right)_0 Q_i | \chi_{0v} \rangle + \frac{1}{2} \sum_{ij} \langle \chi_{0v'} | \left(\frac{\partial^2 \boldsymbol{\alpha}}{\partial Q_i \partial Q_j}\right)_0 Q_i Q_j | \chi_{0v} \rangle + \cdots \tag{10.34}$$

Hence, the Raman selection rules on Δv depend on the relative magnitudes of the derivatives of $\boldsymbol{\alpha}$ in different orders of Q_i. The Raman fundamentals ($\Delta v = \pm 1$) arise from the terms linear in Q_i. Symmetry selection rules may be derived by noting that quantities like $(\partial \boldsymbol{\alpha}/\partial Q_i)_0 Q_i$ and $(\partial^2 \boldsymbol{\alpha}/\partial Q_i \partial Q_j)_0 Q_i Q_j$ transform under point group operations like $\boldsymbol{\alpha}$ itself; any component α_{ij} of $\boldsymbol{\alpha}$ transforms like $x_i x_j$ (i.e., α_{xy} transforms like xy, etc). Hence, Raman fundamentals generally arise only from normal modes Q_i which transform under point group operations like linear combinations of $x_i x_j$ such as $(x^2 - y^2)$, z^2, xy.

Allowed vibrational Raman transitions involve only modes that change the molecular polarizability (the right side of Eq. 10.32 would otherwise reduce into $\mathbf{E}_2 \cdot \boldsymbol{\alpha}(\omega) \cdot \mathbf{E}_1 \langle \chi_{0v'} | \chi_{0v} \rangle$, which is proportional to $\delta_{v'v}$). Cases do arise in which Raman fundamentals are symmetry-allowed, but very weak. For example, alkali halide molecules consist of pairs of oppositely charged ions whose electronic structures are nearly insensitive to the internuclear separation R for R near R_e. The alkali halide molecular polarizability

$$\boldsymbol{\alpha} \simeq \boldsymbol{\alpha}(\mathrm{M}^+) + \boldsymbol{\alpha}(\mathrm{X}^-) \tag{10.35}$$

is thus nearly R-independent, and alkali halides exhibit very weak Raman scattering even though their Raman fundamentals are symmetry-allowed. (Their diatomic vibrational motion is of course totally symmetric in $C_{\infty v}$, and the Σ^+ irreducible representation transforms as both $(x^2 + y^2)$ amd z^2.)

For a nontrivial example of Raman symmetry selection rules, we consider the molecular vibrations in *p*-difluorobenzene. We list in Table 10.1, next to each of the irreducible representations in D_{2h}, the number of vibrations of that species exhibited by *p*-difluorobenzene, the vector and tensor components that transform as that representation, and the corresponding infrared (one-photon) and Raman activity expected for *fundamental* transitions in the pertinent vibrational modes. Table 10.2 gives an analysis of the *p*-difluorobenzene Raman spectra shown in Fig. 10.1. These tables illustrate the point that in centrosymmetric molecules, symmetry-allowed Raman fundamentals appear only in modes with g symmetry, while the infrared-active fundamentals occur only in modes of u

Table 10.1 Molecular vibrations in *p*-difluorobenzene

Irreducible representation	Number of vibrations		IR	Raman
a_g	6	x^2, y^2, z^2		√
b_{1g}	1	xy		√
b_{2g}	3	x		√
b_{3g}	5	yz		√
a_u	2			
b_{1u}	5	z	√	
b_{2u}	5	y	√	
b_{3u}	3	x	√	

Table 10.2 Assignments of *p*-difluorobenzene Raman bands

Liquid			Vapor	
ν (cm^{-1})	Relative intensity	Assignment	ν (cm^{-1})	Relative intensity
1619	(11)	$\nu_2(a_g)$	1615	(3)
1606	(3)	$[2\nu_9(b_{1g})]$, A_g		
1388	(1)	$[2\nu_{16}(b_{2g})]$, A_g		
1285	(3)	$\nu_{25}(b_{3g})$		
1244	(42)	$\nu_3(a_g)$	1257	(30)
1141	(14)	$\nu_4(a_g)$	1140	(5)
1019	(2)	$[\nu_{17}(b_{2g}) + \nu_{26}(b_{3g})]$, B_{1g}	1013	(2)
895	(1)	$[2\nu_6(a_g)]$, A_g	897	(2)
		$[2\nu_{27}(b_{3g})]$, A_g	867	(16)
859	(68)	$\nu_5(a_g)$	859	(73)
840	(89)	$[2\nu_8(a_u)]$, A_g	840	(35)
799	(2)	$\nu_9(b_{1g})$		
636	(23)	$\nu_{26}(b_{3g})$	635	(6)
451	(84)	$\nu_6(a_g)$	450	(37)
376	(47)	$\nu_{17}(b_{2g})$	374	(9)

symmetry. Hence, vibrational modes exhibiting infrared and Raman fundamentals belong to mutually exclusive sets if the molecule has a center of symmetry; this fact has been used as evidence for assigning molecular geometry. Laser Raman spectroscopy has found wide applications in environmental chemistry, biochemistry (e.g., conformational analysis of proteins), and medicine as well as in chemical physics.

A specialized situation called *resonance Raman scattering* arises when the laser frequency ω_1 is tuned close to resonance with one of the molecular eigenstates $|n\rangle = |\psi_{\mathrm{el}}^{n}(q, Q)\, \chi_{nv''}(Q)\rangle$, as shown in Fig. 10.14. The energy denominator $(\omega_{0n} + \omega_{0v,nv''} + \omega_1)$ in Eq. 10.28 becomes very small relative to its value in ordinary Raman scattering, and the transition probability (which is proportional to $|c_m^{\mathrm{R}}|^2$) becomes anomalously large. In this limit, the vibrational energy differences $\omega_{0v,nv''}$ cannot be ignored next to the other terms in the energy denominators, with the result that the transition amplitude $c_m^{\mathrm{R}}(\infty)$ no longer reduces to the symmetrical form (10.30). Hence, the assumptions leading up to Eq. 10.32 (which gives the Raman transition amplitude in terms of a matrix element of $\boldsymbol{\alpha}(\omega)$) fall through: The ordinary Raman selection rules are not applicable to resonance Raman transitions. It turns out that some transitions that are forbidden in ordinary Raman become allowed in resonance Raman spectroscopy (Problem 10.4). Time-resolved resonance Raman (TR^3) scattering has been developed into a useful technique for monitoring the populations of large molecules in electronically excited states (Fig. 10.15). Such excited-state populations might be more conventionally probed by studying the evolution of $S_n \leftarrow S_1$ or $T_n \leftarrow T_1$ one-photon transient absorption spectra on the ns or ps time scale. These transient spectra tend to be featureless (due to spectral congestion) in photobiological molecules and transition metal complexes. TR^3 scattering is a more advantageous probe, because the resulting spectra exhibit sharp vibrational structure similar to that in Fig. 10.1. The enhanced sensitivity inherent in TR^3 can be rendered specific to an excited state of interest by tuning the probe laser frequency ω_1, because each excited state will be uniquely spaced in energy from higher-lying electronic states.

In Chapter 8, we characterized the strengths of one-photon transitions in terms of Einstein coefficients and oscillator strengths. According to Beer's law, a weak light beam with incident intensity I_0 will emerge from an absorptive sample of concentration C and path length l with diminished intensity $I = I_0 \exp(-\alpha Cl)$, where α is the molar absorption coefficient. Beer's law can be recast in the form

$$I = I_0 e^{-N\sigma l} \tag{10.36}$$

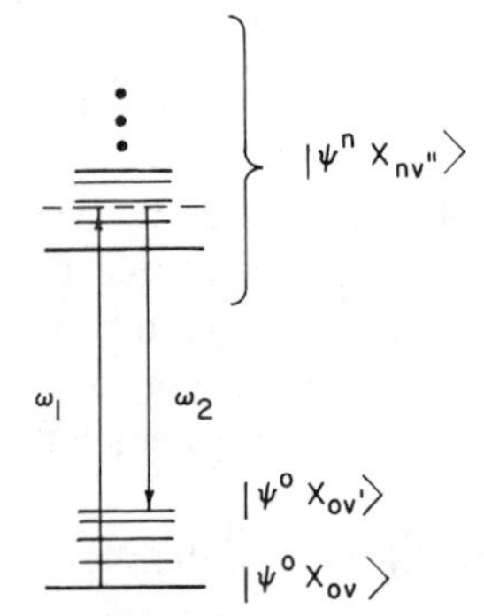

Figure 10.14 Energy level scheme for resonance Raman scattering. The incident photon energy $\hbar\omega_1$ is in near resonance with a vibronic level in some electronically excited state $|\psi^n\rangle$; the energy differences $\hbar(\omega_1 - \omega_2)$ correspond to rotational/vibrational energies in the ground electronic state $|\psi^0\rangle$.

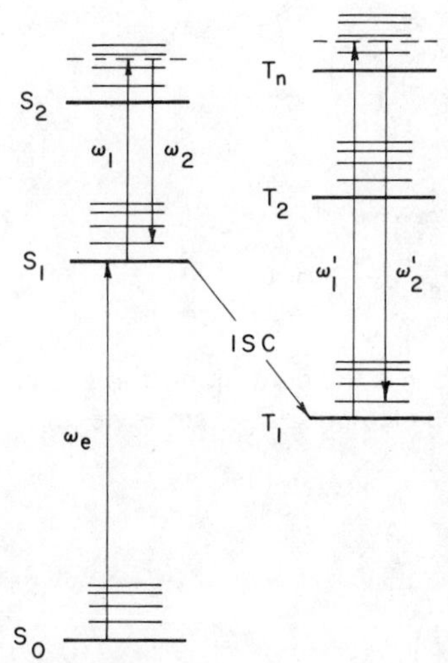

Figure 10.15 Resonance Raman detection of populations in electronically excited states S_1 and T_1 following creation of S_1 state molecules by laser excitation at frequency ω_e. Probing S_1 molecules at frequency ω_1 will generate intense resonance Raman emission at ω_2 if the $S_2 \leftarrow S_1$ transition is E1-allowed, because ω_1 is in near-resonance with the energy separation between vibrationless S_1 and some vibronic level of S_2. Probing at frequency ω_1' will generate similarly intense emission only if appreciable population has accumulated in T_1 by intersystem crossing from S_1, since ω_1' is in resonance with the $T_n - T_1$ energy gap. This excited-state selectivity of resonance Raman scattering has rendered it a useful tool for monitoring time-resolved excited state dynamics.

where N is the molecule number density in cm^{-3} and σ is the absorption *cross section.* This cross section, which has units of area, is related to the absorption coefficient by

$$\begin{aligned} \sigma(\text{Å}^2) &= 1.66 \times 10^{-5}\alpha\,(\text{L mol}^{-1}\text{ cm}^{-1}) \\ &= 3.83 \times 10^{-5}\varepsilon\,(\text{L mol}^{-1}\text{ cm}^{-1}) \end{aligned} \tag{10.37}$$

The physical picture suggested by the concept of a cross section may be appreciated by visualizing the photons as point particles impinging on a sample containing N molecules per cm^3. Each molecule is imagined to have a well-defined cross sectional area σ. A photon is absorbed when it "hits" a molecule, but is transmitted if it traverses the path length l without scoring a hit. Under these conditions, the fraction I/I_0 of photons that are transmitted will be $\exp(-N\sigma l)$. For strongly allowed one-photon transitions, σ is somewhat smaller than the molecular size: The absorption cross section for rhodamine 6G at 5300 Å ($\varepsilon_{\max} = 10^5\ \text{L mol}^{-1}\text{ cm}^{-1}$, Problem 8.2) is 3.8 Å^2.

Raman scattering intensities (which are proportional to $|c_{ml}^{\text{R}}|^2$) are commonly expressed in terms of cross sections. The *differential cross section* $d\sigma/d\Omega_2$ is defined as

$$\frac{d\sigma}{d\Omega_2} = \frac{dN_{\text{sc}}/d\Omega_2}{dN_{\text{inc}}/dA} \tag{10.38}$$

where dN_{inc} is the number of photons $(\mathbf{k}_1, \omega_1)$ which traverse the area element dA normal to wave vector $\mathbf{k}_1$ in the incident beam, and dN_{sc} is the number of photons scattered into the solid angle element $d\Omega_2 = \sin\theta_2 d\theta_2 d\phi_2$ (Fig. 10.16).

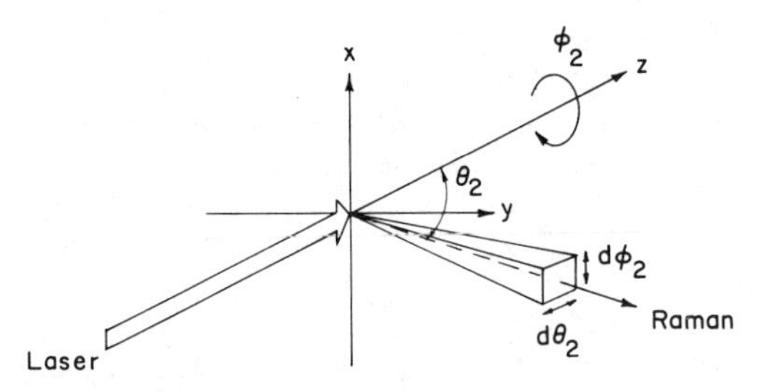

Figure 10.16 Raman scattering geometry. The laser is incident along the z axis, and the Raman emission is scattered into the volume element $d\Omega_2 = \sin\theta_2 d\theta_2 d\phi_2$.

The total Raman cross section

$$\sigma = \int_0^{\pi} \sin\theta_2 d\theta_2 \int_0^{2\pi} \left(\frac{d\sigma}{d\Omega_2}\right) d\phi_2 \tag{10.39}$$

has an interpretation similar to that of the one-photon absorption cross section: σ is related to the fraction I/I_0 of incident photons that remain unscattered after traversing path length l in the sample, via $I/I_0 = \exp(-N\sigma l)$. It may be shown that the differential cross section for Raman scattering is [1]

$$\frac{d\sigma}{d\Omega_2} = \frac{\omega_1\omega_2^3}{(4\pi\varepsilon_0 hc^2)^2}\left|\sum_n \left(\frac{\hat{E}_2\cdot\langle m|\boldsymbol{\mu}|n\rangle\langle n|\boldsymbol{\mu}|k\rangle\cdot\hat{E}_1}{\omega_{kn}+\omega_1} + \frac{\hat{E}_1\cdot\langle m|\boldsymbol{\mu}|n\rangle\langle n|\boldsymbol{\mu}|k\rangle\cdot\hat{E}_2}{\omega_{kn}-\omega_2}\right)\right|^2 \tag{10.40}$$

where $\hat{E}_1$, $\hat{E}_2$ are unit vectors in the directions of polarization of the electric fields associated with photons $(\mathbf{k}_1, \omega_1)$ $(\mathbf{k}_2, \omega_2)$. A special case called Rayleigh scattering occurs when $\omega_1 = \omega_2$ (i.e., the initial and final molecular states are the same). The differential cross section for Rayleigh scattering is obtained by replacing ω_2 by ω_1 in Eq. 10.40, with the result that the cross section becomes proportional to the fourth power of the incident frequency ω_1. This phenomenon is responsible for the inimitable blue color of the cloudless sky, because the shorter wavelengths in the solar spectrum are preferentially scattered by the atmosphere. According to Eq. 10.40, the scattered photon may propagate into any direction $\hat{k}_2$ in general. The angular distribution of Raman scattering (relative intensities of light scattered into different directions $\mathbf{k}_2$) may be obtained by averaging expressions like (10.40) over the orientational distribution of molecules in the sample. For conventional vibrational Raman transitions excited by visible light, a typical cross section $d\sigma/d\Omega_2$ is on the order of 10^{-14} Å^2, with the result that only one photon in 10^9 is scattered in a sample with molecule number density $N = 10^{20}\,\text{cm}^{-3}$ and path length $l = 10\,\text{cm}$. (The cross sections for resonance Raman transitions are, of course, far larger). This is why intense light sources (preferably lasers) are required for Raman spectroscopy. Visible rather than infrared lasers are normally used in vibrational and rotational Raman spectroscopy, because the cross sections (which are proportional to $\omega_1\omega_2^3$) and the photon detector sensitivities are more advantageous in the visible than in the infrared.

Raman line intensities are proportional to the number density N of molecules in the initial state $|k\rangle$, which is in turn proportional to the pertinent Boltzmann factor for that state at thermal equilibrium. Consequently, the relative intensities of a Stokes transition $|k\rangle \rightarrow |m\rangle$ and the corresponding anti-Stokes transition $|m\rangle \rightarrow |k\rangle$ are 1 and $\exp(-\hbar\omega_{mk}/kT)$, respectively. (The factor $\omega_1\omega_2^3$ varies little between the Stokes and anti-Stokes lines, because the Raman frequency shifts are ordinarily small compared to ω_1.) Hence the anti-Stokes Raman transitions (which require molecules in vibrationally excited initial states) are considerably less intense than their Stokes counterparts, particularly when the Raman shift ω_{mk} is large. In much of the current vibrational Raman literature, only the Stokes spectrum is reported (cf. Fig. 10.1).

REFERENCES

1. D. P. Craig and T. Thirunamachandran, *Molecular Quantum Electrodynamics*, Academic, London, 1984.
2. M. D. Levenson and N. Bloembergen, *Phys. Rev. Lett.* **32**: 645 (1974).
3. F. Biraben, B. Cagnac, and G. Grynberg, *Phys. Rev. Lett.* **32**: 643 (1974).
4. D. M. Friedrich and W. M. McClain, *Chem. Phys. Lett.* **32**: 541 (1975).
5. M. M. Sushchinskii, *Raman Spectra of Molecules and Crystals*, Israel Program for Scientific Translations, New York, 1972.

PROBLEMS

1. Two-photon $3s \rightarrow 5s$ absorption is observed in Na vapor at 400 K with counterpropagating laser beams whose frequencies ω_1^0 and ω_2^0 are not quite identical. How large must the frequency difference $|\omega_1^0 - \omega_2^0|$ be so that the Doppler contribution to the linewidth equals the Lorentzian contribution if the $5s$ radiative lifetime is 10 ns?

2. What states in K can be reached by two-photon absorption from $4^2P_{1/2}$ level in the E1 approximation? To what term symbols are the intermediate states restricted?

3. Show that expression (10.40) for the differential cross section in Raman scattering has units of area as required.

4. The acetylene molecule C_2H_2 has five vibrational modes, three nondegenerate and two doubly degenerate (Chapter 6).

(a) Several of this molecule's lowest energy vibrational levels are listed below. The polarizability function is assumed to have the form

$$\boldsymbol{\alpha}(Q_1, \ldots, Q_5) = \boldsymbol{\alpha}_0 + \sum_{i=1}^{5} a_i Q_i + \sum_{ij}^{5} b_{ij} Q_i Q_j + \sum_{ijk}^{5} c_{ijk} Q_i Q_j Q_k$$

with no other terms. Which of the levels will be reached by E1-allowed Raman transition from the 0000°0° level?

v_1	v_2	v_3	$v_4^{l_4}$	$v_5^{l_5}$	Symmetry
0	0	0	1^1	0	Π_g
0	0	0	0	1^1	Π_u
0	0	0	1^1	1^1	Σ_u^+
0	0	0	2^0	1^1	Π_u
0	0	0	0	2^0	Δ_g
0	1	0	0	0	Σ_g^+
0	0	0	0	3^1	Π_u
0	0	0	0	3^3	Φ_u
0	1	0	0	1^1	Π_u
0	0	1	0	0	Σ_u^+
0	1	0	1^1	1^1	Σ_u^+
1	0	0	0	0	Σ_g^+
0	1	0	2^0	1^1	Π_u
3	0	0	0	0	Σ_g^+
3	1	0	0	0	Σ_g^+
0	4	0	0	0	Σ_g^+
0	1	0	3^1	0	Π_g
0	0	0	2^2	0	Δ_g

(b) Consider the hypothetical case in which the laser frequency ω_1 is tuned close to the lowest E1-allowed electronic transition in C_2H_2, so that resonance Raman emission occurs and the polarizability expression (10.32) for the Raman transition probability amplitude is no longer applicable. Which of the vibrational levels listed above can be reached from 0000°0° acetylene by E1 symmetry-allowed resonance Raman transitions, even though they cannot be reached by conventional Raman transitions?

(c) The frequency of mode 2 in acetylene is 1974 cm^{-1}. What will be the approximate ratio of intensities in the Raman fundamentals of this mode in the Stokes and anti-Stokes branches in a 300 K sample?

11

NONLINEAR OPTICS

The discussion of Raman and Rayleigh scattering in Section 10.3 was based on the time-dependent perturbation theory of radiation–matter interactions developed in Chapter 1. The scattered light intensities were found to be linear in the incident laser intensity; the scattered Raman frequencies were shifted from the laser frequency by molecular vibrational/rotational frequencies. Identical conclusions may be reached using a contrasting theory which treats the polarization of bulk media by electromagnetic fields classically. Such a classical theory provides an insightful vehicle for introducing the nonlinear optical phenomena described in this chapter, and so we begin by recasting the familiar Raman and Rayleigh scattering processes in a classical framework.

The total dipole moment $\mathbf{p}$ of a dielectric material contained in volume V is given by the volume integral

$$\mathbf{p} = \int_V \mathbf{P}\, d\mathbf{r} \tag{11.1}$$

of the local dipole moment density $\mathbf{P}$ [1]. Defined in this way, $\mathbf{P}$ (which is normally called the *polarization*) has the same units as electric field. In a material composed of nonpolar molecules or randomly oriented polar molecules, $\mathbf{P}$ vanishes in the absence of perturbing fields. In the presence of an external electric field $\mathbf{E}$, the polarization becomes

$$\mathbf{P} = \varepsilon_0 \boldsymbol{\chi} \cdot \mathbf{E} \tag{11.2}$$

where $\boldsymbol{\chi}$ is the dimensionless *electric susceptibility tensor*. This expression gives the correct zero-field limit $\mathbf{P} = 0$. However, the polarization is not necessarily linear in **E**, because the susceptibility itself may depend on **E**. We shall see that this nonlinearity forms the basis for the phenomena in this chapter. For isotropic media (gases, liquids, and most amorphous solids), the susceptibility reduces to a scalar function $\chi(\mathbf{E})$, and the polarization **P** points in the same direction as **E**. In many crystalline solids, the induced polarization does not point along **E**, and $\boldsymbol{\chi}$ is an anisotropic tensor. Equation 11.2 closely resembles the expression for the dipole moment induced in a molecule by an electric field, $\boldsymbol{\mu}_{\text{ind}} = \boldsymbol{\alpha} \cdot \mathbf{E}$. For atoms and isotropically polarizable molecules at low densities N (expressed in molecules/cm^3), the bulk susceptibility is clearly related to the molecular polarizability by

$$\chi = N\alpha/\varepsilon_0 \tag{11.3}$$

At general number densities where the total field experienced by a molecule may be influenced by dipole moments induced on neighboring molecules, the susceptibility is given instead [1] by $\chi = (N\alpha/\varepsilon_0)/(1 - N\alpha/3\varepsilon_0)$.

We now consider an electromagnetic wave with time dependence $\mathbf{E} = \mathbf{E}_0 \cos \omega_0 t$ incident upon a system of isotropically polarizable molecules. For simplicity, we assume the molecules undergo classical harmonic vibrational motion with frequency ω in some totally symmetric mode Q. The normal coordinate then oscillates as $Q = Q_0 \cos(\omega t + \delta)$, where δ is the vibrational phase and Q_0 is the amplitude. If the molecular polarizability α is linear in Q (as a special case of Eq. 10.33), the vibrational motion will endow the molecule with the oscillating polarizability

$$\alpha = \alpha_0 + \alpha_1 \cos(\omega t + \delta) \tag{11.4}$$

Ignoring the vibrational phases δ (which will be random in an incoherently excited system of vibrating molecules), the polarization induced by the external field will be

$$\begin{aligned} \mathbf{P} &= \varepsilon_0 \boldsymbol{\chi} \cdot \mathbf{E} \\ &= N(\alpha_0 + \alpha_1 \cos \omega t)\mathbf{E}_0 \cos \omega_0 t \\ &= N\mathbf{E}_0 \left\{ \alpha_0 \cos \omega_0 t + \frac{\alpha_1}{2} [\cos(\omega_0 + \omega)t + \cos(\omega_0 - \omega)t] \right\} \end{aligned} \tag{11.5}$$

According to the classical theory of radiation [1], an oscillating dipole moment **p** will emit radiation with an electric field proportional to its second time derivative $\ddot{\mathbf{p}}$. Equations 11.1 and 11.5 then imply that radiation will be scattered at the frequencies ω_0, $\omega_0 - \omega$, and $\omega_0 + \omega$, corresponding to Rayleigh, Stokes Raman, and anti-Stokes Raman scattering, respectively. The scattered electric fields are proportional to $\mathbf{E}_0$, so that the Rayleigh and Raman intensities are linear in the incident laser intensity. Expressions similar to Eq. 11.5 are frequently cited in classical treatments of Raman scattering [2, 3]; they emphasize the central role of Q-dependent molecular polarizability, and they demonstrate heuristically how Raman-scattered light frequencies are shifted by molecular frequencies ω from the laser frequency ω_0.

As lasers with high output powers became accessible to spectroscopists in the 1960s, conspicuous nonlinearities emerged in the polarization $\mathbf{P}(\mathbf{E})$ induced by intense fields. The components P_i of $\mathbf{P}$ may be expanded in powers of components E_j, E_k, E_l of the electric field $\mathbf{E}$ via

$$P_i = \varepsilon_0 \left(\sum_j^3 \chi_{ij}^{(1)} E_j + \sum_{jk}^3 \chi_{ijk}^{(2)} E_j E_k + \sum_{jkl} \chi_{ijkl}^{(3)} E_j E_k E_l + \cdots \right) \tag{11.6}$$

The linear susceptibility $\boldsymbol{\chi}^{(1)}$ gives rise to the Raman and Rayleigh processes treated in Chapter 10; it dominates the polarization in weak fields. As the light intensity is increased, the responses due to the nonlinear susceptibilities $\chi_{ijk}^{(2)}$, $\chi_{ijkl}^{(3)}, \ldots$ gain prominence. A discussion analogous to the one culminating in Eq. 11.5 shows that scattering may occur at frequencies that are multiples of the laser frequency. A process controlled by the second-order nonlinear susceptibility $\chi_{ijk}^{(2)}$ is second-harmonic generation (SHG), whereby two laser photons at frequency ω_0 are converted into a single photon of frequency $2\omega_0$. (A related process is sum-frequency generation, in which laser photons ω_1 and ω_2 are combined into a single photon with frequency $\omega_1 + \omega_2$.) The third-order nonlinear susceptibility $\chi_{ijkl}^{(3)}$ is responsible for third-harmonic generation, $\omega_0 + \omega_0 + \omega_0 \rightarrow 3\omega_0$. It also leads to coherent anti-Stokes Raman scattering (CARS), which is treated in Section 11.3. Generation of nth-harmonic frequencies is governed by the nth-order nonlinear susceptibility $\chi^{(n)}$; ninth-harmonic pulses have been generated by 10.6-μm CO_2 laser pulses in nonlinear media.

Because the scattered light intensities occasioned by the second- and higher order terms in Eq. 11.6 increase nonlinearly with the incident light intensity, higher order contributions to the susceptibility become important at sufficient laser powers. SHG conversion efficiencies of 20% from 1064 to 532 nm are routinely achieved in pulsed Nd^{3+}:YAG lasers, and were unimaginable prior to the invention of lasers.

Explicit formulas for the nonlinear susceptibilities $\chi_{ijk}^{(2)}$, $\chi_{ijkl}^{(3)}$, ... may be derived by working out the coefficients $c_m^{(3)}(t), c_m^{(4)}(t), \ldots$, respectively, in the time-ordered expansion (1.96). Straightforward evaluation of the integrals in the time-ordered expansions rapidly becomes unwieldy, and an efficient diagrammatic technique is developed in Section 11.1 for writing down the contributions to $c_m^{(n)}(t)$ that are pertinent to any multiphoton process of interest. In Sections 11.2 and 11.3, we apply this technique to obtaining the nonlinear susceptibilities for two important nonlinear optical processes, SHG and CARS. Experimental considerations that are unique to such coherent optical phenomena are also discussed.

11.1 DIAGRAMMATIC PERTURBATION THEORY

To illustrate the simplifications introduced by diagrammatic perturbation theory [4], we consider the three-photon processes corresponding to the third-order term in the Dyson expansion of $c_m(t)$,

$$c_m^{(3)}t = \frac{1}{(i\hbar)^3} \sum_{pn} \int_{t_0}^{t} e^{-i\omega_{pm}t_1} \langle m|W(t_1)|p\rangle dt_1 \int_{t_0}^{t_1} e^{-i\omega_{np}t_2} \langle p|W(t_2)|n\rangle dt_2$$
$$\times \int_{t_0}^{t_2} e^{-i\omega_{kn}t_3} \langle n|W(t_3)|k\rangle dt_3 \tag{11.7}$$

We may associate the perturbations $W(t_1)$, $W(t_2)$, and $W(t_3)$ with vector potentials for electromagnetic waves with frequencies ω_1, ω_2, and ω_3 respectively:

$$\langle m|W(t_1)|p\rangle = \frac{i\hbar q}{2mc} (\alpha_{mp} e^{-i\omega_1 t_1} + \bar{\alpha}_{mp} e^{i\omega_1 t_1})$$
$$\langle p|W(t_2)|n\rangle = \frac{i\hbar q}{2mc} (\alpha_{pn} e^{-i\omega_2 t_2} + \bar{\alpha}_{pn} e^{i\omega_2 t_2}) \tag{11.8}$$
$$\langle n|W(t_3)|k\rangle = \frac{i\hbar q}{2mc} (\alpha_{nk} e^{-i\omega_3 t_3} + \bar{\alpha}_{nk} e^{i\omega_3 t_3})$$

Substitution of these matrix elements into Eq. 11.7 then yields terms in $c_m^{(3)}(t)$ proportional to the eight products

$\alpha_{mp}\alpha_{pn}\alpha_{nk}$	$\bar{\alpha}_{mp}\bar{\alpha}_{pn}\alpha_{nk}$
$\alpha_{mp}\alpha_{pn}\bar{\alpha}_{nk}$	$\bar{\alpha}_{mp}\alpha_{pn}\bar{\alpha}_{nk}$
$\alpha_{mp}\bar{\alpha}_{pn}\alpha_{nk}$	$\alpha_{mp}\bar{\alpha}_{pn}\bar{\alpha}_{nk}$
$\bar{\alpha}_{mp}\alpha_{pn}\alpha_{nk}$	$\bar{\alpha}_{mp}\bar{\alpha}_{pn}\bar{\alpha}_{nk}$

These correspond to the eight time-ordered graphs (a) through (h), respectively, in Fig. 11.1. These are only a small fraction of the possible third-order graphs, because the arbitrary assignment of vector potentials to perturbations $W(t_i)$ in Eqs. 11.8 is only one of six permutations of ω_1, ω_2, ω_3 among the $W(t_i)$. Hence, $c_m^{(3)}(t)$ contains 48 time-ordered graphs, of which only eight are shown in Fig. 11.1.

We next evaluate the terms in $c_m^{(3)}(t)$ corresponding to the first two time-ordered graphs. The contribution from graph (a) in the cw limit is

$$\frac{q^3}{8m^3c^3} \sum_{pn} \alpha_{mp}\alpha_{pn}\alpha_{nk} \int_{t_0}^{t} \exp[-i(\omega_{pm} + \omega_1)t_1] dt_1 \int_{t_0}^{t_1} \exp[-i(\omega_{np} + \omega_2)t_2] dt_2$$

$$\times \int_{t_0}^{t_2} \exp[-i(\omega_{kn} + \omega_3)t_3]dt_3$$

$$= \frac{-2\pi q^3}{8m^3c^3} \sum_{pn} \frac{\alpha_{mp}\alpha_{pn}\alpha_{nk}}{(\omega_{kp} + \omega_2 + \omega_3)(\omega_{kn} + \omega_3)} \delta(\omega_{km} + \omega_1 + \omega_2 + \omega_3) \qquad (11.9)$$

The delta function in (11.9) implies the energy conservation $E_m - E_k = \hbar(\omega_1 + \omega_2 + \omega_3)$ pertinent to *three-photon absorption.* This is consistent with graph (a), which shows photons ω_1, ω_2, ω_3 incident at early times, and no photons scattered at long times. The contribution from graph (b) is

$$\frac{q^3}{8m^3c^3} \sum_{pn} \alpha_{mp}\alpha_{pn}\bar{\alpha}_{nk} \int_{t_0}^{t} \exp[-i(\omega_{pm} + \omega_1)t_1]dt_1$$

$$\times \int_{t_0}^{t_1} \exp[-i(\omega_{np} + \omega_2)t_2]dt_2 \int_{t_0}^{t_2} \exp[-i(\omega_{kn} - \omega_3)t_3]dt_3$$

$$= \frac{-2\pi q^3}{8m^3c^3} \sum_{pn} \frac{\alpha_{mp}\alpha_{pn}\bar{\alpha}_{nk}}{(\omega_{kp} + \omega_2 - \omega_3)(\omega_{kn} - \omega_3)} \delta(\omega_{km} + \omega_1 + \omega_2 - \omega_3) \qquad (11.10)$$

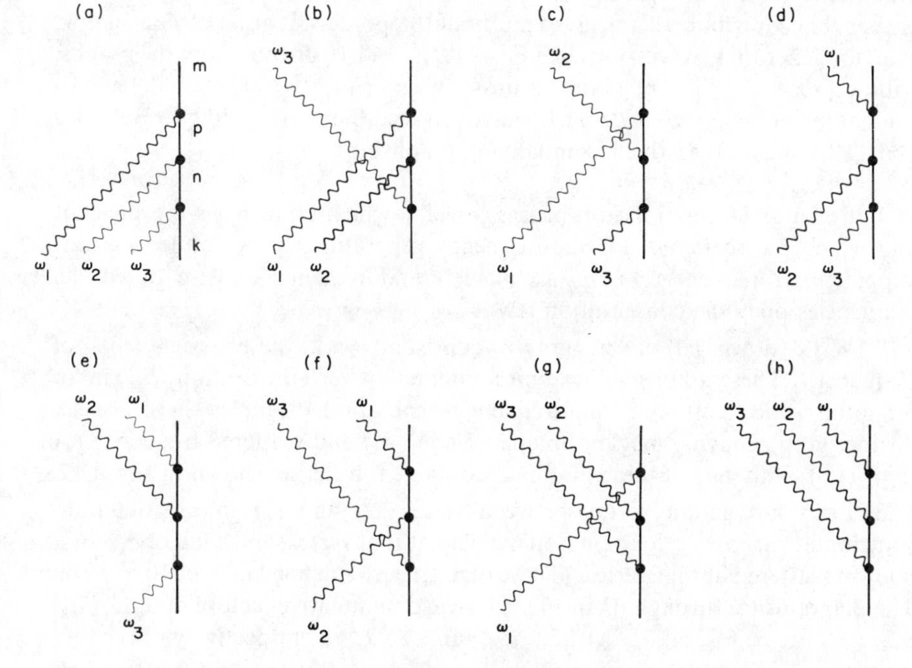

Figure 11.1 Time-ordered graphs representing the eight contributions to $c_m^{(3)}(t)$ in Eq. 11.7 when the perturbation matrix elements are given by Eq. 11.8.

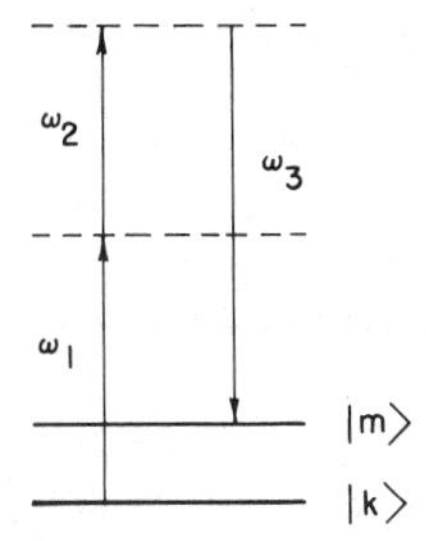

Figure 11.2 Energy level diagram for the process represented by graph (b) in Fig. 11.1. The process is sum-frequency generation when states $|k\rangle$ and $|m\rangle$ are the same, and hyper-Raman scattering when they are different.

The energy conserving condition here is $E_m - E_k = \hbar(\omega_1 + \omega_2 - \omega_3)$, corresponding to absorption of photons ω_1, ω_2 and scattering of photon ω_3 (Fig. 11.2). When the initial and final molecular states $|k\rangle$, $|m\rangle$ are identical, the process is *sum-frequency generation* [5,6], an important gating technique used in time-resolved laser spectroscopies. When the two states differ, the process is *hyper-Raman scattering* [4]; the energy difference $E_m - E_k$ usually corresponds to a rotational/vibrational energy separation in gases, or to phonon frequencies in lattices.

It is clearly tedious to evaluate 48 such integrals. The number of graphs mushrooms as $2^n n!$ with the perturbation order n: the four- and five-photon processes are associated with 384 and 3840 graphs, respectively. The diagrammatic technique's great utility consists in that it quickly isolates those graphs that contribute to any given multiphoton process. It also provides simple rules for generating expressions like Eqs. 11.9 and 11.10 directly from the graphs, without recourse to explicit integration. These rules (which should be self-evident by induction to readers who have retraced the steps leading to Eqs. 11.9 and 11.10 and studied the accompanying graphs) are:

1. For a given multiphoton process, decide which frequencies are incident and which are scattered. In sum-frequency generation, for example, one can stipulate that frequencies ω_1, ω_2 are incident and frequency ω_3 is scattered; the frequencies obey the conservation law $\omega_3 = \omega_1 + \omega_2$.

2. Write down *all* of the graphs consistent with these assignments of frequencies. These graphs will exhibit n interaction vertices (Section 10.1) in an n-photon process. In our sum-frequency generation example, there are six distinct graphs having incident frequencies ω_1, ω_2 and scattered frequency ω_3 (Fig. 11.3). Only one of these graphs is contained in the set shown in Fig. 11.2.

3. Each interaction vertex between states $|p\rangle$ and $|q\rangle$ in any diagram contributes a factor α_{pq} for a photon incident at that vertex, and a factor $\bar{\alpha}_{pq}$ for a photon scattered at that vertex. (These quantities are defined in Eqs. 10.5–10.6.) The diagrams (a) through (f) in Fig. 11.3 yield cumulative factors of $\alpha_{mp}\alpha_{pn}\bar{\alpha}_{nk}$, $\alpha_{mp}\alpha_{pn}\bar{\alpha}_{nk}$, $\alpha_{mp}\bar{\alpha}_{pn}\alpha_{nk}$, $\alpha_{mp}\bar{\alpha}_{pn}\alpha_{nk}$, $\bar{\alpha}_{mp}\alpha_{pn}\alpha_{nk}$, and $\bar{\alpha}_{mp}\alpha_{pn}\alpha_{nk}$, respectively.

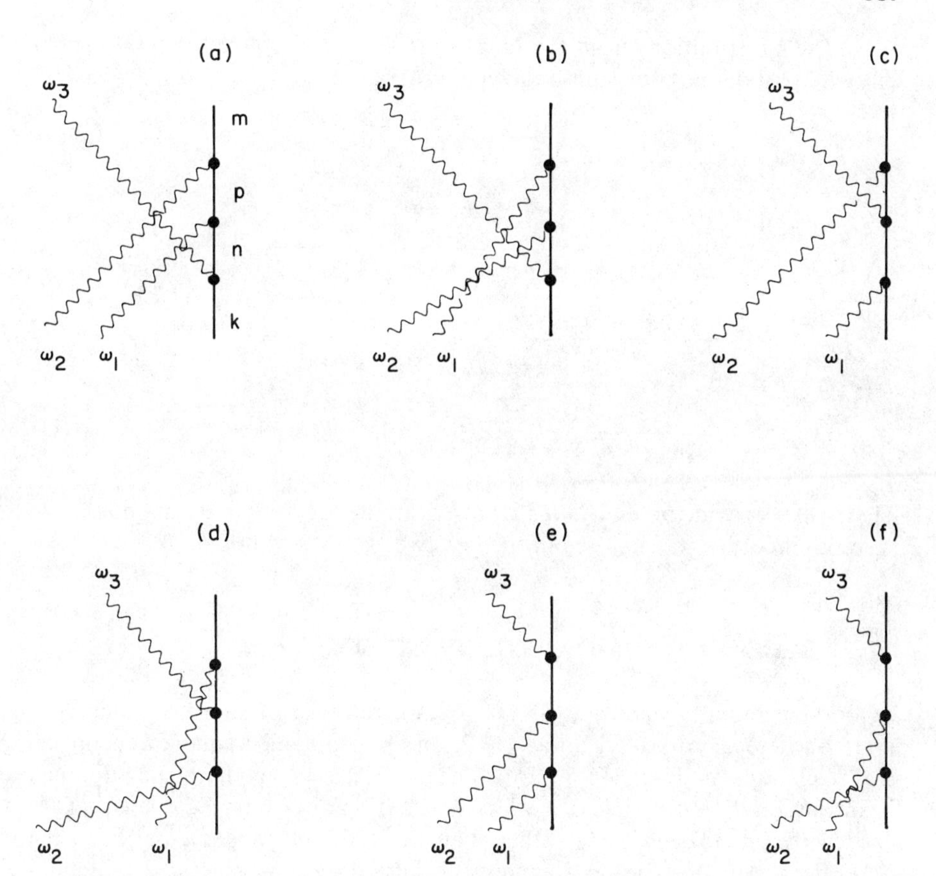

Figure 11.3 The set of time-ordered graphs that can be drawn for the sum-frequency generation process in which ω_1 and ω_2 are the incident frequencies and ω_3 is the scattered frequency. Six graphs result from permuting the three frequencies among the three interaction vertices.

4. Reading from bottom to top of the time line in each graph, each of the first $(n - 1)$ interaction vertices contributes a factor of $(\omega_{kl} + \omega)$ to the energy denominator. Here $\omega_{kl} = (E_k - E_l)/\hbar$, $|l\rangle$ is the state lying above the vertex on the time line, $|k\rangle$ is the initial state, and ω is the total photon energy "absorbed" at all vertices up to and including that vertex. The energy denominators for the six diagrams of Fig. 11.3 are in order

$$(\omega_{kp} + \omega_1 - \omega_3)(\omega_{kn} - \omega_3), \quad (\omega_{kp} + \omega_2 - \omega_3)(\omega_{kn} - \omega_3),$$
$$(\omega_{kp} - \omega_3 + \omega_1)(\omega_{kn} + \omega_1), \quad (\omega_{kp} - \omega_3 + \omega_2)(\omega_{kn} + \omega_2),$$
$$(\omega_{kp} + \omega_2 + \omega_1)(\omega_{kn} + \omega_1), \quad \text{and} \quad (\omega_{kp} + \omega_1 + \omega_2)(\omega_{kn} + \omega_2)$$

5. The contributions from the graphs are summed to give the total probability amplitude. For the sum-frequency generation $\omega_1 + \omega_2 \to \omega_3$, we have

$$c_m(\infty) = \frac{-2\pi q^3}{(2mc)^3} \sum_{pn} \left(\frac{\alpha_{mp}\alpha_{pn}\bar{\alpha}_{nk}}{(\omega_{kp} + \omega_1 - \omega_3)(\omega_{kn} - \omega_3)} + \frac{\alpha_{mp}\alpha_{pn}\bar{\alpha}_{nk}}{(\omega_{kp} - \omega_3 + \omega_2)(\omega_{kn} - \omega_3)} + \frac{\alpha_{mp}\bar{\alpha}_{pn}\alpha_{nk}}{(\omega_{kp} - \omega_3 + \omega_1)(\omega_{kn} + \omega_1)} + \frac{\alpha_{mp}\bar{\alpha}_{pn}\alpha_{nk}}{(\omega_{kp} - \omega_3 + \omega_2)(\omega_{kn} + \omega_2)} + \frac{\bar{\alpha}_{mp}\alpha_{pn}\alpha_{nk}}{(\omega_{kp} + \omega_2 + \omega_1)(\omega_{kn} + \omega_1)} + \frac{\bar{\alpha}_{mp}\alpha_{pn}\alpha_{nk}}{(\omega_{kp} + \omega_1 + \omega_2)(\omega_{kn} + \omega_2)} \right) \quad (11.11)$$

These rules provide an enormous labor-saving device for evaluating nonlinear susceptibilities, as we shall see in the last two sections of this book.

11.2 SECOND-HARMONIC GENERATION

Second-harmonic generation (SHG), the special case of sum-frequency generation where $\omega_1 = \omega_2 \equiv \omega$ and $\omega_3 = 2\omega$, is an invaluable frequency upconversion technique in lasers [5,6]. Most near-UV lasers are frequency-doubled beams originating in visible dye lasers, and Nd^{3+}:YAG-pumped dye lasers are excited by the 532-nm SHG rather than the 1064-nm fundamental from the YAG laser. Autocorrelation diagnoses of pulse durations generated by mode-locked lasers also rely on frequency doubling.

It is clear from Eq. 11.11 that for SHG (in which $|o\rangle$ represents both the initial and final state)

$$c^{(3)}(\infty) = -\frac{4\pi q^3}{(2mc)^3} \sum_{pn} \times \left[\frac{\alpha_{op}\alpha_{pn}\bar{\alpha}_{no}}{(\omega_{op} - \omega)(\omega_{on} - 2\omega)} + \frac{\alpha_{op}\bar{\alpha}_{pn}\alpha_{no}}{(\omega_{op} - \omega)(\omega_{on} +} + \frac{\bar{\alpha}_{op}\alpha_{pn}\alpha_{no}}{(\omega_{op} + 2\omega)(\omega_{on} + \omega)} \right] \quad (11.12)$$

Since the second-order nonlinear susceptibility is proportional to $c^{(3)}(\infty)$, we have in the E1 approximation

$$\chi^{(2)}_{ijk} \propto \sum_{pn} \left[\frac{\langle o|\mu_i|p\rangle\langle p|\mu_j|n\rangle\langle n|\mu_k|o\rangle}{(\omega_{op} - \omega)(\omega_{on} - 2\omega)} + \frac{\langle o|\mu_j|p\rangle\langle p|\mu_i|n\rangle\langle n|\mu_k|o\rangle}{(\omega_{op} - \omega)(\omega_{on} + \omega)} + \frac{\langle o|\mu_j|p\rangle\langle p|\mu_k|n\rangle\langle n|\mu_i|o\rangle}{(\omega_{op} + 2\omega)(\omega_{on} + \omega)} \right] \quad (11.13)$$

where μ_i is the ith Cartesian component of the electric dipole operator. The summations in (11.13) are carried out over all states $|n\rangle$, $|p\rangle$ other than the ground state $|o\rangle$. Efficient frequency-doubling will naturally occur only in materials that are transparent at both ω and 2ω, so only terms for which ω_{po}, $\omega_{no} > 2\omega$ will contribute to $\chi^{(2)}_{ijk}$.

The practical problems associated with SHG may be appreciated by considering a classical theory for wave propagation in the medium. By combining the Maxwell equations (1.37c) and (1.37d), we obtain the homogeneous wave equation [1]

$$\nabla \times \nabla \times \mathbf{E} + \mu_o \varepsilon_o \ddot{\mathbf{E}} = 0 \tag{11.14}$$

for electromagnetic waves propagating in free space. Since $\nabla \times \nabla \times \mathbf{E} \equiv \nabla(\nabla \cdot \mathbf{E}) - \nabla^2 \mathbf{E}$ and since $\nabla \cdot \mathbf{E} = 0$ in vacuum, the wave equation becomes

$$\nabla^2 \mathbf{E} - \mu_o \varepsilon_o \ddot{\mathbf{E}} = 0 \tag{11.15}$$

This has solutions (Section 1.3) of the form

$$\mathbf{E}(\mathbf{r}, t) = \mathbf{E}_1 e^{i(\mathbf{k}_1 \cdot \mathbf{r} - \omega_1 t)} \tag{11.16}$$

When an electromagnetic wave propagates through a frequency-doubling medium, the homogeneous equation (11.15) becomes superseded by [1,5]

$$\nabla^2 \mathbf{E} - \mu_o \varepsilon_o \ddot{E} = \mu_o \ddot{\mathbf{P}} \tag{11.17}$$

where the *source term* in $\ddot{\mathbf{P}}$ reflects the fact that electromagnetic waves are radiated from regions with oscillating polarization $\mathbf{P}$. Owing to the source term, the plane wave (11.16) is not a solution to the wave equation inside the medium. When the optical nonlinearity is dominated by second-order terms due to SHG, the polarization is

$$\begin{aligned} P &= \varepsilon_o \chi E \\ &= \varepsilon_o(\chi^{(1)} E + \chi^{(2)} E^2) \end{aligned} \tag{11.18}$$

where we have assumed that the susceptibility is isotropic to simplify our algebra. The wave equation then becomes

$$\begin{aligned} \nabla^2 E - \mu_o \varepsilon_o \ddot{E} &= \varepsilon_o \mu_o \left(\chi^{(1)} \ddot{E} + \chi^{(2)} \frac{\partial^2 E^2}{\partial t^2} \right) \\ &\equiv \mu_o \ddot{P} \end{aligned} \tag{11.19}$$

The source term in (11.19) now contains two contributions P_1 and P_2 due to the linear and nonlinear susceptibilities $\chi^{(1)}$ and $\chi^{(2)}$, respectively. As a zeroth-order approximation to **E**, we may take the plane wave (11.16). The nonlinear contribution to the source term is then

$$\ddot{P}_2 = -\varepsilon_o \chi^{(2)} E_1^2 \omega_1^2 e^{2i(k_1 r - \omega_1 t)} \tag{11.20}$$

This equation asserts that as the incident wave of frequency ω_1 propagates through the medium it stimulates the radiation of new waves with frequency $\omega_2 = 2\omega_1$ at every point r along the optical path. The SHG electric field dE_2 generated by the incident wave travelling through distance dr near $r = r_o$ will then have an amplitude proportional to [5]

$$\varepsilon_o \chi^{(2)} E_1^2 \omega_1^2 e^{2i(k_1 r_o - \omega_1 t)} dr \tag{11.21}$$

Note that the amplitude varies as E_1^2, so that the SHG intensity is quadratic in the laser intensity. Since the SHG radiation will itself propagate with wavevector $\mathbf{k}_2$, the infinitesimal field generated near r_o will be

$$dE_2 \propto \varepsilon_o \chi^{(2)} E_1^2 \omega_1^2 e^{2i(k_1 r_o - \omega_1 t)} e^{i[k_2(r - r_o)]}$$

after it has propagated from r_0 to some arbitrary point r down the path. The quantity $k_2(r - r_0)$ is the phase change accompanying SHG beam propagation from r_0 to r. The total SHG field observed at point r is the resultant sum

$$E_2 \propto \varepsilon_o \chi^{(2)} E_1^2 \omega_1^2 e^{i(k_2 r - \omega_2 t)} \int_o^r e^{i(2k_1 - k_2) r_o} dr_o$$

$$= \varepsilon_o \chi^{(2)} E_1^2 \omega_1^2 e^{i[k_2 r - \omega_2 t]} \left[\frac{e^{i(2k_1 - k_2) r} - 1}{i(2k_1 - k_2)} \right] \tag{11.22}$$

of interfering waves generated between positions 0 (corresponding to the edge of the medium) and r. The wave vectors k_1, k_2 are related to the fundamental and SHG frequencies ω_1, ω_2 by

$$k_1 = n_1 \omega_1 / c \tag{11.23}$$

and

$$k_2 = n_2 \omega_2 / c = 2 n_2 \omega_1 / c \tag{11.24}$$

where n_1, n_2 are the refractive indices of the medium at frequencies ω_1, ω_2. Since most materials are dispersive ($n_1 \neq n_2$), one ordinarily finds $2k_1 - k_2 \neq 0$. Hence the SHG amplitude will oscillate as $\exp[i(2k_1 - k_2)r] - 1$, with periodicity

$$l_c = \pi / |2k_1 - k_2| \tag{11.25}$$

along the optical path. This oscillation, caused by interference between SHG waves generated at different points along the path, severely limits the attainable SHG intensity unless special provisions are made to achieve the *index-matching* condition $2k_1 - k_2 = 0$. The distance l_c, called the *coherence length*, is typically several wavelengths in ordinary condensed media.

In an index-matched medium, the coherence length becomes infinite, and complete conversion of fundamental into second harmonic becomes theoretically possible. Index matching is not possible in isotropic media with normal dispersion, where $n_2 > n_1$. (In dilute isotropic media, the molecular polarizability α is proportional to $(n^2 - 1)$ according to the Clausius-Masotti equation. It is apparent from Eq. 10.31 that the frequency-dependent polarizability $\alpha(\omega)$ increases when the optical frequency is increased, in the normal dispersion regime where ω is smaller than any frequency ω_{on} for E1-allowed transitions from the ground state $|0\rangle$ to excited state $|n\rangle$ of the medium. It then follows that $n_2 > n_1$ in normally dispersive media.) It is possible in principle to achieve index-matching in anomalously dispersive isotropic media, in which ω is larger than some ω_{on}, and in which n_2 is not necessarily larger than n_1. However, such media are likely to absorb prohibitively at ω_1 (not to mention ω_2). Practical index matching is instead achieved in birefringent crystals [6], in which the refractive indices differ for the two linear polarizations and depend on the direction of propagation. The incident laser fundamental is linearly polarized, and the SHG emerges with orthogonal polarization. The direction of propagation is adjusted by aligning the crystal in a gimbal mount so that the refractive indices at ω_1 and ω_2 become equal for the respective polarizations. Large SHG conversion efficiencies can then be obtained.

Still another criterion for an SHG medium is implicit in Eq. 11.18, where the presence of the nonzero second-order nonlinear susceptibility $\chi^{(2)}$ implies that the polarization $\mathbf{P}$ cannot simply change sign if the electric field $\mathbf{E}$ is reversed in direction. (The inclusion of only odd-order terms varying as $E, E^3, E^5, \ldots$ in $\mathbf{P}$ would ensure that $\mathbf{P}(-\mathbf{E}) = -\mathbf{P}(\mathbf{E})$.) Hence SHG is impossible in any medium for which reversing $\mathbf{E}$ produces an equal but opposite polarization $\mathbf{P}$. Such media include isotropic media (liquids, gases) and crystals belonging to centrosymmetric space groups. For these materials, $\chi^{(2)}$ vanishes by symmetry. By way of contrast, efficient third-harmonic generation is possible in isotropic media, and was demonstrated many years ago in Na vapor. A common SHG crystal for frequency-doubling Nd^{3+}:YAG lasers is potassium dihydrogen phosphate (KDP), which belongs to the noncentrosymmetric space group $\bar{4}2m$.

11.3 COHERENT ANTI-STOKES RAMAN SCATTERING

Coherent anti-Stokes Raman scattering (CARS) is one of several four-photon optical phenomena that can occur when a sample is exposed to two intense laser beams with frequencies ω_1, ω_2. Some of the other phenomena, two of which are shown in Fig. 11.4, are the harmonic generation and frequency-summing

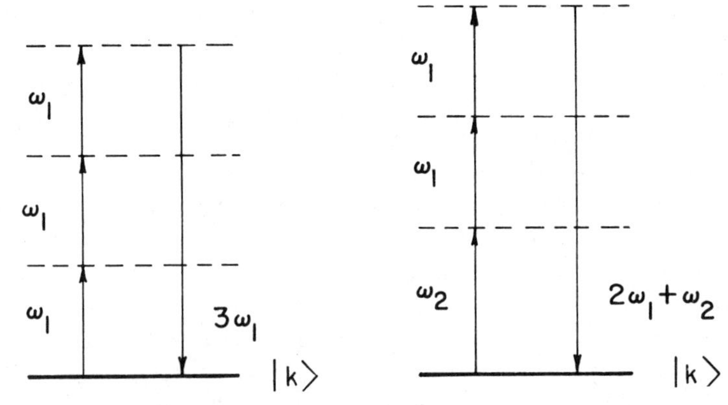

Figure 11.4 Energy level diagrams for third-harmonic generation $\omega_1 + \omega_1 + \omega_1 \rightarrow 3\omega_1$ (left) and frequency summing $\omega_1 + \omega_1 + \omega_2 \rightarrow (2\omega_1 + \omega_2)$ (right), two of the four-photon processes which are possible in a material subjected to two intense beams at frequencies ω_1 and ω_2.

processes which yield the scattered frequencies $3\omega_1$, $2\omega_1 + \omega_2$, $\omega_1 + 2\omega_2$, and $3\omega_2$. In CARS, two photons of frequency ω_1 are absorbed, one photon of frequency ω_2 is scattered via stimulated emission, and a photon at the new frequency $\omega_3 = 2\omega_1 - \omega_2$ is coherently scattered (Fig. 11.5). It is apparent in this figure that when the frequency difference $\omega_1 - \omega_2$ is tuned to match a molecular vibrational/rotational energy level difference, ω_3 becomes identical with an anti-Stokes frequency in the conventional Raman spectrum excited by a laser at ω_1. In this special case (called resonant CARS), the scattered intensity at ω_3 exceeds typical intensities of Stokes bands in ordinary Raman scattering by

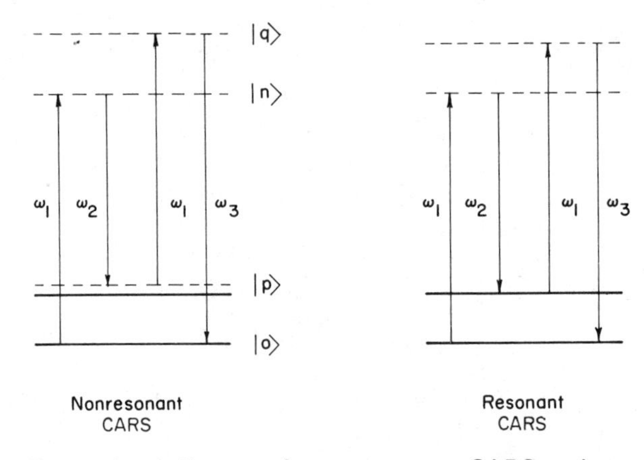

Figure 11.5 Energy level diagrams for nonresonant CARS and resonant CARS. The latter case occurs when $(\omega_1 - \omega_2)$ matches some molecular energy level difference $E_p - E_o$.

several orders of magnitude. Resonant CARS thus offers greater sensitivity and improved spectral resolution over conventional Raman spectroscopy.

As a first step in deriving expressions for CARS transition probabilities, we list in Fig. 11.6 the 12 time-ordered graphs corresponding to absorption of two photons at ω_1 and scattering of photons at ω_2, ω_3. Using the diagrammatic techniques introduced in Section 11.1, we immediately get for the CARS contribution to the fourth-order coefficient in the time-dependent perturbation expansion (1.96)

$$
\begin{aligned}
c^{\text{CARS}}(\infty) = -\frac{2\pi q^4}{(2mc)^4} \sum_{npq} \Bigg[& \frac{\bar{\alpha}_{oq}\bar{\alpha}_{qp}\alpha_{pn}\alpha_{no}}{(\omega_{oq} - \omega_2 + 2\omega_1)(\omega_{op} + 2\omega_1)(\omega_{on} + \omega_1)} \\
& + \frac{\bar{\alpha}_{oq}\bar{\alpha}_{qp}\alpha_{pn}\alpha_{no}}{(\omega_{oq} - \omega_3 + 2\omega_1)(\omega_{op} + 2\omega_1)(\omega_{on} + \omega_1)} \\
& + \frac{\bar{\alpha}_{oq}\alpha_{qp}\bar{\alpha}_{pn}\alpha_{no}}{(\omega_{oq} - \omega_2 + 2\omega_1)(\omega_{op} - \omega_2 + \omega_1)(\omega_{on} + \omega_1)} \\
& + \frac{\bar{\alpha}_{oq}\alpha_{qp}\bar{\alpha}_{pn}\alpha_{no}}{(\omega_{oq} - \omega_3 + 2\omega_1)(\omega_{op} - \omega_3 + \omega_1)(\omega_{on} + \omega_1)} \\
& + \frac{\alpha_{oq}\bar{\alpha}_{qp}\bar{\alpha}_{pn}\alpha_{no}}{(\omega_{oq} - \omega_2 - \omega_3 + \omega_1)(\omega_{op} - \omega_2 + \omega_1)(\omega_{on} + \omega_1)} \\
& + \frac{\alpha_{oq}\bar{\alpha}_{qp}\bar{\alpha}_{pn}\alpha_{no}}{(\omega_{oq} - \omega_2 - \omega_3 + \omega_1)(\omega_{op} - \omega_3 + \omega_1)(\omega_{on} + \omega_1)} \\
& + \frac{\bar{\alpha}_{oq}\alpha_{qp}\alpha_{pn}\bar{\alpha}_{no}}{(\omega_{oq} + 2\omega_1 - \omega_2)(\omega_{op} + \omega_1 - \omega_2)(\omega_{on} - \omega_2)} \\
& + \frac{\bar{\alpha}_{oq}\alpha_{qp}\alpha_{pn}\bar{\alpha}_{no}}{(\omega_{oq} + 2\omega_1 - \omega_3)(\omega_{op} + \omega_1 - \omega_3)(\omega_{on} - \omega_3)} \\
& + \frac{\alpha_{oq}\bar{\alpha}_{qp}\alpha_{pn}\bar{\alpha}_{no}}{(\omega_{oq} - \omega_3 + \omega_1 - \omega_2)(\omega_{op} + \omega_1 - \omega_2)(\omega_{on} - \omega_2)} \\
& + \frac{\alpha_{oq}\bar{\alpha}_{qp}\alpha_{pn}\bar{\alpha}_{no}}{(\omega_{oq} - \omega_2 + \omega_1 - \omega_3)(\omega_{op} + \omega_1 - \omega_3)(\omega_{on} - \omega_3)} \\
& + \frac{\alpha_{oq}\alpha_{qp}\bar{\alpha}_{pn}\bar{\alpha}_{no}}{(\omega_{oq} + \omega_1 - \omega_3 - \omega_2)(\omega_{op} - \omega_3 - \omega_2)(\omega_{on} - \omega_2)} \\
& + \frac{\alpha_{oq}\alpha_{qp}\bar{\alpha}_{pn}\bar{\alpha}_{no}}{(\omega_{oq} + \omega_1 - \omega_2 - \omega_3)(\omega_{op} - \omega_2 - \omega_3)(\omega_{on} - \omega_3)} \Bigg]
\end{aligned}
\tag{11.26}
$$

This coefficient is proportional to the third-order nonlinear susceptibility responsible for CARS. It may be simplified somewhat by defining

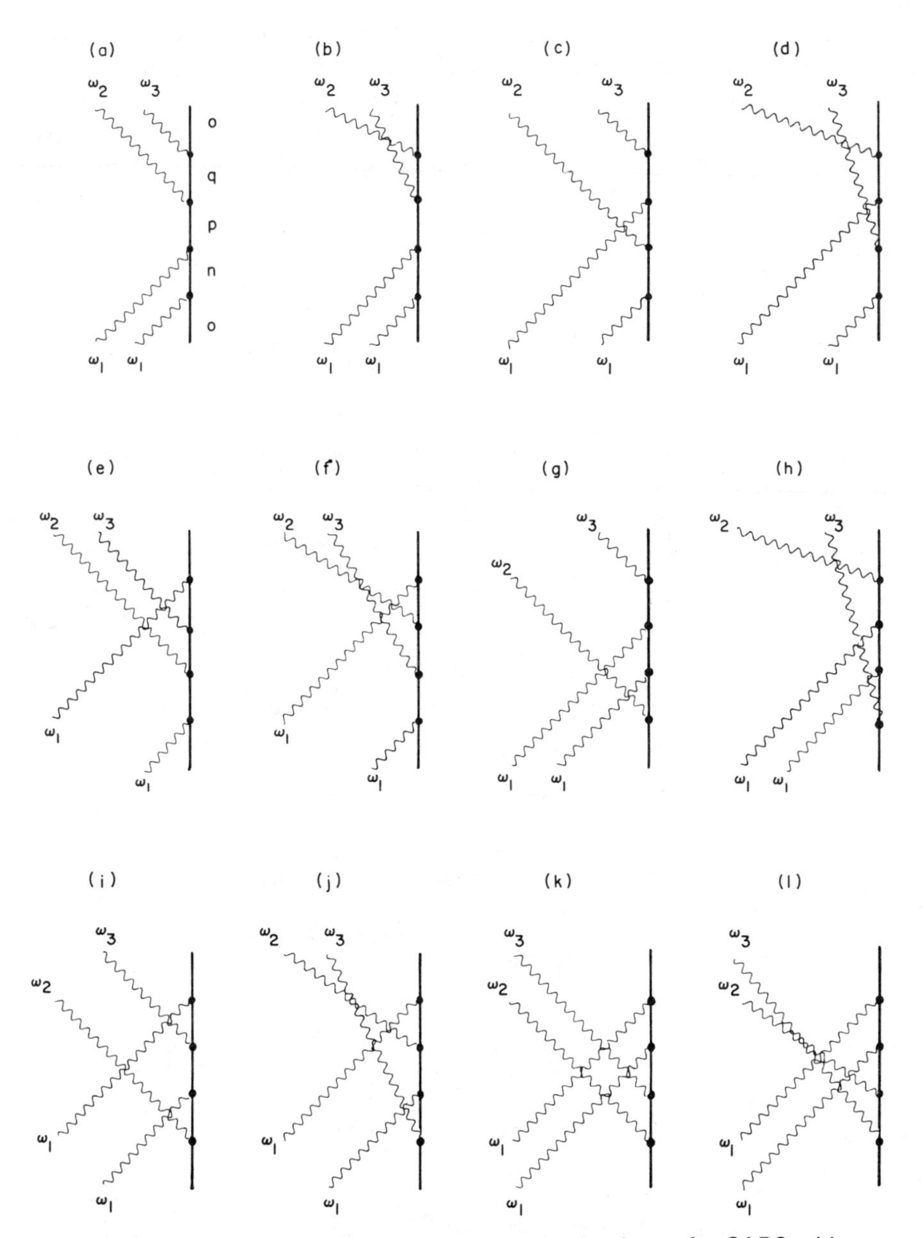

Figure 11.6 The 12 time-ordered graphs that can be drawn for CARS with two incident photons at frequency ω_1 and scattered photons at frequencies ω_2 and ω_3.

$\Delta \equiv \omega_1 - \omega_2 = \omega_3 - \omega_1$. In the E1 approximation, repeated application of the identity $2\omega_1 - \omega_2 - \omega_3 = 0$ then leads to

$$\chi_{ijkl}^{\mathrm{CARS}} = h^{-3} \sum_{npq} \left[\frac{(\mu_j^{on}\mu_l^{np} + \mu_l^{on}\mu_j^{np})\mu_k^{pq}\mu_i^{qo}}{2(\omega_{oq}+\omega_3)(\omega_{op}+2\omega_1)(\omega_{on}+\omega_1)} \right.$$
$$+ \frac{(\mu_j^{on}\mu_l^{np} + \mu_l^{on}\mu_j^{np})\mu_i^{pq}\mu_k^{qo}}{2(\omega_{oq}+\omega_2)(\omega_{op}+2\omega_1)(\omega_{on}+\omega_1)} + \frac{\mu_l^{on}\mu_k^{np}\mu_j^{pq}\mu_i^{qo}}{(\omega_{oq}+\omega_3)(\omega_{op}+\Delta)(\omega_{on}+\omega_1)}$$
$$+ \frac{\mu_j^{on}\mu_i^{np}\mu_l^{pq}\mu_k^{qo}}{(\omega_{oq}+\omega_2)(\omega_{op}-\Delta)(\omega_{on}+\omega_1)} + \frac{\mu_l^{on}\mu_k^{np}\mu_i^{pq}\mu_j^{qo}}{(\omega_{oq}-\omega_1)(\omega_{op}+\Delta)(\omega_{on}+\omega_1)}$$
$$+ \frac{\mu_j^{on}\mu_i^{np}\mu_k^{pq}\mu_l^{qo}}{(\omega_{oq}-\omega_1)(\omega_{op}-\Delta)(\omega_{on}+\omega_1)} + \frac{\mu_k^{on}\mu_l^{np}\mu_j^{pq}\mu_i^{qo}}{(\omega_{oq}+\omega_3)(\omega_{op}+\Delta)(\omega_{on}-\omega_2)}$$
$$+ \frac{\mu_i^{on}\mu_j^{np}\mu_l^{pq}\mu_k^{qo}}{(\omega_{oq}+\omega_2)(\omega_{op}-\Delta)(\omega_{on}-\omega_3)} + \frac{\mu_k^{on}\mu_l^{np}\mu_i^{pq}\mu_j^{qo}}{(\omega_{oq}-\omega_1)(\omega_{op}+\Delta)(\omega_{on}-\omega_2)}$$
$$+ \frac{\mu_i^{on}\mu_j^{np}\mu_k^{pq}\mu_l^{qo}}{(\omega_{oq}-\omega_1)(\omega_{op}-\Delta)(\omega_{on}-\omega_3)} + \frac{\mu_k^{on}\mu_i^{np}(\mu_j^{pq}\mu_l^{qo} + \mu_l^{pq}\mu_j^{qo})}{2(\omega_{oq}-\omega_1)(\omega_{op}-2\omega_1)(\omega_{on}-\omega_2)}$$
$$\left. + \frac{\mu_i^{on}\mu_k^{np}(\mu_l^{pq}\mu_j^{qo} + \mu_j^{pq}\mu_l^{qo})}{2(\omega_{oq}-\omega_1)(\omega_{op}-2\omega_1)(\omega_{on}-\omega_3)} \right] \tag{11.27}$$

where we have used the abbreviation $\mu_i^{on} = \langle o|\mu_i|n\rangle$ and similarly throughout. The terms in Eq. 11.27 have been listed in the same order as the graphs (a)–(l) in Fig. 11.6. When the frequencies ω_1, ω_2 are tuned so that $\hbar(\omega_1 - \omega_2) \equiv \hbar\Delta$ matches some molecular energy level difference $E_p - E_0$ (Fig. 11.5), the terms in $\chi_{ijkl}^{\mathrm{CARS}}$ proportional to $(\omega_{op} + \Delta)^{-1}$ become large. These terms, which correspond to graphs (c), (e), (g), and (i), are responsible for the resonant CARS phenomenon: When the laser frequency ω_2 is swept across the resonance condition $\hbar(\omega_1 - \omega_2) = E_p - E_o$ while ω_1 is held fixed, the CARS intensity peaks sharply at the value of ω_2 at which Stokes Raman scattering off the laser frequency ω_1 would be observed. The remaining (nonresonant) terms in $\chi_{ijkl}^{\mathrm{CARS}}$ contribute a weak background intensity which varies little with ω_2. It may be shown that the total scattering intensity at ω_3 is [4]

$$I_3 = \frac{I_1^2 I_2 k_3^4 N^2}{16\pi^2\varepsilon_0^4 c^2} \sum_{ijkl} |\hat{E}_i(\omega_3)\hat{E}_j(\omega_1)\hat{E}_k(\omega_2)\hat{E}_l(\omega_1)\chi_{ijkl}^{\mathrm{CARS}}|^2 \tag{11.28}$$

Here I_1, I_2 are the incident light intensities at ω_1, ω_2; N is number of scattering molecules; $\hat{E}_i(\omega_3)$ is the projection of the unit electric field vector at frequency ω_3 along Cartesian axis i, and similarly for $\hat{E}_j(\omega_1)$, etc.; and $\chi_{ijkl}^{\mathrm{CARS}}$ is averaged over the molecular orientational distribution. The intensity is proportional to the square of both the number of scattering centers and the third-order nonlinear susceptibility.

Generation of a CARS signal requires that the momentum conservation condition

$$2\mathbf{k}_1 = \mathbf{k}_2 + \mathbf{k}_3 \tag{11.29}$$

be satisfied for the incoming and scattered photons. The magnitude k_i of each wave vector is given by $n_i\omega_i/c$, where n_i is the sample medium's refractive index at frequency ω_i. In dilute gases, where the refractive index dispersion is low (i.e., n_i is relatively insensitive to ω_i), one automatically satisfies (11.29) by using a collinear beam geometry in which all three wave vectors are parallel. The refractive indices depend appreciably on the frequencies ω_i in liquids, however. In this case, index matching may be achieved by crossing the incident beams at an angle θ which achieves momentum conservation (Fig. 11.7). From the law of cosines, the required angle θ is given by

$$\cos\theta = \frac{4k_1^2 - k_3^2 + k_2^2}{4k_1k_2}$$

$$= \frac{4\omega_1\omega_2 n_3^2 - 4\omega_1^2(n_3^2 - n_1^2) - \omega_2^2(n_3^2 - n_2^2)}{4\omega_1\omega_2 n_1 n_2} \tag{11.30}$$

Owing to this momentum conservation, the CARS signal with wave vector $\mathbf{k}_3$ is directionally concentrated in a laser beam with a divergence of typically 10^{-4} steradians. This contrasts with conventional Raman spectroscopy, in which the signal is dispersed over 4π steradians. CARS is thus an advantageous technique for studying vibrational transitions in samples where the scattered signal of interest is accompanied by fluorescence background, a problem frequently encountered in biological systems. Its directional selectivity, combined with the intensity enhancement encountered in resonant CARS, renders it sensitive enough to detect gases at pressures down to $\sim 10^{-10}$ atm. Disadvantages of CARS include the need for a tunable laser to sweep ω_2 (a single-wavelength laser suffices in conventional Raman spectroscopy), and the sensitive alignments required for momentum conservation in condensed samples. The ultimate

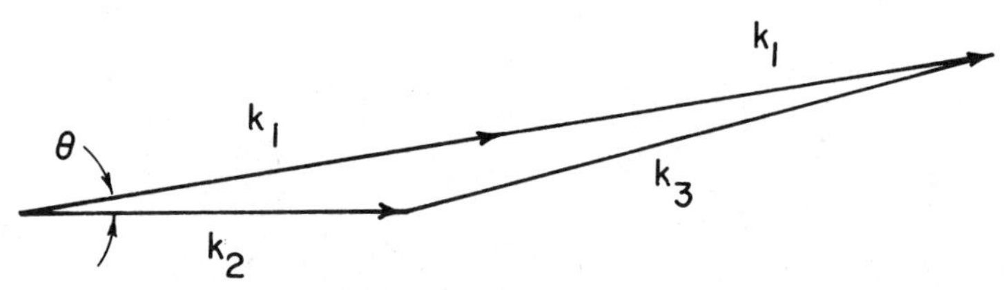

Figure 11.7 Index-matching geometry for conservation of photon momentum in CARS, $2\mathbf{k}_1 = \mathbf{k}_2 + \mathbf{k}_3$. The experimental angle between the laser beams at frequencies ω_1 and ω_2 must be adjusted to the value θ given in Eq. 11.30 for observation of CARS; the scattered anti-Stokes signal emerges in the well-defined direction $\hat{k}_3$.

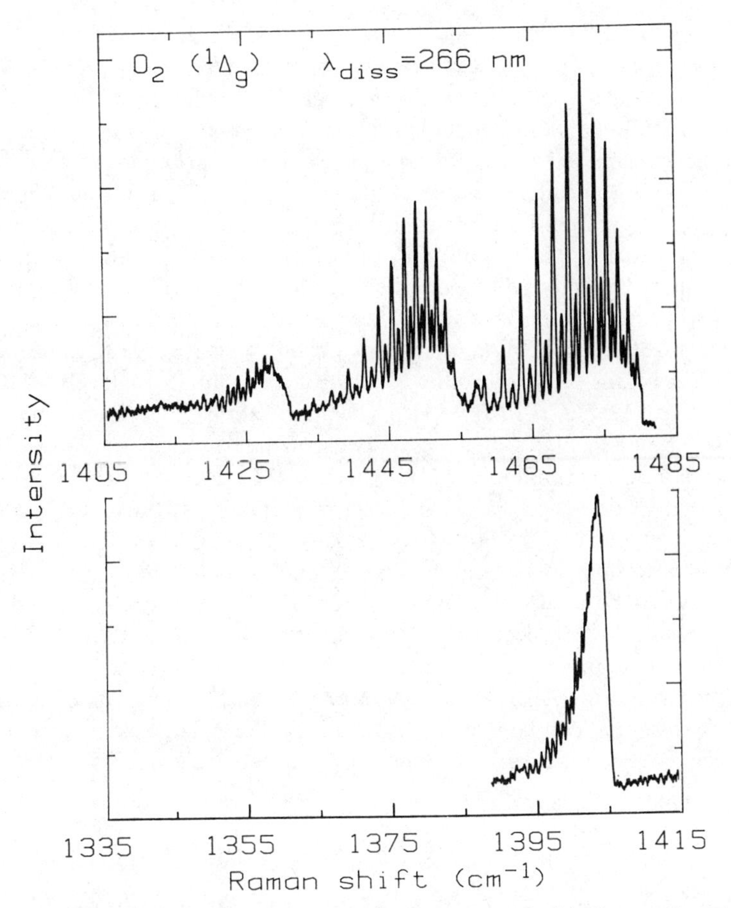

Figure 11.8 Vibrational Q-branch CARS spectrum of $a^1\Delta_gO_2$ produced by O_3 photodissociation at 266 nm. The bands originating from $v'' = 0$, 1, 2, and 3 are centered at 1473, 1450, 1428, and 1403 cm^{-1}, respectively. Fine structure arises from rotational transitions with $\Delta J = 0$. Reproduced by permission from J. J. Valentini, D. P. Gerrity, D. L. Phillips, J. C. Nieh, and K. D. Tabor, *J. Chem. Phys.* **86**; 6745 (1987).

limitation on CARS sensitivity is imposed by background scattering arising from the nonresonant terms in Eq. 11.27.

An example of CARS detection of molecules at low densities is given in Fig. 11.8, which shows the CARS spectrum of $a^1\Delta_g$ O_2 molecules created by photolysis of ozone [7],

$$O_3 \xrightarrow{h\nu} O_2(a^1\Delta_g) + O(^2P)$$

The ozone is photolyzed by a 266-nm (near ultraviolet) laser pulse. The fixed pump frequency ω_1 for CARS detection of the nascent O_2 molecules is provided by 532-nm second-harmonic pulses from a Nd:YAG laser. Tunable pulses (572–578 nm) from a dye laser provide the variable probe frequency ω_2. The Raman shift $\Delta = \omega_1 - \omega_2$ is plotted as the horizontal coordinate in Fig. 11.8. The CARS transitions (which obey the selection rules $\Delta v = 1$, $\Delta J = 0$ between states $|o\rangle$ and $|p\rangle$) originate from vibrational levels $v'' = 0$ through 4 in $a^1\Delta_g$ excited state O_2 molecules produced in the photodissociation. The rotational Q-branch lines appear at different frequencies for different J owing to differences in B'', B' for the lower and upper vibrational levels in each CARS transition. The line intensities may be analyzed to yield the rotational/vibrational state populations in the O_2 protofragments, which reflect on the dynamics of the photodissociation process. (CARS line intensities are not proportional to state populations (cf. Eq. 11.28); peak intensities for transitions connecting levels (v'', J'') and (v', J') vary as $[N(v'', J'') - N(v', J')]^2$.) The alternations in CARS rotational line intensities shown in Fig. 11.8 are *not* a consequence of nuclear spin statistics in $^{16}O_2$, since $a^1\Delta_g$ O_2 (unlike $X^3\Sigma_g^-$ O_2, Section 4.6) can exist in either even- or odd-J levels. Rather, they indicate a propensity for selective O_3 photodissociation into even J levels. Measurement of these state populations by conventional Raman spectroscopy is not feasible, since the initial O_3 pressure is only 1 torr. Laser-induced fluorescence (in which the photofragment molecules are excited to a higher electronic state, and the resulting rotationally resolved fluorescence band intensities are analyzed to determine the state populations) is more sensitive than CARS. However, this technique requires an E1-accessible electronic state that can be reached by a tunable laser. There is no such state in O_2, which begins to absorb strongly only in the vacuum UV. Hence, this example illustrates the generality as well as sensitivity of CARS.

REFERENCES

1. M. H. Nayfeh and M. K. Brussel, *Electricity and Magnetism*, Wiley, New York, 1985.
2. G. W. King, *Spectroscopy and Molecular Structure*, Holt, Rinehart, & Winston, New York, 1964.
3. J. I. Steinfeld, *Molecules and Radiation*, 2d ed., MIT Press, Cambridge, MA, 1985.
4. D. P. Craig and T. Thirunamachandran, *Molecular Quantum Electrodynamics*, Academic, London, 1984.
5. G. C. Baldwin, *An Introduction to Nonlinear Optics*, Plenum, New York, 1969; N. Bloembergen, *Nonlinear Optics*, W. A. Benjamin, Reading, MA, 1965.
6. A. Yariv, *Quantum Electronics*, 2d ed., Wiley, New York, 1975.
7. J. J. Valentini, D. P. Gerrity, D. L. Phillips, J.-C. Nieh, and K. D. Tabor, *J. Chem. Phys.* **86**: 6745 (1987).

Appendix A

FUNDAMENTAL CONSTANTS

Atomic mass unit	amu $= 1.6605655 \times 10^{-27}$ kg
Electron rest mass	$m_e = 9.109534 \times 10^{-31}$ kg
Proton rest mass	$m_p = 1.6726485 \times 10^{-27}$ kg
Elementary charge	$e = 1.6021892 \times 10^{-19}$ coulomb (C)
Speed of light	$c = 2.99792458 \times 10^{8}$ m/s
Permittivity of vacuum	$\varepsilon_0 = 8.85418782 \times 10^{-12}$ $C^2/J \cdot m$
Permeability of vacuum	$\mu_0 = 1.2566370614 \times 10^{-6}$ henry/m
Planck's constant	$h = 6.626176 \times 10^{-34}$ J·s
	$\hbar = 1.0545887 \times 10^{-34}$ J·s
Free electron g factor	$g_e = 2.00231931$
Rydberg constant	$R = 1.0973731 \times 10^{5}$ cm^{-1}
Bohr radius	$a_0 = 0.52917706 \times 10^{-10}$ m
Fine structure constant	$\alpha = 1/137.03604$
Avogadro's number	$N_A = 6.022045 \times 10^{23}$/mol
Boltzmann constant	$k = 1.380662 \times 10^{-23}$ J/K

Data taken from E. R. Cohen and B. N. Taylor, *J. Phys. Chem. Ref. Data* **2**: 663 (1973).

Appendix B

ENERGY CONVERSION FACTORS

$$
\begin{aligned}
1\ \text{J} &= 6.24146 \times 10^{18}\ \text{eV} \\
&= 5.03404 \times 10^{22}\ \text{cm}^{-1} \\
&= 1.43834 \times 10^{23}\ \text{cal/mol}
\end{aligned}
$$

$$
\begin{aligned}
1\ \text{eV} &= 1.60219 \times 10^{-19}\ \text{J} \\
&= 8065.48\ \text{cm}^{-1} \\
&= 2.30450 \times 10^{4}\ \text{cal/mol}
\end{aligned}
$$

$$
\begin{aligned}
1\ \text{cm}^{-1} &= 1.98648 \times 10^{-23}\ \text{J} \\
&= 1.23985 \times 10^{-4}\ \text{eV} \\
&= 2.85724\ \text{cal/mol}
\end{aligned}
$$

$$
\begin{aligned}
1\ \text{cal/mol} &= 6.95246 \times 10^{-24}\ \text{J} \\
&= 4.33934 \times 10^{-5}\ \text{eV} \\
&= 3.49989 \times 10^{-1}\ \text{cm}^{-1}
\end{aligned}
$$

$$
1\ \text{erg} = 10^{-7}\ \text{J}
$$

Appendix C

MULTIPOLE EXPANSIONS OF CHARGE DISTRIBUTIONS

The general problem is to evaluate the electrostatic potential $\phi(\mathbf{r})$ at some point $\mathbf{r}$ due to the presence of a molecular charge distribution $\rho(\mathbf{r})$. According to classical electrostatics, it is given in SI units by

$$\phi(\mathbf{r}) = \frac{1}{4\pi\varepsilon_0} \int \frac{\rho(\mathbf{r}')d\mathbf{r}'}{|\mathbf{r} - \mathbf{r}'|} \tag{C.1}$$

In consequence of the identity (J. D. Jackson, *Classical Electrodynamics*, Wiley, New York, 1962)

$$\frac{1}{|\mathbf{r} - \mathbf{r}'|} \equiv 4\pi \sum_{l=0}^{\infty} \sum_{m=-l}^{l} \frac{1}{2l+1} \frac{r_<^l}{r_>^{l+1}} Y_{lm}^*(\theta', \phi') Y_{lm}(\theta, \phi) \tag{C.2}$$

the charge distribution $\phi(\mathbf{r})$ may be expressed as

$$\phi(\mathbf{r}) = \frac{1}{\varepsilon_0} \sum_{l,m} \frac{1}{2l+1} \left[\int Y_{lm}^*(\theta', \phi') r'^l \rho(\mathbf{r}') d\mathbf{r}' \right] \frac{Y_{lm}(\theta, \phi)}{r^{l+1}} \tag{C.3}$$

where $\mathbf{r} = (r, \theta, \phi)$, $\mathbf{r}' = (r', \theta', \phi')$, and $r_>$ $(r_<)$ is the greater (lesser) of r and r'. Equation C.3 is the well-known *multipole expansion* of the electrostatic potential $\phi(\mathbf{r})$. It is particularly useful for evaluating the electrostatic potential at distances r which are large compared to the distances r' over which the molecular charge distribution is appreciable (Fig. C.1). In this long-range limit, the expansion (C.2) converges rapidly with $r_> = r$ and $r_< = r'$. We may define the *multipole moments*

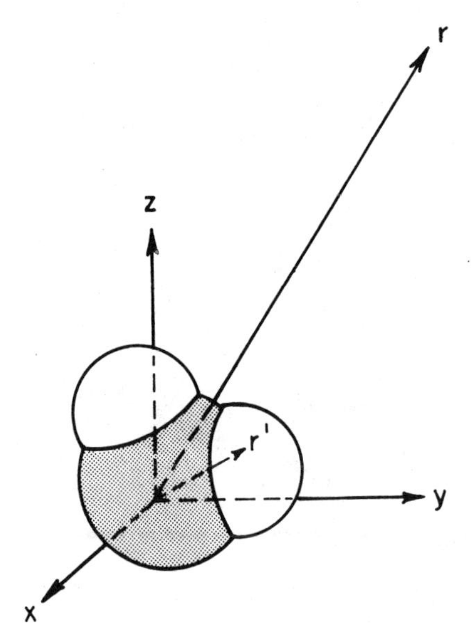

Figure C.1

q_{lm} of the charge distribution as

$$q_{lm} \equiv \int Y^*_{lm}(\theta', \phi') r'^l \rho(\mathbf{r}') d\mathbf{r}' \tag{C.4}$$

To appreciate the physical significance of the q_{lm}, we evaluate some of the lower order moments explicitly. The moment q_{00} is

$$\begin{aligned} q_{00} &= \int Y^*_{00}(\theta', \phi') \rho(\mathbf{r}') d\mathbf{r}' \\ &= \frac{1}{\sqrt{4\pi}} \int \rho(\mathbf{r}') d\mathbf{r}' = q/\sqrt{4\pi} \end{aligned} \tag{C.5}$$

where q is the total molecular charge. In like manner, the moments q_{10} and q_{11} are

$$\begin{aligned} q_{10} &= \int Y^*_{10} r' \rho(\mathbf{r}') d\mathbf{r} = \sqrt{\frac{3}{4\pi}} \int r' \cos\theta' \rho(\mathbf{r}') d\mathbf{r}' \\ &= \sqrt{\frac{3}{4\pi}} \int z' \rho(\mathbf{r}') d\mathbf{r}' = \sqrt{\frac{3}{4\pi}}\, \mu_z \end{aligned} \tag{C.6}$$

and

$$q_{11} = \int Y_{11}^* r' \rho(\mathbf{r}') d\mathbf{r}' = -\sqrt{\frac{3}{8\pi}} \int r' \sin \theta' e^{-i\phi'} \rho(\mathbf{r}') d\mathbf{r}'$$

$$= -\sqrt{\frac{3}{8\pi}} (\mu_x - i\mu_y) \qquad \text{(C.7)}$$

where μ_x, μ_y, μ_z are the Cartesian components of the molecule's electric dipole moment $\boldsymbol{\mu}$. The multipole moment q_{20} is

$$q_{20} = \int Y_{20}^* r'^2 \rho(\mathbf{r}') d\mathbf{r}'$$

$$= \frac{1}{2}\sqrt{\frac{5}{4\pi}} \int (3z'^2 - r'^2) \rho(\mathbf{r}') d\mathbf{r}' \qquad \text{(C.8)}$$

$$= \frac{1}{2}\sqrt{\frac{5}{4\pi}}\, Q_{zz} \qquad \text{(C.8)}$$

where Q_{zz} is the zz component of the molecule's electric quadrupole moment tensor.

Substitution of the moments back into Equation C.3 finally yields the electrostatic potential

$$\phi(\mathbf{r}) = \frac{1}{\varepsilon_0} \sum_{l,m} \frac{1}{2l+1} q_{lm} \frac{Y_{lm}(\theta, \phi)}{r^{l+1}}$$

$$= \frac{1}{4\pi\varepsilon_0} \left(q/r + \boldsymbol{\mu} \cdot \mathbf{r}/r^3 + \frac{1}{2} \sum_{ij} Q_{ij} \frac{r_i r_j}{r^5} + \cdots \right) \qquad \text{(C.9)}$$

which transparently shows the charge, dipole, and quadrupole contributions to $\phi(\mathbf{r})$. Such multipole expansions yield valuable insights into molecule–radiation interactions and long-range intermolecular forces.

Appendix D

BEER'S LAW

Discussions of light absorption in a homogeneous sample with molar concentration C and path length b are usually couched in terms of *Beer's law*, which states that the light intensity I transmitted by such a sample is related to the incident intensity I_0 in a parallel light beam by

$$I = I_0 e^{-\alpha bC} \tag{D.1}$$

Here α is the *molar absorption coefficient*, which has units of L/mol·cm. Experimental spectroscopists often work with the *decadic* absorption coefficient ε, defined by

$$I = I_0(10)^{-\varepsilon bC} \tag{D.2}$$

Clearly $\varepsilon = \alpha/2.303$. Absorption spectra are frequently recorded as the quantity εbC (which is termed the *absorbance* or *optical density*) versus incident light wavelength λ; knowledge of the absorption cell geometry and sample concentration then allows extraction of the wavelength-dependent absorption coefficient $\varepsilon(\lambda)$. Beer's law is applicable only when the incident light intensity I_0 is low enough that the molecular state populations are essentially unperturbed by it; the redistribution of molecules into excited states by the light beam would otherwise materially change ε for different photons within the beam. In addition, Beer's law is strictly valid only in very thin optical samples.

Appendix E

ADDITION OF TWO ANGULAR MOMENTA

Here we consider the nature of the angular momentum states that result from the vector addition of two angular momenta $\mathbf{J}_1$ and $\mathbf{J}_2$ to form a resultant $\mathbf{J}$,

$$\mathbf{J} = \mathbf{J}_1 + \mathbf{J}_2 \tag{E.1}$$

The Cartesian components of $\mathbf{J}_1$ and $\mathbf{J}_2$ obey the usual angular momentum commutation rules

$$[\hat{J}_{1x}, \hat{J}_{1y}] = i\hbar\hat{J}_{1z} \tag{E.2}$$

$$[\hat{J}_{2x}, \hat{J}_{2y}] = i\hbar_{2z} \tag{E.3}$$

$$[\hat{J}_1^2, \hat{J}_{1z}] = 0 \tag{E.4}$$

$$[\hat{J}_2^2, \hat{J}_{2z}] = 0 \tag{E.5}$$

and the cyclic permutations $x \to y, y \to z, z \to x$. Since $\mathbf{J}_1$ and $\mathbf{J}_2$ are independent angular momenta, all components of $\mathbf{J}_1$ commute with all components of $\mathbf{J}_2$,

$$[\hat{J}_{1i}, \hat{J}_{2j}] = 0 \tag{E.6}$$

for all (i, j). Using the commutation rules (E.2) through (E.6), it is easy to show that

$$\begin{aligned}[\hat{J}_x, \hat{J}_y] &\equiv [\hat{J}_{1x} + \hat{J}_{2x}, \hat{J}_{1y} + \hat{J}_{2y}] \\ &= i\hbar(\hat{J}_{1z} + \hat{J}_{2z}) = i\hbar\hat{J}_z \end{aligned} \tag{E.7}$$

and that

$$[\hat{J}_z, \hat{J}^2] = 0 \tag{E.8}$$

Hence, the vector $\mathbf{J} = \mathbf{J}_1 + \mathbf{J}_2$ itself behaves quantum mechanically like an angular momentum. According to (E.8), it is possible to set up a complete set of eigenstates that are simultaneously eigenstates of $\hat{J}^2$ and $\hat{J}_z$.

The question now arises as to whether such eigenstates can also be eigenstates of $\hat{J}_1^2$, $\hat{J}_2^2$, $\hat{J}_{1z}$, and $\hat{J}_{2z}$. It follows from (E.2) through (E.8) that

$$[\hat{J}_1^2, \hat{J}^2] = [\hat{J}_2^2, \hat{J}^2] = 0 \tag{E.9}$$

$$[\hat{J}_1^2, \hat{J}_z] = [\hat{J}_2^2, \hat{J}_z] = 0 \tag{E.10}$$

but that

$$[\hat{J}^2, J_{1z}] = 2i\hbar(J_{1x}J_{2y} - J_{2x}J_{1y}) \neq 0 \tag{E.11}$$

$$[\hat{J}^2, J_{2z}] = 2i\hbar(J_{2x}J_{1y} - J_{1x}J_{2y}) \neq 0 \tag{E.12}$$

Hence two possible commuting sets of observables are

$$J_1^2, J_{1z}, J_2^2, J_{2z}$$

and

$$J_1^2, J_2^2, J^2, J_z$$

The eigenstates $|j_1 m_1 j_2 m_2\rangle$ of the first commuting set obey the eigenvalue equations

$$\hat{J}_1^1 |j_1 m_1 j_2 m_2\rangle = j_1(j_1 + 1)\hbar^2 |j_1 m_1 j_2 m_2\rangle \tag{E.13}$$

$$\hat{J}_{1z} |j_1 m_1 j_2 m_2\rangle = m_1 \hbar |j_1 m_1 j_2 m_2\rangle \tag{E.14}$$

$$\hat{J}_2^2 |j_1 m_1 j_2 m_2\rangle = j_2(j_2 + 1)\hbar^2 |j_1 m_1 j_2 m_2\rangle \tag{E.15}$$

$$\hat{J}_{2z} |j_1 m_1 j_2 m_2\rangle = m_2 \hbar |j_1 m_1 j_2 m_2\rangle \tag{E.16}$$

and are referred to as the *uncoupled representation* of the resultant angular momentum states. Since $\hat{J}^2$ does not commute with $\hat{J}_{1z}$ or $\hat{J}_{2z}$, these uncoupled states are not eigenfunctions of $\hat{J}^2$ in general, and the value of J^2 is indefinite. For the second commuting set of observables, it is possible to construct a complete set of eigenstates $|j_1 j_2 jm\rangle$ which simultaneously obey

$$\hat{J}_1^2 |j_1 j_2 jm\rangle = j_1(j_1 + 1)\hbar^2 |j_1 j_2 jm\rangle \tag{E.17}$$

$$\hat{J}_2^2 |j_1 j_2 jm\rangle = j_2(j_2 + 1)\hbar^2 |j_1 j_2 jm\rangle \tag{E.18}$$

$$\hat{J}^2|j_1 j_2 jm\rangle = j(j+1)\hbar^2|j_1 j_2 jm\rangle \tag{E.19}$$

$$\hat{J}_z|j_1 j_2 jm\rangle = m\hbar|j_1 j_2 jm\rangle \tag{E.20}$$

This is the *coupled representation.* For a given combination of j_1 and j_2, the possible values of the resultant angular momentum quantum number j are given by the Clebsch-Gordan series

$$j = (j_1 + j_2), (j_1 + j_2 - 1), \ldots, |j_1 - j_2| \tag{E.21}$$

(P. W. Atkins, *Molecular Quantum Mechanics*, 2d ed., Oxford Univ. Press, London, 1983). For given j, the quantum number m can assume one of $(2j + 1)$ values ranging between $+m$ and $-m$. The coupled states $|j_1 j_2 jm\rangle$ are not generally eigenstates of $\hat{J}_{1z}$ or $\hat{J}_{2z}$ (cf. Eqs. E.11, E.12) and so the values of J_{1z} and J_{2z} are generally indefinite in these states.

The pictorial *vector models* of the uncoupled and coupled representations (Fig. E.1) embody the physical consequences of the angular momentum commutation rules. In the *uncoupled* representation, the vectors $\mathbf{J}_1$ and $\mathbf{J}_2$ can lie anywhere on their respective cones with their tips on the edges of the cones. Since these cones are invariant to rotations about the z axis, they represent states with fixed $|J_1| = \hbar\sqrt{j_1(j_1 + 1)}$ and fixed $|\mathbf{J}_2| = \hbar\sqrt{j_2(j_2 + 1)}$. The projections of all vectors $\mathbf{J}_1$ on the lower cone have the definite value $m_1\hbar$. Since all

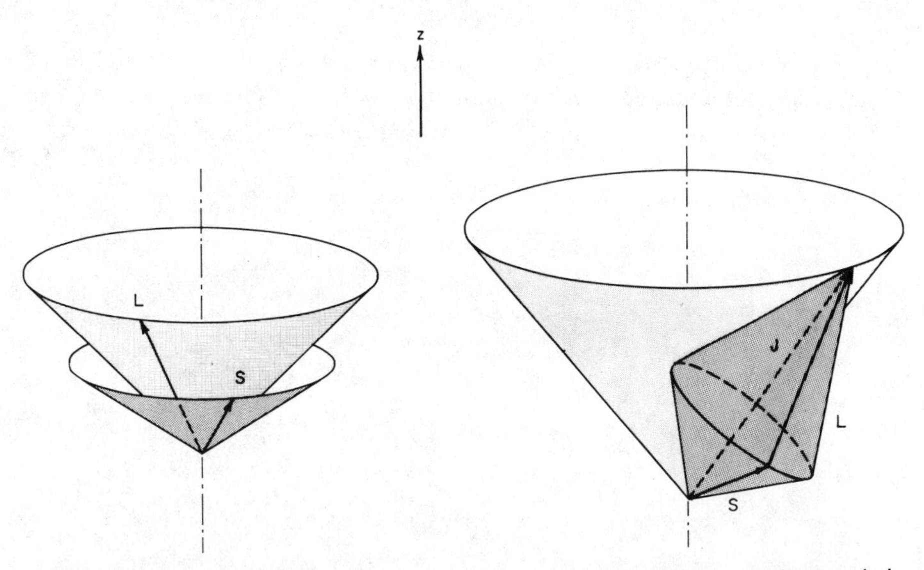

Figure E.1 Vector models for the uncoupled representation (left) and the coupled representation (right). J_1^2, J_{1z}, J_2^2, and J_{2z} are constants of the motion in the uncoupled representation; J_1^2, J_2^2, J^2, and J_z are constants of the motion in the coupled representation.

orientations of $\mathbf{J}_1$ on the cone are equally likely, its projections J_{1x} and J_{1y} along the x and y axes are indefinite. Similar observations apply to $\mathbf{J}_2$: J_{2z} is fixed, while J_{2x} and J_{2y} are indefinite. Since the projections of $\mathbf{J}_1$ and $\mathbf{J}_2$ on the xy plane are uncorrelated (i.e., the phases of $\mathbf{J}_1$ and $\mathbf{J}_2$ fluctuate independently), neither the orientation nor magnitude of their vector sum $\mathbf{J}$ is a constant of the motion in the uncoupled representation. (However, $\mathbf{J}$ does exhibit a definite projection $J_z = J_{1z} + J_{2z} = (m_1 + m_2)\hbar$ along the z axis.)

In the vector model of the *coupled* representation, the vector $\mathbf{J}$ can be found anywhere on the large cone. It exhibits fixed length $|J| = \hbar\sqrt{j(j+1)}$ and fixed projection $J_z = m\hbar$ along the z axis. The individual angular momentum vectors $\mathbf{J}_1$ and $\mathbf{J}_2$ have the fixed lengths $\hbar\sqrt{j_1(j_1+1)}$ and $\hbar\sqrt{j_2(j_2+1)}$, respectively. Since their resultant $|J|$ must also be of constant length, the coupled vector model depicts J_1 and J_2 precessing together, head to tail, to produce a resultant vector $\mathbf{J}$ of fixed length. (In quantum mechanical language, the relative phase of $\mathbf{J}_1$ and $\mathbf{J}_2$ is fixed.) It is clear from the coupled vector model that the motions of $\mathbf{J}_1$ and $\mathbf{J}_2$ on their respective cones do not yield fixed projections J_{1z} and J_{2z} along the z axis; their sum $J_z = m\hbar$ is, of course, definite. Such vector models prove to be useful in discussion of the anomalous Zeeman effect in atoms (Section 2.6) and angular momentum coupling in diatomics (Chapter 4).

Since the uncoupled states $|j_1 m_1 j_2 m_2\rangle$ form a complete set of eigenstates, the coupled states $|j_1 j_2 jm\rangle$ may be expressed as the linear combinations

$$|j_1 j_2 jm\rangle = \sum_{\substack{m_1\\m_2}} |j_1 m_1 j_2 m_2\rangle\langle j_1 m_1 j_2 m_2 | j_1 j_2 jm\rangle \tag{E.22}$$

where the coefficients $\langle j_1 m_1 j_2 m_2 | j_1 j_2 jm\rangle$ are known as the Clebsch-Gordan coefficients. Techniques for obtaining these coefficients using the raising/lowering operators $\hat{J}_{\pm} = \hat{J}_{1\pm} + J_{2\pm}$ are described in Section 2.2.

Appendix F

GROUP CHARACTER TABLES AND DIRECT PRODUCTS

C_2	E	C_2		
A	1	1	z, R_z	x^2, y^2, z^2, xy
B	1	-1	x, y, R_x, R_y	xz, yz

C_3	E	C_3	C_3^2		$\varepsilon = \exp(2\pi i/3)$
A	1	1	1	z, R_z	$x^2 + y^2, z^2$
E	$\begin{cases}1\\1\end{cases}$	ε ε^2	ε^2 ε	$\begin{cases}(x, y)\\(R_x, R_y)\end{cases}$	$\begin{cases}(xz, yz)\\(x^2 - y^2, xy)\end{cases}$

C_4	E	C_2	C_4	C_4^3		
A	1	1	1	1	z, R_z	$x^2 + y^2, z^2$
B	1	1	-1	-1		$x^2 - y^2, xy$
E	$\begin{cases}1\\1\end{cases}$	-1 -1	i $-i$	$-i$ i	$\begin{cases}(x, y)\\(R_x, R_y)\end{cases}$	(xz, yz)

C_5	E	C_5	C_5^2	C_5^3	C_5^4		$\varepsilon = \exp(2\pi i/5)$
A	1	1	1	1	1	z, R_z	$x^2 + y^2, z^2$
E′	$\begin{cases}1\\1\end{cases}$	ε ε^4	ε^2 ε^3	ε^3 ε^2	ε^4 ε	$\begin{cases}(x, y)\\(R_x, R_y)\end{cases}$	(xz, yz)
E″	$\begin{cases}1\\1\end{cases}$	ε^2 ε^3	ε^4 ε	ε ε^4	ε^3 ε^2		$(x^2 - y^2, xy)$

C_6	E	C_6	C_3	C_2	C_3^2	C_6^5		$\varepsilon = \exp(2\pi i/6)$
A	1	1	1	1	1	1	z, R_z	$x^2 + y^2, z^2$
B	1	-1	1	-1	1	-1		
E′	1	ε	ε^2	ε^3	ε^4	ε^5	(x, y)	(xz, yz)
	1	ε^5	ε^4	ε^3	ε^2	ε	(R_x, R_y)	
E″	1	ε^2	ε^4	1	ε^2	ε^4		$(x^2 - y^2, xy)$
	1	ε^4	ε^2	1	ε^4	ε^2		

D_2	E	C_2^z	C_2^y	C_2^x		
A_1	1	1	1	1		x^2, y^2, z^2
B_1	1	1	-1	-1	z, R_z	xy
B_2	1	-1	1	-1	y, R_y	xz
B_3	1	-1	-1	1	x, R_x	yz

D_3	E	$2C_3$	$3C_2'$		
A_1	1	1	1	z, R_z	$x^2 + y^2, z^2$
A_2	1	1	-1		
E	2	-1	0	(x, y) (R_x, R_y)	(xz, yz) $(x^2 - y^2, xy)$

D_4	E	C_2	$2C_4$	$2C_2'$	$2C_2''$		
A_1	1	1	1	1	1		$x^2 + y^2, z^2$
A_2	1	1	1	-1	-1	z, R_z	
B_1	1	1	-1	1	-1		$x^2 - y^2$
B_2	1	1	-1	-1	1		xy
E	2	-2	0	0	0	(x, y) (R_x, R_y)	(xz, yz)

D_5	E	$2C_5$	$2C_5^2$	$5C_2'$		$\phi = 2\pi/5$
A_1	1	1	1	1		$x^2 + y^2, z^2$
A_2	1	1	1	-1	z, R_z	
E_1	2	$2\cos\phi$	$2\cos 2\phi$	0	(x, y) (R_x, R_y)	(xz, yz)
E_2	2	$2\cos 2\phi$	$2\cos 4\phi$	0		$(x^2 - y^2, xy)$

D_6	E	C_2	$2C_3$	$2C_6$	$3C_2'$	$3C_2''$		
A_1	1	1	1	1	1	1		x^2+y^2, z^2
A_2	1	1	1	1	−1	−1	z, R_z	
B_1	1	−1	1	−1	1	−1		
B_2	1	−1	1	−1	−1	1		
E_1	2	−2	−1	1	0	0	$\begin{cases}(x, y)\\(R_x, R_y)\end{cases}$	(xz, yz)
E_2	2	2	−1	−1	0	0		(x^2-y^2, xy)

C_{2v}	E	$C_2(z)$	σ_v	σ_v'		
A_1	1	1	1	1	z	x^2, y^2, z^2
A_2	1	1	−1	−1	R_z	xy
B_1	1	−1	1	−1	x, R_y	xz
B_2	1	−1	−1	1	y, R_x	yz

C_{3v}	E	$2C_3$	$3\sigma_v$		
A_1	1	1	1	z	x^2+y^2, z^2
A_2	1	1	−1	R_z	
E	2	−1	0	$\begin{cases}(x, y)\\(R_x, R_y)\end{cases}$	$\begin{cases}(x^2-y^2, xy)\\(xz, yz)\end{cases}$

C_{4v}	E	C_2	$2C_4$	$2\sigma_v$	$2\sigma_d$		
A_1	1	1	1	1	1	z	x^2+y^2, z^2
A_2	1	1	1	−1	−1	R_z	
B_1	1	1	−1	1	−1		x^2-y^2
B_2	1	1	−1	−1	1		xy
E	2	−2	0	0	0	$\begin{cases}(x, y)\\(R_x, R_y)\end{cases}$	(xz, yz)

C_{5v}	E	$2C_5$	$2C_5^2$	$5\sigma_v$		$\phi = 2\pi/5$
A_1	1	1	1	1	z	x^2+y^2, z^2
A_2	1	1	1	−1	R_z	
E_1	2	$2\cos\phi$	$2\cos 2\phi$	0	$\begin{cases}(x, y)\\(R_x, R_y)\end{cases}$	(xz, yz)
E_2	2	$2\cos 2\phi$	$2\cos 4\phi$	0		(x^2-y^2, xy)

C_{6v}	E	C_2	$2C_3$	$2C_6$	$3\sigma_d$	$3\sigma_v$		
A_1	1	1	1	1	1	1	z	$x^2 + y^2, z^2$
A_2	1	1	1	1	-1	-1	R_z	
B_1	1	-1	1	-1	-1	1		
B_2	1	-1	1	-1	1	-1		
E_1	2	-2	-1	1	0	0	$\begin{cases}(x, y)\\(R_x, R_y)\end{cases}$	(xz, yz)
E_2	2	2	-1	-1	0	0		$(x^2 - y^2, xy)$

C_{1h}	E	σ_h		
A′	1	1	x, y, R_z	x^2, y^2, z^2, xy
A″	1	-1	R_x, R_y, z	xz, yz

C_{2h}	E	C_2	σ_h	i		
A_g	1	1	1	1	R_z	x^2, y^2, z^2, xy
A_u	1	1	-1	-1	z	
B_g	1	-1	-1	1	R_x, R_y	xz, yz
B_u	1	-1	1	-1	x, y	

C_{3h}	E	C_3	C_3^2	σ_h	S_3	$\sigma_h C_3^2$		$\varepsilon = \exp(2\pi i/3)$
A′	1	1	1	1	1	1	R_z	$x^2 + y^2, z^2$
A″	1	1	1	-1	-1	-1	z	
E′	$\begin{cases}1\\1\end{cases}$	$\begin{matrix}\varepsilon\\\varepsilon^2\end{matrix}$	$\begin{matrix}\varepsilon^2\\\varepsilon\end{matrix}$	$\begin{matrix}1\\1\end{matrix}$	$\begin{matrix}\varepsilon\\\varepsilon^2\end{matrix}$	$\begin{matrix}\varepsilon^2\\\varepsilon\end{matrix}$	(x, y)	$(x^2 - y^2, xy)$
E″	$\begin{cases}1\\1\end{cases}$	$\begin{matrix}\varepsilon\\\varepsilon^2\end{matrix}$	$\begin{matrix}\varepsilon^2\\\varepsilon\end{matrix}$	$\begin{matrix}-1\\-1\end{matrix}$	$\begin{matrix}-\varepsilon\\-\varepsilon^2\end{matrix}$	$\begin{matrix}-\varepsilon^2\\-\varepsilon\end{matrix}$	(R_x, R_y)	(xz, yz)

$C_{4h} = i \times C_4 \qquad C_{5h} = \sigma_h \times C_5 \qquad C_{6h} = i \times C_6$

S_2	E	i		
A_g	1	1	R_x, R_y, R_z	$x^2, y^2, z^2, xy, xz, yz$
A_u	1	-1	x, y, z	

S_4	E	C_2	S_4	S_4^3		
A	1	1	1	1	R_z	$x^2 + y^2, z^2$
B	1	1	-1	-1	z	
E	$\begin{cases}1\\1\end{cases}$	$\begin{matrix}-1\\-1\end{matrix}$	$\begin{matrix}i\\-i\end{matrix}$	$\begin{matrix}-i\\i\end{matrix}$	$\begin{cases}(x, y)\\(R_x, R_y)\end{cases}$	$\begin{cases}(xz, yz)\\(x^2 - y^2, xy)\end{cases}$

$S_6 = i \times C_3$

D_{2d}	E	C_2	$2S_4$	$2C_2'$	$2\sigma_d$		
A_1	1	1	1	1	1		$x^2 + y^2, z^2$
A_2	1	1	1	−1	−1	R_z	
B_1	1	1	−1	1	−1		$x^2 - y^2$
B_2	1	1	−1	−1	1	z	xy
E	2	−2	0	0	0	$\{(x, y)$; (R_x, R_y)	(xz, yz)

$D_{3d} = i \times D_3$ $\quad D_{2h} = i \times D_2$

D_{3h}	E	σ_h	$2C_3$	$2S_3$	$3C_2'$	$3\sigma_v$		
A_1'	1	1	1	1	1	1		$x^2 + y^2, z^2$
A_2'	1	1	1	1	−1	−1	R_z	
A_1''	1	−1	1	−1	1	−1		
A_2''	1	−1	1	−1	−1	1	z	
E'	2	2	−1	−1	0	0	(x, y)	$(x^2 - y^2, xy)$
E''	2	−2	−1	1	0	0	(R_x, R_y)	(xz, yz)

$D_{4h} = i \times D_4$ $\quad D_{5h} = \sigma_h \times D_5$ $\quad D_{6h} = i \times D_6$

T	E	$3C_2$	$4C_3$	$4C_3'$	$\varepsilon = \exp(2\pi i/3)$
A	1	1	1	1	
E	$\{1$	1	ε	ε^2	
	$\{1$	1	ε^2	ε	
T	3	−1	0	0	$\{(R_x, R_y, R_z)$; (x, y, z)

T_d	E	$8C_3$	$3C_2$	$6\sigma_d$	$6S_4$	
A_1	1	1	1	1	1	
A_2	1	1	1	−1	−1	
E	2	−1	2	0	0	
T_1	3	0	−1	−1	1	(R_x, R_y, R_z)
T_2	3	0	−1	1	−1	(x, y, z)

O	E	$8C_3$	$3C_2$	$6C_2$	$6C_4$		
A_1	1	1	1	1	1		
A_2	1	1	1	−1	−1		
E	2	−1	2	0	0		$(x^2 - y^2, 3z^2 - r^2)$
T_1	3	0	−1	−1	1	$\begin{cases}(R_x, R_y, R_z)\\(x, y, z)\end{cases}$	
T_2	3	0	−1	1	−1		(xy, yz, zx)

$$O_h = i \times O$$

$C_{\infty v}$	E	$2C_\phi$	σ_v		
Σ^+	1	1	1	z	$x^2 + y^2, z^2$
Σ^-	1	1	−1	R_z	
Π	2	$2 \cos \phi$	0	$(x, y), (R_x, R_y)$	(xz, yz)
Δ	2	$2 \cos 2\phi$	0		$(x^2 - y^2, xy)$

$D_{\infty h}$	E	$2C_\phi$	C_2	i	$2S_\phi$	σ_v		
Σ_g^+	1	1	1	1	1	1		$x^2 + y^2, z^2$
Σ_u^+	1	1	−1	−1	−1	1	z	
Σ_g^-	1	1	−1	1	1	−1	R_z	
Σ_u^-	1	1	1	−1	−1	−1		
Π_g	2	$2 \cos \phi$	0	2	$-2 \cos \phi$	0	(R_x, R_y)	(xz, yz)
Π_u	2	$2 \cos \phi$	0	−2	$2 \cos \phi$	0	(x, y)	
Δ_g	2	$2 \cos 2\phi$	0	2	$2 \cos 2\phi$	0		$(x^2 - y^2, xy)$
Δ_u	2	$2 \cos 2\phi$	0	−2	$-2 \cos 2\phi$	0		

A set of rules for obtaining the *direct product* of two irreducible representations in any point group was set down by E. B. Wilson, Jr., J. C. Decius, and P. C. Cross in their classic book, *Molecular Vibrations*, McGraw-Hill, New York, 1955. The multiplication properties are as follows:

$$\mathrm{A} \otimes \mathrm{A} = \mathrm{B} \otimes \mathrm{B} = \mathrm{A}$$

$$\mathrm{A} \otimes \mathrm{B} = \mathrm{B}$$

$$\mathrm{A} \otimes \mathrm{E} = \mathrm{B} \otimes \mathrm{E} = \mathrm{E}$$

$$\mathrm{A} \otimes \mathrm{T} = \mathrm{B} \otimes \mathrm{T} = \mathrm{T}$$

$$\mathrm{g} \otimes \mathrm{g} = \mathrm{u} \otimes \mathrm{u} = \mathrm{g}$$

$$\mathrm{g} \otimes \mathrm{u} = \mathrm{u}$$

$$' \otimes ' = '' \otimes '' = '$$

$$' \otimes '' = ''$$

$$\mathrm{A} \otimes \mathrm{E}_1 = \mathrm{E}_1 \qquad \mathrm{B} \otimes \mathrm{E}_1 = \mathrm{E}_2$$
$$\mathrm{A} \otimes \mathrm{E}_2 = \mathrm{E}_2 \qquad \mathrm{B} \otimes \mathrm{E}_2 = \mathrm{E}_1$$

For subscripts on A and B representations,

$$1 \otimes 1 = 2 \otimes 2 = 1$$
$$1 \otimes 2 = 2$$

in all groups except D_2 and D_{2h}; for these,

$$1 \otimes 2 = 3$$
$$2 \otimes 3 = 1$$
$$1 \otimes 3 = 2$$

In all groups except C_4, C_{4v}, C_{4h}, D_{2d}, D_4, D_{4h}, and S_4,

$$\mathrm{E}_1 \otimes \mathrm{E}_1 = \mathrm{E}_2 \otimes \mathrm{E}_2 = \mathrm{A}_1 \oplus \mathrm{A}_2 \oplus \mathrm{E}_2$$
$$\mathrm{E}_1 \otimes \mathrm{E}_2 = \mathrm{B}_1 \oplus \mathrm{B}_2 \oplus \mathrm{E}_1$$

In the exceptions noted above,

$$\mathrm{E} \otimes \mathrm{E} = \mathrm{A}_1 \oplus \mathrm{A}_2 \oplus \mathrm{B}_1 \oplus \mathrm{B}_2$$

In the point groups T_d, O, O_h,

$$\mathrm{E} \otimes \mathrm{T}_1 = \mathrm{E} \otimes \mathrm{T}_2 = \mathrm{T}_1 \oplus \mathrm{T}_2$$
$$\mathrm{T}_1 \otimes \mathrm{T}_1 = \mathrm{T}_2 \otimes \mathrm{T}_2 = \mathrm{A}_1 \oplus \mathrm{E} \oplus \mathrm{T}_1 \oplus \mathrm{T}_2$$
$$\mathrm{T}_1 \otimes \mathrm{T}_2 = \mathrm{A}_2 \oplus \mathrm{E} \oplus \mathrm{T}_1 \oplus \mathrm{T}_2$$

In the linear point groups $C_{\infty v}$ and $D_{\infty h}$,

$$\Sigma^+ \otimes \Sigma^+ = \Sigma^- \otimes \Sigma^- = \Sigma^+$$
$$\Sigma^+ \otimes \Sigma^- = \Sigma^-$$
$$\Sigma^+ \otimes \Pi = \Sigma^- \otimes \Pi = \Pi$$
$$\Sigma^+ \otimes \Delta = \Sigma^- \otimes \Delta = \Delta$$
$$\Pi \otimes \Pi = \Sigma^+ \oplus \Sigma^- \oplus \Delta$$
$$\Delta \otimes \Delta = \Sigma^+ \oplus \Sigma^- \oplus \Gamma$$
$$\Pi \otimes \Delta = \Pi \oplus \Phi$$

Appendix G

TRANSFORMATION BETWEEN LABORATORY-FIXED AND CENTER-OF-MASS COORDINATES IN A DIATOMIC MOLECULE

The positions of the nuclei with masses M_A, M_B in a diatomic molecule may be specified by the vectors $\mathbf{R}_A$, $\mathbf{R}_B$ with respect to an arbitrary space-fixed origin (cf. Fig. 3.1). They may also be specified using the vectors

$$\mathbf{R}_{cm} \equiv (M_A\mathbf{R}_A + M_B\mathbf{R}_B)/(M_A + M_B) \tag{3.1}$$

$$\mathbf{R} \equiv \mathbf{R}_B - \mathbf{R}_A \tag{3.2}$$

where $M = (M_A + M_B)$ is the total nuclear mass. The nuclear kinetic energy in the diatomic molecule is

$$T = \tfrac{1}{2}M_A\dot{\mathbf{R}}_A^2 + \tfrac{1}{2}M_B\dot{\mathbf{R}}_B^2 \tag{3.3}$$

Using the inverse

$$\mathbf{R}_A = \mathbf{R}_{cm} - \left(\frac{M_B}{M}\right)\mathbf{R} \tag{G.1}$$

$$\mathbf{R}_B = \mathbf{R}_{cm} + \left(\frac{M_A}{M}\right)\mathbf{R} \tag{G.2}$$

of the laboratory to center-of-mass transformation (Eqs. 3.1–3.2), the kinetic energy becomes

$$T = \tfrac{1}{2}M_A\dot{\mathbf{R}}_{cm}^2 + \tfrac{1}{2}M_B\dot{\mathbf{R}}_{cm}^2 + \tfrac{1}{2}M_A\left(\frac{M_B}{M}\right)^2\dot{\mathbf{R}}^2 + \tfrac{1}{2}M_B\left(\frac{M_A}{M}\right)^2\dot{\mathbf{R}}^2$$

$$= \tfrac{1}{2}M\dot{\mathbf{R}}_{cm}^2 + \tfrac{1}{2}\mu_N\dot{\mathbf{R}}^2 \qquad \text{(G.3)}$$

where $\mu_N = M_A M_B/(M_A + M_B)$ is the nuclear reduced mass of the diatomic molecule. The first term in the kinetic energy is associated with translation of the molecule's center of mass. The second term $\tfrac{1}{2}\mu\dot{\mathbf{R}}^2$ may be separated into $\tfrac{1}{2}\mu\dot{R}^2 + \tfrac{1}{2}\mu R^2\dot{\theta}^2$, representing the kinetic energies of molecular vibration (changes in the length of the internuclear axis) and molecular rotation through an angle θ about an axis perpendicular to the molecular axis $\mathbf{R}$.

AUTHOR INDEX

SUBJECT INDEX